. . . The Heavens and the Earth

... THE HEAVENS AND THE EARTH

A Political History of the Space Age

Walter A. McDougall

Basic Books, Inc., Publishers   New York

Library of Congress Cataloging in Publication Data

McDougall, Walter A., 1946–
  The heavens and the earth.

  Bibliographic notes: p. 466
  Includes index.
  1. Astronautics and state—United States.
  2. Astronautics and state—Soviet Union.
  3. Astronautics—United States—History. 4. Astronautics
  —Soviet Union—History. I. Title.
  TL789.8.U5M34 1985 338.4′76294′0973 84-45314
  ISBN 0-465-02887-X (cloth)
  ISBN 0-465-02888-8 (paper)

To Mac and the Memory of Carol

Contents

PART III

Vanguard and Rearguard: Eisenhower and the Setting of American Space Policy

PART IV

Parabolic Ballad: Khrushchev and the Setting of Soviet Space Policy

PART V

Kennedy, Johnson, and the Technocratic Temptation

PART VI

The Heavens and the Earth: The First Twenty-five Years

Illustrations

Abbreviations Used in Text

AAF	Army Air Forces
ABM	antiballistic missile
ABMA	Army Ballistic Missile Agency
ACDA	Arms Control and Disarmament Agency
AEC	Atomic Energy Commission
AFB	Air Force Base
AFSC	Air Force Systems Command
AID	Act for International Development
AIS	American Interplanetary Society
ARDC	Air Research and Development Command
ARPA	Advanced Research Projects Agency
ARS	American Rocket Society
A-SAT	antisatellite
ASTP	Apollo-Soyuz Test Project
BoB	Bureau of the Budget
C^3I	communications, command, control, and intelligence
CEP	Circular Error Probability
CIA	Central Intelligence Agency
comsat	communication satellite
COPUOS	Committee on the Peaceful Uses of Outer Space
COSPAR	Committee on Space Research
CSAGI	Special Committee for the International Geophysical Year
DDR & E	Director of Defense Research and Engineering
DEW	distant early warning
DoD	Department of Defense
ELDO	European Launch Development Organization

EOR	earth-orbit rendezvous
ESA	European Space Agency
ESRO	European Space Research Organization
FCC	Federal Communications Commission
FY	fiscal year
GALCIT	Guggenheim Aeronautical Laboratory, California Institute of Technology
GDL	Leningrad Gas Dynamics Laboratory
GIRD	Group for the Study of Rocket Propulsion Systems
GNP	gross national product
GOP	Grand Old (Republican) Party
HEW	Health, Education and Welfare
ICBM	intercontinental ballistic missile
ICIC	Commission for Interplanetary Communications
IGY	International Geophysical Year
INTELSAT	International Telecommunications Satellite Consortium
IRBM	intermediate-range ballistic missile
ITU	International Telecommunications Union
JATO	jet-assisted take-off
JCS	Joint Chiefs of Staff
JPL	Jet Propulsion Laboratory
LANDSAT	NASA's Land Survey remote sensing satellite
LOR	lunar-orbit rendezvous
LOX	liquid oxygen
MAD	mutual assured destruction
MIRV	multiple independently targeted reentry vehicles
MOL	Manned Orbiting Laboratory
MRBM	medium-range ballistic missile
MRV	multiple reentry vehicles
NACA	National Advisory Committee for Aeronautics
NAS	National Academy of Sciences
NASA	National Aeronautics and Space Administration
NASC	National Aeronautics and Space Council
NATO	North Atlantic Treaty Organization
NDEA	National Defense Education Act

NRL	Naval Research Laboratory
NSC	National Security Council
NSF	National Science Foundation
OCB	Operations Coordinating Board
ODM	Office of Defense Mobilization
OGPU	Soviet secret police, also NKVD and KGB
OIMS	All-Union Society for the Study of Interplanetary Communications
OSD	Office of the Secretary of Defense
OSRD	Office for Scientific Research and Development
PERT	program evaluation and review technique
PPBS	Planning-Programming-Budget System
PSAC	President's Science Advisory Committee
PSI	pounds per square inch
R & D	research and development
RAND	Research and Development Corporation
RBNS	Research Board for National Security
RNII	Jet Scientific Research Institute
SAC	Strategic Air Command
SALT	Strategic Arms Limitation Talks (or Treaty)
SIOP	Single Integrated Operational Plan
SLBM	submarine-launched ballistic missile
STS	Space Transportation System (the shuttle)
TCP	Technological Capabilities Panel
TLI	trans-lunar insertion
TsAGI	Tsentral'nyi Aero-Gidrodinamichesky Institut
TsBIRP	Central Bureau for the Study of the Problems of Rockets
TVA	Tennessee Valley Authority
UN	United Nations
USAF	U.S. Air Force
USIA	U.S. Information Agency
WSEG	Weapons Systems Evaluation Group

N.B.: For the sake of convenience, all American space vehicles and missions are identified by Arabic numerals (e.g., *Discoverer 13*, *Apollo 8*) and all Soviet vehicles and missions by Roman numerals (e.g., *Sputnik III*, *Salyut VI*). For acronyms of specific spacecraft, see the index.

Preface

I missed the first moon landing. In July 1969 I had the night shift as chief of artillery fire direction in a particularly nasty jungle base in the III Corps region of South Vietnam. A three- or four-day-old copy of *Stars and Stripes* told us of *Apollo 11.* I do not recall that it made much of an impression on us, except maybe to poke our ready sense of irony. Presumably we had more immediate concerns. As a student in high school and college I had, like most Americans, kept up with the space program on television, knowing the tedium of countdowns and the tension of lift-offs and splashdowns. But I was not a "buff." In the hyperactive technological landscape of the time, I think we took the space rocket for granted—the government's equivalent of a Pontiac GTO.

By the time I finished graduate school in 1974, the space program touched me no more than it did anyone else. I consigned Skylab to the "feature section" of my mind, considered the American-Soviet rendezvous a (rather vulgar) punctuation of détente, and noticed the proliferation of unmanned space systems not at all. I wrote lectures on European diplomatic history, tried to adjust to life in California, and busied myself with a book on French foreign policy in the 1920s. Then, late in 1977, my department inquired after my future research plans, pursuant to a review of my performance as a junior professor. My reported interest in Allied economic cooperation during World War I (a likely and under-worked field) found little favor: "more of the same," said the chairman, and my colleagues were unmoved. This was dismaying, to say the least. So, with the boldness that sometimes crystallizes out of confusion, I determined to please myself, to follow my own curiosity no matter how academically outrageous its direction. Two subjects in international relations particularly interested me: international management of nonterritorial regions, like the oceans or Antarctica, and the interplay of international rivalry and technological change. My reconnaissance of these fields then yielded two discoveries: first, that the political history of the early Space Age neatly encompassed both these interests, and second, that the opening gun of the Space Race, *Sputnik I,* had tremendous repercussions in the domestic as well as the international history of our time. This book is the result of those discoveries.

Some say that the events of the 1950s and 1960s are too recent to be susceptible to "serious" historical treatment. I will not argue the point at length, because any cutoff point for history is artificial. There are historians who consider anything after the execution of Charles I to be "journalism." But I appeal to pragmatism—the fact that college students today, for whom JFK and Vietnam are as dim as Truman and Korea were to me, need to learn postwar history from historians. I appeal also to precedent—that historians were writing profitably, though not definitively, on the origins of World War I and the Cold War fifteen or twenty years after the events. It is already a quarter century since *Sputnik I.*

What about sources? Are sufficient materials available on such a recent subject? Suffice to say that my problem, as that of any twentieth-century historian, was too much material, not too little. To be sure, some documents remain classified, but many others have been declassified under the Freedom of Information Act. When the available record does not permit complete confidence, I have resorted to qualifications. More important, I have tried to ask of the history of the Space Age questions whose answers do not hang on the contents of this memo or that dispatch. Soviet sources were a vexing difficulty, but no more so than for any scholar working on post-1917 history. We can hardly dispense with all attempts to understand the Soviet Union because its secrecy is uncongenial to empirical scholarship. In sum, the student of contemporary history may not have access to everything but is in no worse a position than a medievalist, whose total corpus of information is by comparison tiny, arbitrary, and unreliable.

What about perspective? How can we hope to grasp the historical significance of events so close to our own day? This is a dangerous question, for it leads inevitably to quandaries of epistemology—how can any historian claim to "know" anything?—and hermeneutics—how can any perspective be more "right" than any other? I understand the mentality of the 1960s better than will many historians of the future because I shared it, but I cannot know how the changes of our time will play themselves out in decades and centuries to come. The medievalist does know the end of the story he tells (if any history has an end), but has more trouble grasping the mentality, much less the narrative, of the fourteenth century. I can only insist that we have a duty to think historically about the recent past and can be encouraged by the fact that change is taking place on a more compressed scale in our time. The 1960s are already dear, dead days. Future historians will have much more to say, and will surely correct much of what I write here. But the time has also surely come to make a start.

There is a large tableau hanging in the Smithsonian's Hirshhorn Museum in Washington. It is an op-art creation by the Israeli Yaacov Agam, and it consists of dozens of vertical, V-shaped panels, each one painted in black and colored checks on one face and in seemingly random colors on the other face. If one stands at the far left and looks on edge, the colors merge into a grand field of black horizontal lines on a rainbow background. As one moves in front of the tableau and across its expanse, the eye sees an ever-changing checkerboard of colors, yet white boxes and rectangles also appear that were invisible before. *Transparent Rhythms,* the title of the work, percolate throughout. But on the far right, again viewed from on edge, the whole matrix resolves itself to the simplest pattern of all: three brilliant, horizontal spectra, the middle one's polarity reversed. Gone is the chaos, but gone too are the intermediate patterns.

It is the nature of the historian, like the engineer, to solve problems through artificial order. Unfortunately, the historian is inside the checkerboard, but that does not alter the fact that some of its patterns are truer, more enduring and encompassing, than others, whether we can see them or not. That is why history cannot dispense with intuition, no matter how meticulous our empiricism, and why this historian, however cautious and wary of the *trompe l'oeil,* aspires to view history from the edge.

Returning astronauts, abashed by the praise they receive, make a ritual of thanking the anonymous thousands who made their flight possible. Academic authors unfortunately have to name names, but when their book is finally completed despite intermittent research and teaching, they must strain to remember all those who contributed over the years. I hope I have done so.

I came to the Space Age an apprentice. Dr. Monte D. Wright, NASA historian (retired), and Dr. Alex Roland (now of Duke University) took me in. Their expertise was invaluable, and their friendship and occasional skepticism served as encouragement and goad to my endeavor. They and I were also ably served by NASA archivist Lee Saegesser and the staff of the NASA History Office. The dedicated professionals of the Truman, Eisenhower, Kennedy, and Johnson libraries were of great service, especially those, like Martin M. Teasley at Abilene and Martin L. Elzy at Austin, who processed endless requests for declassification with efficiency and good cheer. George Perros, Renee Jaussaud, and the staffs of the Legislative and Modern Military Branches of the National Archives helped me locate and obtain permission to consult previously unseen collections, especially the papers of the Senate Space Committee. Richard Baker, historian of the United States Senate, was an excellent guide in the congressional maze. Walter L. Kraus, historian of the Air Force Systems Command at Andrews Air Force Base (AFB), provided access to declassified but still obscure information on the military space program. I also thank the librarians of the Arms Control and Disarmament Agency, the U.S. Air Force (USAF) History Office, the Milton Eisenhower

Library at Johns Hopkins University, and the Firestone Library at Princeton University.

The most dependable observers of the world of space technology and policy are the analysts of the Science and Technology Division, Congressional Research Service, Library of Congress. Dr. Charles S. Sheldon II (now deceased), who chronicled and interpreted the Soviet space program from its inception, Barbara Luxenberg (now with the Patent Office), and Marcia Smith were all patient and helpful, as was the staff of the Manuscript Division, Library of Congress.

In 1979 I journeyed to Paris and London to gather data on the European space effort and was welcomed by Michel Bourely, Legal Counsel, and the librarians of the European Space Agency, by J. Lawrence Blonstein of EUROSPACE, and the director and staff of the International Institute of Strategic Studies, who provided access to their excellent library and clipping files.

I thank the Committee on Research and the Institute for International Studies of the University of California, Berkeley (especially Director Carl Rosberg and Professor Ernst Haas), for generous financial support during two long sabbaticals. The National Air and Space Museum provided me with office, collegiality, and clerical support on short notice during a critical period of my writing. I am grateful to Drs. Walter Boyne, now director, Paul Hanle, chairman of the Space Science Division, and Allan Needell for a productive and enjoyable four months.

Most of all I am indebted to the Woodrow Wilson International Center for Scholars of the Smithsonian Institution, where I had the privilege to reside as a Fellow during the year in which most of this book was written. Its atmosphere, facilities, location, community of scholars, inspired leadership, and exceptional staff make the Wilson Center unique. It is no exaggeration to say that all of the following contributed, through substantive, technical, or moral support, to the completion of this book: James H. Billington and Prosser Gifford, director and assistant director; William Dunn, business manager; Zdenek David, librarian; Ann Sheffield and Frannie Conklin of the History, Culture, and Society Program; Sam Wells of the International Security Studies Program; Peter Braestrup and Cullen Murphy of the Wilson Quarterly; Bradford Johnson, research assistant; fellow Fellows Loren Graham, Michael Geyer, Arnold Kramish, Franklyn Holzman, and Pete Daniel; Eloise Doane, chief of secretarial services and wizard of the word processors; and the rest of the Wilson Center staff.

Other colleagues whose suggestions contributed to the finished product include C. West Churchman (Berkeley), Leonard David (National Space Institute), Stephen E. Doyle (Office of Technology Assessment), Alain Dupas (European Space Agency), David Ehrenfeld (Rutgers), Arnold Frutkin (NASA, retired), Eilene Galloway (congressional expert and veteran of space law, retired), Wreatham Gathright (State Department),

John Heilbron and Roger Hahn (Berkeley), Melvin Kranzberg (Georgia Tech), John Logsdon (George Washington University), Hans Mark (Deputy Administrator, NASA), Bruce Mazlish (MIT), the Rev. John Bruce Medaris (Maj. Gen., U.S. Army, retired), Barry Posen (Harvard), Robert Post (editor, *Technology and Culture*), Herbert Reis (U.S. delegation, UN), and Reginald Zelnik (Berkeley).

Alex Roland and Charles Maier (Harvard) read the manuscript and made valuable suggestions. My editor at Basic Books, Steven Fraser, as well as Maureen T. Bischoff, Linda Carbone, and Debra Manette, gave me expert guidance at every stage. A Radio Shack TRS-80 Model III typed the manuscript, and a wonderful human being, Grace O'Connell, typed the back material. Unstinting moral support came from William J. Bouwsma, Robert L. Middlekauff, and the Rt. Rev. Robert S. Morse.

I thank you all.

—WALTER A. McDOUGALL
Moraga, California
March 1984

. . . The Heavens and the Earth

Introduction

Three hundred sixty million years ago, we are told, there lived a fish we call *Eusthenopteron*. It frequented the murky shallows of rivers in Pangaea, the vast supercontinent that was later to divide, like a living cell, into North America, Europe, and Asia. Over the course of millennia, climatic changes gradually dried up its rivers. But the Crossopterygians, the transcendental elite to which *Eusthenopteron* belonged, were both stubborn and blessed, hence candidates for metamorphosis. They already sported lobe fins for "walking" on the bottom of their late-Devonian streams and prototype lungs for gulping air in case of foul, stagnant water. In time, *Eusthenopteron's* muscular fins tugged it across mud flats that once, as streambeds, had marked the absolute boundary of its fishy universe. Now the viscous channel revealed itself to be a cradle, and then a platform to a new universe of solids and gases. Animal life had come to the land. In time our analogous friend became an amphibian, *Ichthyostega*, although the new name is a human conceit that *Eusthenopteron* would probably resent.

In A.D. 1961 *Homo sapiens*, in turn, left the realm of solids and gases and lived, for 108 minutes, in outer space. Life again escaped, or by definition extended, the biosphere. The earth's crust and canopy of air became another platform to a new universe as infinite as soil and sky must have seemed to *Eusthenopteron*. The opening of the Space Age was another cleavage, more sharp than blunt, in natural history. It took an era for marine fugitives to populate the land. But by the end of the 1980s some human beings will be constantly in space, if only as scientists, soldier-spies, or telephone repairmen. By the middle of the next century human colonies may be populating earth's neighborhood. Of all the analogies contrived to convey the meaning of the Space Age, therefore, the amphibian adventure of the Devonian period is the most evocative.[1]

Of course, there are gross differences between the two events. *Eusthenopteron* did not contrive to build a "land suit" or "earth rover" filled with water. It adjusted to the land and became, as we would say, a new species. Man fashions his own environment to take with him. Nor did fish come up to survey the sand out of curiosity or lust for power—they did so by instinct, to survive. But these distinctions only illuminate the current human crawl from the cradle. Man transcends his element without ceasing to be man, for he is *Homo faber*, the toolmaker, the technologist. And man explores through idiosyncratic choice, because he is also *Homo pictor*, the symbolist, the dreamer. But tools and dreams, though both products of the imagination, are responses to contrary sides of human nature. Their complex coexistence defines much of human history, and their integration in the human personality has always been a task as mortal as *Eusthenopteron*'s metamorphosis.

When the first artificial satellite, *Sputnik I*, circled the earth in 1957, on-call philosophers of the press and politics contradicted each other, and sometimes themselves, on what the Space Age symbolized. To some it was the newest and most spectacular evidence of mankind's irrepressible, questing nature. To others the promise of space technology and its imponderable social, political, and psychological effects was to *change* man's nature by fostering a global consciousness, material abundance, and, ultimately, in the pre-Leninist vision of rocket theorist Konstantin Tsiolkovsky, "the perfection of human society and its individual members."[2] The confusion is not trivial. Through our technology, our ability to manipulate Nature, we are subcreators, a demiurge. Hence we have never, from Protagoras to Francis Bacon to Tsiolkovsky, been able to separate our thinking about technology from teleology or eschatology. For reason cannot predict whether our tools and dreams, which together permit us to "invent the future," will lead us to perfection or annihilation or unending struggle against Nature and ourselves. Yet those futures whisper their warnings and temptations to us even as we study merely the politics and technology of our own day. As we proceed, therefore, I invite the reader to keep in mind that grander time scale, stretching from the Devonian to the impatient and not-so-distant future, of which all that is contained in this book is only a tick.

Ours is an age of perpetual technological revolution. As most alert undergraduates would attest, history seems to be speeding up. The great "-ations" of the age—centralization, bureaucratization, democratization, secularization, differentiation—are all said to be elements of modernization, and the causal connections among them are the stuff of modern historical debate. Modernization is itself bound up with technology, with industrialization. Depending on how loosely one employs the term, there have been many industrial revolutions in the past two hundred years: that of the factory system and textiles; that of coal, iron, and steam; that of chemicals and electricity; that of oil and the internal combustion engine; that of nuclear energy and jet aircraft; finally that of electronics, computers, and rocketry. What is it that makes the Space Age unique?

From the Sumerian beginnings of agriculture to the twentieth century, technological revolutions have forced adjustments in social, political, and economic institutions. Of course, these institutions may also have inhibited and shaped the use of new tools, but they were largely passive toward invention itself. Their purpose was not to generate the tools in the first place. This has been as true of the modern world as of the ancient. With the rise of the power-state and mercantilist economics in the sixteenth and seventeenth centuries, European governments undertook to foster enterprise, exploration, and industry in a conscious effort to stimulate the wellsprings of their own wealth and power. But with a few precocious exceptions (Prince Henry of Portugal, Charles II's Royal Society, for example), the early modern state sought to exploit existing tools, not create new ones.

The liberal rebellion against mercantilist theory, associated with the followers of Adam Smith, only underscored this passive relationship of the state to new technology. It deemed free enterprise to be the most stimulating environment for private ingenuity and bade the government to remain aloof. To be sure, the state assumed a central role in the nineteenth-century spread of railroads, gas lighting, semaphore and telegraph communications, and agricultural experiments, but self-conscious institutionalized research and development (R & D) emerged only in the late decades of the century. It began in dynamic German industrial firms and was picked up, tentatively before 1914, by governments interested primarily in weapons and public works.[3]

Twentieth-century warfare finally established state-sponsored and -directed R & D as a public duty and necessity. Rapid development of new weaponry, ersatz strategic materials, and more productive manufacturing processes became an imperative of national survival in total war. Even though government research spending dwindled almost to nothing again in the interwar years, the World War I model of command economies in new technology, as well as in investment and distribution generally, was enduring and seductive. Thorstein Veblen, Herbert Hoover, and assorted technocrats invoked it in the United States. The government of Weimar Germany, though financially barren, encouraged a peacetime public and private partnership in basic research. In the new Soviet Union, the Communist Party embraced state-controlled technological change as the partner of ideology in the building of socialism. There the wartime emergency measure became an article of peacetime faith, and the world's first technocracy emerged.[4]

Now "technocracy" is a familiar word meaning "the management of society by technical experts." As such it is an ideal type, for no matter how complicated society's tools or pervasive the technical advisers, politicians and other power-brokers still govern, even in the late twentieth century, even in the USSR. Let us define technocracy therefore as follows: the institutionalization of technological change for state purposes, that is, the state-funded and -managed R & D explosion of our time.

In World War II command technology came of age. Without hesitation and building on the 1914–1918 model, every major belligerent mobilized national talent and resources for science and technology to support the war effort. The four great breakthroughs of those years impressed on the world's imagination as nothing before the possibilities for planned change implicit in command R & D. They also came to shape the military environment of our lifetime. These war babies were the British development of radar, the American atomic bomb, the German ballistic missile, and the American electronic computer.

"What hath man wrought?" was a question variously answered in the aftermath of the first atomic explosions. The existence in the world of horrid political regimes and unprecedented engines of destruction impressed on the wartime generation what a near thing the survival of civilization had been and would be. In the developing Cold War, the dilemma posed by the refinement of nuclear weapons—and of horrid political regimes—was not resolved, and remains so to this day. But the spectacular achievements of wartime R & D also encouraged the belief that conscious application of "Manhattan Project" methods to problems of poverty, health, housing, education, transportation, and communication might eliminate material want and (it was supposed) the material causes of war.[5] Whether or not such hopes were well grounded—whether we can mate our tools with our dreams, but not with our nightmares— demobilization again brought extreme cuts in government research funds, while those that remained went mostly for weapons and atomic power. Only the USSR reified the notion of centralized mobilization of science and technology in peacetime, and even there reality fell far short of the ideal. Political terror and secrecy, bogus scientific theory, similar concentration on military-related research, and general Soviet "backwardness" reinforced majority opinion in the West that state direction of science was self-defeating.

By 1965, in the space of ten years, this proposition was overthrown. We need not call it a revolution, for state involvement in basic and applied research had been growing, however leisurely or unevenly, for a century. Call it instead a saltation, an evolutionary leap in the relationship of the state to the creation of new knowledge. Not only did the Soviets rethink and reaccelerate their R & D machinery in these years, but Western governments came to embrace the model of state-supported, perpetual technological revolution, create national infrastructures for such a program, and quintuple their funding for R & D in support of national goals. What had intervened to spark this saltation was Sputnik and the space technological revolution.*

The political responses to the Sputnik challenge are the subject of this book. For in these years the fundamental relationship between the government and new technology changed as never before in history. No longer did state and society react to new tools and methods, adjusting, regulating, or encouraging their spontaneous development. Rather, states

* See the appendix.

took upon themselves the primary responsibility for generating new technology. This has meant that to the extent revolutionary technologies have profound second-order consequences in the domestic life of societies, by forcing new technologies, *all* governments have become revolutionary, whatever their reasons or ideological pretensions.

The goad that Sputnik became was honed as well by international competition. The first artificial satellite (and the intrinsically more important rocket that launched it) was an incremental and predictable feat of engineering toward which both Superpowers had been nudging since 1945. Yet Sputnik had abrupt, discontinuous effects in politics and ideology because of the volatile historical conjuncture at which it occurred. The United States had assumed the responsibility of free world leader and maintained it on two premises: first, the evident superiority of American liberal institutions, not only in the spiritual realm of freedom, but in the material realm of prosperity; second, the overwhelming American superiority in the technology of mass destruction, shielding those under its umbrella from external aggression. It was axiomatic that the United States was both "better" and mightier than its chief rival. The future belonged to it, at least for the foreseeable American Century.[6]

Sputnik seemed to belie both premises. Not only did the USSR herald its imminent strategic parity through intercontinental ballistic missiles (ICBMs), but this scientific feat of revolutionary implications suggested to a half-informed world that American reliance on the marketplace and the discoordinated efforts of the private sector, corrupted by consumerism, was anachronistic in an age of explosive technological advance. By the dim, reflected light of the Soviet moons, the United States resolved to change. Whatever principles and luxuries of freedom be threatened, whatever sacrifices be called for, the United States *had* to meet this challenge. For the first time since 1814 the American homeland lay under direct foreign threat; its citizens felt that constant fear and pressure that Europeans had lived with for centuries. The global strategic ecumene had closed. And if the USSR now had the capacity to deliver mass destruction to the U.S. heartland, how credible was the American deterrent to its allies? If the Soviet space triumphs that followed in frustrating succession seemed to show that communism was the best path toward rapid modernization, how credible was the appeal of liberal democracy to the underdeveloped nations shedding their colonial status?

In a time of Cold War, decolonization, and indigenous social and racial challenges to the noninterventionist state, Sputnik was a powerful catalyst. Critics accused President Eisenhower of being old-fashioned and out of touch. Clearly a vigorous technical and legislative counter-offensive was needed to regain the initiative in the Cold War. Hence the second Eisenhower administration vacillated, while the opposition incubated new, technocratic ideas. But Ike was not out of touch. He understood the problems of the age perhaps better than his critics among the Best and the Brightest. He feared the economic and moral consequences of a headlong technology and prestige race with the Soviets; he feared the political and social consequences of vastly increased federal powers in education, science, and the economy; he feared, as expressed in his Farewell Address, the assumption of inordinate power and influence by a "military-industrial complex" and a "scientific-technological elite." So his last years were a holding action in which he sharply accelerated the federal R & D "machine" even as he sought to prevent the foreign challenge from being translated into a panicky domestic upheaval.[7]

The technocratic model triumphed under Presidents Kennedy and Johnson. Four months after taking office, Kennedy asked Congress to commit the United States to go to the moon. The decision was a product of the growing technocratic mentality and immediate political trends evident in the reverses in Laos, the Congo, the Bay of Pigs in Cuba, and the flight of Yuri Gagarin, the first man in space. The moon program was a lever by which the young President, who extolled vigor and assaults on The New Frontier, and the nation, which seemed to have lost faith in itself, could find their legs and come to grips with the internal and external challenges of the post-Sputnik world. As Vice-President Johnson capsulized: "Failure to master space means being second best in every aspect, in the crucial arena of our Cold War world. In the eyes of the world first in space means first, period; second in space is second in everything."[8] Space technology was drafted into the cause of national prestige. Later, advanced technology in general was tapped as the vehicle for national and international regeneration.

What Sputnik did, in simultaneously presaging nuclear parity and suggesting Soviet scientific superiority, was to alter the nature of the Cold War. Where it had previously been a military and political struggle in which the United States need only lend aid and comfort to its allies in the front lines, the Cold War now became total, a competition for the loyalty and trust of all peoples fought out in all arenas of social achievement, in which science textbooks and racial harmony were as much tools of foreign policy as missiles and spies. The self-confident administrations of Kennedy and Johnson set out to prove what had previously been taken for granted—the superiority of American institutions. And their chosen weapon was induced technological revolution, followed hard by government control of research, education, economic fine-tuning, and social welfare in all its manifestations. Foreign political and domestic social challenges, it was believed, were equally susceptible to the technological and managerial fix: revolutionary change without revolution, qualitative problems solved with quantitative methods. Under the impact of total Cold War, technocracy came to America. Under the impact of the technological revolutions of the 1960s, exploitation of knowledge and skilled labor replaced the exploitation of raw materials and semiskilled labor as the key factors in economic progress. Forcing more intense development from an already superior scientific and industrial base, the Technological Republic of America must surely outfight and outshine the Ideological Republics of Moscow and Peking.

The R & D saltation in the United States, triggered by foreign pressure, was itself transmitted abroad by the international imperative to accelerate technological, hence social and economic, change elsewhere. After Sputnik, Charles de Gaulle's Fifth Republic dedicated itself to grandeur and independence through science and technology. By the mid-1960s all Western Europe awakened to the "technology gap" that Space Age research and management had apparently opened between the United States and the merely industrial nations. Even the Soviet Union, ruing Khrushchev's braggadocio after its early space triumphs, found itself trailing the United States in space and missile programs. It redoubled its deliberate efforts in command R & D and altered communist doctrine to accommodate the space technological revolution.

By the end of the first decade of the Space Age, tremendous change had occurred in all advanced countries. Now that governments had found the political will, nurtured a climate of expectations, and mobilized national resources for translating dreams into reality, politicians and professors again indulged the utopian notion that man could truly invent his own future. President Kennedy and others hoped that competition among diverse states could be channeled into peaceful pursuits—exploring the cosmos as the moral equivalent of war, conquering disease, desalinating ocean water, developing the postcolonial world. Advocates of space law hoped to preserve tranquil space from Cold War rivalry—the example of a humanity united in space could be transferred back to earth. Perhaps communications or earth-resources satellites would bring understanding between cultures and global economic planning, while reconnaissance satellites erased the distrust that hampered disarmament. If not, then the universal threat posed by futuristic weaponry might conjure a global political will to eliminate the material causes of discord.

Thus the age of man's first steps out of the terrestrial cradle was a time of fear and euphoria both. (Was *Eusthenopteron* glad of the land, or did it dream in its spinal R-complex of a riverine Eden long departed?) In the formative years of the early Space Age, the euphoria faded, the fear remained. There was no transformation of the international system, no revision of priorities toward global "welfare" and cooperation, no metamorphosis in human philosophy and values. Instead, there was only the maturation of the power complex of the R & D State. For the present and foreseeable future, this maturation defines the character of the Space Age in history.

But if sudden acceleration of technological change under the aegis of the "post-modern" state was the main historical product of Sputnik, then did it not flame out in a remarkably short time? Technocracy failed to save (or establish) Third World democracy, or American inner cities, or even U.S. economic competitiveness in basic industries. The Soviet Union lost the race to the moon and failed to "bury" capitalism, as Khrushchev had promised. Nor did Western Europe achieve stable growth, political consensus, or independence from the Superpowers through command R & D. Instead, an international backlash criticized technocracy as tyrannical, antihumane, imperialist, macho, and polluting. Western space budgets dwindled, nuclear power lost its charm, "green" political parties arose in Europe, technology assessment and harassment replaced technology stimulation, and "progressive" social critics talked of exploring inner space rather than outer space. If the R & D revolution defines the Space Age, then was it not rather trivial after all?

Look again at the appendix. The 1970s were certainly a node in public enthusiasm for technology, but they brought no counterrevolution. If public and official wisdom now spoke of "trade-offs" and held that political, military, and social problems were transformed, but not solved, by technology, the process of command invention nevertheless continued.

In the 1970s—and 1980s—the laser emerged as a tool of a thousand uses, military and civilian. Microminiaturization produced pocket calculators and accelerated the evolution of all computers to the point where silicon chips measuring one forty-eighth of a square inch contain 2,250 transistors performing 100,000 calculations per second. Communications satellites passed through four generations as global and regional systems proliferated, and transformed the television industry in North America. New communications satellites (comsats) beckon that will have greater capacity than humanity can use—potentially every individual may have access to spaceborne television, telephone, and data transmissions. On land, fiber optics emerged with the promise to explode the capabilities of conventional circuits—photonics (signals traveling on waves of light rather than electrons) may replace electronics in myriad applications.

In the 1970s the scale of large and small reached literally inconceivable extremes. Molecular beam technology can etch the *Encyclopaedia Britannica* on a postage stamp, or a billion angels on the head of a pin. Layers a single atom thick can be deposited on metals or bunched in order to conduct electricity thirty times faster even than silicon chips. On the other end of the continuum, space-based sensors permit us to detect radiation at distances of tens of billions of light-years (10^{22} miles); the "end" of the universe beckons. Cosmology reeled with the apparent discovery of background radiation from the Big Bang of Creation, pulsars, and possibly black holes. Advanced robotics transformed manufacturing, but were most stunningly displayed in planetary probes like the Viking lander that performed biochemical experiments on Mars directed by radio 150 million miles away. The 1970s also produced the Space Transportation System, the reusable shuttle craft that will open a second Space Age of ineffable potential.

One need not mention technical breakthroughs unconnected with space—the foundations of nuclear fusion or genetic engineering—to conclude that there is more than a "gee-whiz" element in all of this. This is the dream of Trotsky and Mao—continuous revolution—and as Daniel Boorstin, the Librarian of Congress, has observed, there is no technological counterpart to a Restoration or counterrevolution.[9] Styles and moods may shift, but there is no returning to the *ancien régime* of 1957 or 1941. We are on Mr. Toad's Wild Ride, and though we may become "velocitized" and insensitive to our speed, or bored, or content to putter about in the apparently motionless comfort of the back seat, at some point we must dare again to look out the window and perceive that we are careening beyond the very human scales of space and time.

How did all this come about? Are our societies locked into irreversible technological change to the point where human institutions themselves have become "part of the machine?" Or do people, acting through politics, retain their ability to choose which future to invent, or whether to try? If so, can we and our leaders be trusted with such responsibility? What is the relationship between man and his machines?

Most philosophers have assumed technology to be a fundamental human activity, perhaps even the defining human activity that breaks our "Faustian kinship with the worm." Indeed, modern thinkers from Francis Bacon to Herbert Spencer, Auguste Comte, and St. Simon elevated technological progress from its role as an incidental product of intellectual, economic, or military experience to the level of moral imperative.[10] To nineteenth-century philosophers especially, history seemed to be the story of progress, and in the end technology would eliminate material scarcity, erase the cause of the world's ills, and free mankind from the necessities imposed by nature. Even after twentieth-century war and tyranny revealed the dangers of modern technology, many, perhaps most, political philosophers still affirm it as a tool holding great promise for mankind. (Everyone quotes Lord Acton to the effect that "Power tends to corrupt," but few really live by it.) In this sanguine view, social reform, political revolution, "soft" or decentralized technology, technology "for The People," or merely the grafting of scientific advisers and technology assessors on to the body politic may suffice to make technology that "genie in the service of mankind" after all.

Other recent critics of technology challenge the primary assumption that the modern technosystem is a natural social attribute. To them technology represents an extrinsic system in which man himself is entrapped: "machines are in the saddle and ride mankind." Western man came to mechanize and standardize himself and so built a mythology around machines glorifying the values of predictability, efficiency, and utility at the expense of Nature and humanity. Or else the machinery of modernity achieved such dominance that politics, economics, and culture are now themselves *situated in* a "technical milieu." In this melancholy view, the global technosystem we have built is now spontaneously expansive and subject to its own technical morality, of which more power is the only object.[11]

Whether heady or gloomy, such accounts of man's relation to his own technology imply a causal link between new technology and the forms by which people collectively assert power and ideas. Either technology determines social and political forms, or political decisions, cultural values, and social structure shape the creation and diffusion of new technology. But such accounts also imply a holistic model of historical change that leads to endless "chicken and egg" questions.[12] Harvard sociologist Daniel Bell tried a different approach and concluded that we have fallen into a "confusion of realms."[13] He divided human history into three realms, not one: the social structure is the realm of economy *and* technology; the polity, the realm of authority; and culture, the realm of symbolism that explores the existential questions facing all human beings all the time—death, love, loyalty, tragedy. Each realm has distinctive principles and historical rhythms that set it apart one from the others. At any given time, therefore, they are not likely to be "in step" with each other, but radically disjunctive.[14]

In this scheme, there is no question of technology swallowing up political choice and cultural values. First, technology and society are united in the same realm, that is, a people that embraces a new technology *ipso facto* must embrace the new institutions needed to accommodate it. Second, in establishing autonomous realms for politics and culture, Bell helps us to understand why polities choose new technologies at given moments. For our attention is directed away from the technosocial "chicken and egg" realm toward those that harbor the two extremes of human organization: the largest human unit, the international political system with its own laws of competition for power and security; and the smallest human unit, the individual, who dreams and creates, responding to his own culture in an effort to make *his* life, at least, meaningful.

In seeking the origins of the space technological revolution, we arrive at both of these extremes and find them pushing in the same direction, the international imperative and the individual will, the massive power-states and the intrepid pioneers like Tsiolkovsky, Goddard, Korolev, and von Braun.[15] The needs of the former and the dreams of the latter combined, and once the political decisions were made, the spread of Space Age infrastructure and socioeconomic adjustments—what MIT historian Bruce Mazlish termed a "social invention"—was automatic.[16]

What does all this suggest about our prospects of controlling the pace of technological change? Bell argued that technology governs change in human affairs while culture guards continuity. Hence technology is always disruptive and creates a crisis for culture.[17] Similarly, international challenges like Sputnik create a crisis for the domestic technology of a rival state. The need for survival and for the survival of national self-image is an imperative capable of pressing states of whatever social system into revolution from above. To be sure, the state can at our behest turn off the R & D complex and suppress invention, but it cannot stamp out the individual problem solver, whose thoughts are free and whose culture, at least in the West, provides a transcendental validation for the urge to make and do. Nor can the state destroy the capacity of foreign countries to raise challenges. The logical and ironic conclusion is that only a technological hegemony, a Big Brother, so complete as to control the nations of the earth *and* the thoughts and activities of its subjects would suffice to choke off the sources of technological change. In the meantime, the unending race to keep up with foreign military and economic competition threatens to erode the very values that make one's society worth defending in the first place.

That, succinctly stated, is the dilemma of the Space Age and the moral of our story.

PART I

The Genesis of Sputnik

Even before now thought was impatient enough, it wants everything at once, takes each step slowly, because its steps are difficult to correct. Therein lies the tragic position of thinkers.   —ALEXANDER HERZEN, 1849

If we can even now glimpse the infinite potentialities of man, then who can tell what we might expect in some thousands of years, with deeper understanding and knowledge.

There is thus no end to the life, education, and improvement of mankind. Man will progress forever. And if this be so, he must surely achieve immortality.

So push confidently forward, workers of the earth, and remember that every ounce of your efforts is eventually bound to bring a priceless reward.
                    —KONSTANTIN TSIOLKOVSKY, 1911

We must master the highest technology or be crushed.   —V. I. LENIN, 1919

In the period of reconstruction technology decides everything.
                    —JOSEF STALIN, 1931

IN 1881 the People's Will was done. Tsar Alexander II lay mortally wounded beside his magnificent carriage in the Catherine Canal Road. His lower body was blown inside out by a contrivance from the chemistry laboratories of the St. Petersburg Technological Institute, founded by Alexander's father for the scientific modernization, hence viability, of the Russian Empire in the industrial age. There Dmitri I. Mendeleev codified the periodic table of elements in the late 1860s and told his students, "Science and industry, these are my dreams. They are everything today. . . ." It was also there that a generation of angry students harbored "an almost sacramental reverence for chemical mixtures," especially variations on Alfred Nobel's recent brew, dynamite. Nikolai Chernyshevsky expressed it best: "The warmth people need came not from sentimentality, but from a cold match striking the hard surface of social reality and lighting thereby a fire."[1]

Nikolai Kibalchich was the manufacturer of "cold matches"—bombs— for the conspiracy called the People's Will. Captured and sentenced to death after the tsaricide, Kibalchich seemed to lose interest in politics and life itself, but not in chemistry or revolution. He spent his last days in prison hastily designing a rocket-propelled aircraft, discussing it with anyone who would listen, and imploring the guards to see that his notes got to the proper authorities.[2]

When the tsar succumbed to his chemistry students, Konstantin Tsiolkovsky was twenty-four. Raised in provincial obscurity and self-taught, Tsiolkovsky had just mailed some calculations to the St. Petersburg Society for Physics and Chemistry, only to learn that his impressive theses were already well known to the scientific community. Though disappointed, Tsiolkovsky drew confidence from this episode and pressed on into truly original fields of research. Had his isolation and lack of a library not inhibited a normal academic career, he might never have hearkened to the unorthodox call of his intellect and psyche—Tsiolkovsky was obsessed by the conquest of gravity. "It seems to me," he wrote,

that the basic drive to reach out for the sun, to shed the bonds of gravity, has been with me ever since my infancy. Anyway, I distinctly recall that my favorite dream in very early childhood, before I could even read books, was a dim consciousness of a realm devoid of gravity where one could move unhampered anywhere, freer than a bird in flight.[3]

He parodied himself in an essay called "The Gravity Hater" about a "friend" who took gravity to be "his personal, bitterest enemy. He delivered threatening, abusive speeches about it and convincingly, so he imagined, set out to prove its entire worthlessness and the bliss that 'would come to pass' through its abolition." Gravity pressed us down like worms, but a gravity-free environment "would make the poor equal to the rich."[4]

Tsiolkovsky's 1883 diary "Free Space" imagined how the laws of classical mechanics would operate in zero gravity. Houses of any size would not collapse under their own weight; mountains and palaces of any shape and size might stand without props; a man could hold in his hand "a thousand *poods* of earth" or tread on a needle point without being pierced. There would be no up and no down. Motion could take place only by exchange of momentum between bodies. If a man threw a stone, he would himself move in the opposite direction from the toss at a velocity proportional to the masses of stone and man. Hence the most natural way of moving in a gravity-free environment was to throw off matter in the opposite direction. From a psychological hatred of gravity, Tsiolkovsky arrived, through scientific reason, at the principle of rocket flight in space. The practical means of moving in the vacuum of space, without air to provide combustion and lift, and of escaping the earth in the first place, was the rocket.[5] Two decades before the Wright brothers demonstrated heavier-than-air flight within the atmosphere, Tsiolkovsky imagined reactive flight outside of it. By 1903 he had published the mathematics of orbital mechanics and designed a rocket powered by the (precocious) combination of liquid oxygen and liquid hydrogen. By 1911 he claimed to "glimpse the infinite potentialities of man,"[6] and when he died in 1935 the USSR celebrated him as the father of cosmonautics, and was itself at work to become its father*land*.

Modern rocketry and social revolution grew up together in tsarist Russia. There is no anomaly in the fact that the most "backward" of the Great Powers before World War I was the one that fostered violent rebellion against the chains of human authority and the chains of nature. Whether expressed in Kibalchich's chemistry or Tsiolkovsky's dream that spaceflight would bring equality and "the perfection of mankind and its individual members," the will to power over natural constraints and the will to revolution knew no contradiction. To be sure, modern rocketry began as an accessory to spaceflight, a goal of idiosyncratic proselytes. But over time it became first a military necessity and then a symbol of dynamism in an age of competing technocratic systems. The advent of spaceflight in our time, therefore, is not just a tale of the gumption and luck of Russian, German, or American rocketeers, but also of the progress of the idea of command technology as a tool and symbol of the modern scientific state.[7] Above all, it is a tale of the specific ideology of the Soviet Union, the world's first technocracy. For rocketry and revolution planted seeds together as children in old Russia, parted ways for decades as Lenin and Stalin prepared the technocratic soil for their growth, then huddled together again against the storms of Cold War. The harvest sprung skyward in *Sputnik I*, the shot truly heard round the world.

CHAPTER 1

The Human Seed and Social Soil: Rocketry and Revolution

The great pioneers of modern rocketry—Tsiolkovsky, Goddard, Oberth, and their successors Korolev, von Braun, and others—were not inspired primarily by academic or professional interest, financial ambitions, or even patriotic duty, but by the dream of spaceflight. To a man they read the fantasies of Jules Verne, H. G. Wells, and their imitators, and the rocket for them was only a means to an end.[1] "For a long time," wrote Tsiolkovsky,

I thought of the rocket as everybody else did—just as a means of diversion and of petty everyday uses. I do not remember exactly what prompted me to make calculations of its motions. Probably the first seeds of the idea were sown by that great fantastic author Jules Verne—he directed my thought along certain channels, then came a desire, and after that, the work of the mind.[2]

Konstantin Tsiolkovsky was born in Izhevskoye in Ryazan Gubernia. His father, he said, was a failed inventor and philosopher.[3] Soviet writers have also made him out to be a critic and victim of tsarist persecution. Be that as it may, Tsiolkovsky grew up in a depressed but intellectually rich household. He loved mathematics and visionary technology, which provided a world in substitution for the normal society of which partial deafness deprived him. At sixteen he conceived of a spaceship driven by centrifugal force:

I was so worked up that I couldn't sleep all night—I wandered about the streets of Moscow, pondering the profound implications of my discovery. But by morning I saw that my invention had a basic flaw. My disappointment was as strong as my exhilaration had been. . . . Thirty years later, I still have dreams in which I fly up to the stars in my machine, and I feel as excited as on that memorable night.[4]

At twenty-one, Tsiolkovsky became a schoolteacher in Kaluga province. He began experimentation in his own makeshift laboratory that earned him election to the same St. Petersburg Society that had rejected his earlier work. In 1903 he published "Exploration of Cosmic Space with Reactive Devices," in which he set down the principles of rocket motion.

By the turn of the century Tsiolkovsky was no longer alone in his interest in flight and rocketry. I. V. Meshchersky published *The Dynamics of a Point of Variable Mass* in 1897, N. E. Zhukovsky, "the father of Russian aviation," built a wind tunnel at Moscow University in 1902, and engineer D. P. Ryabouchinsky founded the Kuchino Institute of Aerodynamics in 1906. Behind their investigations stood a Russian tradition in mathematics, chemistry, and theoretical physics as advanced as any in the world.[5]

Under the old regime, Tsiolkovsky received only a single grant, 470 rubles, from the Academy of Sciences. So rather than building and testing, he specialized in theory and design, culminating in the idea of the "rocket train," or multistage rocket, which was the high road to earth orbit. As the possibilities became clearer in his mind, his revolutionary imagination leaped higher. Spaceflight would mean liberation from limits; limitlessness meant perfection. In 1911 he gazed into the future with the brazenness of Wells and without the foreboding of a Henry Adams. Where they envisioned the tapping of nuclear energy and terrible wars, Tsiolkovsky saw a multitude of colonies around the earth "like the rings of Saturn," where people could tap solar energy a hundred or thousand times as great as on the surface of the earth. "However, this may not satisfy man either, and having conquered these bases, he may extend his hands for the remaining solar energy which is two billion times greater than that received by earth. . . . It will be necessary to move farther away from earth and become an independent planet—a satellite of the sun and a brother of earth."[6] Once that was accomplished, human society would surely achieve perfection.[7]

Social perfection was also the goal of the Bolshevik Party, which assumed control of Russian destiny in 1917. The Bolsheviks, too, expected perfection to follow from the planned conquest of nature through technology. But the contradictions that had plagued tsarist efforts at rapid industrialization survived under communism: how to promote technological revolution without the social and political effects of such massive change undermining the political monopoly of the regime? The tsars never solved this dilemma. The Bolsheviks at least treated its symptoms—under Stalin—through the exercise of terror. The Russian rocketeers, the human seeds of spaceflight, were thus elevated as exemplary by the Bolsheviks, only to disappear in the "socialist soil" of the Stalinist state—out of sight, but germinating. Stalinist technocracy, seeming to crush the utopian spirit of modern rocketry, was in fact to institutionalize the link between rocketry and revolution.

The primary characteristics of Russian science since Peter the Great have been its reliance on state direction and its ambivalent relationship with the West. In the nineteenth century both Slavophils and Westernizers granted Russian backwardness in science and technology, but Slavophils hoped that new industrial techniques could be used to strengthen the Russian state without sacrifice of the superior moral traditions of Russian culture. If Russia were to survive and prevail in European and world competition, she must first raise herself to the prevailing level of technology. But the very process of importing "alien" knowledge would seem to threaten her cultural distinctiveness or political stability. The "Official Nationality" of the reign of Nicholas I (1825–55)—Autocracy, Orthodoxy, Nationality—was a holistic structure of mutually supporting struts. When the Crimean War revealed the need to graft a fourth— Technology—on to the official triptich, its integrity was compromised. As the "Four Modernizations" of a later (Communist) giant bespeak less the totality of the will to modernize than the hope of limiting change to specific nonpolitical areas, so the tsarist governments in their last fifty years tried to quarantine imported capital and technology lest the attendant social and economic fallout subvert the state. A stress on pure science for native Russians and the class base of the technical intelligentsia were features of the tsarist effort in this regard.[8]

Thanks to their triumphs in space, it is difficult to conceive of a time when Russians were known as brilliant theoreticians but poor engineers. Yet such was the reputation and the fact before World War I. This was the age of Nikolai I. Lobachevsky, Ivan P. Pavlov, Mendeleev, and Tsiolkovsky, all working with the chalkstick in pure fields of research. The Academy of Sciences under Nicholas I made it a matter of policy to direct scholars toward mathematics, classics, and Oriental studies, the better to inhibit diffusion of subversive Western political and economic theory.[9] Disciplining technicians was more difficult. Peter the Great had protected the state from dependence on specialists of uncertain loyalty by drafting the nobility into mandatory state service, but the expanding technical structure of the nineteenth century forced the state to tap a larger social base. So commoners who graduated from technical schools received exemptions from taxation, special ranks, and other badges of noble status.[10] Still, the curricular emphasis on pure science and admin- istration meant that on-site engineers had to come from abroad (negating self-sufficiency) or from tekhniki among uneducated craftsmen (negating the social policy). The increased number of students from lower social origins (especially after 1890) and their underutilization only increased native Russian disaffection with the regime.[11] Most basically, technical students trained in rationalization and efficiency were bound to be critical of the political order no matter what their class origins.

Employed scientists and engineers (as opposed to students) favored the moderate Constitutional Democrats, or Kadet party. Their interest lay in nudging the regime toward more support for research and industry rather than in its utter abolition.[12] As early as 1866 technical experts founded the Russian Technical Society with the aim of coordinating science, technology, industry, and government for the development of the economy. Only in wartime did the reforming technicians score some gains. A Commission for the Study of Scientific Productive Forces, headed by the famed geophysicist and Kadet V. I. Vernadsky, gathered 139 experts to work on applied R & D. Other wartime models of science/industry/government cooperation, such as the Chemical Committee of the Chief Artillery Directorate and the War Industries Committees, pioneered the national mobilization of technological resources, which became a cornerstone of Bolshevik economics.[13] The wartime experience and Bolshevik rhetoric both gave Russian technicians reason to hope that their agenda would finally be adopted after 1917, but both the energy and the disaffection of the technicians would likewise find expression in the new era to come.

From the moment of seizing power the Soviet leadership fell into debate over the proper role of science and technology, and especially scientists and engineers, in the building of Communist society. No previous government in history was so openly and energetically in favor of science, but neither had any modern government been so ideologically opposed to the free exchange of ideas, a presumed prerequisite of scientific progress. No government in history was so committed to material growth as the goal of science and the measure of political legitimacy, yet no modern government had so anathematized "bourgeois" intellectuals and entrepreneurs, the sources of material progress. No government in history so trumpeted its nation's backwardness—a legacy, it insisted, of the criminal lethargy of the old regime—yet no modern government so distrusted the foreign specialists and ideas needed to close the material gap. In their early efforts to apply Marxist theory to practical government in an essentially precapitalist society, the Bolsheviks sought to apply the scientific method to the study of science itself—and came face to face with the aforementioned contradictions. Once again, the problems stemmed from the regime's resistance to change in *some* areas of national life. The Bolsheviks, like the tsars, defined their political system as immutable. Science and technology must be fostered as never before, but somehow must be insulated lest their influence spill over into the realm of ideology and power.

Marxism/Leninism divided society into three strata: the productive forces; the productive relations of people, which, in union with the forces, made up the economic base; and the ideological superstructure. Where did science fit? Surely it contributed, especially through technology, to the economic base. But science also inhabited the realm of abstract ideas, the superstructure. In the early stages of revolution the latter formulation had the upper hand. When the first waves of Red Terror washed over the laboratories, universities, and factories in 1918, academics and capitalist technicians learned that they were "enemies of the people." Some were exiled, others fled, including aircraft designer Igor Sikorsky (father of the helicopter) and George Kistiakowsky (who would help to shape U.S. space policy forty years later).[14] But the persecution of "bourgeois" scientists implied the existence of such a thing as "Marxist science." What distinguished them? Could different versions of objective truth exist even in the hard confines of pure science or machinery?

Marx gave the Bolsheviks little to go on. In *Das Kapital* he seemed to hint that science was a direct productive force. Astronomy, for instance, grew out of the Egyptians' need to predict floods and schedule planting. But he also argued that science came to acquire needs of its own and evolved abstractly without relation to the means of production.[15] In the early 1920s two schools of thought contended in the new Soviet Union. The so-called Mechanists held that dialectical materialism, as the prime mover of history, must be inherent in all growth of knowledge, in bourgeois as well as socialist societies. Hence true science in the West was still dialectical materialist, even if its practitioners were not aware of it. The task of Marxists was simply to reposition Western science in the proper scheme of things. A second school, the Deborinites, feared the consequences of such accommodation. Bourgeois objectivity must be countered lest it poison the class-consciousness of the revolutionary elite. They held that bourgeois and Marxist philosophies were active elements in the class struggle and belonged rightly in the superstructure. Marxist science indeed existed and must be assiduously pursued.[16]

The debate never fully resolved itself in the 1920s. Rather, the threat to the revolution posed by the "brain drain" from Russia after October 1917 eventually determined Leninist policy. Just as the movement for a voluntarist Red Army freed of officers, honors, and bourgeois technicians soon gave way to traditional discipline, conscription, and recruitment of tsarist veterans, so the drive to purge the old technical establishment soon succumbed to the dictates of survival. After the humiliating Treaty of Brest-Litvosk, Lenin interpreted the lessons of the First World War: "The war taught us much, not only that people suffered, but especially the fact that those who have the best technology, organization, and discipline, and the best machines emerge on top. . . . It is necessary to master the highest technology or be crushed."[17]

At the Eighth Party Congress in January 1919, Lenin summarily dropped the ideological approach to the scientific question: "The problem of industry and economic development demands the immediate and widespread use of experts in science and technology whom we have inherited from capitalism, in spite of the fact that they inevitably are impregnated with bourgeois ideas and customs."[18] Staunching the hemorrhage of talent, he declared, "One must spare a great scientist or major specialist in whatever sphere, even if he is a reactionary to the nth degree."[19] Lenin also directed the immediate establishment of a complex of new research institutions, especially for electricity, physics, mining, automotives, fuels, and aviation.

Bolshevism declared itself. The unprecedented founding of governmental institutes, independent of universities, factories, or armies, solely for the promotion of technological progress, helped to reconcile frustrated technicians to the new order. The central place of technological progress in the new Soviet state was then made explicit at the founding of the State Planning Commission, or Gosplan, in December 1920. "In my view," said Lenin, "this is the second program of the Party. We have the [1919] party program. This is the political program. . . . It must be supplemented by a second party program, a plan of work for the creation of our economy and bringing it up to the level of contemporary technology."[20]

Lenin also provided an ideological cover for his pragmatism and opportunism regarding the holdover technicians whatever the outcome of the debate over the place of science in dialectics. All Marxists endorsed the unity of theory and practice. Whoever pursued knowledge must also pursue methods of applying that knowledge to the building of socialism. Similarly, whoever dedicated himself to the revolution must draft all available expertise to its service. Such unity of purpose placed a premium on applied research, satisfying Mechanists, who were ready to enlist bourgeois science into the Marxist cause, as well as Deborinites, who could continue to relegate pure science to the superstructure. But most definitive of all Lenin's formulations was the identification of technological progress with the progress of Bolshevism, as expressed in the dictum "Electrification plus Soviet power equals Socialism." Rapid technological progress was both the purpose and the measure of the revolution.

By the end of 1919 depradations against bourgeois technicians ceased. The Bolshevik regime courted them, offered them political and material encouragement only dreamed of under the tsars, and identified themselves with their goals: rationalization, research, and reconstruction. Down the road, of course, the inevitable conflicts between an unbending political oligarchy and a nonideological but indispensible technical elite would emerge and force confrontation. But for the moment Bolsheviks, scientists, and engineers could celebrate together the founding, in war and revolution, of a powerful new mutation in the evolution of political forms—the world's first technocracy.

For Soviet rocketeers the Bolshevik Revolution was unquestionably a release. In 1918 the new Soviet regime followed Western wartime examples in founding the Central Aerodynamics Institute (TsAGI), then the Zhukovsky Academy of Aeronautics in 1919. Tsiolkovsky was elected to the Soviet Academy the same year and received a pension from the government. War and revolution also spawned books and articles on rockets and space travel. Escapism for the public, they were prophecies for a few and helped to carry the dream of rocketry and revolution into the new Russia. Alexander Bogdanov's *Red Star* (1908, 1918) was the first to mate Sovietism and spaceflight. In his novel a Russian Marxist revolutionary is carried off by Martians to their home, the "Red" planet, where they have achieved a communist paradise including free love, collectivized child-rearing, abolition of money, and, of course, highly advanced technology. After 1917 Bogdanov went on to lecture on behalf of atomic energy, since "the most necessary prerequisite of socialism was a highly concentrated source of energy available to small groups of individuals. . . ."[21]

The 1920s, in the Soviet Union and elsewhere, were the springtime of rocket pioneering.[22] The American Robert Goddard published his treatise on rocketry in 1919 and dedicated his life to experimentation with liquid fuel rocketry. In Weimar Germany, the Transylvanian Hermann Oberth, who read Verne and Wells and turned to spaceflight at the age of fourteen, wrote his thesis on rocketry at Heidelberg. Rejected by the faculty, it was published as *Die Rakete zu den Planetenräumen* (The Rocket into Interplanetary Space) in 1923. Oberth's work on Fritz Lang's silent film classic, *Frau im Mond* (1930), and his 1929 book *Wege zur Raumschiffahrt* (Paths to Space Travel) helped to encourage young Germans to vault the engineering hurdles on the road to fulfillment of the dream.[23] Among those young Germans was Wernher von Braun.

Individual efforts and private organizations proliferated in the Soviet Union as well. But it was there that the state first showed interest in the means (rocketry) and the end (spaceflight). As early as 1920 a disciple of Tsiolkovsky's, F. A. Tsander, met briefly with Lenin and was thrilled by the leader's interest in spaceflight.[24] In 1924 the Soviet regime created a Central Bureau for the Study of the Problems of Rockets (TsBIRP) with these stated objectives:

1. To bring together all persons in the Soviet Union working on the problem.
2. To obtain as soon as possible full information on the progress made in the West.
3. To disseminate and publish correct information about the current position of interplanetary communications (the Soviet rubric for spaceflight).
4. To engage in independent research and to study in particular the military applications of rockets.[25]

Perhaps point four reflected the major concern of the state. But the fact remains that the Soviet Union was the first government to endorse and support the goal of spaceflight. The same year a private All-Union Society for the Study of Interplanetary Communications (OIMS) formed in Moscow. The rough equivalent of the American Interplanetary Society (1926), the German Verein ʿür Raumschiffahrt (VfR) (1927), and the

British Interplanetary Society (1933), the OIMS attracted 150 charter members divided into sections for research, publicity, and publication. In 1927 the OIMS and the TsBIRP cosponsored a Soviet International Exhibition of Rocket Technology in Moscow; subsequent conferences discussed the pragmatic steps needed to move from theory to praxis.[26]

The human raw material was present. In the 1920s Tsiolkovsky entered his most prolific phase, though many of his Soviet-era writings were more speculative than scientific. But Zhukovsky, Meshchersky, and S. A. Chaplygin had by now educated a decade of students in aero- and orbital dynamics. Yuri Kondratyuk embraced rocketry during the war, worked throughout the 1920s, and published his findings in *The Conquest of Interplanetary Space* in 1929. He suggested, among other things, the use of solar power in spacecraft and a separate landing module for lunar and planetary visits. He also explored problems of reentry into the atmosphere. Nikolai A. Rynin's *Interplanetary Flight* was an encyclopedia of astronautical fantasy and theory. In the early 1930s the Soviet state also encouraged translations of foreign literature on rocketry.[27]

Tsander was the leader of the new generation. A Lithuanian graduate of the Riga Polytechnic Institute, he dedicated his life—and children Astra and Mercury—to the cause. Tsander lectured extensively in the USSR on the coming Space Age, and specialized in propellants. At the TsAGI in 1929–30 he built the first Soviet liquid rocket, the Opytnyi Reaktivny-1, powered by gasoline and compressed air. The OR-2 developed 110 pounds of thrust from benzyne and liquid oxygen (LOX). Ten days after its first successful test in March 1933, Tsander died of typhoid fever at the age of forty-six. He left behind his *Problems of Reactive Flight* (1932)—problems to be solved by his youthful collaborator Valentin P. Glushko.[28]

Tsander and the others of the postwar generation, however, were not destined to be Russian Goddards, carrying on in isolation and penury. By 1934 rocketry had already been swallowed up into the belly of the Stalin's Leviathan state. Henceforth rocket research was not to be a private affair, or an avocation of the state apparatus, but an integral part of the Soviet drive for technical supremacy. The link between rocketry and revolution was institutionalized.

In Communist theory, technological progress was virtually equivalent to the march of history. This was why Marx insisted on the mighty progressivism of capitalism. But capitalism slid irretrievably into a monopoly stage as the ruling class inhibited new technology that threatened power structures based on existing means of production. In socialism no such barrier to innovation could theoretically exist: centrally directed R & D, in fact, seemed one of the most seductive features of the Soviet system. Yet technological progress meant much more to the Soviet leadership. It was a principal means of countering the threats of hostile imperialist states, as well as the measuring stick by which communism would prove its superiority to capitalism. In practice, these two elements in Soviet technocracy conflicted from the start. Historian Bruce Parrott has distinguished two abiding tendencies in the Soviet official mind. The first, or traditionalist, viewpoint warned against the hostility of the outside world and Soviet dependence on foreign technology. Since socialism provided a richer soil for innovation, the USSR could and must pursue autarky, sealing off the country from toxic ideas and espionage. But this also required that the Soviets play "catch up" on a massive scale with one hand tied behind their backs.

The second, or nontraditionalist, strain tended to mute the theme of imperialist aggression and warn against undue concentration on military power to the detriment of civilian industry. It argued the good sense of seizing "the benefits of backwardness" (in the words of Stalin's Minister of Trade, Anastas Mikoyan), imitating the successes and avoiding the failures of the West, and importing technology from advanced countries.[29] Soviet R & D policy has wavered between these tendencies throughout its history, depending on the party's current sense of the international climate, internal security, and the lessons of history. At all times, however, the goal was to advance technology at the fastest rate by any means consistent with the political security of the regime. It was only a matter of time before the latter consideration forced a confrontation with the technical intelligentsia.

The "Golden Years" of Soviet science (so called by the Medvedevs and others) from 1922 to 1928 were thus a time of toleration and new resources, when the Soviet Union was most amenable to foreign influence and voluntaristic organizations, when aviation was still in the hands of specialists whose careers dated from before the war, and ideological pressure was modest and indirect.[30] All this fostered a myth that only Stalinist betrayal bent the Soviet twig toward tyranny and xenophobia. Perhaps that is so, but the "golden age" was exceedingly short and was beginning to be undermined almost from the start. It is hard to imagine how a free and inquisitive technical sector could have continued to coexist with a ruling elite defined by ideological purity . . . or how rapid industrialization through expropriation and terror could have occurred in cooperation with an autonomous technical intelligentsia.

But an irony of the "golden age" is that even if it had survived, it could never have launched the Space Age. The tinkerings of a Tsander or even a Korolev could have lasted for decades without issue but for the validating ideology and industrial base established by Stalin. If rocketry was to mature, its theater must move from the excited halls of international conferences and the smoky studies of devotees to the bunkers and test stands of the purposeful state.

In 1923 Josef Stalin won the right to appoint industrial managers throughout the USSR. Following Lenin's death in April 1924, so-called

Old Specialists suddenly found themselves demoted in favor of Red Directors who were less skilled but loyal to Stalin. In December 1925 the Fourteenth Party Congress adopted Stalin's industrial program and urged the comrades to overtake the capitalist world "in a relatively minimal historical period." By then the failure of world revolution was apparent—Soviet defeat in the Battle of Warsaw (1920), failure of Communist agitation in the German Ruhr district under French occupation (1923), the end of the German inflation (1924), and the Locarno Treaties (1925) forced the Party Congress to admit the "partial stabilization of capitalism," which in turn fostered the "partial progress of technology" in the West. This perception, combined with the diplomatic recognition of the USSR by Britain and France in 1924 and the Soviet treaties of cooperation with Weimar Germany (1922, 1926), *should* have justified "nontraditionalists" who emphasized the possibility of trade and coexistence. Instead, Stalin proclaimed in 1927 that "the period of 'peaceful co-existence' is receding into the past. . . ." The USSR must fight to build "Socialism in One Country" and guard against encirclement and imperialist attack. The next year, his power secure, Stalin launched the First Five Year Plan for hothouse industrialization.[31]

The imaginary foreign specter served as a pretext for intimidating opponents of the Five Year Plan. Stalin's 1931 speech on behalf of the plan reminded the party that:

The backward are beaten. But we do not wish to be beaten! No, we do not! The history of Old Russia consisted of being constantly beaten for her backwardness. The Mongol Khans beat her. The Turkish beys beat her. The Swedish feudalists beat her. The Polish-Lithuanian nobles beat her. The Anglo-French capitalists beat her. Everyone beat her—for her backwardness.[32]

Once again, he claimed, capitalist pressures for war were building, but the rapid leap ahead for which the Five Year Plan was blueprint would bring the USSR astride the Western economies in short order.[33]

To be sure, technological superiority was to be a primary legitimizer of Communist authority, but in order to whip the nation to the necessary efforts, the regime constantly had to invoke the threat from more developed, hostile states abroad. Like St. Augustine, who cried, "Lord, make me righteous—but not yet," the Party constantly cried, "We are inevitably superior—but not yet." Throughout the 1930s, when the international situation was worsening steadily, Stalinist propaganda alternated between self-congratulation and exhortation to overcome inferiority if the revolution was to survive.[34] Even as Stalin warned against the USSR becoming "an appendage of the capitalist world economy," he accelerated imports of foreign capital goods. In 1932, 80 percent of all machine tools installed came from abroad.[35] Metallurgy and automotives from the United States, chemical plants from Germany and France, electrical factories from RCA, General Electric, and Vickers—exporting gold and foodstuffs squeezed from newly collectivized farms, Stalin purchased an industrial economy from the imperialists. With the machinery came foreign technicians, and it was that influx that provided the background for the absorption of the technical intelligentsia into the monolithic system.

Late in 1927 the local secret police chief in the Caucasian town of Shakhty reported the activities of a group of "wreckers"—engineers who allegedly conspired to sabotage coal mines at the instigation of their former capitalist bosses and "foreign interests." Stalin seized the moment to signal a radical approach toward experts, the Central Committee identifying a "new form of bourgeois counterrevolution by a small group of specialists who had been especially privileged in the past." In the wave of "indignation" following the Shakhty engineers' trial, the Party encouraged harassment of "bourgeois" and foreign technicians.[36] "Shakhtyites," Stalin announced, "are now ensconced in every branch of our industry. . . ."[37] Stalin then generalized the new approach in the so-called Industrial Party Affair. Eight leading technologists were accused of leading a conspiracy of over 2,000 engineers to take over the government in league with dissidents, emigres, and Western governments.[38] Where Shakhty had legitimized persecution of technicians for purely professional or political shortcomings, the Industrial Party Affair established guilt by association. Entire cadres of regional industries landed in prison if one of their number fell under suspicion. Their alleged crime was usually a tendency toward nonideological technocracy as reflected in below-quota performance, opposition to elements of the Five Year Plan, or sympathy with the "Right Opposition" in the Party associated with Bukharin.[39]

Soviet technocracy meant institutionalization of technological change for state, that is, Communist Party, purposes. Traditional technocracy implied the abolition of politics altogether in favor of public management by pragmatic technicians.[40] The differences were papered over in the 1920s, inviting Old Specialists and foreign observers to misunderstand the Communist message. Stalin determined to disabuse them, proclaiming "technocracy" to be a cover for attempts by specialists to blackmail the state and set themselves up as a political rival to the Party. Criticism of policy on supposedly "objective" grounds became grounds for arrest.[41]

The Industrial Party trial of 1930 was the biggest "show trial" prior to the Great Purges and fixed the role of technicians in the Soviet Union. Far from "co-opting" the state, "the engineer, the organizer of production," in Stalin's words, "would not work as he wished, but as he was ordered, in such a way as to serve the interests of his employers. . . . Of course, the assistance of the technical intelligentsia must be accepted and the latter, in turn, must be assisted. But it must not be thought that the technical intelligentsia can play an independent historical role."[42] In the wake of the industrial trial, the "nontraditionalist" Bukharin and other Old Bolsheviks who favored pure science ("Great practice requires great theory") and a measure of scientific freedom lost their influence and, by 1938, their lives.[43]

The central dilemma of technical progress in a political monopoly remained. It tended to raise up a class in control of the means of production by dint of expertise and therefore a potential rival to the Party. Indeed, such rivalry was even more enduring than that of the capitalists, for technicians could presumably seize power (or at least cause trouble) even in a post-class struggle society. This prospect has obsessed the leaders of every Communist country, from Stalin to Mao-Tse-Tung to Pol Pot. Mao's (and Trotsky's) approach was continuous revolution, Pol Pot's was the elimination of technology and expertise altogether.

Stalin's answer to the potential autonomy of the technical intelligentsia was to discipline it through terror and simultaneously to train a new cadre of Red Experts. The first step toward the latter goal was the corruption of the Academy of Sciences, suddenly accused by the police of being "a center for counterrevolutionary work against Soviet power."[44] The new 1927 Charter of the Academy placed it under the direction of the Council of Peoples Commissars and expanded its membership from forty-four to seventy. The following year the state summarily violated its own charter by adding fifteen more members, charging the academy with discrimination against Communists. The new seats clustered in philosophy and social sciences, ensuring an infusion of Marxism, as well as in engineering, a novelty for the academy. The difference in tactics from the tsars is illuminating—the old regime simply forbade social studies, the Communists took them over, then promoted them. When three Party members failed to be elected in 1929, the commissars intimidated the academy into admitting them and amended the criteria for election to include "socio-economic physiognomy." In 1930 Stalin dispensed with legalisms altogether, and the academy became another arm of the party-state.[45]

Once subordinated to the needs of the Five Year Plan and the education of Red Experts, the academy burgeoned. Personnel grew from 1,000 to 7,000 in the decade after 1927, and surpassed 16,000 by 1940. The budget rose from 3 to 175 million rubles by 1940 (though partly through inflation). A Division of Technological Sciences and four institutes for applied research were meant to harness science to the economy.[46] Technical enrollments quadrupled during the First Five Year Plan to 88,600, whereupon they subsided to between 40,000 and 50,000 per year until the war.[47] The growth of the academy reflected Soviet trends in R & D generally. Lenin's immediate and vigorous efforts to support science and technology seem like feints compared to the explosive thrusts of the first two Five Year Plans. Western experts estimate that Soviet

R & D spending *sextupled* between 1927 and 1932. By the time of the second Five Year Plan (1933–37), the USSR was ready to dispense with most imports. The promise of "reverse engineering" (taking apart an import, modifying it for local conditions, and replicating it at home), the deteriorating international climate, embarrassing defections by trade personnel, and the supposed danger posed by 9,000 foreign technicians inside the country all suggested an abrupt return to autarky and isolationism.[48] During the second Five Year Plan, only 2 percent of gross economic investment came from abroad, travel for scientists and commercial agents virtually ceased, and the campaign against "servility before foreign science" reduced access to foreign technical journals.[49]

All this placed immense burdens on the indigenous R & D sector. Between 1932 and 1935 research funds grew 60 percent more, and the Second Five Year Plan authorized yet another tripling. One-sixth of all technical graduates worked in R & D, and the share of national income devoted to it climbed to twice the percentage invested by U.S. private and public sectors together.[50] The Party Central Committee noted in 1933 that world trends seemed to bear out Communist predictions: the depression was causing capitalist countries to close down research institutes even as the USSR built them up. This indicated a "degradation of scientific thought" in the West and gave the lie to Bukharinite defeatists who spoke of "organized capitalism" maintaining momentum through regulation and research.[51] The USSR seemed to be arriving at the point where, having made up for initial backwardness through imports, it could press ahead with confidence.

Gradually the contradictions of socialist R & D impressed themselves on the leadership. Links between the mushrooming research sector and the managerial sector were weak. Plant managers struggling with short-term quotas had not the resources, time, or incentives to retool for the long haul. Engineers cloistered in institutes were out of touch with industrial needs. The terror that hung over experts discouraged risk-taking; failure could mark one as a "wrecker." Committees sprang up in laboratories to dilute responsibility and erode originality. Circulation of new data among the vast scientific and engineering communities was poor. Finally, and most telling perhaps, was the "mechanistic" pattern of administration employed in civilian R & D. Initiative lay less with the scientists than with the managers, constrained in turn by elaborate bureaucratic rules to minimize waste and maximize control. Information flowed vertically, up and down the command structure, rather than among the scientific teams themselves. It was a system well suited to Soviet political purposes, and before 1941 it worked tolerably well in heavy industry and agriculture where intensified exploitation of labor and existing resources could show results. In dynamic fields of technology, however, an "organic" system is more effective. It permits a loosely defined division of labor and relies on spontaneous cooperation among specialists. Leaders supervise but do not stifle the imagination of under-lings. It is more tolerant of waste, expecting researchers to confront some dead-ends, and permits a horizontal flow of information among research teams.[52] The civilian economy of the USSR, lacking a market mechanism and burdened with political overlay, tended toward mechanistic manage-ment. The result was that technological renewal never measured up to the intense effort put into R & D; or, to put it another way, the regime had to invest especially large sums to achieve the results other countries obtained on less.

In sum, the first Five Year Plans institutionalized the Soviet commitment to technological superiority. But the same political ideology that glorified technology set up barriers to innovation. Rather than liberating the creativity of the people, the regime subjected the technical intelligentsia to a monopoly of patronage more constrictive than had existed under the tsars. The results of indigenous R & D never fulfilled the promise . . . except in one sector. In one sector a measure of organic management did obtain; in one sector political supervision often helped rather than retarded innovation. That sector was defense.

Soviet propaganda argued that the capitalist world was implacable; only its tactics changed over time. When the Allied military invention of 1918 failed to overthrow Bolshevism (which was not its initial purpose),[53] the imperialists turned to ostracism and sabotage, biding their time until internal pressures inherent in capitalism sparked another round of imperialism and war. It could be equally argued that the "imperialists" themselves made it possible for the USSR to consolidate and build its strength. Under American pressure, the Japanese evacuated Siberia in 1922, while the power vacuum in Central Europe sustained by Versailles Treaty restraints on Germany gave the Soviets a *cordon sanitaire* against Western hostility as much as it shielded the successor states against Bolshevism. It even permitted a dozen years of surreptitious German/ Soviet cooperation in weapons development and testing.[54] After 1929, furthermore, the Communist International only fostered renewed hostility by attacking liberal and social democratic parties in the West, thus aiding the rise of fascism.[55]

Nevertheless, the giant steel, automotive, tire, and aircraft plants built with foreign help—and the extra two years purchased with the Nazi-Soviet Pact—provided the Soviets with more tanks and airplanes than the German invaders of 1941, a larger army, and weapons that were not necessarily inferior. Still, the Blitzkrieg shredded the Red Army, conquered most of European Russia, and bequeathed a reputation for primitiveness to Soviet weaponry that lasted until Sputnik. It is important to examine pre-war Soviet R & D, however, without the shadow of 1941, so that the rapid development of aircraft and rocketry after 1945 will not seem so mysterious.

Soviet military research surely suffered some of the same handicaps as the civilian sector: the Great Purges, shortages of raw materials (especially aluminum, tin, and copper), duplication of effort in design bureaus, and an excess of secrecy.[56] But the production figures speak for themselves: by 1939 the Soviet Union was turning out 700 to 800 planes per month, more than Japan and roughly equal to the amount produced by Britain, Germany, or the United States. In the 1920s the Soviets imported airplane engines from BMW, Napier, and FIAT, and the first Soviet airline flew planes from Vickers and DeHavilland. But the First Five Year Plan stressed indigenous construction, and by the mid-1930s, Soviet aviation was on solid ground for expansion. The industry was also well-served by its R & D sector. The TsAGI boasted its own, unusually well equipped, shops and factories, and commanded better materials than civilian labs. Second, military R & D benefitted, rather than suffered, from scrutiny by political leaders. Stalin was said to devote almost daily attention to aviation, and "knew well dozens of plant directors, party organizers, [and] chief designers in the defense sector, and arbitrated disagreements." Third, the powerful position occupied by the consumer agency, the military, liberated weapons designers from production quotas and allowed a greater concentration on quality. Fourth, military R & D was organized more "organically," allowing some competition and local initiative in design strategies.[57]

Soviet "state of the art" aviation went on display in the Spanish Civil War, where Nikolai Polikarpov's I-15 and I-16 fighters appeared to hold their own against German Messerschmidts (Bf-109). In A. S. Yakovlev's (exaggerated) claim, "they really gave it to the Messers."[58] But more impressive (for they were contrived to be so) were the aeronautical prizes captured in the years of the Second Five Year Plan. The records reveal not only the virtuosity of Soviet designers but also the role increasingly played by advanced technology in the political legitimation of Soviet rule. As historian Kendall Bailes has illustrated, legitimacy was a continuing problem for the Bolsheviks, who seized power as a self-proclaimed elite possessed of superior understanding of the laws of history. Stalin had the additional onus of being a usurper within a party of usurpers. He and the Party implicitly justified their rule, therefore, by the obvious "success" of their policies for industrialization, collectivization, and high technology, especially the spectacular achievements of Soviet aviation.[59] Of course, Stalin ruled by terror—but the brutal tyrant, in the age of mass politics, most of all needs to cloak his exercise of power in theatrics and ideology.

The air campaign began in 1933, when Stalin set aside August 18 as Aviation Day. Soviets at once entered competitions, seeking to fly higher, faster, and farther than pilots of other countries. To be sure, the TsAGI engineers may have lobbied the Kremlin on behalf of aviation, but the recent gripping flights of Lindbergh and others, including around-the-world jaunts that crossed the vast Soviet Union itself, must have impressed on Stalin the political benefits flowing from a reputation for aviation leadership.[60] So year by year Stalin's Falcons flew to glory. At one time or another (or more than once) Soviet pilots claimed records for the fastest, longest, and highest flights, the first flight over the Arctic to America, and the first landing at the North Pole. The credit went to Stalin. "He is our father," wrote one pilot. "The aviators of the Soviet Union call Soviet aviation 'Stalinist aviation.' He teaches us, nurtures us, warns us about risks like children who are close to his heart. . . . He is our father."[61] Politburo member L. M. Kaganovich wrote that "Aviation is the highest expression of our achievements. Our aviation is a child of Stalinist industrialization; flyers are our proud falcons, raised lovingly and with care by Stalin."[62] Aviation Day became a much-ballyhooed national festival, posters depicted a dreamy and paternal Stalin with airplanes buzzing his head like King Kong, and aircraft began to fill the skies over Red Square on May Day and the Anniversary of the Revolution as unmistakable proof of the wisdom of the Party and its leader.

The aviation campaign also served to distract attention from the growing terror. Its kickoff in 1933 coincided with the appointment of the first Purge Commission, and the peak of the aviation campaign with the show trials of Kamenev and Zinoviev in 1936 and the peak of arrests in mid-1937. The media saturated their audiences with suspenseful accounts of the flights and human interest stories on the pilots, implicitly comparing the arrested "traitors"—the counterrevolutionary detritus of an earlier age—with the New Soviet Man exemplified by Stalin's Falcons and the new age to come.[63]

Yet the age to come was not what *Pravda* prophesied. The record flights apparently bred complacency and drew attention away from military design at the very time when the Luftwaffe, learning from experience, accelerated innovation. By 1938 the Me109e of World War II fame arrived in Spain and cleared the skies of the more numerous Soviet planes. Stalin's enraged search for scapegoats led to the arrest of his leading designer, A. N. Tupolev, then V. M. Petliakov, V. A. Chizhevsky, and the other chief designers of their day. "Stalin reacted very painfully to our failures in Spain. His dissatisfaction and wrath were directed against those who quite recently had been considered heroes. . . ."[64] The Great Purge, then gaining momentum, swallowed up Marshal Tukhachevsky, the leading patron of advanced technology, as well as the chief of the air force and most of the General Staff.

There were causes other than outclassed fighters for the disaster of 1941. The ruthless purge of the officer corps removed those professionals most likely to learn the lessons of Spain and Finland (1939–40). Soviet strategists, meditating on the nature of the "next war," fancied the massed bombardment doctrines of Guilio Douhet and placed undue emphasis on heavy bombers. In the summer of 1941, armadas of Soviet bombers without fighter support were either caught on the ground or smashed in their efforts to assault German tank columns—Field Marshal Kesselring shook his head at such "infanticide."[65] For these reasons, and not the backwardness of Soviet technology, was the Red Air Force destroyed in 1941.[66]

Stalin turned the USSR into an industrial economy at a frenetic pace, but his domestic "policies," especially the purges, and tactical errors frustrated his long-touted primary task of preparing for invasion. Yet by a perverse logic, the Communist combination of unbridled support for, and terror against, the experts whose creations justified Communist rule helped to prepare the ground for future triumphs. For the aircraft designers were not alone in the support they received, nor were they alone in the gulag—by 1938 they were joined by the rocketeers.

During the 1920s Soviet spaceflight enthusiasts comprised a surviving element of civil society, pursuing ideas, working, and sharing dreams according to personal impulse and not those of the party-state. But the First Five Year Plan submerged rocketry into the state apparatus as thoroughly as every other collective activity, and at just about the same time as their German counterparts were receiving the mixed blessing of Wehrmacht patronage. The voluntary TsBIRP and OIMS groups disbanded, and in their place the state founded the Group for the Study of Rocket Propulsion Systems (GIRD) at Leningrad in 1931. Ancillary GIRDs sprang up in Moscow and elsewhere, but were subsumed, on orders of the Politburo, into the Rocket Research Development Center and then, at the end of 1933, the Jet Scientific Research Institute (RNII). Although the budgets and manpower of these organizations cannot be guessed, they must have benefited from the sixfold increase in R & D spending, permitting the GIRD and RNII to field separate teams for research into engines, vehicles, and rocket aircraft. But the most important work on rocketry lay outside even these burgeoning organizations. In 1928 the Military Revolutionary Council upgraded a research team working on solid fuel rocketry into the Leningrad Gas Dynamics Laboratory (GDL) under the aegis of Tukhachevsky.[67]

During the brief life of the GIRD Tsander built his OR-1 and OR-2 and a flying model, the GIRD-X. After his death the men of GIRD carried on, delighting in the stable flight of a small rocket to 1,300 feet above Muscovite suburban forests. But this sort of wildcatting was coming to an end. The subsequent death of Tsiolkovsky (1935) and the arrest of his pupil Boris Stechkin in the first wave of specialist-baiting marked a new era. Its emerging leaders were Glushko and Sergei P. Korolev, the heads of the GDL and RNII, respectively.

Korolev was born in 1907 in the Ukraine. He began his aviation career as a test pilot, and even after turning to design insisted on climbing into cockpits and risking his life. His brilliance earned him the leadership of GIRD's design and production section—but Korolev's goal in life was not rocketry per se but spaceflight. The Ministry of Defense published his *Rocket Flight into the Stratosphere* in 1934, and Tukhachevsky bestowed upon him the deputy-directorship of the RNII. Glushko was a year older than Korolev and the leading designer of liquid fuel rocket engines. At sixteen he published an article entitled "The Conquest of the Moon by the Earth" and by 1930 was teamed with Tsander. He never had the chance to build large-scale prototypes before the war, but his laboratory testing and theoretical work made breakthroughs in propellants, pumps, ignition, shockless combustion, and regenerative cooling (in which the pipe feeding the frigid LOX curled around and cooled the combustion chamber).[68] While putting good ideas into practice is a far cry from dreaming them up, many of the advances that went into the German V-2 were nonetheless on drawing boards at the GDL in the 1930s.

Glushko, Korolev, and their teams were expected to produce weapons. This divergence led to various dead-ends such as flying bombs, rocket gliders, and rocket planes. But Korolev also seemed to believe—as his U.S. Air Force rivals did in the 1950s—that winged rocket craft, not ballistic missiles, were destined to launch the Space Age.[69] As it happened, his own genius for missile design defeated his expectations. But the work done on rocket-assisted aircraft by the RNII and GDL laid the basis for Soviet jet planes, initial designs for which were complete as early as 1941.[70] From 1936 to 1938 Korolev oversaw a series of rocket tests and winged rocket flights that taught the RNII much about stabilization. By 1939 it launched the world's first two-stage rocket with a ramjet engine that reached 500 miles per hour. The RNII also tested air-to-surface, surface-to-air, and surface-to-surface missiles, the latter to a range of almost thirteen miles.[71]

Korolev's progress in the 1930s, given his management duties and wide array of assignments, was even more extraordinary for the fact that from 1938 onward he himself was a citizen of the gulag. After 1937 "Tukhachevsky" became a synonym for death, and when the Great Purge swept the lower ranks of the marshal's "empire," the entire RNII became suspect. According to defector Leonid Vladimirov, the majority of the important people in the organization did not survive. If so, Korolev was an exception—but to his misfortune, he was reassigned to Tupolev's bureau just in time to follow him to prison in 1938.[72]

It seems mean-spirited to inquire after the effect of institutional terror on the *productivity* of intellectual workers rather than on their minds and hearts and health. Yet the Soviet record suggests that human beings are more adaptive than one might imagine—and that the courage and self-discipline of ordinary people, far from ensuring the bankruptcy of totalitarianism, sometimes work to its advantage. Mass arrests of aircraft engineers (*samizdat* author G. S. Ozerov believed that over 300 were in the camps after 1938) was the *ultima ratio* of Stalinist technocracy. Experts were essential, yet their expertise could not be permitted to form the basis for political rivalry. Intimidation, reorganization, and reeducation could only do so much. Communist logic pointed finally to the solution of the 1930s: the *sharaga*, or work camp, in which the technical intelligentsia was literally a prisoner of the Party. Did not this solution defeat its own purpose by destroying the creativity on which the Party depended? The answer seems to be not in the short run, and not categorically.

Prison camps for technical workers dated from the first purges of Old Specialists after 1930.[73] But there is no evidence that the NKVD (secret police) preplanned the resettlement of scientists and engineers in a politically safe environment. The sophisticated *sharagas*, rather, postdated the dragnet of scientific personnel in 1937–38. Staffing the Tupolevskaya Sharaga, an aircraft design prison under Tupolev, required a scouring by the NKVD of other far-flung camps and prisons to locate and transfer aeronautical specialists. Once reunited, the aircraft designers exhibited an understandable camaraderie and resumed their mental work on behalf of their own masters, despite a miserable climate, meager rations, and the presence of armed guards. We have two main sources concerning life in the *sharagas*, the samizdat by Oserov and Alexander Solzhenitsyn's novel *The First Circle*.[74] Oserov describes his arrival in Tupolevskaya Sharaga:

We were taken to the dining room . . . heads turned to our direction, sudden exclamations, people ran to us. There were so many well-known, friendly faces. At the tables we can see Tupolev, Petliakov, Miasishchev, Neman, Korolev, Putilov, Chizhevsky, Makarov, and many others—the elite, the cream of Russian aircraft technology. It was impossible to conceive that they had all been arrested, that they were all prisoners—this meant a catastrophe for Soviet aviation![75]

In fact, it did not. The prisoners labored on, for something. The freedom to continue their work may have been salvation enough, even if their labors served their captors. Patriotism may have motivated many, since their work, especially after June 1941, was for the survival of the Motherland. Some experts were promised release in return for successful projects. All faced transfer to hard labor camps as the price of indigence or insubordination. The work camps also enjoyed, ironically, the patronage of the state police, who were probably better able even than the military to procure men and materiel in the chaotic years of purge and war. Secrecy and isolation surely hampered the prison R & D effort, but perhaps little more than on the "outside." Cold and poor nutrition must have sapped the inmates' stamina, especially those unlucky enough to have done some hard labor. Korolev himself is said to have spent some months in the arctic Kolyma mining camp.[76] But by 1942 conditions at the *sharagas* may have been no more spartan than anywhere else in the USSR. Better, certainly, to be at Tupolevskaya Sharaga than back at the GDL in besieged Leningrad! And those engaged in military work probably received better treatment; according to Medvedev, civilian scientists drew minimal rations and most did not survive the war.[77]

How long could the loyalty of scientists and engineers be maintained under these conditions? How long until, like the inmates of Orwell's *Animal Farm*, they forgot that any other life existed? In the short run, the risky days of purge and war, Stalin had the best of both worlds: scientific productivity and political security. In Tupolev's camp three departments worked on military aircraft, especially high-altitude fighters and dive bombers, that helped the Red Air Force draw even with the Luftwaffe. Georgy Langemak (reprieved from a 1937 death sentence) designed the most effective Soviet novelty of the war, the Katyusha, or "Stalin's Organ," which fired an array of truck-mounted rockets up to three miles and was death on fieldworks and tanks. Other prison design bureaus produced new tanks, artillery, and locomotives, and the first Soviet jet plane.

What was the overall impact of the purges on Soviet technological progress? The purported liquidation of thousands of engineers must have been deleterious—what contributions might have been made, or could only have been made by individuals who disappeared from history? But in pure numbers of engineers the purges may not have hurt. R & D can only absorb so much labor and resources, especially in the early stages, and the tremendous flood of graduates and funds in the 1930s may have choked some projects. After the German invasion the entire R & D complex suffered severe disruption and loss of funding in any case.[78] The descent of the elite of Soviet aviation into gulag, therefore, may not have inhibited the development of rocketry. Defector and rocket expert Grigory A. Tokady (also known as Tokaty-Tokaev) thought it necessary in the early 1960s to explain not why the USSR was precocious in space technology, but why it had been retarded! And this, he wrote, was the result of the war, not the purges. The main industrial centers of the Soviet Union were overrun or evacuated in 1941, scarce resources went for immediate military needs, and the rocketeers were deflected from spaceflight into quickly realizable weapons systems.[79] From this slant, Russian rocket research could be considered "on schedule" in 1938 and still progressing in 1941. If the subsequent genealogy of spacefaring rockets came to branch off the German rather than the Soviet trunk, it was due above all to the different ways in which World War II shaped the capabilities and perceived needs of Nazi Germany as opposed to the Soviet Union.

Stalinization forged, by main force, the industrial base and the R & D complex necessary for indigenous technological advance in military fields.

It also identified the person of Stalin and the regime with the purposeful conquest of the skies. The seeds of spaceflight disappeared into the soil of Stalinist technocracy, but by 1941 the soil was sufficiently fertilized—at great human cost, to be sure—for germination to occur. Perhaps Stalinist rule then inhibited the growth of rocketry by its contribution to the outbreak and disastrous defeats of the war. But the course of that same war, by ridding Russia of the invader and saving the technocratic Soviet state, by winning a share of the German technical inheritance for the Soviet engineers, and by blowing in a new international storm in which long-range rocketry took on surpassing importance, also put the Soviet drive for spaceflight back on schedule.

CHAPTER 2

Political Rains and First Fruit:
The Cold War and Sputnik

A week after the Allied invasion of Normandy, a gothic devilment buzzed its way across the English Channel. It was an unmanned, winged cylinder twenty-seven feet long, powered by a jet engine on top of the fuselage that developed 110 pounds of thrust. This *Vergeltungswaffe-1*— so dubbed by Adolf Hitler—was the ancestor of the cruise missiles of the 1980s. It carried only a ton of explosives and flew low and slowly enough for interception. Its successor, the V-2 (or A-4, as its designers called it) was even less cost-effective since its unit cost was ten times that of V-1. But the V-2 was invulnerable, the first medium-range ballistic missile. By investing the dwindling resources of the Nazi Empire in these technical adventures, which, without atomic warheads, could only stoke the determination of the enemy, Hitler did achieve a vengeance of sorts. He hastened the day when staggering costs and numbing fear accompanied the efforts of his conquerors to refine the V-2's offspring into engines of terrible destruction.

A detailed history of the German rocket program lies beyond our scope. The subject has been well covered elsewhere, and more important for our purposes is the fact that the German program ended abruptly in 1945.[1] The political environment of the birth of the Space Age was the Cold War to come, not the war just ended. To be sure, the German legacy permitted the Soviets to traverse quickly that terrain of practical experience the war had denied them, and it pushed the United States into the rocket field before it might otherwise have entered. But the V-2 represented few theoretical breakthroughs unfamiliar to Soviet rocketeers. Only a pure determinist could designate the V-2 a sine qua non of the origins of the Space Age in our time. What the German engineers did, with their clever fabrication of what seemed even in World War II a "baroque arsenal," was to prod their enemies to the East and West into premature fear and rivalry, and to make themselves and their blueprints the most prized spoil of war.

Of all Stalin's imprints on Soviet and world history, his technological bequest is perhaps the least appreciated. For if the Soviets' determined drive for nuclear and rocket weapons after the war derived from Cold War competition, it expressed as well the continuous commitment to Soviet technological primacy promised ever since 1917. When, after Stalin's death, Russian rocketry and Soviet technocracy yielded their first fruits in the birth of the Space Age, the world voiced astonishment. However glaring the faults of communism, the might of technocracy stood revealed.

The German Sixth Army, encircled at Stalingrad, surrendered in February 1943, and the Wehrmacht went on the strategic defensive. Casting about for technological fixes to the dilemma of a Blitzkrieg turned war of attrition, Hitler restored top priority to an extraordinary new weapon called the A-4.

The A-4 entered production as the V-2. A British air raid on August 17, 1943, suggested the wisdom of transferring manufacture to the Mittelwerk factory, carved from a slope of the Harz Mountains near Nordhausen in Thuringia. By 1945 nearly 900 V-2s per month emerged from the eerie underground assembly line, manned by slave labor. The production network of the V-2 was also a prototype of the national integration of brain power and materiel characteristic of the technocratic state.[2]

The Rocket Team harbored a hidden agenda, of course: spaceflight— the Gestapo even arrested von Braun in February 1944 on the charge that he was not really interested in the military needs of the Fatherland and planned to flee to England.[3] Indeed, von Braun's team indulged their dreams of larger versions of the V-2 with the potential for orbital flight. Designations A-5 through A-8 were upgraded V-2s, but the A-9 and A-10 were of another order of magnitude. They were to comprise a multistage rocket, the first stage developing 400,000 pounds of thrust, and designed for a range of 3,200 miles. It was the first ICBM on paper, but it was also a spaceship. The A-11 was visualized as a third stage capable of boosting a pilot into earth orbit. Finally, there was the speculative A-12, boasting a first stage with 2.5 million pounds of thrust and capable of orbiting a 60,000-pound payload.[4]

Soviet intelligence seems to have followed German rocket research closely during the war. Defector Tokady testified that the Bureau of New Technology in the Ministry of Aircraft Production collated all data from open and clandestine sources, and when Tokady arrived in occupied Germany in 1945 as aeronautical consultant to the Red Army, he was given complete dossiers on German personnel and facilities.[5] Soviet agents in England must also have reported all they could discover on the V-2, which London's Operation Backfire sought to reverse-engineer from flight data, wreckage, and espionage.[6] Prior Soviet knowledge of German rocketry even suggests that the Kremlin's desire to reach Peenemünde before the Western Allies influenced Red Army operations.

THE A-4

In 1929 the Ordnance Ballistic Section of the German army assigned Walter Dornberger to develop a liquid fuel rocket of longer range than any existing gun, a sobering assignment, given that the Big Berthas of World War I fired projectiles sixty-five miles. Dornberger visited the "rocketport" of the amateur Verein für Raumschiffahrt in Berlin, set young Wernher von Braun to work completing his doctorate, and together they recruited the Rocket Team. Just as in the Soviet Union, the rocketeers did not find state support—the state found them, and at a propitious moment. "The more time I have to think about it," wrote Willy Ley, "the more I have arrived at the conclusion that the VfR progressed as far as any club can progress. ... Experimentation had reached a state where continuation would have been too expensive for any organization except a millionaires' club."[7]

Von Braun and Dornberger chose for their lonely, spacious test site a sweep of sandy coast on the Usedom Peninsula beyond the mouth of the Peene River. But by the time Peenemünde opened in the fall of 1939, the Wehrmacht was rolling over Poland, and Hitler decided the big rockets would not be needed. Von Braun and Dornberger pressed on, with reduced budgets, toward a prototype of their majestic A-4, the first medium-range ballistic missile, standing 46.1 feet high. It was a single-stage rocket powered by LOX and alcohol, developing a thrust of 56,000 pounds, a payload of 2,200 pounds, and a velocity of 3,500 miles per hour while inertially guided by gyroscopes and leveling pendulums to its target 200 miles distant. The first A-4 flight test finally took place in June 1942. It failed, and so did the next. But the third bird, in October, rose from the Baltic dunes in a stable and gentle arc fifty miles high until it passed out of sight en route to the impact area 119 miles downrange. Dornberger's team watched in exultation—like the Alamogordo physicists three years later, they attended in the delivery room as a new Power was born. But where the elemental blast of the atomic bomb rendered its makers diminished, apprehensive, in a sense imprisoned, the elegant, finned cylinder of the A-4 was a metaphor of liberation, defying gravity as it soared aloft with little hint—after the first moments—of the brute force it contained. An aspiring and creative thing, it had brushed the sleeves of space.

At the Yalta conference in February 1945 Marshal Zhukov was stunned to have his plans for advance on Berlin vetoed by Stalin himself: "You are wasting your time. We must consolidate on the Oder and then turn all possible forces north, to Pomerania, to join with Rokossovsky and smash the enemy's 'Vistula' group."[8] Stalin's motives are unknown, but his order redirected the Soviet advance on a line for Peenemünde. The Soviets also secured, with no apparent opposition, both Peenemünde and Nordhausen in the occupation zones drawn up by the European Advisory Commission.

On May 5, 1945, Major Anatole Vavilov stormed the Baltic test site with infantry from Rokossovsky's Second White Russian Army. He met no resistance and took no prisoners—the place was deserted and mostly in ruins. As for V-2 production facilities, when the Soviets finally took possession of the Mittelwerk in July they found it stripped of everything but odd scraps and charred units in railway cars outside. Stalin's reaction was predictable: "This is absolutely intolerable. We defeated Nazi armies; we occupied Berlin and Peenemünde, but the Americans got the rocket engineers. What could be more revolting and more inexcusable? How and why was this allowed to happen?"[9]

The answer was that the German Rocket Team and prescient American officers willed it to happen before the Red Army was in a position to prevent it. As early as mid-January 1945 von Braun took responsibility for the safety of his people, who voted unanimously to flee Peenemünde and go in search of the U.S. Army. As one member of the Rocket Team put it: "We despise the French; we are mortally afraid of the Soviets; we do not believe the British can afford us; so that leaves the Americans."[10] And so, in a harrowing exodus through the stricken Reich, von Braun started south in February with 525 people and thirteen years' worth of documentation hunched in boxcars. Reaching Bleicherode in Brunswick, von Braun buried his paper treasure in an abandoned mine shaft some miles to the north. The team then commandeered a train with forged SS orders and made their way through chaos and air raids to Bavaria. On May 2 a small party went to look for Americans, found an unsuspecting private guarding the road, and approached with hands in the air. The bewildered soldier leveled his rifle as one man stepped forward and said in accented English: "My name is Magnus von Braun. My brother invented the V-2. We want to surrender."[11]

Imagine, too, the dumb wonder of the American lieutenant who veered off from the drive into Nordhausen on April 11 and bumped into the Mittelwerk. There, on railway cars leading into the bowels of the earth, were gigantic rockets lined up like imports from Mars. And inside, a gutted mountain, bizarre machinery, slaves like living skeletons: a scene from Flash Gordon. Nearby they found the workers' camp and thousands of corpses stacked here and there as garbage awaiting pickup. If too weak to work, slaves were left to expire—150 per day—human sacrifices on the altar of machines and power, another logical conclusion, like Stalin's *sharagas*, of totalitarian technocracy.[12]

When word reached the American command of the capture of the Mittelwerk, the Ordnance Department decreed Special Mission V-2: get your hands on a hundred operational V-2s ready for transport to a new White Sands Proving Ground in New Mexico. The task was herculean enough, since few fully assembled and undamaged V-2s remained. What was worse, the Red Army was scheduled to occupy the region in a matter of weeks. Troops hastily gathered up one hundred of every part that looked likely, threw them into impounded freight cars, and hauled everything away into the American zone. In time, after much travail, the ingredients for the hundred V-2s landed on railroad sidings in the New Mexican desert against the day when the von Braun team would arrive to sort them out.

There remained the priceless Peenemünde files, buried in a mine 300 miles to the north. Desperately locating trucks while local miners sweated to reach the cache, the men of Special Mission V-2 disinterred the documents and beat it back to the American zone.[13] Thus, when the Soviets arrived, they found the Nazi dynasty's richest graves already robbed. In July the Americans coordinated their roundup of scientific personnel and offered contracts for work in the United States. After assurances were given that their families would follow, 115 German scientists departed in September for the New World. In this way the Cold War for "intellectual reparations" in the form of Nazi scientists and their secret weapons began before the political Cold War was apparent. At Potsdam in July the Big Three powers agreed to share German scientific facilities, but it was a sham. The Soviets were already carting off entire laboratories, while the German rocketeers, a wind tunnel, and other spoils were en route to the United States. What Winston Churchill called the Wizard War in technical one-upmanship and espionage did not stop on V-E day.[14]

Soviet pursuit of scientific booty seems to have been more premeditated, but less successful.[15] The discovery of a bare Peenemünde kicked off an immediate effort to locate the missing brains. Already in May, Peenemünde veteran Helmut Gröttrup was approached by a member of the Soviet Special Technical Commission in charge of reconstituting the V-2 production line. In August a German engineer told the Americans that "the Russians intend to develop a big rocket with a normal range of 3000 miles and they are needing specialists with knowledge of the theory of flight mechanics and control equipment. . . . The Russians set big prices for getting over to Russian area Professor von Braun and Dr. Steinhoff." They also broadcast over Radio Leipzig to anyone connected with Peenemünde, guaranteeing good wages and personal safety.[16]

All they got were the rank and file of the V-2 program, engineers and minor technicians scattered over the eastern zone. "When I arrived at Peenemünde," wrote Tokady,

there was hardly a German sufficiently competent to talk about the V-2 and other big stuff. There were many, almost all, claiming to be V-2 experts . . . [but they] talked and talked, and displayed the typical characteristics of a second-rater. . . . not only in Peenemünde, but also in all Soviet-occupied Germany, we found not a single leading V-2 expert.[17]

Only one major designer went over to the Soviet side, Gröttrup. His motives are unclear but appear to have been personal rather than political. Having worked in the shadows of the great, perhaps he preferred to lead the Soviet-affiliated missile program rather than remain anonymous in the West. By mid-1946 Gröttrup commanded over 5,000 workers, and new V-2s rolled off the line again in September. They were then static-fired at a German test facility under the direction of ... Glushko. At the same time Korolev, now free after seven years in prison, was deep into designs for larger V-2s and proudly concluding that the Soviet rocketeers really had little to learn from the Germans. "What were our impressions of Peenemünde?" recalled Tokady.

This is an extremely interesting question, and I would like to answer it frankly. We were quite clear on three things: (1) in the field of original ideas and rocket theories, the USSR was not behind Germany; (2) in the field of practical technology of rockets of the V-2 caliber, we were definitely behind the Germans; (3) having seen and studied Peenemünde, we came to the conclusion that there were in the USSR rocket engineers as able and gifted as elsewhere.[18]

Whether the Soviets would have been more impressed had they captured von Braun and his papers is another question.[19] Nevertheless, Gröttrup labored on at Bleicherode until October 1946. Then, late one night, after a drinking party hosted by his Soviet military shepherd, Gröttrup received a hysterical call from his wife. They were all to be rounded up and sent to the Soviet Union at once. Six thousand German technicians, including 200 rocket engineers, and their families, left Germany on twelve hours' notice for a seven-year stint on the steppes.[20]

Paranoid or not, Stalin must have thought his postwar prospects delectable. By spring of 1945 his armies were overrunning Central Europe, the farthest westward thrust of Russian power since Napoleonic days. This century's enemy, Germany, lay beaten and divided. The only check on Soviet policy was the Allied army, but it was already shrinking as U.S. forces shifted to the Pacific. The capitalist powers were also divided among themselves. At Yalta Roosevelt had made no secret of his distaste for British colonialism and was eager to purchase Soviet help against Japan.[21] Indeed, a war that began so badly for the USSR was ending in such a way as to permit the achievement of Stalin's most ambitious war aims.[22]

To be sure, matters could have been even better had the Allied force in Normandy not broken out so swiftly to compete for occupation of Germany. For all his recriminations to Western leaders about their delay in opening the second front, Stalin must have been disappointed at the crossing of the Rhine and "rooted" for the Germans to hold out in the West. (Churchill, for his part, hoped that the Red Army would not reach Berlin and the Danubian Plain.) Still, the revolution had passed its sternest test, broken imperialist encirclement, and reopened the field for Communist expansion for the first time since 1921.

Then, suddenly, by the end of summer, the mood in the Kremlin darkened. The Americans had built and used an atomic bomb.

A mountain of literature exists on the collapse of the wartime alliance and the descent into Cold War. Interpretations range from the stridently anti-Soviet to the stridently anti-American, and the debate, as historian Charles Maier has observed, often seems an extension of the Cold War itself.[23] There are those who see Soviet expansionism after World War II as a case of traditional Russian imperialism and those who attribute it to the global mission of communism. Others hold that imposition of friendly governments on their borders was a defensive reflex by the Soviets and did not indicate implacable hostility or unbounded ambition. Still others explain Soviet behavior with reference to a rapacious Stalin, or to the inner dynamics of the Stalinist state. On the other side are those who blame the United States. To them Stalin was a "traditional" Russian statesman with whom the Americans could have found a *modus vivendi* but for the death of Roosevelt and the irascibility of Truman and his advisers. New Left authors postulate an ideological or commercial imperialism determining American hostility toward the USSR, and argue that the use of the atomic bomb against Japan was a ploy to intimidate the Soviets and force an "American" peace. Finally, there are the historians with a "longer perspective" who consider it almost inevitable that two great states thrust into world leadership, each with its own culture, ideology, interests, and foreign policy traditions, each threatened militarily by the other, should fall into mistrust and rivalry for a time.[24]

In all these views, however, technological change is a dependent variable. The new Superpowers presumably derived political goals, be they obnoxious or benign, from some impulse or another, and then applied military technology to their achievement. This is excusable—in an age of nuclear arms the notion that technological change is an independent variable seems too terrible to entertain. But the opposite hypothesis must at least be considered. For of all the things that made the United States and the USSR distrustful in the moment of victory, perhaps the greatest was the fact that one was not only Communist, but a technocracy, taking for granted its destiny of technological superiority— and the other was not. Regardless of the political climate, the Soviet Union was always in a race, and Stalin had already determined to live in a world with but one nuclear power as briefly as possible.

The USSR supported nuclear research before the war with no less vigor than other countries. V. I. Vernadsky founded the Radium Institute in 1922, Peter Kapitsa and Abram I. Ioffe won international reputations in the 1930s, and the Soviets began Europe's first cyclotron in Leningrad in 1937. When Otto Hahn and Fritz Strassmann discovered nuclear fission in 1938, Soviet scientists quickly explored the theory of chain reaction, including the requirements for explosive conditions in a critical mass of fissionable material. The Presidium of the Academy of Sciences then ordered construction of a more powerful cyclotron to be completed in 1941 and sought access to uranium deposits in the Belgian Congo. Just prior to the invasion, two of I. V. Kurchatov's students proved the possibility of spontaneous fission, and others explored ways of producing U-235 and heavy water in quantity.[25] But just as the war stimulated German rocketry while stifling Soviet work, so it pumped huge sums of money and talent into the British and American atomic programs while stopping the Soviets in their tracks. Whether or not Stalin was aware of the possibility of a uranium bomb in 1941, for the moment the USSR could do nothing.[26] Scientific institutions were evacuated from European Russia before the German armies, and the Leningrad cyclotron gathered rust.[27]

A twenty-eight-year-old colleague of Kurchatov, G. N. Flyorov, relit the atomic fuse. In a letter to Stalin he noted the secrecy that had fallen over American research: "It is essential not to lose any time in building the uranium bomb."[28] A new laboratory directed by Kurchatov emerged in late 1942. Its mission was to build the bomb. The key factor in the decision, according to historian David Holloway, was Soviet knowledge of German and American work rather than any instinct about the postwar environment. But when Soviet agents pierced Germany in 1945, they hunted nuclear physicists as assiduously as rocket engineers. Most had fled, but Gustav Hertz, like Gröttrup, chose the USSR.[29] Again like Gröttrup, he was to find the Soviet specialists at least as capable as himself.

What did Stalin know or guess about the Manhattan Project and the implications of its possible success? If his spies were accurate, he would have heard that some American scientists placed little hope in the project as late as the turn of 1945 and expected at best a single bomb of half a kiloton (equal to 500 tons of TNT).[30] Such an increment in destruction would hardly change the world or repay the investment. But the Americans got a much bigger boom on July 16, 1945, at Alamogordo. President Truman then chose his moment, at Potsdam, to mention casually "that we have a new weapon of unusual destructive force." Stalin replied that "he was glad to hear of it and hoped we would make 'good use' of it against the Japanese." Truman and Churchill concluded that he had not understood that it was a reference to the atomic bomb.[31]

Stalin's breezy reply makes one wonder if he already knew of the Alamogordo test. But the Soviet spy ring in Canada did not report it until August 9, and British Communist spy Klaus Fuchs not until September.[32] Perhaps other sources relayed the news more quickly, for Marshal Zhukov's memoirs of Potsdam suggest neither ignorance nor underestimation of the bomb:

On returning to his quarters after this meeting, Stalin, in my presence, told Molotov about the conversation with Truman.

"They're raising the price," said Molotov.

Stalin gave them a loud laugh. "Let them. We'll have a talk with Kurchatov today about speeding up our work."

I realized that they were talking about the creation of the atomic bomb.[33]

But the Potsdam mystery is academic. On August 6, 1945, the fireball over Hiroshima spoke for itself. Communist journals in the West like the *Daily Worker* and *L'Humanité* at first applauded the bomb as a means of hastening Japan's surrender. In a few weeks they changed their line and attacked the American atomic monopoly. Modest Rubinshtein, a leading agitprop expert on technical affairs, reported that "The American reactionary press insists that the United States must keep the method of production of atomic bombs a secret in the expectation of future war." But, he warned, the monopoly would not last for long.[34]

In mid-August the People's Commissar for Munitions received a puzzling summons to the Kremlin. When Kurchatov appeared as well, "at once it became clear to everyone what the conversation would be about": "A single demand of you, comrades," said Stalin. "Provide us with atomic weapons in the shortest possible time. You know that Hiroshima has shaken the whole world. The equilibrium has been destroyed. Provide the bomb—it will remove a great danger from us."[35]

The equilibrium had been destroyed! Such equilibrium as we Westerners detected after the war consisted of the huge Red Army looming over Western Europe, balanced by American air power and the atomic bomb. Did the Soviets view matters differently? Did their new conventional military dominance in Eurasia serve simply to create a "balance" that was immediately upset by the bomb? The American and British ambassadors both sensed such a feeling in Moscow toward the end of 1945. "Suddenly the atomic bomb appeared," wrote Averill Harriman, "and they recognized that it was an offset to the power of the Red Army. This must have revived their old feeling of insecurity."[36] Sir Archibald Clark Kerr:

> There was great exaltation. Russia could be made safe at last. . . . She could stretch out her hand and take most of what she needed and perhaps more. It was an exquisite moment, all the more so because this resounding success under their guidance justified at last their faith in the permanence of their system.
>
> . . . Then plump came the Atomic Bomb. At a blow the balance which had now seemed set was rudely shaken. Russia was balked by the West when everything seemed to be within her grasp. The three hundred divisions were shorn of much of their value.[37]

One does not have to believe that Truman deliberately hoped to intimidate the Soviets to grasp how the Soviets must have seen their objectives threatened and their power diminished. Stalin gave Kurchatov authority to coordinate research, plan factories, and mount expeditions to prospect for uranium. How long, assuming all-out support, did Kurchatov think it would take to build a bomb? Five years, he predicted.[38]

The USSR exploded an atomic bomb in August 1949, a year "early." Soviet secrecy, pride, distrust of the West, and especially the ideological commitment to technological superiority all militated in favor of a crash program and against international controls on nuclear weapons. Even if one could assume away the rapid degeneration in relations between the Superpowers, it is hard to imagine the Soviets renouncing their drive for the A-bomb, once Hiroshima was history.[39] The Soviet world view made unacceptable any world in which the capitalists possessed superior military technology. It gave them no choice but to press military technology as far and as fast as possible.

What of the Grand Alliance, the one hiatus in the official Leninist line on imperialism and foreign policy, during which British monarchy, American democracy, and Soviet communism fought side by side, and even the Kremlin spoke of the Great Patriotic War against the Teutons? Such nationalism served well during the emergency, but the official Leninist view could not be set aside indefinitely without the dictatorship losing its legitimacy. After the war Soviet theoreticians quickly returned to orthodoxy on the hostility of the capitalists, the inevitability of new wars, the necessity for strict Party leadership, and the role of technological supremacy as a measure of the success, legitimacy, and security of the revolution.

The reversion was evident in the fate of the Soviet Institute of the World Economy and World Politics, charged with describing and inter- preting the global changes resulting from the war. Among its findings was the undeniable productive might revealed by the supposedly decadent American economy. The explanation of American dynamism seemed to lie in the growing government intervention that "eliminated monopolies which threatened the war effort" and made the United States "capable of enormous development" even in the postwar period.[40] If capitalism had changed its spots, what should the USSR do in response? Debate on the question began at the time of the first Soviet request for American reconstruction loans (January 3, 1945), and just a few months prior to the commission of a new Five Year Plan. Traditionalists held that monopoly capitalists still dominated Western policy. The postwar period would bring inflation, unemployment, and technical stagnation, all of which could only increase pressures for war, with the USSR now the sole target. But nontraditionalists at the institute, led by Evgeny Varga, thought that state regulation not only restored Western dynamism by mitigating the contradictions of capitalism but also served to mellow Western foreign policy. Varga predicted political moderation from the Americans, argued for East-West cooperation in reconstruction, and hoped for a postwar world not divided into economic blocs.[41]

Cooperate with capitalism or return to autarky? For the first time in a decade something like open debate broke out on Soviet domestic and foreign policies. Molotov, Mikoyan, and Kaganovich took the traditionalist view. The USSR "must equal the achievements of contemporary world technology. . . . We will have atomic energy, and much else." But several Politburo members disagreed, including Andrei Zhdanov, who stressed consumer industries in the wake of wartime sacrifice.[42] Stalin appeared to choose moderation. His speech on the Five Year Plan counted Britain and the United States among the "peaceloving states," promised rapid reconstruction and a consumer orientation, and sent delegates to the March meetings of the new World Bank and International Monetary Fund. But he also said that war and victory had justified his harsh policies of the 1930s and called on Soviet scientists "not only to overtake but to surpass in the near future the achievement of science beyond the borders of our country." Only this would insure the USSR against "all sorts of accidents."[43] Thus before the Cold War was "declared" and his own diplomats haggled for loans without political strings, Stalin launched the greatest crash military program in the history of the regime.

The new Five Year Plan called for an annual R & D budget four times higher than the record figure allotted (but not implemented) in 1941. The 1946 budget set aside 6,300 million rubles, growing to 9,000 million by 1950.[44] The U.S. government, by comparison, spent $3,850 million on R & D during the whole of World War II, of which $2,000 million went into the Manhattan Project. In 1950 the U.S. government allotted $1,083 million. This figure does not include the large sums spent by private industry, while the 9-billion Soviet figure for 1950 was partly a product of inflation. But it is fair to conclude that the Soviet budget for military research was several times greater than the American, and perhaps six times greater as a percentage of gross national product (GNP).[45]

By 1947 whatever ambiguity existed in Stalin's assessment of the international scene disappeared. The Marshall Plan forced him to choose either to integrate the world economy by accepting Western aid and conditions or to cultivate his Eastern European garden and recover in isolation. As in early 1946 the Varga thesis surfaced for discussion, but this time it would be more accurate to say that it provided an occasion for the reinterment of nontraditional views on capitalism.[46] So the old contradictions in Soviet technology policy survived the war years. By definition the USSR had a superior potential for R & D but was still, by historical accident, temporarily behind. Great efforts must be made to catch up, but not through international mechanisms that carried unacceptable risks. Capitalism was still ahead in the technology race, and even more dynamic than before, yet it was still so economically sterile that it must soon launch wars to stave off collapse! Varga had tried to erase this inconsistency from Soviet theory and had also born witness to the fact, as had leading Stalinists in 1945, that "capitalist encirclement" had been broken. By February 1946 Kaganovich was insisting that it, too, remained.[47]

The attack on Varga signaled the end of the flirtation with international cooperation. Stalin snubbed the Marshall Plan, imposed bilateral trade treaties on the Eastern European states, restored the Communist International (now called Cominform) in September 1947, and agreed with Churchill that the world was divided into two camps. Zhdanov, a bellwether, admonished the USSR to resist imperialism in all its forms and invoked the analogy of Munich and Appeasement three years before Truman did the same.[48] Varga himself held out bravely until the Berlin Crisis of 1948 and formation of the North Atlantic Treaty Organization (NATO) washed away his last handholds. He publically confessed "a whole chain of errors of a reformist tendency" and "departure from a Leninist-Stalinist evaluation of modern imperialism."[49]

Would a more accommodating Western policy have altered the outcome of the Soviet debate? This question lies beyond our scope, but two facts stand out. First, the Soviets could not have entered into meaningful cooperation without relinquishing the myth of conflict between "world camps" and socialist superiority. Second, the Soviets chose to race in new fields of military technology almost at once. Their large standing army might intimidate Eastern Europe, but it was also the drapery to cover a new "window of vulnerability" until they could pull even in technology.[50] By tripling the R & D budget with peacetime crash programs in atomic, aviation, and rocket technology, the Kremlin all but announced its estimate of the dangers of the postwar world and its intention to restore the "balance" upset by the bomb. The diplomatic breakdown that followed seemed only to confirm the wisdom of decisions already made concerning military R & D. Those decisions and others concerning what sort of hardware might impress an adversary skulking in the safety of another hemisphere "kicked" the USSR into the last leg of the pathway to space.

Grigory Tokady was chief of the Aerodynamics Laboratory of the Moscow Military Air Academy and a leading expert in rocketry. As technical adviser to the Soviet occupation army he had the mission of locating data and personnel and otherwise aiding Gröttrup, Glushko, and Korolev to revive the V-2 production line and an affiliated design bureau. What a luscious treat for the Soviet engineers to be turned loose on all that sweet technology after their wartime fast! But the V-2, as the Chief of the Soviet Air Forces confided to Tokady, was not enough. "They were good to frighten England, but should there be an American-Soviet war, they would be useless; what we really need are long range, reliable rockets capable of hitting target areas on the American continent. This is an aim that should dominate the minds and efforts of your rocket group."[51] So, in October 1946, Gröttrup's Germans and their Soviet patrons departed for the East.

They found in the USSR the foundation of a vast new R & D complex. By the end of 1945 Commissar of Armaments D. F. Ustinov chaired a sixty-man Scientific Council to advise on military rocketry. In June 1946 the new Academy of Artillery Sciences set up a department for rocketry and radar under A. A. Blagonravov. Then in April 1947 Stalin called for Tokady himself to brief him on a project that had turned up several times in captured German files. It was Eugen Sänger's antipodal bomber, a piloted, winged rocket to reach an altitude of 160 miles and "skip" on the top of the atmosphere halfway around the world. Answering the summons, Tokady bounced from one NKVD sentry to another like a ball bearing through a pegboard, and came to rest in the Kremlin office of the deputy prime minister. "You know this book?" asked the minister, holding a translation of the secret proposal that Sänger had peddled without success to Nazi leaders. Tokady knew it well, and soon found himself before the Politburo weapons expert, G. M. Malenkov. Sänger's work was theoretical, said Tokady with professional caution. It was not at all certain that such an engine could be built or metals developed to resist the heat of combustion and reentry. Malenkov insisted that "the flying bomb is an outmoded weapon now. . . . The point is that the V-2 is good for 400 kilometres, and no more. And, after all, we have no intention of making war on Poland. Our vital need is for machines which can fly across oceans!" Were the British and Americans pursuing the Sänger Project? Tokady thought it possible: "If it be true that the Americans are so greatly concerned with rocket weapons that they have transformed Texas into a vast Peenemünde, as is often said, it is hardly possible that they have overlooked Sänger's plan. They have combed German scientific centers pretty thoroughly."[52]

The next day Tokady appeared before the Politburo. The important thing, he concluded, was intensive research, whether or not it resulted in hardware. Stalin paced and puffed on his pipe. "Certainly research is necessary," he replied. "But we still need Sänger planes, and their construction should be our immediate objective." Such planes, added Malenkov, could cross the Atlantic and return in one hop. "So they would," said Stalin, "and their possession would make it easier for us to talk to the gentleman-shopkeeper Harry Truman, and keep him pinned down where we want him. Tokaev, we wish you to exploit Sänger's ideas in every way." The Council of Ministers hastily drafted a decree for a special commission comprising Colonel-General I. A. Serov, Tokady, Academician Mstislav V. Keldysh, M. A. Kishkin from the aviation ministry, and Vasily I. Stalin, an air force major general.[53]

The Sänger Project proved to be premature. But the Politburo's interest in intercontinental delivery systems before they even possessed the atomic bomb was a turning point. By the end of 1947, according to engineer A. G. Kostikov, "everybody wanted to design a trans-Atlantic rocket."[54]

Postwar conditions for R & D were better than at any time in Soviet history. European Russia had twice been overrun and scorched, 30 or 40 million citizens were dead of purge or war. Yet during these years after 1945 a threefold increase in R & D spending, construction of modern laboratories and proving grounds, and incessant government pressure for results made for frenetic progress. The Academy of Sciences pushed pure and applied research in atomic energy, radar, jet and rocket technology, electronics and semiconductors, calculating devices (computers), and combustion theory. In 1946 the Gosplan established a department for technology to plot not only five-year plans but yearly and evenly quarterly schedules for R & D and installation of new technology. The Council of Ministers resolved to double or triple the salaries of scientific workers, who suddenly found themselves a privileged class.[55] The research *sharaga* lived on—according to one estimate 15 percent of the top Soviet scientists were still in camps[56]—but the infrastructure for a new, expanded assault on the technological frontier was in place.

Rocket test ranges opened at Kapustin Yar, seventy-five miles into the steppes east of Stalingrad, and at Tyuratam, a railhead in remote Kazakhstan. The Gröttrup team worked in isolation on an island in Lake Seilger, 150 miles from Moscow, before their transfer to Kapustin Yar. There they supervised test launches of V-2s, consulted on the short-lived Sänger Project, and designed new rockets. Their R-10, with greater thrust than the V-2 and a detachable warhead, never entered production, and the multistage R-12 intermediate-range ballistic missile (IRBM) called for stage separation that was still beyond current technique. So Gröttrup's R-14, a full-blooded atomic bomb carrier designed to send a 6,600-pound warhead 1,800 miles, was a single-stage, finless monster fueled by the alcohol-LOX brew so favored by Peenemünde veterans and stabilized in flight by a novel system of swiveling nozzles. It was the most advanced design in the world of 1949. The Soviet Scientific-Technical Council whisked away all the plans for the R-12 and the Germans never saw them again. Instead they were put to work on antiaircraft missiles and trained gaggles of postgraduates who in time took over routine design work. Finally, on November 22, 1953 (ten years to the day before the JFK assassination), they received orders to pack for home, as abrupt as the initial summons of seven years before.[57]

The handling of the Gröttrup team is illustrative of Soviet borrowing. Always behind, the Soviets are constantly tempted to tap foreign hardware and talent. Always prone to secrecy, in part to cover their own backwardness, they are discreet and anxious to patriate foreign skills as rapidly as possible. The Gröttrup team was also "second string," and for this, for politics, and for pride, the Soviet engineers made only partial use of them. Historians Frederick Ordway and Mitchell Sharpe conclude that the Germans contributed in specialized fields (twelve experts in guidance were held back from the 1953 repatriation) and in the "systems engineering" approach to rocket design.[58] Even if Glushko and Korolev had little to learn from the Germans in engineering, the managerial techniques of Peenemünde may have found their way to Tyuratam via the Gröttrup team.

Meanwhile, according to Leonid Vladimirov, the ghosts of the *sharaga* still haunted Korolev at the Tyuratam rocket oasis. His old camp warden, V. N. Chalomei, reportedly stole the credit for Korolev's wartime inventions and tried to get him back after 1946. Tyuratam became a divided fiefdom with Glushko, Korolev, and L. A. Voskresensky in one compound, Mikhail K. Yangel and Chalomei, their bitter rivals, in another.[59] Whether such disunity inhibits or stimulates performance is a question no R & D manager has fully resolved. But Tyuratam produced. By 1949 its Pobeda, or T-1, an all-Soviet upgrade of the V-2 with a range of 550 miles, was in production and supplying the first rocket units of the Red Army.[60] The T-2, an IRBM, was under construction by 1952. Design competition for an all-out ICBM, therefore, must have been underway about this time, and journalist Michael Stoiko reports that Korolev's blueprints for the ICBM that launched Sputnik won approval in 1954,[61] the same year that ICBM development became a top priority in the United States.

In plans for imminent construction of a world-girdling rocket to deliver the newly made atomic bombs, Soviet technological maturity was at hand. But the mid-1950s were also a time of rebirth, or remembering, of what rocketry had once been all about. From Tsiolkovsky to Tsander to Korolev, rockets were about spaceflight. In the early 1930s the Russian technical revolutionaries fell into the hands of a Soviet state whose *raison d'être* was to play forcing house of technological change. From the mid-1930s to the mid-1950s (with the exception of Stalin's Eagles) military might had perforce sole emphasis. But the Soviet expectation of imminent nuclear parity had a side effect. In establishing unprecedented might, it resurrected glory. After a hiatus of two decades, the rocketeers and their patrons in the Kremlin rummaged again in that corner of their imagination that harbored the dream of spaceflight.

When the Soviets sported an atomic bomb, the United States responded, after much debate, with sharply increased defense spending and a program to build a fusion, or thermonuclear, or hydrogen bomb of far greater destructive force. Outbreak of the Korean War in June 1950 hardened American resolve. The Soviets reacted by doubling the Red Army to 5.8 million men by 1955 and, without pausing to admire their atomic bombs, pushed on at once for their huskier offspring. In August 1953 they exploded the first thermonuclear device and tested a deliverable H-bomb in November 1955. The corresponding American dates were November 1952 and May 1956.[62] The H-bomb "race," like the ICBM race, probably began even before the Americans held their hand-wringing debate.

Stalin's death in March 1953 still seemed to presage a thaw. For the first time since the 1920s the collective wits of the Politburo did not have to square their opinions with Stalin on pain of torture and death. Malenkov, Mikoyan, and others approved a negotiated end to the Korean conflict and claimed that Soviet nuclear capacity enabled true peaceful coexistence.[63] Civilian technology might now become the main arena of competition with capitalism. After 1950, R & D spending did level off. A ceiling on the military budget as well could free up investment in light industry and consumer goods. But such views clashed as always with the mythology of capitalist hostility and socialist superiority. Tentative feints toward consumption, trade, and exchange of ideas with the West exposed their advocates to charges of being "soft on imperialism." Even with Stalin gone, Eastern Europe secured, and the H-bomb in Soviet hands, the Politburo still upheld traditionalist assumptions. Klement Voroshilov reasserted the reality of encirclement; Khrushchev, Bulganin, Molotov, and Kaganovich urged more military spending in light of imperialist belligerence in Korea, the effort to rearm West Germany, and the U.S. buildup. "We cannot assume," said Nikolai Bulganin, "that the imperialists expend enormous material and financial resources only to frighten us."[64] Malenkov repented of his consumerism, military R & D rose as part of overall 15 to 16 percent annual increases from 1953 to 1956, and missile expenditures jumped as Tyuratam moved to prototype production of its giant rockets.

So the arms race was not about to end. What of Stalin's other bequests, the rule of terror and the Cold War? Nikita Khrushchev dealt with them before the Twentieth Party Congress in February 1956. There he lectured to an astonished audience at a special midnight session "on the cult of personality and its consequences." He recounted the terrors, tortures, and errors of Stalinist rule, the leader's blunders prior to the Nazi invasion, his collapse in the critical months following, his responsibility for agricultural and diplomatic disasters, and, above all, his attacks on the procedures and loyal personnel of the Communist Party. Stalinism became an official aberration, not a natural expression of doctrine, and Soviet historiography and political vocabulary metamorphosed overnight.[65]

Subsequent events revealed more continuity than Khrushchev's broadside suggested. The wave of de-Stalinization did little to "liberalize" the Soviet bloc. Intramural Party terror and wholesale judicial murder subsided, but the police state lived on, while outbreaks occasioned by de-Stalinization in East Germany, Poland, and Hungary triggered new repression. Khrushchev proceeded to foster his own personality cult, and his airing of Stalin's mistakes did not prevent blunders of his own that brought his own downfall eight years later. The element of continuity most pertinent to us was Khrushchev's adoption of the technocratic myth and his personal identification with Soviet heroism and futurism. Like Stalin, he struggled to establish his legitimacy against Politburo members with better claims to the succession. Like Stalin, he did so in part by styling himself the personal patron of high technology and the theorist most in touch with the historical laws of his age.

The power struggle after Stalin's death focused on military policy. Khrushchev sided with the majority in the 1953 plan that favored bigger conventional forces, and he courted war hero and supreme commander Georgi K. Zhukov. This alliance helped to save him in June 1957 when Malenkov, Molotov, and Kaganovich conspired to oust him from power. But Khrushchev's military plans, as events proved, did not include old Stalinist generals. Having used Zhukov against his political rivals, Khrushchev would use Stalin's maturing missiles against Zhukov to establish both his own monopoly of power and a new age in military strategy.[66]

The dawn of the missile age, which illuminated the 1956 Party Congress, made the whole world appear differently to those, like Khrushchev, with eyes to see. The postwar Soviet agenda had included consolidation of Eastern Europe and a headlong drive for nuclear parity. The Berlin anomaly still rankled the Kremlin, but otherwise the fulfillment of this agenda was in sight. What lay ahead in Soviet foreign policy? Where were the new opportunities in the coming age of mutual nuclear deterrence? Khrushchev's answers to these questions, also delivered to the Twentieth Party Congress, fundamentally revised Leninist dogma on foreign policy. For an age was upon them when the Soviets could put to rest the old fears of capitalist encirclement and bargain with their adversaries as equals. That equality in turn freed the USSR to compete in other ways, economic and political, and in regions beyond its own cordon. And the USSR was free to do so just as the neutralist "Third World" was coming into existence. This compelling chain of logic, beginning with missiles, seemed to prove that worldwide initiative was finally passing to the socialist camp.

Khrushchev's report to the Party Congress boasted of spectacular postwar recovery, achieved with "complete self-sufficiency"—that is, no Marshall Plan handouts—while the capitalist world, though growing, could never abolish its endemic overproduction, unemployment, and inflation. In foreign policy, Khrushchev denounced the "so-called 'Cold War' launched against the countries of the socialist camp" as well as the bloody war that "was launched" in Korea. The inspirers of the Cold War, he explained, alleged that their military blocs were for protection against the "Communist threat," but this was "plain hypocrisy. . . . Now the slogan of 'anti-Communism' is again being used as a smokescreen to hide the pretensions of a particular power to world domination." Thanks to courageous efforts by Communist parties, working-class leaders, and antimilitarist movements in the West, influential circles were beginning to "sober up" and "admit that the socialist camp is invincible." Why? Because thanks to Soviet weapons breakthroughs, the atomic arm of the West was now useless. But another challenge had also arisen to imperialism: the "national liberation struggle of the colonial peoples." The disintegration of the empires was the "universal historic event of the postwar period" and was significant to the USSR, for the peoples of former colonies would not be truly free until they achieved economic autonomy, which meant an association with the socialist camp. Of course, new imperialist rivalries were evident: South Vietnam, for instance, was "passing from the hands of the French to those of the USA," and the Cold War itself was a means of instigating war hysteria and thus justifying imperialist expansion.

Despite these provocations, boasted Khrushchev, the USSR was dedicated to the principles of peaceful coexistence, including mutual respect for territorial integrity and sovereignty, nonaggression, noninterference in domestic affairs, equality and mutual advantage, and economic cooperation. "Of what purpose is war to us?" he asked in conclusion. ". . . Our faith in the victory of Communism is based on the fact that the socialist way of production has decisive advantages over the capitalist." True, Marxist doctrine spoke of the inevitability of war, but that was worked out in a period when imperialism was all-embracing and the socialist forces weak. "But at the present time the situation has changed fundamentally. The world camp of socialism has arisen and become a powerful force. . . . We must still exercise the greatest vigilance. . . . But there is no fatal inevitability of wars."

War between capitalism and socialism no longer inevitable! Did this mean stalemate, the arresting of the revolution? No, for "in connection with the radical changes in the world arena, new prospects are also opening up with regard to the transition of countries and nations to socialism." The Party's tasks were to pursue "peaceful co-existence, strengthen inter-communist ties, and tighten bonds of friendship with the new nations, improve relations with the West, but keep a vigilant eye, and preserve Soviet defense at the level of modern military technique and science."[67]

At the Twentieth Congress Khrushchev perceived a new Cold War. His address was a kind of Communist *Rerum Novarum* for foreign policy. The Cold War would continue, but the material environment reversed the correlation of forces. Soviet nuclear and missile power wiped out at a blow the vulnerability of the socialist camp, encirclement by the imperialists, and the inevitability of war. Competition would shift to other spheres: economic productivity, scientific progress, and influence in the underdeveloped nations, whose struggles for national liberation were the second arrow in the socialist quiver. (Indeed, as soon as October 1956 the new correlation of forces would become manifest in the Anglo-French retreat from Suez following a Soviet threat of "rocket attacks.")

In all these ways the coming dawn of the space and missile age meant a new and better world for the Soviet Union and for its new leader: a deterrent to imperialist war; an amulet of attraction for the elites in the postcolonial world; a technological revolution with which Khrushchev could personally identify; a justification for moving against the conservative, Stalinist military leadership; an indicator of Soviet superiority in science and technology. For all these reasons the prospect of the USSR leading the world in the peaceful as well as military uses of rocketry beckoned irresistibly. New frontiers were opening up for Soviet power, and the Twentieth Party Congress passed the baton to a new post-Stalinist leader in touch with the times. Times of accelerating change—but also continuity perceptible in a technocratic, totalitarian state whose legitimacy and international appeal rested on its material promise of a glorious future, hence on regular palpable indicators that that future was still in healthy gestation.

After his swearing-in as President, Harry Truman was stunned to learn of the nature and progress of the Manhattan Project. Roosevelt had kept him in the dark. Similarly, when Khrushchev and his colleagues were briefed on rocket development after Stalin's death, they were flabbergasted. "Korolev came to the Politburo," wrote Khrushchev in his memoirs,

to report on his work. I don't want to exaggerate, but I'd say we gawked at what he showed us as if we were sheep seeing a new gate for the first time. When he showed us one of his rockets, we thought it looked like nothing but a huge, cigar-shaped tube, and we didn't believe it would fly. Korolev took us on a tour of the launching pad and tried to explain to us how a rocket worked. We were like peasants in a marketplace. . . . We had absolute confidence in Comrade Korolev. When he expounded his ideas, you could see passion burning in his eyes, and his reports were always models of clarity. He had unlimited energy and determination, and he was a brilliant organizer.[68]

In the year following Stalin's death the ICBM was apparently approved, and high-level indications of interest in spaceflight reappeared after twenty years. The president of the Academy of Sciences, A. N. Nesmeianov, announced to the World Peace Council that "Science has reached a state at which it is feasible to send a stratoplane to the moon, to create an artificial satellite of the earth."[69] Several articles appeared in 1954 concerning interplanetary communications, an aeroclub began a cosmonautics division, a biography of Tsiolkovsky was commissioned, and a Tsiolkovsky prize was instituted to honor work in rocketry. Such indications of mild public interest were no more than occurred in the United States, but in the USSR they signaled official interest as well. More telling was the Soviet response to recommendations by the organizers of the International Geophysical Year (IGY) that attempts be made to place

artificial satellites in orbit about the earth. The Soviet Academy of
Sciences named a blue-ribbon Commission for Interplanetary Communications (ICIC) chaired by Academician Leonid I. Sedov. Its stated
purpose was this:

> The problem of realizing interplanetary communications is undoubtedly one
> of the most important tasks among those which mankind has to solve on the
> way to conquering nature. The successful solution of this task will become
> possible only as a result of the active participation of many scientific and
> technological collectives. It is precisely for the unification and guidance of those
> collective efforts of research workers that the permanent ICIC has been established.
> . . . One of the immediate tasks of the ICIC is to organize work concerned with
> building an automatic laboratory for scientific research in space. . . .[70]

Moscow radio reported that a team of scientists had been formed to
build the satellite. Another academician declared satellites a possibility
in June 1955 and believed tackling the problems of spaceflight to be
extremely urgent. On July 30, 1955—a day after a similar American
announcement—the Kremlin revealed that the USSR planned to
launch satellites during the IGY. Sedov predicted one in two years.
Reentry problems were under study as well, he said, and a multistage
rocket would be used in the first attempt.[71]

Soviet officials, therefore, while avoiding premature boasting, did not
hide their intentions. It was just that few took them seriously. Meanwhile,
in remote and secret isolation, Korolev pieced together the world's first
ICBM. In mid-1953 the Ministry for Medium Machine Building was
established—a dummy name for the missile plants (whose political liaison
included a rising Party official named Leonid Brezhnev)—and in June
1955 a new test range arose at Tyuratam (where the new Party secretary
for Kazakhstan, Brezhnev again, took an interest).[72] Throughout 1955
and 1956 Sedov, Blagonravov, and others predicted the coming of the
Space Age. Soviet scientists captivated the First International Conference
on Rockets and Guided Missiles in 1956 with tales of high-altitude
experiments and dogs launched sixty-eight miles high at g-forces five
times normal. There was no doubt, they said, that human rocket flight
was possible.[73]

The IGY began on July 1, 1957. Soviet predictions of a satellite became
a weekly occurrence as Korolev put his giant rocket to the test. The
metallurgists had never succeeded in finding an alloy to withstand the
heat produced by very large rocket engines, so Korolev's solution was a
"cluster of clusters"—twenty separate engines in a central core and four
great skirts, developing 1.1 million pounds of thrust on kerosene and
LOX. Presumably built to carry the primitive two-ton atomic bombs of
the early 1950s on the 4,000-mile run to the United States, the R-7 was
all bulk, short and splayed like a mechanical Cossack in billowing
pantaloons, only three times as high as thick, and only twice as tall as a

V-2. The first R-7 (*semyorka*, or "ol' number seven" to the rocketeers) exploded on ignition in the late spring of 1957. When more failures followed, Korolev's team came under criticism—his rival Chalomei sowing discord. But on August 3 the Soviet ICBM roared off the pad and flew 100 degrees of longitude to the east, into the Pacific Ocean near the Kamchatka Peninsula. After a second success, Moscow announced to the world on August 27, 1957, its possession of a proven ICBM. According to Korolev, it was only then that final approval of a satellite attempt descended from the capital.[74] On September 17, the centennial of Tsiolkovsky's birth, the government promised the world that a satellite was coming soon. On the first of October, it announced the radio frequency on which the satellite would broadcast.

Three evenings later space scientists from various IGY countries talked shop and sipped vodka at the Soviet Embassy in Washington. The hosts disingenuously resisted casual probes from their American colleagues as to the date of their first attempt.[75] A Russian emigré even teased his ex-countrymen: "Poor Tsiolkovsky is turning in his grave. His hundredth birthday has passed without even one Russian satellite in orbit. Under the Tsar we would have had several of them long before now and would have celebrated the anniversary with a flight to the moon." One Soviet guest took offense. Before he returned to Moscow two days hence, he said, the emigré would eat his words.[76]

A hemisphere away Korolev, who had slept little for weeks, fidgeted in his concrete bunker, built by slave labor, at Tyuratam. All evening there had been delays in the countdown, frustration, and suspense—the aggravations that have taught us spectators why engineers and test pilots must be so maddeningly equable. Now, in the darkest, chilliest hour of night, the measured pace of seconds, no longer corresponding much to human heartbeats, finally signaled the moment of ignition. Soviet historian Evgeny Riabchikov recounts:

> The clear tones of a bugle were heard above the noise of the machines on the pad. Blinding flames swirled about, and a deep rolling thunder was heard. The silvery rocket was instantly enveloped in clouds of vapor. Its glittering, shapely body seemed to quiver and slowly rise up from the launch pad. A raging flame burst forth and its candle dispelled the darkness of night on the steppe. So fierce was the glare that silhouettes of the work towers, machines, and people were clearly outlined. . . .
>
> "She's off! Our baby is off!" People embraced, kissed, waved their arms excitedly, and sang. Someone began to dance, while all the others kept shouting, "She's off! Our baby is off!"

The rocket disappeared. Everyone rushed to the radio receivers. The satellite's first signals, from the moment of its separation from the booster, were recorded on tape for its anxious family below: ". . . beep, beep, beep. . . ."[77]

At the IGY gathering in Washington, a Soviet embassy official called Walter Sullivan to the telephone. It was the *New York Times* Washington bureau. Sullivan scratched a message and handed it to Lloyd V. Berkner, who clapped his hands and called for silence. "Radio Moscow has just reported that the Russians have placed a satellite in orbit 900 km. above the earth."[78]

Premier Khrushchev had just returned to Moscow from his dacha in the Crimea. "When the satellite was launched," he recalled, "they phoned me that the rocket had taken the right course and that the satellite was already revolving around the earth. I congratulated the entire group of engineers and technicians on this outstanding achievement and calmly went to bed."[79] It was left to the official announcement the next day to set the tone for seven years of propaganda from a triumphant Soviet, and increasingly Khrushchevian, technocracy: "Artificial earth satellites will pave the way for space travel, and it seems that the present generation will witness how the freed and conscious labor of the people of the new socialist society turns even the most daring of mankind's dreams into reality."[80] In the weeks and months to come, Khrushchev and lesser spokesmen would point to the first Sputnik, "companion" or "fellow traveler," as proof of the Soviet ability to deliver hydrogen bombs at will, proof of the inevitability of Soviet scientific and techno-logical leadership, proof of the superiority of communism as a model for backward nations, proof of the dynamic leadership of the Soviet premier. At the fortieth anniversary of the revolution in November 1957, Khru-shchev predicted that the Soviet Union would surpass the United States in per-capita economic output in fifteen years.

Russian rocketry and revolution embraced again. Only this time the revolutionary flames leaped the oceans, found crackling timber in the United States, and then spread around the world on the strength of the promise not of Marxist dialectic but of Leninist technocracy. It is not too fanciful to suggest that the fires of "ol' number seven" were themselves kindled by the bombs astride the carriage of Tsar Alexander II.

Conclusion

How had the Soviets come so far so fast? How was it that human penetration of space arrived as "early" as 1957? The fact that the first satellites were the feats of a closed, totalitarian society obscures most of the details even as it illuminates the whole. The drive for spaceflight was in the nature of the Soviet beast just as the urge to explore, discover, and overcome nature is part of the nature of man. Communism is strong because it expresses a part, but only a part, of human reality. But the totalitarian nature of the regime means that we have no documents by which to trace the technical progress of the engineers or the industrial capacity supporting R & D. Some facts are known, however, and some inferences can be drawn.

First, spaceflight was not premature. The Soviets showed an unexpected capability in guidance technology and an impressively large rocket. But shooting a satellite into a rough orbital trajectory is not the same as pinpointing an ICBM to its target or positioning a communications satellite; and any garden variety multistage rocket or a big, simple single-stage rocket is sufficient to accelerate a small orb to orbital velocity. The rocket teams in both Superpowers protested that they could have launched a satellite years earlier if left to do so without military or political interference.[1] But the genius of the engineers was only a necessary, not a sufficient, condition. The characteristics of the Soviet regime and the advent of nuclear weapons provided the nourishment and climate sufficient for the space technological revolution to occur. Those characteristics included an ideology of foreign relations that ensured distrust and competition whatever the diplomatic settlement after World War II. They included a self-definition that compelled Soviet leadership to exert maximum effort to equal and surpass the technological achievements of the capitalist states, and a concentration on science and R & D unique in the world. They included a materialistic progressivism that linked the legitimacy of the Party and of its leader to their capacity for inventing the future and conquering nature. In these ways the Soviet Union was especially suited to open the age of spaceflight.

What of the barriers to science and technology in the USSR? Did not the same totalitarianism that glorified technology also stifle its progress? This is demonstrably the case in numerous areas of applied science. The contradictions of socialism were everywhere apparent in Soviet efforts to encourage imaginative research yet retain central direction and ideological control over the technical intelligentsia; in Soviet efforts to hasten the spread of new technology throughout industry and agriculture yet suppress the market mechanisms that give incentive for innovation; in Soviet efforts to borrow knowledge from abroad yet erect barriers to the flow of people and ideas across borders. But the contradictions that inhibited domestic prosperity were less troubling to the leadership than those that might inhibit military technology. For the Soviets leaned on promise and appearance, not on reality—it was more important for the USSR to seem invincible and progressive than for it to be so. This perhaps hints at the ultimate contradiction of Soviet technocracy: founded as a brave new society in which politics were to serve rapid technological change and the material needs of mankind, it transformed itself in short order to one in which technology serves politics. The Soviet Union—as has been often observed—can channel national efforts into specific areas of political importance to the hierarchs. The shortcomings of secrecy, terror, and mismanagement can be balanced by massive investment and concentration.

Was Sputnik the triumph, therefore, of the Soviet Union, or was it really a Russian accomplishment in the spirit of Tsiolkovsky, achieved as much in spite of the Soviet as because of it? Or did the spirit of the Russian dreamers of spaceflight merge with the spirit of the Bolshevik revolution—a spirit of rebellion against nature and the limitations imposed by mankind's own biology and the physical laws of his world? Tsiolkovsky yearned to kill gravity and perfect mankind, themes recurrent in the writings of spaceflight enthusiasts—technological and psychic and social emancipation all rolled into one. After the revolution, Tsiolkovsky became a devoted Bolshevik, and before his death in 1935 he wrote to the Central Committee of the Communist Party: "All my works on aviation, rocketry, and space travel I hereby bequeath to the Party of the Bolsheviks and the Soviet Government—the real leaders in the advancement of civilization. I am convinced they will bring these works to a successful conclusion."[2]

An old man's dreams notwithstanding, can it be argued on more solid grounds that Sputnik was a Soviet triumph, or would another regime in Russia have accomplished the same? In his remarkable study of Soviet technology before 1941, Kendall Bailes lists eight characteristics of Soviet R & D: (1) tension between the need to borrow foreign technology and the desire to foster native creativity; (2) lack of a competitive stimulus, and a planning system that inhibited innovation by production managers; (3) inhibitions due to terror because of the risks perceived for failure; (4) tensions created by the conflict between professionalization of R & D and the Party's attempts to make innovation a "mass movement" stimulating creativity in the working class; (5) relative abundance of unskilled labor and relative scarcity of skilled workers; (6) strong tradition in pure research and high status for pure science in comparison to technical work; (7) organizational separation of R & D and production; (8) tendency of R & D personnel to show less concern for economic criteria than technical performance.[3]

Six of these features are negative, the other two are neutral. Six can also be styled as Russian; they existed (though sometimes for different reasons) before 1917. Only two are specifically Communist: terror and the effort to involve workers themselves in innovation. The latter was not a factor in high-technology defense industries, so that leaves terror as the sole element peculiar to Soviet R & D. And as we have seen, terror was in no way decisive. Thousands of specialists died in the purges and camps, but Korolev and his colleagues, at least, labored on in *sharagas*, designing weapons that contributed materially to Soviet survival and occupation of Central Europe. Much was made of the American capture of the best Germans and the guts of the Mittelwerk factory. But the fact is that if the Red Army had been six months behind in its advance, and the Western Allies had occupied Germany up to the Oder River, the Soviets might have had a difficult time exploiting any of the achievements of Peenemünde. What slowed Soviet rocketry in the 1930s and 1940s was not terror, but war. The German inheritance put the Soviets back on schedule.

The other Soviet constraints on R & D—lack of economic competition and undue neglect of economic criteria—were less relevant to defense technology. The Russian overemphasis on pure science, which contributed to the imaginative beginnings of cosmonautics under the tsars, was a barrier to reification that the Soviets effectively broke down. Similarly, the efforts to train a new elite ameliorated the shortage of engineers and skilled technicians. Stalinist industrialization and unprecedented state support for R & D created the national base mandatory for advanced rocketry, while the Soviet regime's ideology and perceived strategic needs provided its justification. In sum, the science of cosmonautics and the genius of the men who furthered it were Russian. But they owed the rapidity of their success to Bolshevik myth and the command economy forged after 1917 by the world's first technocracy.

Sputnik was a famous victory, an expression of much that is good in Russian culture. It was also an expression of much that must be labeled bad in Soviet practice—the distortion of technology, purchased at such a price, into a cold tool of the state. And this observation already suggests part of the answer to the next question we must consider: why the supremely technological, but not yet technocratic, United States was second into outer space.

PART II

Modern Arms

and Free Men: America

Before Sputnik

Shall we expect some transatlantic military giant to step the Ocean and crush us at a blow? Never! All the armies of Europe, Asia, and Africa combined, with all the treasure of the earth (our own excepted) in their military chest; with a Buonaparte for a commander, could not by force take a drink from the Ohio or make a track on the Blue Ridge, in a trial of a thousand years.

—ABRAHAM LINCOLN, 1838

Under these conditions, Soviet possession of such [multimegaton] weapons and delivery capabilities would place the U.S. in danger of surprise attack and possible defeat. . . . Should we arrive at a condition where the contest is drawn and neither contestant can derive military advantage, we need not assume that this state is unchangeable. . . . We see no certainty, however, that the condition of stalemate can be changed through science and technology.

—Killian Panel Report, 1955

Progress is our most important product.　　　　　—General Electric Slogan

For progress there is no cure.　　　　　—JOHN VON NEUMANN, 1955

THE GLORY and the drama in Europe were over in June 1945, but retrospection on the European war not yet possible. Disclosure of Nazi death camps, generals' accounts of high strategy, the first shivers of Cold War at Potsdam, and the atomic bomb all lay in the future. Thus it was in gloomy frustration that reporters trooped in to Allied headquarters on a rainy day in Paris—the journalistic action lay elsewhere. They were greeted by John A. Keck, a Pittsburgh engineer turned chief of intelligence on German weapons, with the words: "This will make Buck Rogers seem as if he lived in the Gay Nineties." A first summary of what German scientists were up to when V-E Day interrupted them was about to be made public.

The reporters listened with wry skepticism. Surely these tales told more of Nazi dementia than of the imminent future of technology! The Germans, it seemed, had fantasized about launching V-2s from submarines, antiaircraft rockets guided to within ten feet of their targets, infrared scopes enabling riflemen to spot targets at night, IRBMs to bombard targets 1,800 miles distant. Researchers at Hillersleben even conceived of a space station 5,100 miles above the earth to be equipped with a "sun gun," or huge, reflecting mirror—to transmit power to earth or incinerate cities and boil oceans in a flash. The Germans insisted that such a thing was possible, perhaps in fifty years. Turning aside reporters' ridicule, Colonel Keck and his staff insisted that none of this was a laughing matter: "We were impressed with their practical engineering minds and their distaste for the fantastic."[1]

The terrible responsibility that falls on victors in a great war is to identify and expunge the causes of the ordeal to prevent its happening again. After 1713 the Great Powers manned a ring of fortresses around France to remove temptation from the Bourbon kings; after 1815 they suppressed revolution and conscript armies, and exiled the Disturber Napoleon to St. Helena; after 1918 they crusaded against monarchical militarism and imperialism. But what had escaped Pandora's Box this time? Could the chilling weapons forged in World War II exist safely in a world cleansed of fascism? Or were the technologies and technologists themselves the incubus? To defeat the enemy, it is said, you must become the enemy. As the Manhattan Project neared consummation and Allied agents scoured Germany for brains and hardware, the victory over Hitler appeared woefully inadequate. Consider the thoughts of George Fielding Eliot, military correspondent of the New York *Herald Tribune.* In April 1945 he voiced his discomfort with the fruits of the war: robot bombs, rockets, jet planes, possibly a postwar arms race. Scientists "hidden away in a garret" might soon be tinkering with death rays or atomic bombs, and the day might come when an "anemic professor" could touch a button and kill a thousand men a thousand miles away. Eliot called for strict controls on military science and thought it wise if "the leading men of science of Germany and Japan, who have devoted their lives to contriving new weapons and new methods of slaughter, were confined on some distant island—South Georgia, for example—down near the Antarctic Circle. . . ."[2] But this fear of technology itself, apart from those who may wield it, was inconceivable to the Soviet official mind, while Americans would learn to live with the fear of terrible technology in hopes of allaying a greater fear—destruction at the hands of a hostile, transatlantic power—from which they had been uniquely free since 1814.

For a century and a half the American experiment had run its course in isolation from the Eastern Hemisphere. George Washington warned that gratuitous involvement with the rivalries of Europe would corrupt the new republic. The Monroe Doctrine quarantined the New World from European predations lest they entwine the young United States in wicked contests for power. Southern secession was ominous, too, since the existence of two or more strong nations on the continent would re-create the balance-of-power system that impelled states toward centralized rule, standing armies, heavy taxation, fickle alliance systems, and all the ills of the garrison state. Manifest Destiny later ensured that a multipolar international system would never arise in North America. For the blessed United States, foreign war remained a discretionary activity and peacetime a reality, not just an armed truce. After World War I Wilsonian Democrats and Eastern Republicans granted that the United States must now help to manage the world political economy. But no one took this to mean that the United States must remain a nation in arms, maneuvering constantly for military and diplomatic advantage. Rather, the United States remained essentially defensive with respect to the Eastern Hemisphere—until 1945. Within another year Hitler might have had his "New York rocket." Within another year the United States did have the atomic bomb. In a few more years the rocket and bomb were married, and isolation ceased to be even an option for Americans. The global strategic ecumene had closed.

Would postwar Americans want to live in the world their victory had created? As early as August 1944 Chicago physicist Arthur H. Compton warned that the war would not truly be over until firm international control of the coming atomic weapons was in place.[3] But after Hiroshima, civilian and military leaders came to refer to the bomb as the "ultimate tool of peace," the guarantee that the kind of war just ended would never recur.[4] By 1947 negotiations with the USSR for control of atomic energy proved barren, Germany's "mad scientists" had supplemented the indigenous talent of the Superpowers, and their leaders anticipated the day when the technical progeny of wartime research could suffice to deter or destroy a future enemy. Compton was right: advanced weaponry was not controlled, hence World War II never ended for the United States, just as World War I had never ended for Russia.

Unlike the Soviet Union, however, the United States never blindly embraced power in whatever guise. To be sure, Americans love technology; we are, in Daniel Boorstin's phrase, the Republic of Technology. But the federal government never took upon itself the role of compelling the creation of new knowledge and power any more than it sought to concentrate all political, judicial, or police power in its hands. Rather, the Founding Fathers practiced what Boorstin calls "political technology" in drafting a constitution designed to limit and balance the exercise of federal power. Alexander Hamilton and others dissented, but most Americans of the late eighteenth century preferred to rely on the creativity of individuals rather than on a directed economy. "What federalism was in the world of politics, technology would be in the minutiae of individual life. While ideology fenced in, federalism—and technology—tried out."[5]

Certainly the United States had much in common with the USSR. They were both "continental super-states" made possible, in the words of Isaac Asimov, by the railroad and the telegraph. Both were born of revolutions inspired by ideologies of progress, faith in the works of man, and patriotism rooted in common ideas, values, and experience, rather than tradition, "holy soil," or ethnicity. Both regimes came of age internationally in the world wars and prevailed because of geographical expanse, remoteness, and unprecedented mobilization of technical resources. And both emerged from isolationism to find themselves assigned the role of defining from scratch what it meant to be a global, or "super," power.

Yet American and Soviet traditions diverged in the relationship between the state and technological change. American laissez-faire demanded tremendous faith in technology precisely because it was not to be cradled in the arms of a wise and beneficent bureaucracy. The Soviets extolled technology, but just as obviously feared it. Technology was the means of production and represented the stored labor of workers. In private hands it became an instrument of class oppression, and thus rightfully belonged to the instrument of proletarian rule, the party-state. Once in power, the party-state jealously guarded its monopoly by controlling all technology, the generation of new technology, the science it was based on, and ultimately the very freedom to think, create, make, and do that is essential to the humanity of its citizens. In the USSR there was never any official suggestion that the goals of society and the march of technology could conflict. What promoted technology promoted communism, and vice versa.

The United States, on the other hand, was materialistic, but not materialist.[6] Technology was not an end in itself, but a means of serving higher American values—life, liberty, and the pursuit of happiness. Over time, two sorts of challenges arose to the reigning American culture of technology. The first was represented by all those who ever asserted that technology or "modernity" were offensive to other values. They include everyone who ever eulogized the bucolic life, pushed westward to escape the city and its "prison of conveniences," struck against automated factories, mourned the Tennessee Valley Authority's effects on Appalachian culture, or protested nuclear power plants. Such people are either idealized or mocked, but they are still free to act and are sometimes effective. The other challenge to the reigning culture came from would-be technocrats, Marxist or otherwise, who advocated far greater public promotion of technological change. Born of the Progressive Movement at the turn of the century or arriving from Europe itself, people like Herbert Croly and Thorstein Veblen admired European research institutes and economic intervention, and wanted to bring social management through science and technology to the United States. Author of the "economic interpretation of the American Revolution," Charles A. Beard, went so far as to deny the relevance of our "eighteenth century" ideas of government to the world of 1929. Technological revolution, he wrote, "has emphasized as never before the role of government as a stabilizer of civilization," and confronted it with bewildering complexities. But "technology has brought with it a procedure helpful in solving the problems it has created; namely, the scientific method," which "promises to work a revolution in politics. . . . It punctures classical oratory—conservative as well as radical. . . . Disputes about democracy, therefore, creak with rust."[7]

Neither Beard's generation of progressives, nor the depression and global war to come, closed off "disputes about democracy" in favor of manipulation of society by scientific experts. Certainly the war and bomb drove home Beard's truth that technological revolution demanded adaptation, but the United States by no means succumbed to technocracy in the decade after 1945. Instead, American impulses against federal control of education and research, unconstrained military spending, and social engineering remained forceful under Truman and Eisenhower. Historians argue about the origins of the huge military buildup of 1950, the decision to build the H-bomb, and other such turning points. But what is surprising is not that such decisions were made by the leaders of a world power, but that heated debate over them occurred at all. Policy disputes in the early Cold War demonstrate precisely the abiding concern of protesters and presidents alike for values other than power and technology: decentralization, the market economy, academic freedom, fiscal prudence, all of which competed with the demands of national defense.

From 1945 to 1960 Truman and Eisenhower searched for a means of deterring the Cold War enemy without the United States itself becoming another garrison state. But the means they hit upon—nuclear deterrence—chained American defense to rapid technological change at the very moment when Soviet technocracy committed itself to making the missile revolution as quickly as possible. Ironically, the presidential efforts to keep a lid on spending, to keep open the option of a negotiated arms freeze, to preserve the United States as a symbol of free inquiry and international cooperation, and to cope with a secret, singleminded, technocratic adversary without giving in to paranoia all contributed to the failure of the United States to be first into space. And that celebrated failure did more than anything to defeat Truman's and Eisenhower's hopes of adjusting to the Cold War without transforming American government.

CHAPTER 3

Bashful Behemoth: Technology, the State, and the Birth of Deterrence

American science, according to historian Daniel Greenberg, grew up an orphan. For all the dedication of amateur scientists like Washington, Jefferson, and Franklin, the young Republic rejected federal responsibility for the funding and direction of science and technology.[1] Congress saw fit not to imitate the royal academies and universities of Europe, dithered for years before accepting the Smithsonian bequest in the 1840s, and left education in the hands of the states and private institutions. Regulation of technology fell to the federal government only when interstate commerce or federal lands were involved. The U.S. government had an important hand in internal improvements, marine charting, and the survey of the continent in the nineteenth century, but it was a light and reluctant hand, as Alexis de Tocqueville observed:

> The American government does not interfere in everything, it is true, as ours does. It makes no claim to oversee everything and carry everything out; it gives no subsidies, does not encourage trade, and does not patronize literature and the arts. Where great works of public utility are concerned, it seldom leaves them to the care of private persons. . . . But it is important to observe that there is no rule about the matter. The activity of companies, of [towns], and of private people is in a thousand ways in competition with that of the State.[2]

The Civil War spawned the first major experiments in federal patronage: the National Academy of Sciences (NAS), chartered to mobilize expertise for the Union Army, and the Morrill Act of 1862, which ceded land to the states for agricultural schools. Industrial take-off in the following decades increased demand for research in such fields as electricity, metallurgy, and chemicals. But rather than the federal government underwriting the search for new knowledge, an American pattern matured in which research became the business of diffuse private entities: dynamic corporations like those of Edison and Bell; technical schools such as MIT

(1861), Texas A & M (1876), Georgia Tech (1885), CalTech (1891), the Illinois Institute of Technology (1892); and philanthropic foundations like those of Rockefeller and Carnegie. The only exceptions were the army and navy arsenals, which pioneered standardization, interchangeable parts, and assembly-line techniques that became known as the American System of manufacture.[3] Still, by 1914 the United States clearly trailed Europe in the nascent processes of command technology.

The First World War widened the gap. Compared to the European belligerents, the United States made only hesitant feints toward the mobilization of science.[4] The NAS set up a National Research Council, but the military services made little use of it, preferring to believe that élan and generalship were still decisive in modern war.[5] After the armistice, such R & D programs as did exist were cut back, while American firms shied from defense contracting that offered little profit and exposed them to charges of being "merchants of death."

The seminal exception in this period was the National Advisory Committee for Aeronautics (NACA), established as a rider to the Naval Appropriations Bill of 1915. Conceived by Charles D. Walcott of the Smithsonian Institution, who feared that the land of the Wright brothers might fall hopelessly behind the European belligerents in aviation, the NACA began with a $5,000 budget and the unpaid services of twelve presidential appointees. Obscure, humble, and poor—its peak peacetime expenditure was $3.1 million in 1940—the NACA succeeded in keeping American aviation abreast of the latest technology and made occasional breakthroughs of its own. The Samuel P. Langley Memorial Laboratory at Hampton, Virginia, completed the world's first full-scale wind tunnel in 1931 and pioneered streamlined cowling to increase speed, the optimal placement of engines, retractable landing gear, low-drag air foil, and techniques to prevent stalls and spins. By 1936, when Soviet and German fighters dueled in Spain, NACA won permission to expand, and in 1939 it broke ground at Moffett Field, California, for what became the Ames Aeronautical Laboratory. Nevertheless, the NACA, with a total staff of 523 souls, was surpassed in the 1930s by the government labs of totalitarian Europe.[6]

In the 1920s Thorstein Veblen and others made technocracy a vogue among intellectuals, and Herbert Hoover, an engineer himself, lobbied for federal support of R & D. They shared the Progressive notion, itself a form of Social Darwinism, that science was the key to rational progress and that civilization evolved at the pace of its creation of new knowledge. Upon his inauguration as President, Hoover named a Research Committee on Social Trends, but his campaign for a private fund to promote research died in the Depression. Roosevelt's New Dealers fared little better. A Science Advisory Board arose in 1933 under MIT President Compton, but no money could be found for its plan to provide emergency relief to scientists, and the board itself dissolved.[7] FDR's National

Resources Coordinating Board estimated in 1935–36 that American universities spent perhaps $50 million on research, of which $6 million, mostly for agriculture, came from Washington. The entire federal R & D budget for each of the years before World War II averaged around $70 million (with another $50 million in "emergency funds" for social science and statistics). A spare $15 million was the province of the War and Navy departments. Private industry spent an estimated $100 million, for a grand total of American R & D of $264 million per year at the end of the 1930s.[8] This compares poorly with the 1,651 million rubles of the 1941 Soviet plan.

How can we account for this aloofness toward public finance of science and technology? Certainly the United States was not antitechnological, backward, or naive. The country had participated, albeit briefly, in the mobilization of 1914 to 1918 and had lived through ten years of unprecedented economic distress. Nor can the explanation lie in conservative leadership: both Hoover and Roosevelt were among the biggest boosters of research, engineering, and federal intervention. The answer must lie instead in American notions of liberty that affirmed the existence of powerful, autonomous institutions embodying values deemed worthy of protection from a grasping central government. The university harbored academic freedom, the corporation and entrepreneur economic freedom. It was not the business of government to assume a superior wisdom about the allocation of funds for R & D, nor to use taxpayers' money to impinge on scholarship or underwrite private enterprise.

The American pattern of privately funded and executed R & D distinguished American ideology from the Soviet. It was also a luxury, bestowed by the strategic isolation and natural wealth of North America and the inventiveness and vigor of its people, business, and universities. American economic and military needs were met for 150 years without centralized mobilization of intellectual resources. But the international imperative, imposing a nondiscretionary, continuous involvement in the global struggle for power and security, broke the pattern. In December 1941 the United States suddenly lost its luxury to dispense with policy for science and technology, and the postwar period, this time, brought no respite. The history of policy for science ever since has been one of struggle to reconcile abiding American values with the need to meet the real and perceived challenges from abroad.

Given the economics and ideology of American R & D prior to World War II, American rocketry progressed to the stage of large-scale experimentation five to ten years later than the Soviet or German. It began similarly enough when another "pragmatic dreamer," inspired by tales of spaceflight, began singlemindedly to elaborate the physics and technology of rocket flight. Robert Goddard published his thesis with the aid of a Smithsonian grant, joined the faculty of Clark University, and was the first to experiment with liquid fuel rockets in 1926. Jealous and secretive, Goddard soon hunted for a spot far from New England where he could work in liberty from reporters, neighbors, and competing rocketeers. With help from Charles Lindbergh and the Daniel Guggenheim Fund for aeronautics, he moved to the state destined to be associated with futuristic weapons research, New Mexico. Goddard was so aloof that his progress in liquid fuel rocketry, which earned him eighty-three patents, had meager influence on contemporaries, who had little choice but to dismiss him as a sort of crackpot.

Goddard insisted on using only the highest energy fuels. Liquification of hydrogen was still beyond the means of commercial industry, so he settled on gasoline. But only LOX would do for the oxidizer, despite its expense, danger, and the difficulty of making a pump work at −298° F. So Goddard designed a pressurized feed and, after much trial and error, integrated it with fuel pumps, igniter, combustion chamber, and nozzle into a working rocket. Over the course of fifteen years at his ranch near Roswell, Goddard conducted over a hundred static tests and forty-eight flight tests of ever larger rockets. His final specimens were twenty-two feet long, carried 250 pounds of LOX and gasoline, and developed an average thrust of 825 pounds. By amateur standards his achievements glittered, but his lone-wolf operation was anachronistic by 1940, while his secrecy meant that his later writings went uncirculated.[9]

Like the Europeans, Goddard and his contemporaries in rocketry grew up on Verne and Wells. But American popular culture, less respectful of literary conventions, raised science fiction to a genre. Edgar Rice Burroughs capitalized on the spurious discovery of "canals" on Mars by Percival Lowell (after Italian astronomer Giovanni V. Schiaparelli) in his Barzoom tales of Martian civilization. Hugo Gernsback and his protégé David Lasser, the editor of *Wonder Stories,* founded the American Interplanetary Society (AIS). Lasser also argued the practicality of spaceflight in a nonfiction book, *The Conquest of Space,* in 1931. The AIS then corresponded with French aircraft industrialist and rocketeer, Robert Esnault-Pelterie, and the German VfR, and financed liquid fuel rocket work on a farm in New Jersey. Other freebooters included Harry W. Bull of Syracuse University, the Cleveland Rocket Society under Ernst Loebell, and Robert Truax, who built small rockets in his spare time at the U.S. Naval Academy.[10]

The European and American paths diverged in the mid-1930s when the state apparatus drafted German and Soviet rocketeers for military research. The American pioneers found no sponsors. Annapolis observed Truax but offered no support. Army Ordnance pursued lackadaisical work on solid fuel rockets with an eye toward air-to-air weapons, but the project lapsed when its instigator, Leslie A. Skinner, shipped out to Hawaii in 1938. Frank J. Malina and the brilliant Hungarian aerodynamicist Theodor von Kármán did extensive groundwork on little financial and no moral support at Caltech's Guggenheim Aeronautical Laboratory (GALCIT). Ridicule and indifference even moved the AIS to change its name to the American Rocket Society (ARS) in hopes that dropping "interplanetary" might boost its credibility.[11]

The turnaround dated, not surprisingly, from 1941. The military planned for war, and government began to grasp the admonition of Marxist physicist J. D. Bernal that "A national economy, integrated through science and continually advancing by means of scientific research and development, is the basic need of the new era which we are now entering.[12] When General "Hap" Arnold, commander of the Army Air Corps, asked the NAS in 1939 to sponsor R & D on small rockets to aid in the takeoff of heavily loaded planes from short runways, the GALCIT received the first $1,000 of the millions that would make it, by 1944, Caltech's Jet Propulsion Laboratory (JPL).[13] (Even now, skepticism about rockets led von Kármán to choose the word jet for the expanded installation.) Suddenly Goddard was rediscovered, ARS data were collected and digested, and the U.S. government entered the rocket business. In the last two years before the war, GALCIT and the new Rocket Ordnance Section began testing jet-assisted takeoff (JATO) and the antitank rocket weapon that became the bazooka. Four members of the ARS founded Reaction Motors, Inc. (later a division of Thiokol Chemical), the first private firm devoted to rocketry, in 1941. By the end of the war, enterprising aircraft, electrical, and chemical firms, and government laboratories had gained the experience to absorb the achievements of the Peenemünde team almost as quickly as their counterparts in the institutes and prison camps of the USSR. The pace of their progress in long-range rocketry, however, would depend on how the U.S. government chose to assimilate the R & D explosion of World War II.

Even before Pearl Harbor it was apparent to most American officials that the virtuosity and scale of a nation's research went far to determine its performance in modern war. By the end of the war there were no dissenters. Radar was followed by electronic countermeasures for air and sea combat, infrared bombsights, the DUKW amphibious vehicle, the bazooka, the proximity fuze for artillery, and a plethora of other devices that made the war as much a competition in brainpower and ingenuity as of numbers, productivity, and morale. The American scientific effort was especially impressive for its voluntarism and ad hoc organization. And it was associated above all with the name of Vannevar Bush.

. An electrical engineer from New England, veteran of antisubmarine research in World War I, among the first to experiment in computer design, a vice-president of MIT, head of the Carnegie Institution, and a chairman of the NACA, Bush was uniquely suited for leadership in wartime science. While German armies overran Western Europe in June 1940, Bush gained entry to the Oval Office through a sympathetic

acquaintance, Harry Hopkins, and urged F
nation's scientific talent. He also impressed or
maintain autonomy for scientists lest their work
by military supervisors. Roosevelt obliged by
Research Committee, which evolved a yea
Scientific Research and Development (OSR
Conant. The OSRD, they decided, would
programs to universities, deemphasizing fed
tapping the talents of society as a whole. Th
for results and deadlines, but retained a measure of independence from
public supervision. Banking on the patriotism of private citizens and
institutions, and the hunger of universities for long-denied federal
subsidy, Bush established the practice of state-funded but privately
executed R & D. In a matter of months, patterns that had characterized
American research throughout its history were undone. Over fifty universities and industrial firms received contracts of $1 million or more
during the war.

Other models were in the making as well. The War and Navy
departments created their own "in-house" research capabilities almost
out of nothing, the largest and most fateful being General Leslie R.
Groves's Manhattan District with its nuclear laboratories at Oak Ridge,
Tennessee, and Los Alamos, New Mexico. To fulfill the government's
needs—and seize the chance to expand at federal expense—the University
of California and MIT founded Radiation Laboratories, Harvard a center
for radar research, Caltech the JPL, the University of Chicago its nuclear
"Metallurgical Laboratory." Professors across the land suddenly found
themselves in demand by a government eager to spend seemingly
inexhaustible funds for work both challenging and patriotic. The total
federal budget for R & D almost quadrupled from 1940 ($74 million) to
1943 ($280 million), then grew five and one-half times again by 1945
($1.59 billion). The share of the federal budget devoted to R & D more
than doubled over the course of the war even as federal spending
increased tenfold.[14]

The role of R & D in making the United States a qualitative as well as
quantitative "arsenal of democracy" made it an article of faith by midwar
that the federal government would continue to subsidize research beyond
the victory. But as soon as deliberation began on how to do it, the
inherent contradictions between technocracy and democracy created
muddles that have never been fully resolved. The first three issues were
how to organize federally funded scientific research, how to control
atomic energy, and how to coordinate and unify the armed services. The
bargains among conflicting agencies and reconciliation of national interests
in these three areas eventuated in a postwar regime for research that no
one had intended: a regime in which the military dominated despite the
opposite intent of all concerned, a regime soon pinched for funds despite

al recognition of the importance of R & D, a regime divided
ealous and disorganized interests despite the hope for coherent,
al direction.

ne obvious solution to ongoing federal R & D, and the one favored
y Bush, was to extend something like the OSRD into peacetime in
combination with relaxation of wartime secrecy.[15] But the difficulties of
such a plan came to light in the proposed legislation of Senator Harley
M. Kilgore (D., W.Va.), chairman of the Subcommittee on War Mobilization. Kilgore, a populist New Dealer whose faith in governmental
activism proved embarrassing even to members of his own party, was
so impressed by war work of civilian scientists that he wanted a new
agency to direct postwar research. It would be led by a presidential
appointee, governed by a board of eight civilian specialists and nine
cabinet secretaries. Kilgore lumped together basic and applied research,
insisted that all patents from sponsored work belong to the government,
and preached a vague utilitarianism: "to do something for the betterment
of humanity."[16] Such harnessing of science for the solution of human
ills was fetching—but it seemed little different from the Marxist technocracies of Bernal or Bukharin. Bush skewered the Kilgore bill. Congress
would demand visible results from public R & D, hence the meshing of
pure and applied research ensured the impoverishment of the former.
Government monopoly of patents ensured nonparticipation by private
industry. Political linkage of science and government ensured the corruption of the research community. Bush insisted that while the public must
pay, the scientists must be free of political interference. Unfortunately,
such a scheme was unconstitutional: representatives of the people could
not vote funds for activities over which they had no control.

The threads of policy grew more snarled when military R & D entered
the frame. In mid-1944 the service secretaries, Henry L. Stimson and
James Forrestal, appointed a committee to consider a postwar civilian
agency for the direction of military research. But its idea for a Research
Board for National Security (RBNS) troubled the president of the NAS,
Frank B. Jewett, who disapproved of the proliferation of agencies and
feared the RBNS would be vulnerable to pressure from Congress, the
military, and their contractors. He preferred to insulate the new board
within the NAS. But even this stopgap failed to survive the scrutiny of
the most powerful and pervasive agency of government, the Bureau of
the Budget (BoB). How could the military services and NAS take it upon
themselves to form a powerful new body without the approval of White
House or Congress? Budget director Harold D. Smith prescribed that
control of military R & D "must at all times be lodged solely within the
framework of the government"; that the OSRD must stay in harness
until a postwar agency was created by law; that the RBNS be restricted
to an advisory role. Bush's original concern was to prevent government
from running science; by 1945 Smith and President Truman feared the

opposite: "We cannot let this outfit run the government."[17]

In July Bush published his acclaimed treatise, *Science, The Endless Frontier.*[18] It instructed the public on the importance and the complexities of national science policy but solved nothing. His call for a National Research Foundation to include an advisory committee for defense prompted two more bills in Congress suggesting different governing formulas and patent policies. Bush retreated to a strictly advisory RBNS, with R & D contracts let by the military itself. But even this arrangement could not work—neither the services nor the NAS could hire personnel for an agency unauthorized by Congress. In February 1946 the RBNS was terminated and the army and navy began awarding funds and contracts directly to American universities.[19]

From afar science policy appeared the high road to power and progress. It turned out to be a mire through which Congress slogged for four years as the dilemma of planning science in a democracy became frustratingly clear. R & D was too important to be entrusted to *anyone*—the military, the scientists, or the politicians. Government support of R & D *without* political control meant old-style *technocracie*, government by elite experts; government support of R & D *with* political control meant new-style technocracy and the obviation of scientific and economic freedom. In 1950 the muddle "resolved" itself in favor of the President. Congress created the National Science Foundation (NSF), beholden to congressional committees and the BoB for funding and oversight. And when the Senate, upset with the financial burden attending the Korean War, voted only $225,000 to the NSF, the dreams of Bush and Kilgore alike for steady, substantial federal support for research evaporated.[20] In the meantime, the army and navy used their clearance to deal directly with the private sector, a clearance they had never sought. By 1948 the new National Military Establishment accounted for 62 percent of all federal R & D.[21]

A major arena of postwar research that did not fall to the military was atomic energy. But here again the scientists won an empty triumph, federal control prevailed, and another paradigm emerged: the agency set up for intensive R & D in one specific, strategic field of technology. As early as July 1944 Bush, Conant, and Irvin Stewart sketched out a domestic control bill for atomic research. The President would appoint eight civilian and four military commissioners to oversee all handling of nuclear materials, research, and production plants. But Bush and other scientists fretted about a government monopoly. Controls were necessary, to be sure, but not a political imprimatur over private research. The army proposed a compromise, promising "a policy of minimum interference. . . ."[22] Then came Hiroshima. Truman's brief statement of August 6 told the American people that a single bomb with the force of more than 20,000 tons of TNT had been dropped on a Japanese city and opened "a new era in man's understanding of nature's forces." The technical details, however, could not be divulged "pending further examination of possible methods of protecting us and the rest of the world from the danger of sudden destruction." Hardly a comforting message. Truman asked Congress to establish a commission to control this new force, while he sought to make of it a "forceful influence for world peace."[23]

A week later the Government Printing Office released Henry D. Smyth's *General Account of the Development of Methods of Using Atomic Energy for Military Purposes*, which told the world of the Manhattan Project. It concluded ominously that "The ultimate responsibility for the nation's policy on the questions raised by atomic energy rested with its citizens. . . . Now the great political and social questions that might affect all mankind for generations were open for the people to debate and decide through their elected representatives." The scientists would explain, but the people must decide.[24]

Who wanted such responsibility? And how could "the people" be expected to exercise it? What they heard in the media in the wake of Hiroshima was a mixture of euphoria and hysteria: the American atomic monopoly meant peace for all time, but failure to share and control atomic power meant Armageddon; atomic energy could bring an economic millennium and an end to all causes of war, or else the destruction of the world. Inevitably the struggle over postwar management of this new technology fell not to the public but to factions within the government, while scientists played the role not of neutral experts but of major external antagonists. Chicago nuclear physicists rallied quickly to oppose any plan that promoted secrecy and exclusivity at the expense of freedom of information and international cooperation, and lobbied to prevent the "railroading" of the army bill.

By all accounts atomic energy was a revolutionary technology that justified abandonment of old patterns of research. But in favor of what? Unprecedented control and secrecy, or unprecedented cooperation and openness? Atomic bombs and long-range bombers meant that the next war would offer no grace period such as that after Pearl Harbor, but would be decided by the weapons existing at the outset. Vigorous military R & D was therefore essential.[25] On the other hand, these were "forces of nature too dangerous to fit into any of our usual concepts," and thus could not be entrusted to the military.[26] What institutional arrangement was called for?

The army draft entered Congress as the May-Johnson bill of October 1945. Congressman Andrew J. May (D., Ky.), chairman of the House Military Affairs Committee, and Senator Edwin C. Johnson (D., Colo.), hoped to rush the bill through committee and into law without the confusion attending legislation for research in general. Instead they helped to kick off a national protest. Newspapers denounced the haste, which seemed to confirm that May-Johnson was a bid for military control. Chicago physicist Herbert L. Anderson, previously aloof from his protesting colleagues, spoke for many when he implored, "The war is won. Let us be free again."[27] The BoB and Office of War Mobilization and Reconversion also thought the bill overemphasized military applications. Truman himself withdrew endorsement. The Chicagoans then enlisted law professor Edward H. Levi to draft alternative legislation, while young physicist John A. Simpson, Jr., hectored congressmen and joined forces with women's, religious, and United Nations groups.[28] One by one scientists came forward to denounce military control. Harold Urey got carried away and called the May-Johnson "the first totalitarian bill ever written by Congress. . . . You can call it either a Communist bill or a Nazi bill, whichever you think is the worse."[29] Scientists around the country formed an Independent Citizens Committee to publicize the dangers of nuclear war and plead for a "spirit of world security."[30]

The chairman of the Special Committee on Atomic Energy, Senator Brien McMahon (D., Conn.), introduced an alternative bill in December 1945. It called for an Atomic Energy Commission (AEC) under the exclusive control of five civilian commissioners appointed by the President from civilian life, freedom of information in basic science, and a patent policy ensuring rewards for private investors. It also forbade any weapons R & D in violation of hoped-for international agreements and kept all fissionable materials under AEC control. The bill received immediate support from scientists and journalists, while General Groves, so recently a hero, became a symbol of military secrecy and arrogance. These were also the months of investigation into American unpreparedness before Pearl Harbor. The military seemed to be in full retreat.

Then, in mid-February 1946, Canadian agents cracked a Communist spy ring that had passed atomic secrets to Moscow. The public mood shifted overnight. Secretary of War Robert P. Patterson attacked the McMahon bill, asking how the army and navy could be excluded from a matter of the highest importance to national security. Senator Arthur Vandenberg (R., Mich.) proposed an amendment providing a military applications advisory board to consult on all matters relative to national defense. Now McMahon was on the defensive. Such a procedure, he countered, would give the military a veto power over atomic policy and "a position of authority in our national affairs unprecedented in our history."[31] But assorted opponents threatened to knock McMahon back to square one. Congressmen charged that the AEC's proposed powers were unconstitutional, that the bill authorized a give-away of vital secrets to foreign countries, that the whole concept of a state monopoly in a technology was radical. Clare Boothe Luce (R., N.Y.) echoed Harold Urey's earlier hyperbole by declaring the proposed commission "not even socialistic, it is a commissariat!"[32]

On July 18, 1946, the McMahon bill came before the Senate. Behind the scenes, the bill's supporters, the army, and the Chicagoans tacitly agreed to avoid a floor fight, although McMahon himself refused to budge. The Vandenberg amendment had made the bill palatable, if not delectable, to most senators, and it passed by voice vote on a quiet Saturday afternoon. The House was another story. Opponents of various aspects of the bill rose with amendments to virtually every section, one of which abolished AEC authority to "educate the world" on the danger and promise of atomic energy. When the latter was passed, one congressman voiced his disgust: "It looks like isolationism is again in the saddle. I take it that most of you want no part in our international problems." The McMahon Act then passed the House 265 to 79 and was signed into law by President Truman on August 1, 1946.[33]

Atomic energy, unlike governmental R & D as a whole, came to rest in civilian hands. Deemed too important to be left to the generals, it instead spawned a novel agency empowered to develop and direct the use of a specific technology. Congress also made an institutional adjustment by creating the Joint Atomic Energy Committee to oversee the AEC and its works. But such control, from the point of view of dissenting scientists and "globalists," was rendered anodyne from the start. For if national survival depended on the United States remaining in the forefront of military technology, the commission, however constituted, could hardly deny the military the weapons it requested. The solution was by no means a mirror image of the R & D philosophy and organization of the Soviet Union, but it was the first important peacetime compromise on the American road to technocracy. That is, it *would* be, pending the outcome of diplomatic efforts to ban nuclear weapons entirely.

"We are here to make a choice between the quick and the dead," announced Bernard Baruch to the delegates of the United Nations' (UN) own Atomic Energy Commission. He was there as President Truman's personal representative to appeal to the nations for support of a British-American plan to ensure that atomic energy would henceforth be used only for peaceful purposes. The Baruch plan of June 1946 called for an international authority to police all stages of development and use of atomic energy. Once an adequate system of control was in place, the United States would dismantle its existing weapons. A critical *sine qua non* was that declaration of violations and sanctions must not be subject to the *liberum veto* of UN Security Council members.

Andrei A. Gromyko countered with the Soviets' plan. (By now their own atomic program had been going full bore for over a year.) It called for a convention prohibiting all manufacture and use of atomic weapons. All existing weapons must be destroyed within three months of the convention's effective date. Measures to ensure observance would follow, but there could be no tampering with the right of veto. The Australian delegate explained to Gromyko that he was asking the United States to halt production, destroy existing weapons, and reveal to the world its exclusive information, all in return for a paper promise that others would not take advantage of those unilateral acts. Gromyko explained that the United States was asking all other countries to renounce weapons development, reveal to the world their uranium resources and the state of their own research, before the United States had given up its own arsenal. Gromyko's philosophy of disarmament also flew in the face of American internationalists: the very existence of the UN, he said, depended not on world government but on national sovereignty; imposition of inspection or sanctions by majority vote was incompatible with the independence of nations. What were the details of the Soviet plan, then? asked the Canadians and Australians. A week of insistence yielded nothing—there was no Soviet plan, only the demand for abolition of all existing (i.e., American) weapons.[34]

UN committee work droned on for the rest of the year. American and third country representatives worked on schedules for the various stages of dismantlement, information-sharing, and inspection that might satisfy the Soviets, but Gromyko's December proposals were essentially the same as June's. At the final vote on the American plan, every nation endorsed the Working Committee's report with the exception of the Soviet Union and Poland. Baruch resigned, and control of atomic energy was never again a real possibility.

Was it ever a real possibility? Was there a chance to stave off the arms race before the quakes of the late 1940s made the crevice of distrust unbridgeable? Whatever the demerits of the Baruch plan, it was still an unprecedented offer of unilateral disarmament by a Great Power with emerging global interests and justifiable suspicions about its potential adversary. But was it enough? Secretary of Commerce Henry Wallace thought not. This revolutionary technology required a revolutionary diplomacy. Instead both powers played politics more or less as usual, neither willing to take the first step, relinquish an advantage, or take a risk for real peace.

Some distinctions, nonetheless, can be made. For all the trumpeting of the Hearst Press against giving away atomic secrets, American opinion strongly favored international control, while scientists and politicians were beginning to sense the damage a peacetime arms race might inflict on American values. The United States had far more to lose domestically from an all-out R & D race than did the USSR. To be sure, the Truman administration was not as trusting as it could have been. The Baruch plan, with its gradual stages toward a system of controls and inspection, permitted the United States to keep its monopoly until Moscow had proven its goodwill. On the other hand, the Soviet plan, with its insistence on unilateral American disarmament and the sharing of information without inspection, only invited an imminent Soviet nuclear monopoly. That Truman's demarche was cautious is understandable; that it was sincere is beyond question. The same cannot be said for the

Soviets. An all-out R & D race was an expression of, not a threat to, Communist values. In foreign policy, a ban on nuclear weapons would surely strengthen the Soviet strategic position, given the weight of their conventional arms, but it would also (under the Baruch plan) open their country to international inspection, which it is hard to imagine Stalin—paranoid, secretive, brutal, insecure about Soviet backwardness, and traumatized by his ally Hitler's sneak attack—ever accepting. Certainly Gromyko's position afforded no compromise or elaboration, as would have befit an earnest attempt at agreement. Rather than assuming that a more forthcoming American position might have altered Stalin's policy, it is more reasonable to conclude that the nuclear arms race, from Moscow's point of view, had already begun.

The U.S. government spent a greater proportion of national product on defense in the years of mid-World War II than at any time before or since. The army and navy became world-girdling empires of men and machines, crowding out their enemies in a two-front, two-ocean war, and tangling with each other in a matrix of overlapping theaters, missions, and jurisdictions. At stake in their rivalry was not only glory, but the future of the services themselves. After 1918 the country had disarmed to the point of impotence, and the military lived in genteel poverty until the sneak attack arrived that weakness had invited. This time, the officers determined, the nation must flex its strength even after victory, and the leverage of each service in the shakedowns to come would depend heavily on the precedents each was able to set during the war. When dreams of a peace conference, a truly United Nations, a peaceful world under the condominium of the Big Three powers all died like prairie grass in a single day of summer, the Truman administration began to grope for a strategy. How it decided to "fight" a Cold War would go far to determine not only the postwar military structure but the role of government in the processes of technological change.

No domain was more confused, contested, and crucial to the future of American military branches than that of air power. Land warfare belonged to the army and the sea to the navy; to whom did the sky belong? While both services needed air arms for reconnaissance and tactical support, neither had a natural claim to long-range bombing. Yet control of that mission would go far to determine which service—navy, army, or the latter's obstreperous offspring, the Army Air Forces (AAF)—would become bearer of the nation's sword, replete with big budgets, glamour, and technical dynamism, and which would shrink back into a glorified constabulary or coast guard.

This "roles and missions" controversy enveloped the new field of rocketry. Like aircraft, rockets, or "guided missiles," could be put to all sorts of uses: ground-to-air, air-to-air, tactical artillery, perhaps long-range bombardment. Like aircraft, rocketry forced the military services, no less than Congress or the White House, to invent management techniques for large-scale, state-funded R & D. Should a new technology be assigned to one service on grounds of efficiency, or should technologies be divided among the services according to their relevance to assigned missions? From 1943 to 1947 the fierce struggle over military unification ended in a more or less understood division of labor. But it failed to yield guidelines on rocket development, with the result that by the late 1940s, when Stalin had fashioned a unified, intense missile program, the United States permitted its momentum to dissipate through disinterest and disunity. The Truman administration's economies also crippled apparently nonessential programs such as rockets. But those same econ-omies, in the context of Cold War rearmament, dictated a strategy based on strategic technological superiority. Such were these "years the locust hath eaten," during which the United States groped for ways to be a liberal democracy and a Superpower at the same time.

Early in the war the AAF Services division assigned guided missile work to the Special Weapons Group at Wright Field in Dayton, Ohio. By 1945 squabbles within the AAF and the potential revealed in the V-1 and V-2 enticed two Assistant Chiefs of Air Staff, the Air Commu-nications Officer, and the Army Service Forces all to claim jurisdictions in rocketry. "The development characteristics, control, and capabilities of these missiles," concluded the Chief of Staff's office, "have not been developed to the point where definite assignment of the operational employment to a major command can be determined without jeopardizing future development."[35] Its only guidelines dated from October 1944, to the effect that all missiles dropped from aircraft and all those with aerodynamic qualities (i.e., cruise missiles of the V-1 variety) should fall to the AAF; ground-launched ballistic missiles were the business of the Army Service Forces.[36] Such a division, based on the nature of the technologies rather than their potential missions, would place all long-range ballistic missiles in the army, like artillery, rather than in the AAF, like bombers.

After V-J Day the services quickly expanded their missile programs. All sent teams to occupied Germany, allocated funds for peacetime R & D, and grabbed for advantages as the interservice dogfight entered its critical phase. The army ground forces welcomed the planned unifi-cation for a single Department of Defense as a means of getting rid of their increasingly dominant air branch and as a way of containing the growth of the marines. The AAF naturally promoted it with vigor as a means of achieving autonomy and equality as a third service—the United States Air Force. Only the navy opposed unification, in part because the concentration of air power in a separate service stood to take from the navy its roles as the nation's first line of defense and vehicle for the global projection of power. On the other hand, the army and navy felt a common interest in warding off the air force claim to all flying weapons.

From 1945 to 1947 the three pushed their infant R & D programs, promoted divisions of labor favorable to themselves, and concocted scenarios of "next wars" that suggested the decisiveness of their own capabilities.[37]

As historian Daniel Yergin has argued, the roles and missions controversy and the developing Cold War mentality were inseparable. The services had to demonstrate their worthiness in terms of an enemy, and the only plausible candidate after 1945 was the Soviet Union.[38] But the navy could scarcely penetrate Soviet waters, and no one wanted to maintain land forces comparable to the Red Army. That left air power as the only deterrent to renewed Soviet expansion—as well as the only means by which the USSR might assault North America. General Carl Spaatz instructed Congress that the new military frontier of the United States was in the sky. "The next war will be preponderantly an air war. . . . Attacks can now come across the Arctic regions, as well as across the oceans, and strike deep . . . into the heart of the country. No section will be immune. The Pearl Harbor of a future war might well be Chicago, or Detroit, or Pittsburgh, or even Washington."[39] General James Doolittle drove this point home by the simple expedient of discarding the standard Mercator world map for a polar projection: the Soviet Union hovered over North America in a great crescent at what would soon become bomber range.[40] Secretary of the Navy James Forrestal in turn prompted his admirals to stump for "supercarriers" to accommodate the big bombers needed to carry A-bombs to the Soviet heartland. But the AAF was better able to draw on the logic of the geostrategic situation and its own image as the future-oriented service. It also pioneered new techniques of R & D and built strong ties with the aircraft industry in search of support for its world view.

The appeal to private industry was a novelty in interservice fights. After Congress failed to create a unified R & D agency, each branch came to manage its own R & D. But the navy and army preferred their traditional arsenal system that involved private industry and universities primarily as providers of components and basic science. The air force and the aviation industry, on the other hand, realized at war's end that each depended on the size and vigor of the other. But aviation was sick. After tremendous expansion during the war, the industry threatened to collapse without government support. Reconversion consequently posed another novel problem of policy suggesting that separation between the public and private sectors was obsolete. Prior to 1945 American military hardware, if it did not come from armories, at least came from industries with large civilian markets. Reconversion simply meant helping factories move from tanks to autos, military radios to home radios, or boots to loafers. But the civilian market was insufficient to sustain the smaller aviation firms that mushroomed during the war, or even to keep the larger firms in the black. Yet the future security of the nation might depend on the ability of U.S. aviation to expand production rapidly and perform the R & D required to stay at the state of the art.

Wartime expansion of aviation exceeded anyone's expectations. In 1939 Roosevelt called for an air corps of 30,000 planes, total. By January 1942 he demanded new production of 125,000 planes per *year*. This level proved unnecessary, but in 1944, 95,274 aircraft rolled out of American factories, including 16,334 four-engine bombers. Pounds of airframe produced (a measure similar to naval tonnage) leaped from 20.3 million (1940) to 915.0 million (1944). The government financed plant expansion, and prudent planning left the industry a reserve of $117 million. Still, the collapse of the military market in 1945 doomed dozens of subcontractors and research firms, while the twelve "majors" lost an aggregate of $35 million in 1946 and $115 million in 1947. Hundreds of thousands of production workers lost their jobs. Far from cracking down on "war profiteers," the country had a vital interest in subsidizing this most technological division of its arsenal of democracy.[41]

On the day of his return from Potsdam (and of the second atomic strike, on Nagasaki), President Truman acknowledged the dangers of peace for American aviation. "It is vital," he wrote, "to the welfare of our people that this nation maintain development work and the nucleus of a producing aircraft industry capable of rapid expansion to keep the peace and meet any emergency."[42] Aviation R & D, accordingly, was singled out for support in fiscal year (FY) 1946 and rose to $500 million in FY 1947. The AAF and aircraft industry likewise embraced Vannevar Bush's dictum that "The whole practice of warfare was being revised by the laboratories." Anticipating both the contraction crisis and the new importance of R & D, industrialist Donald Douglas approached the AAF in January 1946 with a plan for joint industry-government coordination of R & D with long-range strategic planning—in short, the think tank. Project RAND (coined by Arthur Raymond from Research *and* development), founded with Douglas Aviation and soon spun off as a nonprofit corporation, was the novel result.[43]

Much of the federal money went for rockets and missiles. Army Ordnance, commanding the services of the von Braun team, successfully fired its first V-2 in May 1946 at White Sands Proving Ground. The AAF approved no less than twenty-six missile projects and collaborated with six aircraft firms on surface-to-surface missiles alone: North American, Martin, Curtiss-Wright, Republic, Northrop, and Consolidated Vultee (later Convair). The latter company received contracts to study two missiles with a range of 5,000 miles, one a subsonic jet of the V-1 type, the other a ballistic missile. Northrop touted its own ICBM design. But the AAF, unimpressed by the accuracy of the V-2 and inclined to think in terms of winged, air-breathing vehicles, placed its hopes in cruise missiles, especially the Northrop Snark, conceived in January 1946 to carry a 5,000-pound warhead between continents.

This postwar R & D boom made dead letters of the previous divisions of labor. If, as AAF R & D chief Curtis LeMay wrote, "The long range future of the AAF lies in the field of guided missiles," then rocketry was vital to all the services in the roles and missions controversy.[44] In February 1946 both services received orders to revise the October 1944 memo on missile R & D, while the AAF studied possible arrangements for its coordination. They included, in the eyes of the air staff, a single government agency similar to the Manhattan District, an independent War Department division, or a single command within the AAF. Needless to say, the AAF preferred the third possibility, though it graciously offered Army Ordnance Command the responsibility for tactical "battle-field" missiles.[45] Ordnance retorted that rockets were artillery, not aircraft, and made common cause with the navy to forestall an AAF monopoly. In October 1946 Assistant Secretary of War (Air), Stuart Symington, sought to end the debilitating struggle by assigning all guided missile work to the AAF. But even this was not the final word, for the unification process was also nearing a climax, and the eventual separation of the army ground and air forces would provide Army Ordnance and the navy with a final chance to argue their case.

The framework for the eventual solution to the reorganization of the U.S. military establishment was the Eberstadt Report. Prepared under the auspices of the recalcitrant navy in September 1945, most of its recommendations found their way into the National Defense Act of 1947. These included the formation of a National Security Council to ensure that the United States may become "an alert, smoothly-working and efficient machine" drawing on all military, political, and economic assets to deter an enemy; the formal establishment of the Joint Chiefs of Staff (JCS) under a rotating chairman; a National Security Resources Board to manage industrial preparedness; a large military role in directing R & D; and a Central Intelligence Agency.[46]

The one measure Eberstadt did *not* recommend was abolition of the War and Navy departments in favor of a single Department of Defense. Secretary of the Navy Forrestal seconded this omission. But once congressional intent was clear, he fell back to fighting for a measure of autonomy for the army, navy, and new air force. A compromise to this effect, combined with Forrestal's appointment as the first Secretary of Defense, reconciled the admirals to the inevitable. On July 25, 1947, Congress passed this greatest military reform in American history—a response to the new circumstances of a global power that would never again know peacetime as formerly understood, in an age of scientific/technological revolution. It seemed to impose a layer of centralized, civilian control over the competing services. In fact, the act only changed the locus of interservice rivalry from the strategic councils of the military itself to the halls of committees and agencies with power over the budget and the offices of civilian strategists, scientists, and industrialists. The act also unleashed the young and hungry U.S. Air Force (USAF), already honing its talents and forming alliances with industry and Congress in preparation for the coming age of technocracy.

In the wake of unification, however, a small victory went to the army. Yet another new defense directive divided up missile work not according to roles or missions or technical data but simply according to scale. "Strategic" missile programs fell to the USAF; "tactical" missiles to the army. This decree might have sparked a lively debate on where to draw the line but for the fact that the ultimate arbiter, the budget makers, had just brought down the ax on all parties. If the apparent requirements of global power and technological fertility had changed the face of American research in a few brief years, they had also stunned the administration and Congress with their cost. Battered by postwar inflation, Truman's America was not yet ready to bear those burdens. The budget bonanza ended, the promising postwar start in American rocketry slowed to a crawl, and speculative R & D became subject to rigid proofs of its relevance to an increasingly unified national military strategy. Before examining the effect of the 1948 budget crunch on missile research, therefore, we must consider the evolution of American strategy in the first years of the atomic age.

The fundamental data whence national strategy is derived are the vital national interests as defined by the leadership and the potential threats to those interests as mounted by foreign states. American ambassador Walter Bedell Smith explained to Stalin in April 1946: "We are faced in America, as in the USSR, with the responsibility of making important, long-range decisions on our future military policy, and these decisions will depend to a large extent on what our people believe to be the policies of the Soviet Union."[47] Why did the American government come to view the USSR as an aggressive state threatening vital American interests? Certainly the apparent lessons of history (futility of appeasement as evidenced by Munich, totalitarian cynicism and expansionism as evidenced by the Nazi-Soviet Pact, dangers of being unprepared as evidenced by Pearl Harbor) and the impression made by the Soviets' own rhetoric (superiority of communism, inevitability of world revolution, Leninist prescriptions for amoral exploitation of diplomacy, force, agitation, and propaganda) combined to make most Americans sensitive to and distrustful of the USSR. Then the Communist subversion of Poland and other East European states (hauntingly reminiscent of Hitler's salami tactics), disputes over administration in Germany, failure to control atomic energy, apparent Soviet unwillingness to demobilize, Soviet-supported destabilization of Greece and Turkey, the Communist coup d'etat in democratic Czechoslovakia, and the Berlin blockade followed each other with head-throbbing rhythm in the late 1940s and hardened the American inclination to assume the worst about Soviet intentions.[48]

A Soviet apologia could point to American provocations as well and perhaps exonerate the Stalinist regime of some of the mischief attributed to it. A principal historical issue, therefore, has been whether the United States adopted a policy of air-atomic power in response to objective military requirements, or whether American groups desirous of such a strategy (e.g., the air force, weapons scientists, industry) promoted a paranoid view of the USSR for their own purposes. But given that the United States, for whatever reasons, assumed responsibility for guiding events in the Eastern Hemisphere according to its own values and interests, and that the Soviet Union was the only major opponent of American policies, then the American strategic problem was inexorable: how to deploy American resources in such a way as to minimize the Soviet threat, and how to do so at the least cost to American taxpayers, institutions, and domestic values, including civilian control, a small standing army, an open society, and a free market economy? To be sure, interest groups and institutions had their own motives, but the evolution of U.S. strategy in the postwar years is striking for its tentativeness, confusion, and quest for the least drastic option. In the end, Truman resigned himself to the fact that the *least* drastic option was heavy reliance on nuclear weapons.

Chief of Staff George C. Marshall stated the dilemma of postwar American strategy as early as 1943: "We are trying to avoid war, but at the same time we have to carefully avoid a financial effect on our economy that would be as disastrous as a war might well be. . . . I think the maintenance of a sizeable ground expeditionary force probably impracticable. . . . Having air power will be the quickest remedy."[49] In 1945 the AAF had grown to 243 air groups and 2.4 million men. How many groups and airmen would be needed in peacetime? Three weeks after Hiroshima, AAF generals hit on the magic number of seventy groups, including twenty-five bomber groups, and 400,000 men. Their mission was already espoused as one of deterrence as well as counterattack. Hap Arnold advised Marshall that "we must . . . secure our nation by developing and maintaining those weapons, forces, and techniques required to pose a warning to aggressors in order to deter them from launching a modern, devastating war." In testimony before the House Committee on Military Affairs, a parade of air generals took the same line. The mighty air force could strike back at an attacker but, more important, deter him from attacking in the first place.[50]

Atomic bombs only seemed to strengthen the AAF case for deterrence. A blue-ribbon board named for General Spaatz and including generals Ira C. Eaker, Hoyt Vandenberg, and Lauris Norstad, reported in October 1945 that while the current paucity and expense of atomic warheads made them an accessory, rather than a centerpiece, of air strategy, nuclear technology would develop quickly: "any serious compromise of research and development ca⌐ be the sounding of the death knell of this country."[51] General Arnold responded by making LeMay the chief of R & D and organizing the AAF in March 1946 into its postwar triad of the Strategic Air Command, Tactical Air Command, and Air Defense Command. General Eisenhower told Baruch in June that the atomic bomb in American hands was a deterrent to aggression in the world. General Spaatz insisted that the "hysterical" demobilization of the army made the atomic bomb an essential part of U.S. strength even if the American monopoly was transitory: "Any step in the near future to prohibit atomic explosives would have a grave and adverse military effect on the United States."[52]

Yet, for all that, the atomic bomb had not found a place in U.S. strategy. War plan "Pincher" (June 1946) viewed the bomb as a distinct advantage, but could not integrate its use due to secrecy concerning its numbers and destructive force. The AAF plan "Makefast" made no allowance at all for atomic weapons. In any event, the military had little notion of how bombs could possibly prevent a Soviet takeover of Western Europe. Initial thoughts came to rest on bombing of Soviet industrial targets, but they in turn were too dispersed for an atomic offensive to be decisive.[53]

Even if an air-atomic strike had been deemed a war-winning capability, neither the bombs nor delivery systems were there to use. Spaatz estimated that there were only a dozen or so warheads in the arsenal during his time as Air Force Commander, February 1946 to April 1948. When the AEC took over the Hanford, Washington, reactors in January 1947, it found plutonium being manufactured "at a fraction of its wartime rate and the resulting bombs were still considered 'laboratory weapons.' "[54] The delivery problem was just as acute. By August 1946 the AAF had shrunk to fifty-two groups, only six of its twenty-six B-29 groups had aircraft, and these were based in the Southwest where they threatened no one but Canada and Mexico. Two years later the United States possessed only thirty-two B-29s converted for atomic bombs.[55] When President Truman thought to inquire about the size of the atomic arsenal (which he apparently did only in April 1947), he was shocked to find it a fraction of his expectations.[56]

In mid-1947 the crises that sparked the Marshall Plan and Truman Doctrine also inspired the first investigations of American atomic requirements. With the results of the Bikini Atoll tests in hand, the JCS concluded that atomic bombs could indeed nullify any nation's military effort and demolish its social and economic structures. But given the current scarcity of bombs, they must be used against population centers and only in extraordinary circumstances against military targets. The JCS also urged Congress to define "aggressive acts" in such a way as to permit a preemptive strike if an enemy should achieve an atomic capability.[57] In October the Joint Strategic Survey Committee estimated U.S. atomic needs, and the JCS "placed an order" with the AEC: 400

warheads as destructive as the Nagasaki bomb, to be delivered by 1953. These bombs, dropped on 100 urban targets, would be adequate to "kill a nation" and would suffice until such time as an enemy also acquired nuclear weapons. In other words, 400 bombs were judged to offset Soviet conventional superiority.[58]

Most important of all the 1947 studies was that of the President's Air Policy Commission, chaired by lawyer Thomas Finletter. In the course of more than 350 meetings and interviews, the commission asked to be apprised of the current American war plan. After "quite a run-around" from the Pentagon, Finletter complained to Truman, who directed Generals Eisenhower and Vandenberg and Admirals William D. Leahy and Chester W. Nimitz to testify. But they, too, obfuscated until Eisenhower turned to his colleagues and confessed: "Gentlemen, these five civilian gentlemen are just patriotic American citizens trying to do something they've been asked to do by the President. I think we really owe it to them to tell them that there is no war plan."[59]

Nineteen forty-seven—and no operative war plan for the new global power! But that was the point: how could any nation fulfill a truly global role? Forrestal testified that

you cannot talk about American security without talking Europe, the Middle East, the freedom and security of the sea lanes, and the hundreds of millions of underfed, frustrated human beings throughout the world. . . . It would do no good for us to be a Sparta in this particular hemisphere and have chaos prevailing elsewhere in the world. . . . Had Athens . . . a little less philosophy, and Athens had a few more shields, it might have been a good combination.[60]

The military could not be asked to find a solution to everything, however intimidating the atomic bomb. A coherent strategy required a frame of assumptions, not only about one's own intentions and the enemy's capabilities but about one's own *capabilities* and the enemy's *intentions*, before analysis could be brought to bear.

At this juncture in 1947, when the Truman administration was most in need of a framework for coping with the enigmatic Soviets, George Kennan's "The Sources of Soviet Conduct" appeared in *Foreign Affairs*. The article seemed to do for Americans what Leninist theory on imperialism did for the Soviets—it simplified. The USSR, wrote Kennan, was not just another imperial power, but the bearer of a messianic religion, an incarnate ideology for which coexistence in the long run was a contradiction in terms. The USSR expanded its influence by every possible means, its sole concern being to fill "every nook and cranny available to it in the basin of world power." The only response to such a challenge was "containment."[61] Soon the author of this explosive piece was feted and honored and assigned by the State Department to educate American diplomats and generals on the nature of the enemy. In a very few years Kennan came to regret his invention of the political technology of containment, just as some scientists regretted inventing the bomb. But Kennan's portrait of the Soviets, and the strategy of containment he proposed, entered the minds of men preoccupied with the need for a strategy and accustomed to think of strategy in military terms. Containment, deterrence, and the "sources of Soviet conduct" meshed like the pieces of a jigsaw puzzle.

The Finletter Commission reported its findings in a report entitled "Survival in the Air Age" on New Year's Day 1948. Its solution was deterrence, built on an air arm so strong "that other nations will hesitate to attack us or our vital interests because of the violence of the counterattack they would have to face...." The danger was that the American people might refuse the financial burden necessary for military procurement and the underwriting of a "permanently enlarged aircraft industry."[62] Finance was indeed the hinge on which all else turned. The $37 billion budget for FY 1947 was down 60 percent from the wartime peak, but still four times the size of prewar budgets. Inflation and unprecedented spending caused Truman's BoB Director, James E. Webb, to slash AAF budgets by half a billion dollars in 1947. But the lower the budget ceiling, the smaller the conventional military and the greater the reliance on atomic bombs. When the JCS removed its embarrassment by preparing a new war plan, "Half Moon," in May 1948, it called for an offensive in Europe, a defensive stand in Asia, and "a powerful air offensive designed to exploit the destructive and psychological power of atomic weapons...."[63] Truman asked the Chiefs to draft an alternate plan eschewing the use of nuclear weapons, but his own budgetary policies doomed conventional options. According to the JCS, the $14.4 billion ceiling on the FY 1950 defense budget would not permit the United States to retain even a foothold in Europe in case of war.

The atomic option was open to two sorts of criticisms. The first, suggested briefly by the navy, was that it was immoral. A democratic society ought not to plan strategies based on annihilation of civilian populations. The other was that it would not work. The Harmon Report of May 1949 predicted that an atomic attack on Soviet cities would not induce a surrender, destroy communism, or weaken the hold of the Moscow regime. By validating Communist propaganda, it might even strengthen the regime and steel the people's will to resist. Nor would atomic attack seriously impair the ability of the Red Army to advance rapidly in Western Europe, the Middle or Far East.[64]

But what was the alternative? Truman's FY 1951 budget lowered the military ceiling by another billion dollars. The new Secretary of Defense, Louis Johnson, then upheld a two-to-one vote in the JCS against the navy's supercarrier and, in the Admirals' Revolt to follow, sacked the leading critics of an air-atomic strategy. The National Security Council (NSC) declared deterrence to be national policy, and Truman finally and wistfully concluded that international control of the bomb was not to be. In July 1949 he told a top-secret meeting: "I am of the opinion we'll never obtain international control. Since we can't obtain international control we must be strongest in atomic weapons."[65] The JCS asked for a substantial increase in the nuclear stockpile and approved LeMay's request for rapid deployment of the long-range B-36. The Harmon Report may have sown doubts about the value of A-bombs in war-fighting, but as a relatively cheap tool of deterrence, they seemed to fill the bill as shield for Athens. "The only war you really win," General Vandenberg explained, "is the war that never starts."[66]

From 1945 to 1949 American leaders searched for a counterweight to Soviet conventional might. Few preferred to rely on an atom-armed Strategic Air Command (SAC)—but that reliance was dictated by geography, technology, and finance. In the space of five years, the United States adjusted to the fact that its global military responsibilities had not ended with the coming of peace but would extend into an indefinite future. In hopes of insulating the civilian economy and society and forestalling the garrison state, Truman opted for military technocracy— a strategy of deterrence based on the presumed superiority of the United States in weapons technology. But no sooner had that decision been made, marking a profound break with American tradition, than the Soviet crash program produced an atomic bomb on September 23, 1949, and the USSR emerged as a serious technical as well as political rival. Dependence on air-atomic power—technical supremacy—would prove neither cheap nor innocuous after all. The United States had made a sober commitment to military technocracy in hopes of avoiding a more pervasive militarization. But the Soviet A-bomb, together with the Communist victory in China and other challenges, persuaded the bashful American behemoth that it had still not done enough. After the A-bomb would come the race for the H-bomb, then the race for long-range rockets, and after that—a race for space.

CHAPTER 4

While Waiting for Technocracy: The ICBM and the First American Space Program

Truman's stubborn campaign against inflation and growth in federal spending, begun in late 1946 and continued (with the exception of a FY 1949 aircraft procurement program) until the Korean War, stunted the growth of the rocketry seedlings strewn hither and yon by the military services just after the war. In this first conflict between the demands of international power and the demands of domestic economy, the latter won, and the first precocious steps by the armed services toward an ICBM and a satellite program were not followed up. But these were also the years in which Soviet researchers vaulted ahead in pursuit of their capitalist technological rival. The United States commanded far greater national wealth and a far larger industrial base, but was still fishing for the lowest adequate level of national mobilization. Better atomic bombs and long-range bombers were accepted first, in accordance with deterrence, then a restoration of conventional strength, in response to Korea. But the rocketeers, though firmly planted in the military-industrial establishment, were left to cool their heels until Washington became reconciled to the next compromise with technocracy. Surprising Soviet progress in rocketry, and the advent of the hydrogen bomb, served finally to resurrect the ICBM, while the desperate American need for hard intelligence on Soviet capabilities served to convince the highest councils of government that spaceflight, too, was not folly but of tremendous potential importance to American security.

In FY 1945 the AAF spent $3.7 million, or a mere 2.6 percent of its R & D, on missiles. The following year the figure leaped to $28.8 million and 14.6 percent for twenty-six missile programs, and the AAF projected a doubling of this sum to $75.7 million for FY 1947. Instead, the President's austerity plan cut back missile R & D to $22 million. Eleven programs died at once, including Convair's Project MX-774, the 5,000-

mile ballistic missile.[1] Funding for all rocketry, army and navy as well as AAF, fell to $58 million in FY 1947. Thereafter funding showed an upward trend, but did not reach serious proportions until FY 1951.[2] The sort of long-range ballistic rocketry required for satellite launching disappeared.

The reasons for cancelling MX-774 seemed valid enough to a government not yet committed to an all-out technology race. The Air Materiel Command judged that MX-774 "does not promise any tangible results in the next eight to ten years." Technical barriers to IRBMs and ICBMs also gave pause. Available fuels lacked the specific impulse necessary to boost five-ton atomic warheads around the world, and no one knew yet how the bombs might survive the heat of reentry into the atmosphere in any case. It was assumed that a 5,000-mile missile "would amount to little more than a thorough study for some time to come."[3] Besides, the United States had a substantial lead in long-range bombers and access to air bases close to Soviet borders. The United States did not need an ICBM as badly as the Soviets.

The decision to shelve the ICBM reflected at least four mentalities current at the time: the need for rigorous economy, which dictated that scarce funds be put into bigger bombers and eventually jet aircraft; the assumption of American superiority in aviation; the preference of "blue-sky" air officers for manned bombers; and scientific pessimism about the technical problems. Vannevar Bush reflected the last trait in December 1945: "I say technically I don't think anybody in the world knows how to do such a thing [build an accurate ICBM] and I feel confident it will not be done for a long period of time to come."[4] By 1949 he had decided that a long-range atomic missile might in fact prove possible, but that its cost would be "astronomical."[5] Of course, Bush also admitted in the same book that he was "not much of a prophet";[6] in this case, as in others, some soldiers were right and some scientists were wrong. But in the budget-cutting context of the late 1940s, the B-36 won out over the ICBM, and the Pentagon concentrated on boosting the operational strength of the SAC, not on long-term R & D.

The $2.3 million spent on MX-774 was not a total loss. It bought the first studies of ICBM design and engineering, static tests of a main rocket engine, and helped to create a pool of expertise among industrial contractors that could be tapped in the future. Convair engineers progressed beyond V-2 technology with swiveling engines and a separable nose cone. They also dispensed with interior fuel tanks and simply used the outer "thin skin" of the rocket as the fuel container, permitting more volume at lower weight. The Convair team, led by Karel Bosshart, was so encouraged by its progress in two short years that it convinced management to continue design studies through 1950 at company expense. The new USAF tried to sneak the project in through the back door by billing Convair missiles as high-altitude research vehicles. The

R & D Board of the Department of Defense (DoD) said nix.[7]

High-altitude research was a promising use for sizable rockets, but the navy had already cornered it. In 1946 its Bureau of Ordnance and Naval Research Laboratory (NRL) founded the Viking program to probe the upper atmosphere for scientific purposes. The Viking was based on the V-2 and a small upper stage called the Aerobee, built by Martin and Aerojet Engineering. Six Vikings flew from the decks of ships and the NACA test range at Wallops Island, Virginia, through 1950. The navy also fired a fully fueled V-2 from the deck of the carrier *Midway* in September 1947. Operation Pushover followed, in which the navy purposely exploded two V-2s on deck to determine the ability of ships to withstand rocket accidents.

During the lean years, long-range ballistic research continued, unobtrusively and on a shoestring, under the von Braun team in New Mexico. Although the struggle for control of missile work had left the long-range programs in the air force, which then abandoned them, the army recovered its freedom to tinker with V-2s. Project Hermes, a collaboration between Army Ordnance and General Electric, modified the German rockets for larger payloads. Project Bumper added a pencil-thin upper stage, the WAC Corporal, which soared to record heights of 250 miles. When the store of V-2s began to run out, the army planned an expanded program of home-made missiles. In November 1950 the von Braun team set up shop at a new Ordnance Guided Missile Center at Redstone Arsenal, Huntsville, Alabama. Limited by DoD directives and budget to tactical weapons, the Germans began work on the 500-mile-range Redstone, built around the liquid fuel engine developed by North American Aviation for the Navaho cruise missile. The first Redstone, roughly equivalent to the Soviets' 1949 Pobeda, was fired in August 1953—thus the United States had fallen perhaps four years behind the USSR.

The story of early American satellite projects parallels almost exactly that of the ICBM: a brief flurry of enthusiasm after the war, followed by budget cuts and cancellations, followed after some years by sudden revival in reaction to Soviet progress. Indeed, it must have been with some irony that the Peenemünde veterans recalled choosing the Americans in part because the British "couldn't afford them." Stuck away in Huntsville, with a threadbare budget, restricted to rockets with a 200-mile range, von Braun's team counted their best years passing by with little appreciable progress toward outer space. Their only consolation was that public fascination with space travel rose as sharply after the Second World War as it did after the First.

Why the boom in science fiction? It has often been suggested that the atomic bomb was responsible, creating at once an appetite for vicarious scientific adventure and a need to externalize fear. Be that as it may,

Hollywood mass-produced low-budget thrillers premised on technological nightmares: atomic mutations, giant insects, visitors from outer space. Beginning in the midsummer of 1947 the American public also began to see unidentified flying objects, kicking off a flying saucer "epidemic" of such proportions that the air force launched a special investigation and began compiling thousands of case studies that, in the end, satisfied no one.[8] Science fiction books and magazines rebounded from the wartime slump (with its paper shortage) to reach a circulation by 1949–53 double the prewar peak and seven times the wartime trough. In 1951 *Life* magazine estimated the science fiction readers in the United States at over 2 million.[9] In retrospect, it was all a form of cultural anticipation. Science fiction writer Arthur C. Clarke has observed that virtually every technological breakthrough in the contemporary world was imagined a generation before by some novelist. After V-2s and atomic bombs, any fantasy seemed credible. But it is also true that these popular cultural expressions had no effect on American missile and space policies.

Rather, the practical thinkers, who could not be dismissed as a lunatic fringe, were the ones to seed American military and scientific communities with notions of the coming Space Age. Their ideas involved nothing more than extrapolations of existing rocket technology and the require-ments of the age of intercontinental weapons already on the horizon. Symptomatic of the practical drift of space enthusiasts in the United States after 1945 was the change in the membership of the American Rocket Society. Begun in the 1930s as a clique of "space cadets," the ARS in 1946 drew 59 percent of its membership from business and another 10 percent from the military and government. It came to associate more and more closely with the aviation industry and, by 1959, with 12,000 members, would be a thoroughly professional club and lobby.[10]

Von Braun himself undertook to make spaceflight respectable in a series of articles for *Collier's* magazine in 1952. A Walt Disney film, while trading on the sensational nature of the subject, also presented spaceflight in a "down-to-earth" way. When Disneyland opened in 1955, the Tomorrowland "moon rocket" was among its most popular rides, in part because of its verisimilitude. For all that, most Americans still considered space travel something for the medium-to-distant future. Even professional prophets—people involved in rocketry, space science, and science fiction—were cautious in their expectations. All but one of such experts polled in 1953 believed that interplanetary travel would someday occur, but only a quarter predicted a manned lunar visit by 1969. Twelve visionaries did not expect a manned lunar voyage until the twenty-first century, if at all. The most realistic response was that of German rocketeer Willy Ley: unmanned spaceflight would be achieved "three to five years after initiation of the project."[11] The technology was there; only the technocratic will, at least in the United States, was lacking.

The reason why the United States waited so long to join the race for space was *not* that it never occurred to anyone. Just three days after their surrender to the U.S. Army, the Peenemünde engineers sat down to brief agents of the Naval Technical Mission at the little Bavarian town of Kochel where Peenemünde's wind tunnel had been located. Among the Americans were Clark Millikan and Hsue-shen Tsien, the brilliant Chinese aerodynamicist from Caltech. They listened with growing appreciation as the Germans told of the possibilities opened by their rocketry: artificial earth satellites, manned space stations, interplanetary voyages. Their report, "Survey of Development of Liquid Rockets in Germany and Their Future Prospects," communicated the excitement to the Navy Bureau of Aeronautics in Washington.[12] Lieutenant R. P. Haviland took upon himself the task of researching the technical feasibility of artificial satellites, then elaborated the possibilities of such vehicles for science, communications, mapping, and meteorology, and submitted findings in Project Rex of August 1945. Captain Lloyd Berkner and Commander Harvey Hall were convinced. By October Hall had formed a committee on space rocketry, and by the end of the year the Navy Bureau endorsed an Earth Satellite Vehicle Program.[13]

Robert Truax, already an "old timer" in naval rocketry, and his colleagues settled on an uncomplicated path to space, a single-stage rocket fueled by the high-energy combination of LOX and liquid hydrogen (LH_2). Homer J. Stewart and Frank Malina of JPL confirmed the feasibility of the navy plan, but preferred a two-stage booster for its decided advantage in propellant-to-weight ratio. Encouraged, the Navy Bureau let contracts to Aerojet Engineering to design the required engine and set up a pilot plant for manufacturing liquid hydrogen. North American Phillips was put to work on a solar-powered engine to run the satellite systems, and navy engineers tackled the guidance and attitude controls. It was a full-fledged satellite program in microcosm. But costs rose to the $8 million range for design work alone. The project had to outgrow the Navy Bureau. In the spring of 1946, its satellite committee cast about for collaborators.[14]

The AAF was the likeliest candidate. Although "the obvious military, or purely naval applications in themselves, may not appear at this time to warrant the expenditure," R & D chief LeMay agreed to look into a joint satellite venture. Meanwhile, Commander Hall tried repeatedly to meet with Vannevar Bush, only to hear that Bush was familiar with the satellite proposal and "preferred not to get involved in any discussions about it at this time." In fact, he was as skeptical of satellites as of the ICBM, and found it necessary to mock in print

some eminent military men, exhilarated perhaps by a short immersion in matters scientific, [who] have publicly asserted that we are [interested in high-trajectory guided missiles spanning thousands of miles]. We have been regaled by scary articles, . . . we even have the exposition of missiles fired so fast that they leave the earth and proceed around it indefinitely as satellites, like the moon, for some vaguely specified military purposes.[15]

The AAF, however, turned to its new advisory body, the RAND Corporation, for an independent opinion on the prospect and value of an earth satellite. The report, released on May 2, 1946, under the auspices of Douglas Aircraft, was exceptional for its foresight. It considered two configurations for a space booster: a four-stage rocket based on the German combination of LOX and alcohol, and a two-stage vehicle fueled by LOX and LH_2. It also delved into the problems of structural weight, trajectories, and stabilization. The authors confined themselves to "conservative and realistic engineering" and counted on no breakthroughs. Still, they concluded that a satellite vehicle was quite possible.[16]

What was required to launch an artificial satellite, RAND explained, was acceleration of a rocket to 17,000 miles per hour. At this velocity the upper stage would not return to earth but would revolve about the earth as inertia ("centrifugal force") matched the force of gravity. Such a vehicle would complete an orbit every one and one-half hours. Satellites would "undoubtedly prove to be of great military value," but the report emphasized scientific applications, including study of cosmic rays, gravitation, earth magnetism, astronomy, meteorology, and the upper atmosphere. Scientific instruments, it was assumed, required a payload of 500 pounds and twenty cubic feet, hence the concentration on more efficient multistage boosters. The report predicted that an initial satellite vehicle would take about $150 million and five years to build.[17]

To be sure, RAND could not predict at present what specific payoff might justify that effort: "The crystal ball is cloudy." But an analogy with early aviation was justified. "We can see no more clearly all the utility and implications of spaceships than the Wright Brothers could see flights of B-29s bombing Japan and air transports circling the globe." Two things were clear, according to RAND: satellites would become one of the most potent scientific tools of the twentieth century, and achievement of a satellite by the United States "would inflame the imagination of mankind, and would probably produce repercussions in the world comparable to the explosion of the atomic bomb." A satellite also "offers an observation aircraft which cannot be brought down," and could function as a communications relay station. Finally, it would be the first step toward journeys to Mars and Venus. "Who would be so bold as to say that this might not come within our time?"[18]

The RAND study so intrigued the AAF that at subsequent meetings with the navy, AAF officers claimed that their space thinking was as advanced as anyone's and tried to take over the project. They argued, as in all the other R & D quarrels, that this was a matter of strategic aviation, their natural responsibility, while the navy, as always, argued that the unknowns were great and division of responsibility premature.

But the navy's generous bid for a joint program, hence a stronger budgetary position for satellite research, came to nothing. Rear Admiral Leslie Stevens then requested the Joint R & D Board to coordinate an Earth Satellite Vehicle Program, but the chaos attending the 1947 military unification swallowed it up. Eventually, a reconstituted board assigned the problem to its Committee on Guided Missiles, which referred it to its Technical Evaluation Group. At that subterranean level, in March 1948, it was decided that, while satellites may be feasible, no "military or scientific utility commensurate with the expected cost of such a vehicle" had been demonstrated. The navy, squeezed by Truman's budget cuts, began transferring satellite funds to more pressing projects, until Admiral Stevens was told on June 22, 1948, that the Earth Satellite Vehicle Project was cancelled.[19]

How quickly quelled was the excitement generated by the German inheritance! World-circling rockets and satellites sketched on paper in 1946 were buried in closed files by 1948. Even the USAF missile program shrunk to four small projects, none of them long-range ballistic rockets. But the very next year new changes were in the works; the postwar lull was ending. For the Soviet A-bomb forced an immediate reassessment of American strategy that split the government and physics community and climaxed in another technical breakthrough—the hydrogen bomb—that restored in a flash the glitter of long-range rocketry.

The debate on whether the United States ought to pursue the "super-bomb"—an H-bomb based on nuclear fusion rather than fission—forced American leaders to face up to the prospect of permanent peacetime technocracy.[20] Would this country henceforth base its security on force-fed technological revolution, even if it meant an ongoing race with the Soviets? Physicists Edward Teller and Ernest O. Lawrence, AEC Commissioners Lewis L. Strauss and Gordon Dean, Senator McMahon himself and assorted military leaders thought so. But AEC chairman David Lilienthal and protesting scientists led by Robert Oppenheimer warned of the consequences of a race that could never be "won." In November 1949 Truman appointed a special committee that split two against two on rapid development of the H-bomb, with Secretary of State Dean Acheson's the deciding vote. Acheson could not ignore the argument that "If we let Russia get the 'super' first, catastrophe becomes all but certain—whereas if we get it first, there exists a chance of saving ourselves."[21] He accepted a compromise suggested by strategic adviser Paul Nitze whereby the United States would go ahead with research to prove the feasibility of the H-bomb while conducting a comprehensive strategic review.

Meanwhile, theorists on both sides debated "the nature of the beast," the Soviet Union. Was Stalinist Russia an incorrigibly ambitious yet cautious power chary of war and liable to be influenced by American behavior, as Kennan now argued? Or was the USSR firmly and irrevocably committed to military superiority, if only to intimidate the West while it expanded its empire step by step, as Nitze believed? Heretofore, the military chiefs had never considered the chilling possibility that the USSR might bid for supremacy in both conventional *and* high-technology armaments. In February 1950 General Herbert B. Loper of the AEC Military Liaison Committee suggested (correctly) that the Soviets may have embarked on a determined nuclear program as early as 1943 and may already be in the latter stages of H-bomb development. Loper cautioned that his speculations were of a "fantastic order," yet they galvanized the JCS, which hastily recommended a crash program for the "superbomb."[22]

There remained the moral argument. Kennan insisted that the purposes of a democracy could never be fulfilled by ever more terrible deterrents. Many atomic scientists, appalled by the consequences of their earlier research, sought to expiate themselves through opposition to this next escalation. The Joint Chiefs retorted that "it is folly to argue whether one weapon is more immoral than another" and warned that unilateral renunciation by the United States would logically result in a dangerous realignment of world powers and the loss of the Eastern Hemisphere to Soviet rule.[23] This was a risk with which the President would not associate his name. Acting on the Acheson compromise and JCS urging, Truman gave the go-ahead for the H-bomb on January 31, 1950.

The Nitze committee reported in March. It recommended a "rapid and sustained build-up of the political, economic, and military strength of the free world. . . ." Only the United States had the wherewithal to balance the power of an adversary that, unlike previous expansionist powers, was "animated by a new fanatic faith, antithetical to our own, and seeks to impose its absolute authority over the rest of the world." Given the Soviet fission bomb, the United States could not expect any lasting abatement of the crisis, pending a change in the nature of the Soviet system itself. The current American lead in atomic weapons would, by current trends, disappear by 1954. The only course of action for Americans was to summon their courage and intelligence, and face up to costs and dangers: "Budgetary considerations will need to be subordinated to the stark fact that our very independence as a nation may be at stake."[24] Three months later, on June 2, 1950, the armies of Communist North Korea launched a blitzkrieg campaign to conquer the South. The Nitze report entered history as NSC-68 and became national policy in September. American defense spending tripled.

The Korean War effectively ended the long postwar debate on the proper American posture vis-à-vis the Soviet Union. Hopeful Americans would continue to look for "change in the nature of the Soviet system" and the Eisenhower administration would renew the struggle to keep military spending and technology under control, but Korea established the reality of the Communist threat and the necessity for American primacy in strategic arms. The budgetary wraps came off, and missile men in the military, Congress, and industry launched a publicity campaign on behalf of the ICBM. General Donald L. Putt, the inspiration for the new Air Research and Development Command (ARDC) at Wright Air Force Base (AFB), senatorial friends like Lyndon B. Johnson (D., Tex.), and journalist friends like the *New York Times'*s Hanson Baldwin began to bring pressure on the White House. They charged ignorance and confusion in R & D policy and warned of Soviet progress in guided missiles. In response, Truman appointed the president of Chrysler, K. T. Keller, to serve as a special adviser on missiles. The press reacted to the appointment as if it meant a change of policy rather than a measure to defuse criticism. Keller was touted as the czar of push-button warfare, the head of a new Manhattan Project for missiles that would absorb not the current $30 million per year but over $3 billion.[25]

In fact, Keller merely conducted a rapid survey and recommended that rocket R & D be coordinated by the Office of the Secretary of Defense. His final report granted that missiles might become important weapons but that overenthusiasm could also do harm.[26] After Sputnik, when all parties demanded explanations for the U.S. lag in missiles, ex-President Truman defended himself by insisting that he had "called in a top industrial engineer . . . with instructions to knock heads together whenever it was necessary to break through bottlenecks [and] assured him I would back him to the hilt. . . ."[27] The Keller appointment was indeed made on the basis of a report (from USAF Undersecretary John A. McCone) that recommended a Manhattan District–style mobilization. But Keller either had no stomach for it—he continued to function at Chrysler—or judged it unnecessary. In any event, Truman was unconcerned when, after eleven months of study, Keller had not asked to give him a single briefing.[28]

The first missile czar, passive though he was, did take American rocketry off "hold." In the spring of 1951, when the USAF offered up a new ICBM project called the Atlas, Keller bought it. With an initial allocation of $500,000, scarcely more than a retainer for the contractors, the USAF was back in the ICBM business. To be sure, technical hurdles still made it seem a reckless bet. The inordinately rigorous specifications issued by ARDC—accuracy to 0.01 degree over 5,000 miles with a 10,000-pound payload—made Convair's problems "enormous, but not insuperable."[29] Still, Atlas commanded only $14 million in FY 1954, three years after its inauguration.

The technical drawbacks disappeared only with the proof of concept of the hydrogen bomb, more efficient and a thousand times more powerful than its atomic progenitor. The key to creating a thermonuclear reaction, the process that powers the stars, was the booster principle— using a small fission reaction to trigger nuclear fusion in deuterium or tritium, forms of hydrogen with extra neutrons. As General Loper suspected, the Soviets jumped without pausing to the next stage in nuclear technology, but American physicists, led by Teller and Stanislaw Ulam, had to await the outcome of the policy debate before fashioning a fusion device. They first chose liquid deuterium as their thermonuclear fuel, the simplest solution from an engineering standpoint, and exploded the first American hydrogen device in the MIKE test on November 1, 1952, at Eniwetok Atoll in the Marshall Islands.

MIKE was a huge success, yielding the equivalent of 10 megatons of TNT, close to the predicted figure of a thousand times more than the Hiroshima bomb. But it was not a bomb. Liquid deuterium had to be cooled to temperatures below −250° C. To deliver such a device to the target required a "flying refrigerator" of impossible bulk and weight. The solution, arrived at both by Andrei Sakharov for the USSR, and by Teller and Ulam, was a dry precipitate of deuterium, the saltlike substance lithium-deuteride. It was more difficult to explode than liquid deuterium was to "burn," but it was the secret to making H-bombs relatively small and light. The Soviets apparently achieved this first, as evidenced by their "Joe-4" explosion of August 1953. Though Joe-4 had a smaller yield than MIKE, the Soviets boasted that it was a true bomb, whereas MIKE was "only a very cumbersome and untransportable structure based on principles unsuitable for producing a weapon."[30] On March 1 of the following year the United States pulled even. The first test in Operation CASTLE, designated BRAVO, yielded 15 megatons that atomized Bikini Atoll. More important, it was readily adaptable for delivery by aircraft— or by rockets.

Bureaucratic resistance to ICBMs, financial competition between bombers and missiles, and the favoritism to manned bombers shown by "blue-sky" USAF generals could now be swept aside. In anticipation of improved warheads, a review committee chaired by Millikan eased specifications on the ICBM to a circular error probability (CEP) of one mile and a payload weight of 3,000 pounds.[31] Another committee, directed by the famous mathematician John von Neumann, urged the acceleration of missile programs in February 1954, and the RAND Corporation found ICBMs to be technically feasible far sooner than the twelve years that remained in existing schedules. The main reasons were the hydrogen warhead, which permitted a smaller, less accurate rocket; an improved guidance system devised by Stark Draper; and the "blunt body" reentry vehicle designed by the NACA. Finally, reports of Soviet programs in rocketry prompted RAND to conclude at long last that the United States was in a race for the ICBM, and currently was losing.[32]

The new evidence favoring a crash ICBM program carried the day in the hands of an aggressive, irascible civilian named Trevor Gardner. Appointed a special assistant in R & D by the Secretary of the Air Force, he found the source of the long American snooze in big rocketry in the impossible specifications laid down for ICBM performance. Armed with the RAND and von Neumann reports, Gardner leapfrogged the USAF bureaucracy and persuaded Secretary Harold E. Talbott, to accelerate the Atlas project in the belief that an operational missile could be completed in just five years. The Air Council believed him, established an autonomous vice-command within ARDC for the sole purpose of managing Atlas, and tapped Major General James McCormack for the job. When he retired in ill health, it fell to a brigadier, Bernard A. Schriever, to take command of the new Western Development Division and build an American ICBM.

Of Schriever it could be said, "thou art come to the kingdom for such a time as this." Born in Germany—his parents expatriating in World War I—Schriever grew up in Texas, trained as an engineer at Texas A & M and Stanford, and joined the army in 1931. After a stint as test pilot at Wright Field, he took a degree in advanced aeronautical engineering and went on to fly sixty-three combat missions in the Pacific. Schriever then spent three years as chief scientific liaison in the AAF and had just been named Assistant to the Commander of ARDC when the call came to head up the Western Development Division with its headquarters in Inglewood and test facilities at Point Arguello, near Lompoc, California. In the course of the next few years, this energetic and eclectic test pilot/ bomber pilot/administrator/engineer/military politician would derive and implement techniques for the management of large systems, expand the USAF model of contract R & D, and help to make systems integration an American science to be emulated by the world. Schriever quickly signed on Simon Ramo and Dean E. Wooldridge, two von Neumann committee members, to provide overall technical coordination, thus overriding the contractors and the USAF R & D bureaucracy. Finally, he won the following order from the Deputy Chief of Staff, Materiel: "The Atlas program will be reoriented and accelerated to the maximum extent that technological development will permit. . . . The Atlas will be given the highest program priority in the Air Force. Processing of any aspect of this program will be given precedence over any others in the Air Force."[33] An American ICBM was finally to be a matter of engineering alone. The date was June 21, 1954.

As with the ICBM, so with earth satellites. When the first projects for spaceflight disappeared because their sponsors could not justify the cost, the brash young USAF got the message: it must somehow demonstrate a military mission for satellites. In 1948 this was not yet possible, as Deputy Chief of Staff, Materiel, reported:

. . . launching of an Earth Satellite Vehicle is technically, although not economically, possible. The passage of time, with accompanying technical progress, will gradually bring the cost of such a missile within feasible bounds. It seems therefore imperative, in order that the USAF maintain its present position in aeronautics and prepare for a future role in astronautics, that a USAF policy regarding Earth Satellite Vehicles be promulgated.[34]

The USAF also assigned RAND the task of "continuing studies of the potential military utility of earth satellites—including work on the use of such devices for cold war politico-psychological advantage for communications and for purposes of observation."[35] The directive was superfluous—RAND had its satellite design "reviewed, brought up to date, and compared with the latest Navy satellite proposal" already, but a formal contract in 1948 ensured that RAND would "keep abreast of the art" and extend the satellite vehicle study to include that on the long-range rocket.[36]

RAND accordingly reported again in October 1950. This time its researchers delved at length into the political and military implications of earth satellites. Few documents demonstrate so clearly the exceptional nature of this first strategic "think tank." Its job was to divine the future and, by predicting and recommending, to help define it as well. At a time when the Soviets were proceeding full tilt on missiles, but giving little thought to the implications of space technology, the Americans were dragging their feet on missiles but, thanks to RAND, glimpsing with prescience the effects of the opening of the Space Age. The differing concentrations were crucial, for the developmental lag and the theoretical lead of the United States were responsible *both* for the United States finishing second in the satellite race *and* for the fact that the eventual American space program was much more suited to national strategic needs than was the Soviet. The RAND document of October 1950, more than any other, deserves to be considered the birth certificate of American space policy.[37]

Following an introduction, the RAND report asked "What would be the peacetime and wartime utility of a satellite vehicle program for United States national defense?" Satellites were not weapons, but they could hardly be more relevant to national security, given their primary function as future tools of *strategic and meteorological reconnaissance.* Satellites could gather data of high military value not available from any other source; they would be a "novel and unconventional instrument of reconnaissance." That unconventional nature, in turn, would ensure their perception as a factor in the existing balance of strength, hence they must also be dealt with in terms of their politico-psychological effects. Because of the political implications of spaceflight, RAND perceived, what the U.S. government *says* about it is just as important as what it *does.* The launching of satellites could not be kept secret, hence the political handling of satellites became paramount. The successful launching of a satellite would be a spectacular event causing a worldwide sensation. While the reassurance of friendly and even neutral nations as to American strength would be salutary, the response of the USSR might be dangerous. Recent Soviet press releases clipped by RAND gave an indication. In December 1947 Modest Rubinshtein attacked American use of "Hitlerite ideas and technicians" in its rocket research and spoke of a "sect" urging the United States to go beyond the "ruthless, ghoulish doctrine of the Hitler marauders." The Americans were speaking of the "fantastic" idea of earth satellites, which another Soviet writer referred to as an "instrument of blackmail." As for ICBMs becoming a "grand strategic weapon," this was a wild utopia and a deliberate bluff.[38] Such propaganda depicting the United States as a mad militarism made it advisable to downplay the military potential of satellites and counter the expected Soviet reaction by stressing the peaceful aspects of this "remarkable technological advance." It might be theoretically possible to keep a satellite secret, according to RAND, but that would maximize the negative sensation upon its eventual discovery. Advance publicity was preferable, especially if it stressed that the satellite was not in any sense a weapon.[39]

The principal political problem that RAND expected to be posed by an American satellite was the reaction of allies and adversaries when they discovered this significantly enhanced American capability for secret reconnaissance. It could be assumed that the Soviets would treat such spying from space as a terrible threat. "Fear of loss of secrecy is constant and intense. A picture of the outside world as engaged in penetrating Soviet secrets is likely to be highly anxiety-provoking." The Soviets would surely consider satellite reconnaissance an attack upon their secrecy and therefore illegal. But was it illegal? The question was wide open. At present, RAND reported, overflight of a nonassenting nation, that is, violation of its air space, was contrary to international law. But did air space have an upward limit? The Chicago Convention on Civil Aviation of 1944 affirmed national sovereignty but also promoted the right of innocent passage. The USSR never adhered to it and insisted that "the air space above land is as much territory of the state as the land itself." It was very doubtful that the USSR would accept any "vertical limitation" on its sovereignty or accept that any passage of a spacecraft over its territory might be innocent. Rather, orbiting a satellite over the Soviet Union might be construed by the Kremlin as an act of aggression.[40]

In these few pages the RAND Corporation spelled out the central political problem attending the birth of the Space Age. The new Superpowers were locked in Cold War. One of the contestants was an open society, the other secret and closed. A great premium was thus attached to reliable surveillance techniques by the open society. Reconnaissance satellites offered such a technique. But just as important as developing such technology was establishing the legal right to use it. If the USAF was able to convince the masters of the budget that space-based photography was actually possible (no one could gainsay its importance), how would the Soviets respond to its use? RAND allowed that "the satellite would put the Russian leadership in an awkward position." They would surely consider it an attack, but the instrument would be beyond the range of retaliation. One possible response would be a Soviet appeal to international law, which could result in the United States being found in the wrong if evidence was presented concerning the gathering and transmission of photographic data. However, a harmless American satellite would probably not be found to violate sovereignty. The Soviets might also respond with force or threats, for instance against any neighboring states housing American tracking stations. Or they could initiate a war of nerves with unpredictable consequences. Finally, they might consider satellite overflight to be justification for military reprisal and, while "all out war" would probably not result, "the line dividing 'peace' and 'war' may well be blurred."[41]

An American satellite program, therefore, must reckon with Soviet reactions, and American policy must be charted in advance. The enormous benefits of satellite reconnaissance were clear, noted RAND, whatever the future of U.S.-Soviet relations, for such reconnaissance would be imperative in a climate of hostility or of "settlement" as long as the USSR remained a closed society. How then should the United States proceed? "Our objective," opened the report's concluding section, "is to reduce the effectiveness of any Soviet counteraction that might interfere with the satellite reconnaissance operation before significant intelligence results are secured. Perhaps the best way to minimize the risk of countermeasures would be to launch an 'experimental' satellite on an equatorial orbit." Thus the first satellite would not cross Soviet territory and could test the issue of "freedom of space" in the best political environment. If results were satisfactory, the decision could then be taken to proceed with a second "work" satellite. Of course, an American satellite might provoke the Soviets to begin a satellite program of their own, but there would be little reason for them to do so given "the ease with which they can find out information about United States targets in other ways."[42]

The RAND satellite report was released to the USAF under wraps on October 4, 1950, seven years to the day before *Sputnik I.*

Douglas Aircraft took "all measures necessary to keep the program rolling," and was rewarded with signs of "increasing receptiveness" in the USAF. Project Feedback was implemented to study design of reconnaissance satellites. Westinghouse, RCA, and Lawrence R. Hafstad, recently of the DoD's R & D Board, joined the team.[43] A technical study, roughly defining hardware specifications, followed in April 1951. Two more years of research, in which North American, Bendix, Allis-Chalmers, and Vitro Corporation also participated, permitted RAND to sketch in detail the booster, spacecraft, payload, and subsystems needed to do the job. A final report then made its way up USAF channels and provoked another year of study and debate, during which time the Atlas ICBM, the only possible booster for a USAF satellite, achieved top priority.

On March 16, 1955, the USAF secretly circulated General Operation Requirement #90 (SA-2C). It briefed appropriate American industrial firms on the parameters of Project WS-117L, "a strategic satellite system." They included the ability to attain a precise, predicted orbit; to be stabilized on three axes with a "high-pointing accuracy"; to maintain a given attitude for disturbing torques; to receive and execute commands sent from the ground; and to transmit information to ground stations.[44] This was no "quick and dirty" orbiting beeper, but a large, sophisticated spacecraft integrating the most advanced technology from a dozen fields of American industry.

It was a paragon of peacetime command technology . . . and the first American space program.

CHAPTER 5

The Satellite Decision

When Dwight D. Eisenhower hammered the Democrats in 1952 on Korea, communism, and corruption, it was natural to find Democrat George Kennan a cynical critic of the general in politics. But Eisenhower knew as well as Kennan the dangers for the United States in a militarized Cold War. He became President intent on disengaging from Korea, avoiding "brushfire" wars in the future, slashing the defense budget, and reining in the generals and admirals whose ever-growing demands for new hardware threatened the integrity of the Treasury. He also intended, to a degree consistent with American obligations, to demilitarize the Cold War through arms control. Yet Kennan, voicing the prevailing attitudes of the intelligentsia, thought Eisenhower lacked the brains to be a successful president and sneered at a public that would vote for him: "[Eisenhower] incorporated, in personality, manner, and appearance, all that Americans liked to picture as the national virtues. He was the nation's number one Boy Scout."[1]

What Ike understood, and Kennan seemed not to, was that military power is an essential component of *political* competition.[2] Whether or not the Soviet leaders ever considered invading Western Europe,[3] their ability to do so, their rapid development of nuclear weapons, their intimidation and agitprop were all powerful influences on the behavior of every government in those "nooks and crannies" of world power. What held NATO together and emboldened other nations to resist Communist blandishments was belief in the countervailing power of the United States. The problem facing Truman, then Eisenhower, was how to maintain an image of military resolve without undermining American values, institutions, and economic health. In his own dogged efforts, Truman came to rely on the atomic shield, because it was internationally impressive, domestically unobtrusive, and, above all, cheap. Eisenhower shouldered the shield, but he also inherited a ground war in Asia, a spiraling defense budget, and economic policies such as wage and price controls that seemed to stifle growth in the name of combatting inflation. Eisenhower's increased reliance on nuclear strength, combined with arms control initiatives and a lower defense budget, made for a natural, even "liberal" progression from Truman's policy.[4]

Unfortunately for Ike, he was destined to feel the drum beat of the international imperative even more relentlessly than his predecessor. It fell on him to cope with Soviet hydrogen bombs, long-range bombers, and missiles. And as he hoped to cut spending but was no more willing than Truman to risk falling behind the USSR in nuclear arms, Eisenhower felt desperately the need for accurate intelligence about Soviet progress. The U-2 spyplane was a stopgap, space-based reconnaissance the solution. But, as the RAND report had explained, political preparation for reconnaissance satellites was as important as the technical. So Eisenhower's administration came to have two priorities in missile and space policy in the mid-1950s. The first was to make up for lost time in R & D leading to American missiles; the second was to ease into the Space Age in such a way as to preserve American hopes for penetrating the Iron Curtain once and for all. Each consideration contributed, in its own way, to the tardy timing of the first American space satellite.

Eisenhower's defense policy took form during the presidential transition. After his promised trip to Korea, the President-elect met with John Foster Dulles, George Humphrey, and Joseph Dodge, designated heads of the State Department, Treasury, and Budget, aboard the cruiser *Helena*. There they pondered the revolution in government that had occurred in their adult lifetime. From 1932 to 1952 the federal budget had grown from less than $4 billion per year to $85.5 billion, of which 57.2 percent went to the Pentagon. Such spending levels, thought the *Helena* passengers, endangered the economy as much as inadequate arms would endanger free world security. "The relationship between economic and military strength," said Eisenhower, "is intimate and indivisible." For the conflict with the USSR was not, in his view, building up to a hot war for which one must prepare at all costs. Rather the United States should prepare for the "long haul," for it could lose only by spending itself into bankruptcy. Hence the government must balance essential military force with a healthy economy in what the administration came to refer to as "The Great Equation."[5]

The determination to trim military spending stemmed most directly from the ordeal in Korea. If the United States was obliged to respond to Communist offensives anywhere, anytime, then American vitality would bleed away. But unmistakable strategic superiority, and rhetoric indicating a willingness to use it, might deter the enemy from such adventures, return diplomatic initiative to the United States, and still permit lower budgets. This strategy, known as the New Look, was embodied in NSC-162/2 and approved in October 1953. The United States would depend in the first instance on indigenous forces to resist Communist attacks, but back them up with tactical air and sea power, possibly including nuclear weapons, and finally, if necessary, the ultimate deterrent of "massive retaliatory power" to be applied "by means and at places of our own choosing." NSC-162/2 called at the same time for demobilization of a quarter of all men under arms and a drop in military spending of 30 percent over four years! The only service to be spared was the USAF, which provided "more bang for the buck."[6]

The New Look followed logically from weapons decisions made under Truman. In fact, Truman's cabinet planned a similar trend as soon as the fighting could be stopped in Korea. But the New Look obviously increased American dependence on technological leadership at a time when the USSR had already demonstrated a thermonuclear potential and was proceeding rapidly with transcontinental bombers and missiles. The United States no longer enjoyed much breathing room even if "massive retaliation" (simply a more categorical form of deterrence) did persuade Moscow to refrain from local adventures. Thus the Eisenhower policies, designed to minimize the impact of the Cold War on domestic life, also pushed the country further along the road to technocracy. The services, especially the USAF, were justified in requesting ever greater sums for R & D, and the ills of a militarized, centrally directed command economy would creep in anyway. The new President could only hope that a basis for arms control could be found before the illness was too far advanced.

After three months in office, and the first frustrating efforts to trim the federal budget, Eisenhower opened his mind on all this to the American Society of Newspaper Editors:

Every gun that is made, every warship launched, every rocket fired signifies, in the final sense, a theft from those who hunger and are not fed, those who are cold and are not clothed.

This world in arms is not spending money alone.

It is spending the sweat of its laborers, the genius of its scientists, the hopes of its children.

The cost of one modern heavy bomber is this: a modern brick school in more than 30 cities.

It is two electric power plants, each serving a town of 60,000 population.

It is two fine, fully equipped hospitals.

It is some 50 miles of concrete highway.

We pay for a single fighter plane with a half million bushels of wheat.

We pay for a single destroyer with new homes that could have housed more than 8000 people. . . .

This is not a way of life at all, in any true sense. Under the cloud of threatening war, it is humanity hanging from a cross of iron.[7]

The President then challenged the leaders of the Soviet Union (Stalin had died weeks before) to say plainly what they were willing to do to satisfy the world's hunger for peace. Eisenhower proposed a limitation on the size of all military forces; a limit on the proportion of strategic materials devoted to military purposes; international control of atomic energy; limitation or prohibition of other sorts of very destructive weapons; and adequate safeguards including a practical system of inspection under the UN.[8]

Throughout all the tortuous and sterile negotiations on arms control and disarmament since 1946, the propaganda, posturing, and positions of the Superpowers had not changed much.[9] The Americans insisted on a trustworthy system of inspection prior to agreement on arms limitation; the Soviets repeated their call for prior abolition of nuclear weapons and "general and complete disarmament," while refusing to allow inspection teams to penetrate their wall of secrecy. Eisenhower's graphic depiction not of the horrors of nuclear war but of the waste of deterrence had no measurable effect on the Kremlin, still preoccupied with post-Stalin struggles. Only the growing outcry against nuclear fallout from weapons tests jolted the UN Subcommittee on Disarmament and elicited a first hint, in September 1954, that the USSR might consider a formula for on-site inspection. Eisenhower responded in March 1955 by appointing Harold Stassen a Special Assistant to the President on Disarmament, and a summit conference was planned for Geneva in the summer.

How far would Eisenhower have gone toward the abolition of nuclear weapons, had rapid progress been made toward a comprehensive test ban? Studies from later in his administration suggest that he hoped a test ban might lead to a limitation or "freeze" on deployment of strategic systems. But had the Superpowers outlawed nuclear weapons entirely, the United States would have been thrown on to the other horn of its strategic dilemma: how to balance the Soviet superiority in conventional arms without becoming a garrison state. Whatever his deepest hopes, Eisenhower's New Look and the disarmament initiatives both placed a premium on trustworthy surveillance of the Soviet Union. If arms *competition* continued, the U.S. military must be kept from going overboard in its demands for weapons, and that required reliable information on Soviet deployment and R & D. If arms *control* was possible, how much more progress might be made if the United States did not have to demand elaborate on-site inspection! In short, whatever the administration's intentions, the United States needed to spy on the Soviet Union more than the otherwise devious Soviets needed to spy on the Americans.

In March 1954 Eisenhower summoned the Office of Defense Mobilization's Science Advisory Committee and apprised its members of the growing danger faced by the United States. Modern weapons made it easier for a closed dictatorship to gain the advantage of surprise over an open society. It was imperative that the best minds in the country attend to the technological problem of preventing another Pearl Harbor. The result was the Technological Capabilities Panel (TCP) Report, or "Killian Report," or "Surprise Attack Study." Its authors included James F. Killian, president of MIT, Lee A. DuBridge, president of Caltech, James B. Fisk, later president of Bell Labs, James Phinney Baxter III, president of

Williams College, Edwin H. Land, inventor of the Polaroid camera, James H. Doolittle, chairman of the NACA, and over forty scientists and engineers.

Reporting to the NSC in a "full-dress" secret session, the Killian panel presented their findings in terms of a timetable. Period I (late 1954 to 1955) was characterized by an American air-atomic advantage but also a vulnerability to surprise attack due to the lack of a reliable early warning system, inadequate air defense, and a growing Soviet bomber force. Neither side was in a position to mount a "decisive" attack, defined as one that would eliminate the ability to strike back and/or reduce political and cultural life to chaos. Period II (1956–57 to 1958–60) would bring a very great offensive advantage to the United States, given the New Look buildup in the nuclear stockpile and SAC. The Soviets would also be testing new bombers and missiles, but the American operational bomber force would be overwhelming. This was the period in which the United States could best undertake diplomatic initiatives to the benefit of the free world. Period III (1958–60 to ?) would bring a rapid increase in Soviet jet bombers, which could be dangerous if the USSR achieved this stage as early as 1958 and U.S. defenses were not in place. At some point this phase would blend into Period IV, during which both sides would have the capability to destroy the other even in retaliation. The single most important variable was the early achievement of ICBMs by either side. Once mutual stalemate arrived, it could conceivably be altered and one side reclaim the advantage. *"We see no certainty, however, that the condition of stalemate can be changed through science and technology."*[10]

The panel recommended the highest national priority for the USAF ICBM program, an IRBM suitable for land or shipboard launch, rapid construction of a distant early warning (DEW) line in the arctic, a strong and balanced research program on the interception and destruction of ballistic missiles, a greater application of science and technology to methods of fighting peripheral wars, and especially an increase in intelligence capabilities. "We *must* find ways," wrote Edwin Land, "to increase the number of hard facts upon which our intelligence estimates are based, to provide better strategic warning, to minimize surprise in the kind of attack, and to reduce the danger of gross overestimation of the threat. To this end, we recommend adoption of a vigorous program for the extensive use, in many intelligence procedures, of the most advanced knowledge in science and technology."[11]

American intelligence had indeed been a frustrating business heretofore. But clandestine sources clearly suggested that the Soviets had gone far beyond "improving the V-2." The Gröttrup Germans told tales upon returning that were disquieting as much for what they did not reveal as for what they did: the Germans had been kept in the dark; Soviet engineers felt confident enough to proceed on their own; and they showed special interest in long-range rockets and guidance systems. The Central Intelligence Agency (CIA) proceeded to place a radar station in Samsun, Turkey, to monitor Soviet flights from the range at Kapustin Yar. But rockets launched from this site were impacting some 1,500 miles away in the Central Asian desert—they were IRBMs. The ICBM program, if it had reached the testing stage, must be deeper inside the country and out of radar range. It was then that the Killian Panel insisted that the most advanced technology be applied to intelligence gathering.

In short, the United States needed a spy plane. The job was given to Lockheed's incomparable Skunk Works, a design bureau under Kelly Johnson that specialized in brainstorming emergencies with a small team of imaginative engineers. In just eighty days, with twenty-three designers on the job, the Skunk Works created the U-2, an extremely high-flying photo-reconnaissance plane invulnerable (at the time) to antiaircraft. The unlikely looking U-2s, with wings so long they draped on the ground upon landing like a child's toy, began criss-crossing the USSR in June 1956. The following spring, a U-2 flying out of Peshawar, Pakistan, returned with photographs of Tyuratam, the Soviet ICBM test facility in Kazakhstan, after having crossed the entire expanse of Western Russia and landed at Bodö in northern Norway.[12]

Blatant violation of Soviet airspace was a risky, hit-and-miss means of espionage. If continuous surveillance of Soviet installations and exact targeting of Soviet bases were to be assured, the solution was to spy from outer space. Camera-toting satellites, circling the earth south to north in a polar orbit, could view the entire surface of the earth as it rotated below, return to any location in a few days' time, home in on suspicious areas, and do it all under the legal cover of freedom of space— if such legal cover could be established. Given the simultaneous USAF go-ahead for Project WS-117L and Land's insistence on applying the most advanced knowledge of science and technology to intelligence, there is little doubt that portions of the Killian Report that remain classified to this day include the recommendation that the highest national priority attach to the development and operation of reconnaissance satellites.

The Killian Report noted that early achievement of the ICBM was the "single most important variable" in its timetable, then curiously dropped the subject. It is likely that sections still classified deal with them as well. But the Soviet program was not a total mystery, even before the flights of the U-2. The Basic National Security Policy papers of December 1954 and January 1955 opened with the staccato warning that the United States would soon be in grave peril.

The Soviet guided missile program, over the next few years, will bring increasingly longer-range missiles into production. Assuming an intensive effort, the USSR may develop roughly by 1963 (1960 at the earliest) operational intercontinental ballistic missiles. The U.S. program for missiles of this type should approximate this timetable, provided that intensive effort continues. There is no known defense against such missiles at this time.[13]

Against this background of worried preparation for the missile revolution, the struggle to hold down military spending, hopes for arms control agreements, and concern for the rapid achievement and legal protection of spy satellites, the Eisenhower administration came to address the satellite question. The legal risk elucidated by the RAND Corporation in 1950 stemmed from the likelihood of vigorous Soviet protest against the hoped-for right of satellite overflight. How could the United States establish the "freedom of space?" How to finesse a satellite into orbit without anyone objecting and thus set a legal precedent for subsequent space activities? The suggestion made by RAND was to start with an innocuous, nonmilitary, scientific "test" satellite launched on a trajectory that avoided the Soviet Union. By the time the Killian Panel reported and the USAF began its final review of WS-117L, the occasion for such an "innocuous" solution had presented itself. On October 4, 1954, the Special Committee for the International Geophysical Year (CSAGI) recommended that governments try to launch earth satellites in the interest of global science.

The IGY idea sprang from an informal gathering of scientists at the home of James Van Allen in Silver Spring, Maryland, in 1950. Lloyd Berkner, S. Fred Singer, J. Wallace Joyce, and the Briton Sydney Chapman were among the circle that sat discussing ways of coordinating high-altitude research around the world. Why not, thought Berkner, hold another International Polar Year such as those staged by the nations in 1882 and again in 1932? Reducing the interval between such cooperative exercises from fifty to twenty-five years was especially felicitous, since 1957–58 would be a period of maximum solar activity. They took their idea to the International Council of Scientific Unions, which expanded its purview from the poles to the whole earth, and won the support of sixty-seven nations for an IGY. At the end of 1952 the NAS appointed a U.S. National Committee for the IGY that succeeded, under Hugh Odishaw, in talking the White House into boosting NSF funding for FY 1955 to $13 million, of which $2 million was marked for IGY preparation. That October, while awaiting the opening of the CSAGI meeting in Rome, Berkner and ten of his associates in space science spent half a night cataloguing the technical problems and scientific rewards of launching a satellite. They agreed unanimously to recommend a satellite project to the CSAGI as a major goal of the IGY.[14]

Although official approval and funding were yet to be achieved, candidates lined up for the honor of launching the United States into the Space Age, for the dream had never died. In 1952 Aristid V. Grosse, Temple University physicist and Manhattan Project veteran, made a study of the "satellite problem" for President Truman. He consulted with von Braun and eventually proposed the orbiting of an inflatable balloon that would appear to the naked eye as an "American Star" rising in the West. The report stressed the importance of satellites for science, military observation, and especially psychological competition with the Soviet bloc: ". . . the satellite would have the enormous advantage of influencing the minds of millions of people the world over during the so-called period of 'cold war' or during the peace years preceding a possible World War III." Furthermore, "it should not be excluded that the Politbureau might like to take the lead in the development of a satellite," which "would be a serious blow to the technical and engineering prestige of America the world over." The Grosse report vanished into White House files.[15]

In late 1954 the American Rocket Society also lobbied the NSF for funds to start a satellite program.[16] But the most credible candidate for the task of opening the Space Age was Wernher von Braun. His Redstone, although a modest entry only 34 percent more powerful than the V-2 (and 30 percent lighter!), was now in the testing stage. George Hoover of the Air Branch, Office of Naval Research, realized that the Redstone now made possible the early satellite hopes of the Navy Bureau of Aeronautics. Hoover gently approached the Germans at Huntsville, while Frederick C. Durant III, former president of the ARS, peddled to army, navy, and civilian representatives a plan for a five-pound satellite launched atop a Redstone augmented by "strap-on" solid rockets. He got no action, but the Office of Naval Research and Redstone Arsenal did agree to propose a joint project called Orbiter. The army would supply the booster and the navy the satellite, tracking, and data analysis. They billed it as a "no-cost" satellite, since it could be done with existing technology.[17] Von Braun's own 1954 report, "A Minimum Satellite Vehicle," asked for only $100,000—a tiny price to pay given that "a man-made satellite, no matter how humble (five pounds) would be a scientific achievement of tremendous impact." Since it could be realized in few years with technology already available, "it is only logical to assume that other countries could do the same. *It would be a blow to U.S. prestige if we did not do it first.*"[18]

Meanwhile, the presidents of the NAS and NSF, Detlev Bronk and Alan Waterman, piloted the IGY proposal through the federal government. Because of its international nature, not only the NSF, National Science Board, and Science Advisory Committee, but the CIA, State, NSC, BoB, and the White House all had to clear the project. Waterman and Bronk briefed State in March 1955, then took their case to the White House, where the President showed an interest. But at these higher levels of government, considerations other than science and prestige already weighed heavily. For unbeknownst to the IGY people, there were now *two* goals of high national importance—establishing legality for satellite overflight as well as being first into space—and *three* proposals to open the Space Age—Project Orbiter, the IGY proposal, and WS-117L. Donald Quarles, Assistant Secretary of Defense for R & D, had his staff analyze the whole problem from a military perspective, routed it through the Special Assistant to the President on Governmental Operations, Nelson A. Rockefeller, and thence to the NSC. The product was NSC-5520, subject to special security precautions and limited distribution.

"The U.S. is believed to have the technical capability to establish successfully a small scientific satellite of the earth in the near future," began the report. "If a decision to embark on such a program is made promptly, the U.S. will probably be able to establish and track such a satellite within the period 1957–58." The Technological Capabilities Panel and the Science Advisory Committee recommended an immediate start for a "very small" satellite, but also a reexamination of the principles and/or practices of international law regarding "Freedom of Space." The report also noted that the USSR was now working on a satellite program of its own. Following a portion of the document that is still classified, it was recognized that "[C]onsiderable prestige and psychological benefits will accrue to the nation which first is successful in launching a satellite." If the Soviets were first, it could have important repercussions on the determination of Free World countries to resist Communist threats. "Furthermore, a small scientific satellite will provide a test of the principle of 'Freedom of Space.' The implications of this principle are being studied in the Executive Branch. However, preliminary studies indicate that there is no obstacle under international law to the launching of such a satellite." The IGY presented "an excellent opportunity," and the United States should emphasize the peaceful purposes of its first satellite. However, care must be taken not to prejudice U.S. freedom of action to proceed with satellite programs outside the IGY.[19]

The report then asked the NSC to take the following actions: initiate a small, scientific satellite program aimed at launching by 1958; endeavor to launch a satellite under international auspices such as the IGY, in order to demonstrate peaceful purposes; but do so without implying that only international scientific satellites were permissible or that prior consent was needed from any nation over which the satellite may pass, and without impeding other (missile) programs. Rockefeller approved the report and appended his own thoughts. The achievement of a satellite, he believed, "will symbolize scientific and technological advancement to peoples everywhere. The stake of prestige that is involved makes this a race that we cannot afford to lose." But the legal questions were so new and uncertain that "it is highly important that the U.S. effort be initiated under auspices that are least vulnerable to effective criticism." Since the Soviets themselves had announced plans to launch a satellite, the United States could use this to good effect against a concerted Communist effort to denounce an American project as evil or threatening. Rockefeller concluded by urging especially that exploratory work continue on more "sophisticated" satellites referred to in a portion of the report that is still classified.[20]

The NSC, therefore, gave indubitable primacy to protection of the reconnaissance satellite program, and its approval of an IGY satellite on May 26 stipulated that its peaceful nature be stressed and that the project not interfere with military programs. The resulting public announcement, made by White House Press Secretary James Hagerty on July 28, accorded with the plan to proceed with maximum publicity of the scientific, international character of the program:

> On behalf of the President, I am now announcing that the President has approved plans by this country for going ahead with the launching of small, earth-circling satellites as part of the United States participation in the International Geophysical Year. . . . The President expressed personal gratification that the American program will provide scientists of all nations this important and unique opportunity for the advancement of science.[21]

The United States had pledged itself to open the Space Age during the IGY. But another grave decision remained: which rocket would do the honors? Again, the double goals of the American program demanded a delicate solution: the administration wanted to be first but also sought to protect "freedom of space" and avoid interfering with the priority missile programs. The choice of the satellite booster fell to Assistant Secretary Quarles. He and Waterman agreed in June that some branch of the DoD had to supply the rocket but the IGY National Committee should take responsibility for the satellite. Quarles then named an advisory group under Homer Stewart of JPL to debate the best method of launch.

The entire Stewart Committee met for the first time in July at the Pentagon.[22] After a briefing by Quarles and RAND representatives, it reconvened at the NRL for the official proposal of a "scientific satellite program," then turned to the vying military services. The committee hosted USAF and army delegations, visited the Martin plant building the navy's Viking sounding rocket, and heard von Braun's briefing on the Redstone. The harried month of fact-finding left the committee divided between the army's Project Orbiter and an NRL scheme based on an upgraded Viking. The USAF, anxious to establish its role in the coming Space Age, regaled all and sundry with their "World Series" satellite proposal based on the huge Atlas ICBM, but could not deny that the mission would interfere with its military duties nor promise that the Atlas could perform before the end of the IGY.[23] The army proposal was likelier. Though based on the "Minimum Satellite Vehicle" plan of 1954, Project Orbiter now entailed the use of Sergeant solid-fuel rockets as second, third, and fourth stages atop the Redstone. Its estimated cost of

$17.7 million fell below the $20 million NSC ceiling. The NRL's Viking, by contrast, was much smaller than the Redstone, and its upper stages would have to be designed from scratch. But the NRL impressed the panel with its plans for the scientific components and electronics and the necessary radio tracking system dubbed "Minitrack."

The committee's decision was a near thing. Project Orbiter had the better booster, NRL the better satellite and support. Given the scientific backgrounds of the committee members, they could be expected to lean toward the latter. But the vote taken on August 3, 1955, split three for Viking and two for Redstone. Robert McMath, a University of Michigan astronomer, was ill and not present, but later admitted that he favored the army proposal. The remaining two members, pleading ignorance of guided missiles, went along with the majority. In this way, it seems, the United States came to place its hopes for priority in space on a scientific, nonmilitary program based on a slim, experimental first-stage rocket and three entirely new upper stages, expected to coax a grapefruit-sized satellite to orbital velocity, before the end of 1958.[24]

In retrospect, the decision seems disastrous. The Viking booster, though ultimately successful in producing satellites during the IGY, failed to beat the Soviets or even the hamstrung army team. Was the illness of a committee member or the elegance of NRL's satellite design the sole cause of this fateful twist in American history? Or was it, as Stewart suggested privately in 1960, the ethnic origins of the Huntsville Germans? Some members were uncomfortable with the notion that the U.S. satellite would ride on a descendant of the Nazi V-2. Some Redstone personnel also blamed von Braun's imperious demeanor for the army defeat, but this seems unlikely given his history of smooth and even charismatic bearing before the authorities funding his projects.[25]

It is far more likely that the political insistence on the civilian character of the satellite program, one divorced as much as possible from the military, had a major bearing on the committee's choice. Presumably, none of the committee members had seen the NSC report or Rockefeller's minutes—they may not have been fully sensitive to the "freedom of space" issue or the importance of being first into space. But the jury *had* been briefed by RAND and Quarles, and was instructed to keep in mind the importance of a nonmilitary, scientific image for the enterprise.[26] The Viking was a scientific rocket, constructed by private industry for high-altitude research. The Redstone was a military missile under development by an army arsenal. Whether or not the Stewart committee knew it, their decision was ideal from the political, if not technical, standpoint.

When news of the decision reached Huntsville, there was anger and consternation. Major General Leslie Simon protested vigorously enough to force a new hearing on Project Orbiter. Modifying the Redstone, he insisted, was a far simpler matter than designing a new four-stage Viking. Only the Redstone had a chance of beating the Soviets, who might be ready to launch as early as January 1957. Finally, the satellite job would not interfere with weapons development, since the Redstone was well along in testing and was soon to enter production at Chrysler Corporation. The NRL and Martin Company made assurances of their own and promised to support the program in "the aggressive fashion necessary to achieve a satellite at the earliest practicable date." The Stewart Committee stuck to its previous decision for Viking, Quarles's Advisory Group on Special Capabilities concurred, and Quarles himself backed them up.[27]

This final imprimatur at the level of the Assistant Secretary of Defense strongly suggests that the decision for the NRL proposal, called Project Vanguard in the navy's September contract with Martin, was more than technical. Although the Stewart Committee had been responsible for choosing a means to launch satellites "during the IGY," there was little doubt that the Redstone promised a satellite soonest. Furthermore, according to CIA Chief Allen Dulles, U.S. intelligence services knew of Soviet progress and had a good notion of when the Soviets might launch.[28] If being first was the primary consideration in U.S. satellite policy, the DoD could have overridden its advisory committees. But speed was *not* the primary consideration; in the end, assuring the strongest civilian flavor in the project was more important. A satellite launched with a military rocket designed by the Peenemünde team, no matter how small or scientific, might provoke a Soviet challenge to the legality of satellite overflight. That the principals knew the army rocket was the best bet is demonstrated by the fact that Stewart himself went to Huntsville and told von Braun to keep his modified Redstone available as an ace in the hole. And when another variant of the Redstone, the Jupiter-C, went into service to test nose cones, or "reentry vehicles," for nuclear warheads, the army sent inspectors to Cape Canaveral to ensure that all Jupiter upper stages were dummies. Von Braun was *not* to be allowed to launch a satellite "accidentally" by boosting a nose cone into orbit.[29]

The first phase of American policy for the coming Space Age, begun in 1950 at the RAND Corporation, came to an end with the decision in favor of Vanguard in the fall of 1955. Occupied by the need to keep abreast of the USSR in long-range rocketry, the Eisenhower administration put the ICBM on a crash basis. Absorbed by the need to monitor Soviet R & D and deployment whether arms race or arms control obtained, it also gave priority to the USAF spy satellite program, two and one-half years before the Space Age opened. Worried about the legal and political delicacy of satellite overflight, it seized the IGY opportunity to initiate an unobtrusive scientific satellite program under civilian auspices. Finally, the administration was advised of the propagandistic value of being first into space. Of all these critical policy areas, however, the last had the lowest priority. For there were two ways the legal path could be cleared for reconnaissance satellites. One was if the United States got away with an initial small satellite orbiting above the nations of the earth "for the advancement of science"—and had no one object to it. The other way was if the Soviet Union launched first.

The second solution was less desirable, but it was not worth taking every measure to prevent.

If the United States ever regained something like "normalcy" in the postwar era, it was limited to the four years between the Korean truce and Sputnik, when the economy expanded, the budget stabilized, the military shrank, and the country was at peace. Americans dared to hope that the emergencies were over, that the depressions and wars that seemed to punish the virtuous were past, and that individuals, freed from national controls and international crises, might again claim their own lives according to the American dream.

The President encouraged such hopes. He provided a pause in the social change and growth of government dating from the New Deal and especially from 1941. Thanks to nuclear deterrence, Americans might be free to place their brains, discipline, and initiative at the service of private goals, and so replenish the national reserves of wealth and productivity, depleted during the decades of "abnormalcy." But the USSR had not gone away and still yearned to puncture the American nuclear umbrella. A by-product of that effort was the first space race, although Eisenhower chose not to inform the nation of that fact for reasons of principle and reasons of state. The administration was neither ignorant nor complacent. Its goals were to stay abreast of Soviet progress; seek negotiations to curb, or at least manage, the arms race; and ease the United States into the missile age without panic. Judging protection of the secret reconnaissance satellite program to be more important than the public prestige satellite program, however, Eisenhower failed to foresee how a Soviet space triumph would reinforce challenges at home, not only to his leadership, but to all the values he hoped to restore to the core of American life.

Due to the Korean War and Truman's rearmament, the military's share of federal spending peaked in FY 1953 at 57 percent. Adding in the costs of atomic research, foreign military assistance, and other defense-related programs, the figure approached two-thirds of the entire budget. By FY 1955 Eisenhower had rolled back defense spending by $8 billion, or 20 percent. Believing the "old-time religion" that military spending was not a spur to the economy but a dead loss, Ike released resources and thousands of young men for civilian pursuits and was rewarded with the steady growth at low inflation that economists soon took to be normal. Sociologists, in turn, pointed to increased social mobility, especially for the many white ethnics sent to college under the GI Bill, to endorse the notion of the United States as a "classless" society in which the great majority of citizens could aspire to middle-class status and adopt middle-class values. This was the age of the "end of ideology," the celebration of "consensus," the nurturing of "the vital center."[30] The "Red-baiting" senator Joseph McCarthy (R., Wisc.) was repudiated. Eisenhower succeeded in isolating the fading extremists on the Right, even as Stalinist tyranny had discredited extremists on the Left.

In foreign affairs, too, these seemed to be years of relaxation. Stalin was dead and the new leadership seemed willing to translate peaceful coexistence into practical agreements. The Austrian State Treaty of 1955 brought the first retreat of Soviet armed forces in Europe, and the Geneva Conference raised hopes that the Kremlin might accede to on-site inspection for a nuclear test ban. Nineteen fifty-six brought new troubles: civil war in Algeria, the Suez crisis, and Soviet suppression of the Hungarian revolution. Finally, Democratic candidate Adlai Stevenson's clumsy call for a unilateral cessation of nuclear testing damaged the administration's delicate efforts to close a deal with Moscow.[31] But none of these events imperiled the United States or required financial or military sacrifice. Despite talk of "rolling back" Soviet power, Eisenhower and Dulles preferred pacification to confrontation, as in Indochina in 1954. Most Americans had no stomach for crusades to liberate captive nations, but neither were they eager to lower the nuclear shield through unilateral disarmament. Ike won reelection in 1956 by a larger majority than in 1952.

For all that, the later 1950s were a seed time of ideas, discontents, and change—at home and abroad—that would define the Eisenhower second term as a massive rearguard action in defense of concepts of the role of government in the United States, the role of the United States in the world, and the role of technology in pursuit of political goals that Americans had previously taken for granted. For these were the years in which Herbert Marcuse denounced American society from a Marxist-Freudian slant, Paul Goodman explained rebellious "youth culture" in terms of alienation and disillusionment with the establishment, C. Wright Mills purported to expose the "power elite" that really ran the country, and John Kenneth Galbraith critiqued American consumerism and called for rapid expansion of the public sector.[32] To such critics the United States *was* a class society: a few ruled, most were manipulated into working for "two cars and a barbeque," and those on the bottom (especially blacks, just discovered by the intellectuals) had no hope of participating in the mainstream of life—a life that, the critics said, was not worth living anyway. Existentialism became a fad, the postwar resurgence of churches ended, baby-boom children tuned into rhythm and blues broadcast from the inner city, and the reigning social problem was white juvenile delinquency.

In 1954 the Supreme Court ruled against racial segregation in schools, and by 1957 Eisenhower was sending federal troops to Little Rock, while

Democrats were stymied on the "race question" by their Southern wing. Then the economy finally dipped into recession made less tolerable by the new expectation of stable growth, and liberal economists chanted for deficit-spending to stabilize demand. By the late 1950s the propositions gained currency that the United States was not democratic; perpetuated maldistribution of wealth and opportunity; alienated the poor, the black, the worker, and ultimately the female (Simone de Beauvoir's *The Second Sex* appeared in English in 1952); and veiled those evils under a creed of individualism and federalism ("states' rights"). Gradually, the notions that (1) these inequities were correctable, and (2) were correctable primarily through exercise of federal power and money overflowed the intelligentsia and trickled down, in vulgar forms, to the real and imagined victims themselves.

In foreign policy, too, seeds took root in these years. The backward nations (later the "underdeveloped, developing, less-developed, and Third" world) emerged from the disintegration of the colonial empires. "Neutralist" regimes appeared, such as Gamal Abdel Nasser's and Ho Chi Minh's—nationalist, anti-Western, rhetorically or actually socialist. In 1957 the United States and Israel nervously watched Egypt and Syria experiment with a United Arab Republic with the Soviet Union in attendance; in 1958 U.S. marines went to Lebanon; in 1959–60 Fidel Castro won power in Cuba. But overshadowing all these sprouts of the 1950s was the technological one, the impending missile revolution. From post-Sputnik hindsight, U.S. missile programs seemed "a mess," confused, underfunded, riddled with interservice rivalries. There was truth in these charges, but the confusion and inefficiency were perhaps no more than could be expected to attend the birth of a new technological age. Eisenhower could have avoided these criticisms only by more profligate spending, more centralized direction of R & D, more alarming rhetoric to justify a "race posture." All this was precisely what he hoped to prevent. His plans miscarried; that does not make them unworthy. Thanks to Sputnik, the missile age would not ease in but open with a tidal force that swept along every domestic and international trend that Ike had hoped to contain.

The federal budget held steady through the years of the unannounced satellite race. The military services inveighed against the "artificial" spending ceilings imposed on the Pentagon, but Eisenhower still clutched to the Great Equation. The Joint Chiefs, Ike remarked dryly, "don't know much about fighting inflation. This country can choke itself to death piling up military expenditures just as surely as it can defeat itself by not spending enough for protection." Still, Eisenhower had to respond to new Soviet strategic capabilities in the wake of hydrogen bomb testing and the alleged "bomber gap" of 1954. The result was a revised defense policy, inevitably dubbed the "New New Look," which downgraded massive retaliation in favor of simple deterrence and thus upgraded the maintenance of a capability for limited war—but without big budget increases.[33] The military share of the budget fell to 50 percent; total R & D spending, up from $1.8 to $3.1 billion during the Korean War, grew slowly, and was still under $5 billion in FY 1958. A far greater share of those sums, however, flowed to the Atlas, whose budget climbed from $14 million in FY 1954 to $515 million two years later and $2.1 billion as it moved into testing in FY 1958. Other ballistic missiles, including the Redstone, actually suffered cuts in funding until FY 1957 established an upward trend. But an R & D program can only absorb so much money in its early stages, and American security was not endangered by Eisenhower's ceilings. Soviet rocket progress was closely monitored, while B-47s and, after 1956, B-52 intercontinental jet bombers came on line. The Killian Report accurately predicted that the late 1950s would be the time of maximum U.S. superiority.

The Killian Panel had also noted that the fleeting years of superiority would provide the optimal window for diplomatic initiatives. Thus Eisenhower made several proposals designed again to soften the transition to the missile age. "Open Skies," authored by Rockefeller (and inspired by a young Henry Kissinger) and presented to the Soviets at Geneva, was such a device. The United States, said Eisenhower, was prepared to turn over the blueprints of all its bases and armed forces and to permit regular and frequent inspection flights over U.S. territory in return for the same privileges inside the USSR. Eyeballing the Soviet delegation, he entreated, "I only wish that God would give me some means of convincing you of our sincerity and loyalty in making this proposal." As if on cue, a thunderstorm broke with a flash and doused the lights in the hall.[34] Khrushchev later denounced "Open Skies" as a transparent espionage device, and refused to take it seriously.[35]

American contacts in 1956 and 1957 dealt mostly with a nuclear test ban. But the President's State of the Union Address, opening his second term in January 1957, expanded the field. The United States, he said, was "willing to enter any reliable agreement which would reverse the trend toward ever more devastating nuclear weapons; reciprocally provide against the possibility of surprise attack; mutually control the outer space missile and satellite development; and make feasible a lower level of armaments and armed forces and an easier burden of military expenditures."[36] A few days later Henry Cabot Lodge submitted a memorandum to the UN General Assembly suggesting a plan of controls whereby "future development in outer space would be directed exclusively to peaceful purposes and scientific purposes" by bringing "the testing of [satellites and missiles] under international inspection and participation."[37]

These were the first proposals ever made for control of space technology. They predated Sputnik itself. Did the Eisenhower administration not take spy satellites seriously after all? On the contrary, no hope was more abiding than that of "opening up" the Soviet Union. If it could be done voluntarily in the context of arms control, Eisenhower was even willing to forego a purely national space program. But if that was not possible, then the Soviet Union must be "opened up" by other, clandestine means, and meanwhile a U.S. commitment to the peaceful uses of space and "Open Skies" was on the record to support the later claim that spaceborne reconnaissance was itself a peaceful activity. In July Dulles iterated American "willingness to cooperate in the working out of a system which would insure that outer-space missiles be used exclusively for peaceful purposes and scientific purposes."[38]

Khrushchev's dismissal of "Open Skies" and the Superpowers' unwillingness to talk disarmament on each other's terms forced Eisenhower to prepare for the imminent missile age. Thus the middle 1950s, far from being a period of "complacency," were the most dynamic and imaginative years in the history of American military R & D. More new starts and technical leaps occurred in the years before 1960 than in any comparable span. Every space booster and every strategic missile in the American arsenal, prior to the Space Shuttle and the Trident submarine-launched ballistic missile (SLBM) of the 1970s, date from these years. In this way, too, the mid-to-late 1950s were a seed-time.

Atlas made rapid progress after 1955. Staging was still a challenge in big rocketry, so Schriever's team, like Korolev's, found a way around it. All the Atlas engines were fired at launch, with two of the boosters dropping off later. The Atlas also had the "thin-skinned" pressurized airframe developed for the MX-774. Finally, NACA engineers solved the reentry problem by going beyond the "blunt body" principle to the ablative nose cone, which radiated heat away and sheltered the warhead upon reentry. The army proved it out with the Jupiter, and the USAF adopted it for the Atlas. Schriever also pioneered the concurrency system, by which major systems were not tested individually so that faults could be easily traced, but all at once. This increased risks but vastly accelerated development. Concurrency made systems management and interface the trickiest chore in R & D.

Nevertheless, the Atlas, again like Korolev's R-7, was far from a serviceable weapon. It took too long to fuel and fire and could not stand up to the high acceleration take-offs needed for a missile force under attack. It was, so to speak, a "zero-generation" ICBM. Hence in May 1955 the USAF was authorized to build a second ICBM, the Titan, which drew on many Atlas components but was to be a genuine thick-skinned booster with a second stage ignited in flight. Titan was also fueled by a storable hypergolic mixture.[39]

American IRBM programs also took shape in 1955. The Killian Panel recommended, as part of its coherent plan for accelerating missile development, a single intermediate-range rocket for the use of both the army and navy. Thus were born the Army Ballistic Missile Agency

(ABMA) at Huntsville, commanded by Major General John Bruce Medaris, and the Navy Special Projects Office. The joint missile was known as the Jupiter, a derivative of the Redstone. But fuel storage, launching, and tracking of a large, multistage liquid fuel rocket would all be headaches aboard ship. What the navy really needed for a piece of the nuclear action was a small, solid-fuel missile capable (it was hoped) of launch by submarines. In July 1956 the navy won permission to back out of Jupiter, and within three months the solid-fuel Polaris joined the Atlas and spy-satellite programs at the top of the national priority list. Thanks to AEC breakthroughs in warhead design and the army's inertial guidance system, the Polaris was visualized in its final form as early as 1957. The German wartime dream of submarine-launched missiles was to come true, thanks to the imagination and efficiency of the navy bureaucracy and Eisenhower's DoD.[40]

Finally, the USAF entered the IRBM sweepstakes with Thor. In obedience to Killian recommendations, the DoD shut its eyes to the apparent untidiness of parallel programs. Douglas Aircraft won the Thor contract in December 1955, and the missile entered testing only four months behind the Jupiter in January 1957. All these military missiles, with the exception of Polaris, were liquid-fueled and took a dangerously long time to fire. But USAF planners, impressed by the specifications of the Polaris, already peered into the second generation. In early 1956 the Western Development Division sold the Pentagon its plan for a solid-fueled, intercontinental rocket that could be launched within sixty seconds of an alert. Hence its name—the Minuteman. With a range of 6,000 miles, lighter and cheaper than Atlas or Titan, it could make up in numbers what it gave up in warhead size. By the early 1960s Minuteman would join the Polaris and B-52 as the third leg of the U.S. strategic triad. Titan, Atlas, and Thor would already be obsolete as weapons. Instead, they made up the efficient, diversified stable of military and civilian space boosters, received serial improvements and upgrades, and survived into the 1980s.[41]

While all this missile progress went forward, the American space program was consigned to Project Vanguard. And Vanguard stumbled in the starting blocks. Delays in assigning formal authority to the NRL to direct the project forced it to take ad hoc decisions on how large a staff to reassign, how to draw lines of authority, how to ensure an orderly flow of money, how to handle security classification, and how to define its relationship with the contractor, Martin. The latter protested its subordination to the research lab and—ominously—insisted that it would never have given assurances about a quick satellite success if it had known that a parade of NRL watchdogs would peer over the shoulders of its own Baltimore engineers. Martin executives bristled at the appointment of John P. Hagen, superintendent of NRL astronomy, and Milton W. Rosen as project director and technical assistant for Vanguard. To make matters worse, Martin promptly won the contract for the airframe of the new Titan and transferred its top designers and much of its work force to the military missile. In the minds of the DoD and the contractor, Vanguard became a second-string project.[42]

Money was also a problem. The NSC had approved $20 million for the IGY satellite program, but before Vanguard was even underway the NRL upped its estimate to $28.8 million. Governmental procedure also required "incremental financing," so that the first number on the contract was only $2 million, and Martin and the NRL would have to petition for their funds every step along the way. When the USAF then grudgingly permitted the Navy Lab to use its Missile Test Center at Cape Canaveral on the condition that the Vanguard people build their own facilities, cost estimates leaped again to $43.8 million, then $63 million in June 1956. Emergency funds from the NAS and NSF kept the program going until the annoyed Eisenhower himself ordered the BoB to find the needed money.[43]

Vanguard's problems stemmed in part from its weak position as a civilian competitor of high-priority military programs. But its civilian character was its very *raison d'être*, and pains were taken to protect this public image. "The Earth-satellite program," explained Vanguard's Richard W. Porter to Clifford Furnas, Assistant Secretary of Defense for R & D, and others at Washington's Cosmos Club, "should be thought of as an IGY project in which the DoD is cooperating, rather than as a DoD project."[44] When NAS scientists hosted some Soviets just prior to the opening of the IGY on July 1, 1957, John Hagen casually spoke of the Naval Research Lab satellite. A junior staffer quickly corrected him: "The National Academy's satellite, Dr. Hagen."[45]

Vanguard's first stage tested out well in December 1956. But just as its army rivals predicted, the upper stages played hob with the timetable. The full configuration, TV-2, lingered in Baltimore until a nasty letter from Hagen prompted its delivery to Florida. It was found to be contaminated with filings, metal chips, dirt, and technical bugs. Static firings could not commence until August 1957, whereupon the first four all failed. By then the Soviets had already fired two ICBMs successfully and were predicting a satellite at any time.

A year before, on September 20, 1956, the army's Jupiter-C had soared to a record altitude of 682 miles, far into outer space, and returned to earth 3,355 miles downrange from Cape Canaveral. Its top speed was about 13,000 miles per hour. A live upper stage, assuming it fired properly, could easily have reached orbital velocity. In January 1957 the Thor underwent its first test. A contaminant in the LOX caused a loss of thrust and the rocket exploded. On March 1 the Jupiter roared brilliantly for over a minute before excessive heat in the tail section caused it to veer off course. The second Thor test was apparently successful, but erroneous readings caused the safety officer to send the destruct command. A week later, on April 26, the Jupiter again went out of control due to sloshing propellants. On May 21 a third Thor was destroyed on the pad due to overpressurization in the LOX tank. Jupiter flew successfully on a shorter range at the end of May. Then Atlas tried for the first time. A random valve malfunction shut down the engines, and the safety officer destroyed it. On August 8 the Jupiter pushed the revolutionary ablative nose cone 300 miles high and 1,200 miles downrange. It was the first fully successful demonstration of the heat-shield concept for reentry vehicles. Three weeks later the Jupiter succeeded again. On September 20, 1957, the fifth Thor also succeeded, impacting 1,300 miles down-range.[46]

As October 1957 arrived, the United States was marginally behind the Soviets in ICBM development and comfortably ahead in guidance technology, warhead design, and solid-fuel technology. By the time either side could deploy significant numbers of long-range missiles, American superiority would be clear. In size the United States had nothing to compare with the Russian ICBM, but the Jupiter-C probably had a 50 percent or better chance of placing a satellite into orbit any time after September 1956.

At Huntsville they knew the score. By the summer of 1956 General Medaris, von Braun, and the others suspected that time was running out on Vanguard. The Soviet ICBM test in August confirmed their hunch. When Medaris received word in late September that the new Secretary of Defense, Neil McElroy, would be coming down the following week, "our whole organization was thoroughly fired up with the necessity of giving Mr. McElroy the full and complete story. . . ."[47] He arrived on the afternoon of October 4. Secretary of the Army Wilbur M. Brucker and Generals Lyman Lemnitzer and James Gavin were also present and in the midst of cocktails when the public relations officer rushed up to Medaris and gasped the news. The Soviets had just put up a satellite.

There was an instant of stunned silence. Then von Braun started to talk as if he had been vaccinated with a victrola needle. In his driving urgency to unburden his feelings, the words tumbled over one another. "We knew they were going to do it! Vanguard will never make it. We have the hardware on the shelf. For God's sake, turn us loose and let us do something. We can put up a satellite in sixty days, Mr. McElroy! Just give us the green light and sixty days!"[48]

Von Braun's more cautious commander interjected, "No, Wernher, ninety days." But Medaris promised McElroy a 99 percent chance of success in at least one of two shots: "When you get back to Washington and all hell breaks loose, tell them we've got the hardware down here to put up a satellite any time."[49]

But Washington knew what Huntsville had been doing. It was Huntsville that knew not what Washington had been thinking.

Conclusion

All hell did break loose. Sputnik was a sharp slap to American pride, but worse, it suggested Soviet technical and military parity with the West, which in turn undermined the assumptions on which free world defense was based. To those in the know, the limited importance of the Soviet satellite and the true proportions of military might were clear. But to Eisenhower's opponents, ranging from hawkish senators to civil rights activists, critics of Republican economics to pushers of federal aid to education, Sputnik was an opportunity to sell their programs as cures to the presumed ailments of American life that contributed to the "loss" of the space race. From October 1957 to the end of his term, Eisenhower was under siege, and with him the public values he championed. Thus Sputnik was the greatest defeat Eisenhower could have suffered, and it wiped out much of five years' efforts to meet the Cold War challenge without America, in his view, ceasing to be America.

From 1789 to 1941 the U.S. government stood relatively aloof from science and technology. Americans loved machines but worshipped them no more than they worshipped central government. Abiding minorities, from Thoreau to Woody Guthrie, defined liberty and happiness in terms other than materialism and power. Even for the mass of Americans technology was a means, not an end, and not something to be directed by the federal government. Creation and application of new knowledge were properly left to private persons and institutions, plotting the nation's progress in a marketplace of ideas and techniques.

A rival, technocratic tendency challenged this aloofness in the twentieth century. Social scientists and others believed that private initiative neither fostered science adequately nor ensured the most beneficial use of technology. Inspired by Marxist or war economy models from Europe, they had some successes—perhaps the TVA was the most famous—but they did not carry the day. As late as the 1950s Eisenhower could win cheers for denouncing a planned expansion of the TVA as "creeping socialism." The only ally of technocracy was war. The NAS and land grant colleges dated from the Civil War; the National Research Council and NACA from World War I; the OSRD and Manhattan District from World War II; the AEC, NSF, and military R & D offices from the Cold War. Atomic energy especially frightened Americans with the prospect of what damage science could do in the wrong hands even as it infatuated them with the possibilities of planned change. Not trusting themselves, or other nations, Americans had no choice but to trust their government.

The United States was by no means a technocracy in the postwar decade. The Truman and Eisenhower administrations continued to define most government spending as a necessary evil. But they also had to suffer massive federal spending and command R & D in the arena of national defense, for soon after the victory in 1945 the United States fell heir to responsibilities stretching out toward the indefinite future. Many civilian and military leaders may have coveted the power that went with those responsibilities, but many others who did not still granted the need for military force sufficient to contain an outspokenly hostile and demonstrably expansionist technocratic rival. Five years of debate on how best to do so ended with a reluctant decision to rely on superior technology in order to avoid turning the whole country into a garrison state.

Eisenhower ratified this decision in the New Look. Smaller defense budgets meant greater reliance on strategic forces—B-47s and B-52s— while the AEC perfected small, more efficient thermonuclear warheads. Under this umbrella Americans enjoyed their first taste of postwar stability and, predictably, indulged in the relaxation, temptations, and ultimate discontents of "normal" life. But during those years two trends hinted at new turbulence to come. The first was the revival of social engineering among intellectuals increasingly sensitive to the gap between the promise and reality of American life. Denouncing middle-class materialism, they nonetheless demanded that the government extend the fruits of that materialism to the underprivileged. While their critiques were often poignant, their solutions dictated ever broader federal intervention in domestic life. The other trend was international and technological—the missile revolution. The Soviets showed off long-range bombers above Red Square and progressed rapidly toward an ICBM. Although reliable data on Soviet R & D were still scanty, the Killian Panel sketched the trends with some accuracy. The United States need not fear for five to eight years, barring a suicidal attack. But by the end of that period the United States must have prepared for the missile age.

Eisenhower's considerations were twofold: match Soviet capabilities and ensure a credible deterrent, but do so without the military going "hog-wild." So beneath budgetary ceilings, Eisenhower approved a tableau of missile programs sufficient to secure the nation from future Soviet rocket-rattling. It was not imperative that the United States be first to do this or that, only that it be prepared to deploy missiles in equal or greater numbers at a higher level of guidance, survivability, and reliability. At the same time, Eisenhower hoped that diplomatic breakthroughs might put a lid on the arms race and guard American society from the vapors of a boiling technocracy.

The most important ingredient of a balanced and moderate missile program was reliable information about the doings of the secretive Soviet Union. It would prevent overreaction to rumors of Soviet weapons, permit verification of arms control agreements, and provide targeting data in case of war. This critical need turned spaceflight, a by-product of missile research, from an expensive fantasy into a practical bargain. However much their lull in rocket research forced the Americans to play "catch up" in space technology, their advanced thinking on the applications of spaceflight meant that the American space program, when it came, would be better and more quickly adapted to national needs. Observation satellites were the greatest prize, but they could not have been more delicate from the standpoints of international law, diplomacy, and strategy. How could the United States ensure the legality of satellite overflight? As early as 1950 the RAND Corporation suggested that a small, civilian, scientific satellite would have the best chance of setting a precedent for "freedom of space." When the IGY proposed just such a program and placed it in an international setting, the need and opportunity found each other.

Officials and scientists charged with deciding on the mode of launch then faced a fateful decision. The only extant missile up to the job was the army's Redstone/Jupiter. The navy proposal had other important merits. But the Stewart committee knew the importance of keeping the project free of military involvement, while Assistant Secretary Quarles was aware of the reconnaissance satellite program and the legal imponderables, even risks, surrounding it. If he had been motivated above all by the desire to beat the Soviets he could have overruled the Stewart committee in favor of the Redstone. He chose not to do so. To be sure, RAND, Rockefeller, and the NSC were all aware of the likely impact of the first satellite. But being first was not the only consideration. The tragedy is that Eisenhower, or perhaps just his DoD advisers, failed to imagine the use that would be made of a Soviet "surprise" by those who yearned to overthrow his policies, his party, and his philosophy of government.

Four days after *Sputnik I*, poor Quarles was called on the White House carpet. "There was no doubt," he confessed, "that the Redstone, had it been used, could have orbited a satellite a year or more ago." Ike said that when this information reached Congress they would surely ask why such action was not taken. The President "recalled, however, that timing was never given too much importance in our own program, which was tied to the IGY, and confirmed that, in order for all the scientists to be able to look at the instrument, it had to be kept away from military secrets."

Quarles then accentuated the positive: ". . . the Russians have in fact done us a good turn, unintentionally, in establishing the concept of freedom of international space. . . . The President then looked ahead five years, and asked about a reconnaissance vehicle."[1]

PART III

Vanguard and Rearguard:

Eisenhower and the

Setting of American

Space Policy

America is very beautiful, and very impressive. The living standard is remarkably high. But it is very obvious that the average American cares only for his car, his home, and his refrigerator. He has no sense at all for the nation. . . . He also has no sense for great ideas which take as long as a number of years to achieve. . . . Russians do! Well, you certainly know what I mean, because you are a former German.

—LEONID SEDOV to ERNST STUHLINGER, October 7, 1957

If America ever crashes, it will be in a two-tone convertible.

—BERNARD BARUCH, October 16, 1957

How shall a democracy grapple with a totalitarian state without compromising its fundamental principles? . . . How can we compete against such a Spartan system with our unwieldy, clumsy Government which was designed to best prevent the accumulation of excessive power?

—VICTOR GILINSKY, November 8, 1957

There is much more to science than its function in strengthening our defense, and much more to our defense than the part played by science. The peaceful contributions of science—to healing, to enriching life, to freeing the spirit— these are [its] most important products. . . . And the spiritual powers of a nation—its underlying religious faith, its self-reliance, its capacity for intelligent sacrifice—these are the most important stones in any defense structure.

—DWIGHT D. EISENHOWER, November 7, 1957

BY Thanksgiving weekend of 1959 seven years had passed since Eisenhower and his new appointees determined to wage Cold War and still uphold the principles of limited government that, in their view, were the wellsprings of American power. Of the drafters of the Great Equation, Dulles was dead by 1959, and Dodge, Humphrey, and Charles Wilson had left government. Only Ike, now sixty-eight, remained. He had endured seven budget cycles, two recessions, the Soviet H-bomb, an alleged "bomber gap," and Sputnik. And still his budget for FY 1960 showed a billion-dollar surplus. It came to $92.2 billion, an increase of only 20 percent during seven years when the gross national product (GNP) rose nearly 25 percent in constant dollars, average industrial wages 34 percent, teachers' salaries 50 percent, and private investment 45 percent.[1] The economy was sound, national defense adequate, and the future unmortgaged. If the President took the long view that Thursday morning, he probably did give silent thanks. But the short-term view surely made him apoplectic. A lengthy steelworkers' strike, interrupted by a Taft-Hartley injunction, remained unsettled. Khrushchev's recent visit failed to advance arms control. Americans were jerking again to the electric prods of their journalistic shepherds, who outdid *Tass* in finding political and even moral significance in the Soviet "moonshots" of the autumn. And, as always, another budget was coming due.

In the national press Eisenhower appeared as an affable but simple old man who did not govern so much as preside over the Cabinet and NSC. "Assistant President" Sherman Adams, it was said, did the strong-arming. But now Adams, too, was gone, victim of the vicuña coat scandal the previous year. Given their assumptions of presidential sloth and senility, some journalists were hard put to account for anything getting done in the White House. Those privy to Eisenhower's inner meetings knew differently, although fifteen years would pass before their testimony found an audience. They saw Ike as a dominant boss with a quick temper, a shrewd politician who kept the practical and ethical bases for action before himself and his subordinates. Thus Robert Anderson (treasury), Christian Herter (state), Thomas Gates (defense), and the others were not surprised that Thanksgiving weekend to hear their chief open a budget discussion by asking: "What is the true problem which faces Western Civilization?" No one spoke; the President clearly intended to answer this one himself.

The question is whether free government can continue to exist in the world, in view of the demands made by government and peoples on free economies, while simultaneously facing the continuing threat posed by a centrally controlled, hostile, atheistic, and growing economy? . . .

We have got to meet the [Soviet threat] by keeping our economy absolutely healthy. Without the health and expansion of our economy, nothing we can do in the long run, domestically or in the foreign field, can help. We are the world's banker. If our money goes bad, the whole free world's position will collapse or be badly shaken. . . .

We must get the Federal Government out of every unnecessary activity. We can refuse to do things too rapidly. Humanity has existed for a long time. Suddenly we seem to have an hysterical approach, in health and welfare programs, in grants to the states, in space research. We want to cure every ill in two years, in five years, by putting in a lot of money. To my mind, this is the wrong attack.[2]

Seven months later, in June 1960, Eisenhower's second term was counting down, and he wanted to bequeath a budget—FY 1962 this time—"that is fine and decent and in accord with the Administration's philosophy." When he entered office, Ike reminisced, the budget was something like $73 billion. He had pledged to cut spending, regulation, and taxes—critics charged that "the Republicans will ruin the country." Yet he had succeeded, he recalled, in spite of ridicule and the opposition's "collectivist spirit." But then came the missile programs and the Sputniks, and "an almost hysterical fear among some elements of the country," then demands that government "make life happy in a sort of cradle-to-grave security." All this jeopardized the "basic values of self-dependence, self-confidence, courage, and a readiness to take a risk."[3]

In Eisenhower's view, government could not legislate happiness or "cause" economic growth, or remove the risk from human life without removing its freedom and meaning as well. Eisenhower told his Cabinet the story of a man so sated with pleasures in the afterlife that he asked to transfer to hell—only to learn he was already there. Ike was not contemptuous of the poor, but of those in the elite—industrialists, politicians, bureaucrats, educators, small businessmen, sectional interests—all of whom sought subsidies to remove their risks or build their power. State-managed economics were doubly dangerous: they did not work and at the same time corrupted the private sector, which was the true source of wealth. The art of governance involved meeting unavoidable responsibilities without compromising the liberty and dynamism of civil society.

Sputnik posed a great challenge to Eisenhower's presidency precisely because it seemed to belie this philosophy. As a foreign threat with military overtones, it was clearly the government's business. As a blow to U.S. credibility, it seemed to demand a response in kind. As a technocratic accomplishment, involving the integration of new science and engineering under the aegis of the state, it called into question the assumptions behind U.S. military, economic, and educational policy—every means by which the mobilization of brainpower is achieved. As an arcane technical feat, it suggested new dependence on a clique of experts, whom the people's representatives had no choice but to trust. All told, Sputnik threatened to undercut Eisenhower's efforts to usher in the missile age without succumbing to centralized mobilization and planning.

Instead, Ike was forced to adjust to an age of technological competition with civilian and military space programs, elevation of scientific advice to the apex of government, federal aid to education and greatly enlarged subsidies for basic research, reform of the Pentagon to enhance central authority, especially in R & D. In these ways, the Eisenhower administration joined the vanguard of the Space Age. But it was equally in the rearguard, for Eisenhower placed his stamp on every post-Sputnik policy, lest the genii of state-funded science, technology, and social mobilization get out of hand. Still, it was a losing battle. Soviet threats and achievements, especially in space, kept up the pressure on the United States until the emergency measures taken by Ike became permanent and his opponents made a virtue of necessity by embracing the technocratic method.

American space policy dates from the last Eisenhower years. In all things it aimed at sufficiency, not universal superiority vis-à-vis the Soviet Union, and not at all at achievement of some self-generated ideal. To train X thousands of engineers, to reach the moon by 19XX, to plant X numbers of missiles in silos regardless of Soviet deployments, to plan for economic growth of X percent without unemployment or inflation— these were not the assignments of a free society but the dictates of a command economy. Eisenhower refused to accept that the missile and Space Age necessarily meant an age of technocracy. Whether or not he was quixotic, or his principles obsolete in an age of computers and satellites, he stood for four years like Janus, in the vanguard and the rearguard.

CHAPTER 6

"A New Era of History"
and a Media Riot

Senate Majority Leader Lyndon B. Johnson presided over his Texas ranch on October 4, 1957. Among his guests was Gerald Siegel, prior counsel to the Senate Democratic Policy Committee, future vice president of the Washington *Post*, but current applicant for a job with Mrs. Johnson's Austin television station. Together the guests heard the news of the Soviet satellite, "and simultaneously," wrote Johnson, "a new era of history dawned over the world." After dinner the party strolled in the dark along the road to the Pedernales River, their eyes drawn upward. "In the Open West you learn to live closely with the sky. It is a part of your life. But now, somehow, in some new way, the sky seemed almost alien. I also remember the profound shock of realizing that it might be possible for another nation to achieve technological superiority over this great country of ours."[1]

Johnson returned to the house and telephone. Siegel volunteered to stay up for another term, Lady Bird's television job now beside the point. This was a national emergency, and LBJ was the sort of man who had to act at once.[2] The ideal forum for an inquiry was the Preparedness Subcommittee of the Senate Armed Services Committee, created by Johnson's patron, Richard Russell of Georgia, as a showcase for the young Texan in 1950. Patterned on the committee that had propelled Truman's career, the subcommittee filed a hundred reports on aspects of defense during the Korean War. When the Republicans won control of the Senate in 1952, it fell into limbo, but the 1954 elections restored both it and Johnson, now Majority Leader, to the limelight.[3]

But there were other considerations that stayed Johnson's hand for a month after Sputnik. What did he, or Russell, or anyone, know about outer space? Sensational hearings might rebound badly if the White House was able to account for its space and missile policies and make the inquiry seem unpatriotic and exploitative. Eisenhower was much beloved, and attacking him in the defense realm was risky, as the 1954 "bomber gap" fracas had revealed. So it was not certain that Senate

Democrats would launch what became the famous "Johnson hearings" until a wave of public hysteria, the administration's evident confusion, and the body blow of *Sputnik II* on November 3 made a conspicuous inquiry into the causes of the United States' "missile mess" a no-lose strategem for the Democrats.

As it happened, the public outcry after Sputnik was ear-splitting. No event since Pearl Harbor set off such repercussions in public life. But public knowledge of issues as complex as those raised by the opening of the Space Age was no more or less than what the news media imparted. It is tempting, in studying press coverage of the Sputniks, to attribute a manipulative intent to certain journalists, generals, and politicians; the administration was probably right in charging that some pundits exaggerated the danger of the Soviet satellite and kept that danger before the public in order to marshal support for their own agendas. But it is also true that the journalists who forged the boilerplate were as ignorant of space as the public, and had to educate themselves even as they reported. And unlike the public, the editorialist or congressman was obliged to analyze, conclude, and interpret almost from the first instant.

The outcry over Sputnik was lengthy, loud, and imposing, precisely because all these features, sinister and righteous, combined to raise the volume. Sputnik challenged the assumptions of American military and fiscal policy, and thus seemed to have scary implications for American security and prosperity. It involved a romantic but eerie enterprise—space travel—that Americans had come to associate, thanks to Hollywood and science fiction, with sudden and irresistible horrors. It was clearly connected with nuclear missiles. Finally, it lent itself to opponents of the administration as the seal of alliance between military-industrial "hawks" on the one hand and social-educational activists on the other. Therefore, the response to Sputnik was not just random clamor or a manipulated panic, but the chaotic product of several waves, their crests and troughs overlapping to reinforce alarm one week and confused inertia the next. One day Sputnik seemed to challenge the viability of American life; the next it only illustrated the worst aspects of the Communist way of life. One day it required Americans to overhaul their institutions, the next that they cling to them and the principles they embodied more dearly. To wit:

On October 5 Senator Alexander Wiley (R., Wisc.) saw "nothing to worry us" and thought Sputnik salutary for it would "keep us on our toes." But his colleague Styles Bridges (R., N.H.) called for an "immediate revision of national psychological and diplomatic approaches. . . ." The United States must be less concerned with the "height of the tail fin in the new car and be more prepared to shed blood, sweat, and tears if this country and the free world are to survive." Senators Henry Jackson (D., Wash.), Stuart Symington (D., Mo.), and Russell warned of severe danger from the demonstrated Soviet missile capacity and blasted the White

House for withholding the truth.[4] Editorialists, following the State Department itself, acknowledged on the sixth and seventh that Sputnik was a propaganda victory without parallel, and linked it, for the first time, to competition for influence in the underdeveloped world.[5] Should the United States respond in kind to such feats? Or did Sputnik only show that "in a totalitarian country scientists are told what to do. They can be quickly mobilized and their mass effort directed at any single objective. . . ." Perhaps, but the Dallas *News* thought "some advantages of tight, totalitarian control will be helpful to our democratic processes." More shields for Athens. The New York *Herald Tribune* attacked American "hamstringing" of scientists and the "starving of R & D."[6] Senator Mike Mansfield (D., Mont.) called for subsidy of scientific talent from the second year of high school, Jacob Javits (R., N.Y.) for an acceleration of the overall defense effort.[7] But if some urged the United States to become more like the USSR, the Washington *Evening Star* wondered if the Soviets might now become more like us. A nation so scientifically advanced could not expect to maintain thought control over its citizens.[8]

By the middle of the following week (Sputnik occurred on a Friday), the nation's pundits began to grasp that the satellite confirmed the Soviet claim to an ICBM. This meant looming strategic stalemate in which the Soviets might "win the peace without ever having to make war." According to French general Pierre Gaullois, "The spectacular Russian thunderbolt might terrify the already weak-kneed allies of the U.S. into pressing us to yield to Soviet demands all over the world or 'go it alone.' " NATO officials were thrown into a "worse state of flux" than after the Russian H-bomb.[9] Several more days and the alleged causes of the humiliation of the United States began to surface: interservice rivalry, underfunding, complacency, disparagement of "egghead" scientists, inferior education, lack of imagination in a White House presided over by a semiretired golfer, and a general lethargic consumerism. Walter Lippmann expected that continued Soviet progress would radically alter the world balance of power.[10]

The opening of the Space Age! It was a big story, the biggest. Between moralizing editorials, networks and newspapers discussed the scientific principles behind it all. What was a satellite? What made it "stay up"? What were apogee, perigee, inclination? How could one spot Sputnik in the night sky? The public learned as if from a rookie professor, who kept one chapter ahead in the textbook. The uses to which satellites might be put went unreported, the real connections between satellite and missile forces were lost on the reporters, the fact that Sputnik was far more an engineering triumph than a scientific one was an especially fateful misapprehension. On October 14, the *New York Times* did report the existence of the USAF reconnaissance satellite program, but the article went little noticed. Instead, the public heard sensational assaults on American materialism and contempt for science.

October 13, 1957. Courtesy of the *Sacramento Bee*.

What was the impact of this sort of coverage? A Gallup Poll conducted between October 11 and 14 gave little evidence of panic. Half the sample thought the Soviet satellite a "serious blow to U.S. prestige," but 46 percent thought it decidedly not! Over half showed surprise that the Soviets beat the United States, but 44 percent did not, and fully 61 percent thought satellites would "more likely be used for good purposes than bad." Another poll had 49 percent believing that the USSR was moving ahead in missiles, but 32 percent did not, and a wise (or apathetic) 19 percent had no opinion.[11] On what basis were citizens to answer such a question? U.S. missile programs were secret or underreported, while the Soviet satellites were bombshells that, in Sherman Adams's words, "sail over our heads and land on the front pages of every American newspaper."[12] Perhaps a portion of the public was reassured when those in the know, like General Omar Bradley, Admiral Arleigh Burke, or Curtis ("It's just a hunk of iron") LeMay explained the state of missile R & D and the primitive, passive nature of Sputnik, but the newspapers tended to be incredulous. If the polls can be trusted, one may conclude that the media were not responding to a grassroots movement when they played up the space story and guessed its ominous meaning. In its initial stages, the national response to Sputnik was rather an aimless, agitated "media riot."

A national primer on the mechanics of spaceflight arrived on millions of doorsteps two weeks after the fact in the pages of *Life* magazine. But among the scientific stories and charts was editorial material that *instructed* the American people to panic and told them that their wiser neighbors already had. Their "growing familiarity with Sputnik did nothing to soothe Americans' shock"; Soviet scientists were as good as any in the world; the Eisenhower administration was so stingy that the security of the nation was compromised. "Let us not pretend that Sputnik is anything but a defeat for the United States," said *Life*. There *was* a space race, and the first heat was already lost. "We must revise our naive attitude toward basic research. The armed forces must understand that money spent on background research is not money thrown away. . . . We must give much more aid and encouragement to our educational institutions in turning out scientists and engineers. . . . We must change our public attitude toward science and scientists." Sputnik was "really and truly 'the shot heard round the world,' " and in an environment that "may be altered more in the next twenty years than in the last two hundred," the United States must reaffirm its commitment to liberty, but not lag in weapons against communism.[13]

The product of the first frenzied month of the Space Age, though inadvertent, was a new symbolism that had more reality for mass politics than did the actual data on Sputnik, Soviet and American defense budgets, missile progress, and science and education. As one cartoonist illustrated, things looked different by the light of the Soviet moon. It was irrelevant that the new perceptions may have obscured rather than enlightened—even a respected war hero and President could not suppress a new political symbolism so at variance to the one that had defined his political lifetime.

The month of *Sputnik I* also taught the administration the truth of Rockefeller's admonition, "This is a race we cannot afford to lose."[14] But Sputnik could not be undone. Instead, the Cabinet fought to regain its balance and control of public opinion. On the morning of October 8, at the White House meeting when Quarles admitted a Redstone could have beaten the Soviets, Ike asked if there was anything about Sputnik that invalidated American R & D programs. No, answered Detlev Bronk, "we can't always go changing our program in reaction to everything the Russians do." Nor was the military situation affected, for the government had known of the Soviet ICBM since August. What then of the long-range significance of Sputnik? The Office of Defense Mobilization (ODM) Science Advisory Committee, meeting Eisenhower on the fifteenth, was equally unruffled. The United States was still ahead in science. Prudence demanded only that American science be given some encouragement, and a permanent science adviser be placed in the White House to ensure that the United States maintained its lead.[15]

In public, the administration offered reassurances in hopes of heading off a stampede on the Treasury. But past logic—the need for fiscal restraint over the long haul—was illogic under the new symbolism. The Republican public relations effort appeared clumsy and substantiated charges that Ike was out of touch. Hagerty explained on the day after Sputnik that the government had not viewed the satellite program as a race, but rather as a scientific contribution to the IGY.[16] But given its "evident" importance, asked the critics, why was it *not* viewed as a race? Dulles then fed Hagerty a more subtle account—that the satellite *was* of considerable importance, but that Sputnik involved "no basic discovery, and the value of a satellite to mankind will for a long time be highly problematical." The Soviets won, said Dulles, because of their takeover of German rockets, scientists, and facilities, and because "Despotic societies which can command the activities and resources of all their people can often produce spectacular accomplishments. These, however, do not prove that freedom is not the best way."[17] Sputnik was invigorating! "We live in exciting times," thought Dulles. "Can the free world prevail in the face of the most formidable, despotic, materialistic, atheistic society that the world has ever known?" If the United States, too, became a garrison state, of course it could, but "the question is whether we can surpass [the USSR] and still retain the essentials of freedom."[18]

Such subtleties were lost on the newsmen, but in their defense it must be said that the facts seemed to belie the administration's version. The best German rocketeers had fallen into *American* hands. If Sputnik was just a "neat scientific trick" (as Defense Secretary Wilson put it), why was the DoD so anxious to insist that U.S. missile programs were proceeding rapidly? Eisenhower himself claimed that Sputnik "does not rouse my apprehensions, not one iota. . . . They have put one small ball into the air." Yet across town McElroy announced urgent studies on the removal of bottlenecks in U.S. missile programs.[19] These statements were not necessarily contradictory, but they seemed so to the press. Journalists rarely admit that they themselves can become actors in a play they purport only to review, that actions (e.g., removing bottlenecks) may be subjective responses to a bad press rather than objective reactions to a real crisis.

In the meantime, "I told you so's" and expressions of alarms flooded the White House. Defense analyst Ernst Steinhoff informed Quarles that RAND had predicted the Soviet satellite six months before, that the army could launch one anytime, that Sputnik was hurting U.S. prestige more than any sum of foreign aid could make good. Symington insisted on a special session of Congress. Jimmy Doolittle of the NACA demanded immediate acceleration of missile programs. Javits wanted a Manhattan Project for missiles.[20] Academic opinion was nearly unanimous in favoring sudden, substantial support for education. The National Planning Association saw a shift in global power and called on the government to help Americans transcend materialism, make the sacrifices necessary to defend Western civilization, and demonstrate to uncommitted nations that socialism need not mean subordination to Moscow.[21]

Two administration figures who recovered quickly from the drubbing, and seemed to grasp the new symbolism, were Vice President Richard Nixon and UN Ambassador Henry Cabot Lodge. The latter urged Ike to muzzle everyone but himself, allay the impression that the government was cumbersome compared to the Soviet, court Congress, and seize the high ground with a claim to inside knowledge.[22] Nixon made the first major administration speech on the challenge of the Space Age in San Francisco. "There has been a lot of loose talk," he began, but Sputnik made no difference in the military balance. The West was stronger and would outproduce a slave economy in the long run. The ICBM was dangerous, he granted, and it would be foolish to brush off Sputnik as a "stunt." But it would do a signal service if it strengthened American resolve and the country responded intelligently. The ultimate threat, served by Sputnik, was Communist penetration of Asia and Africa, a threat the United States must meet with vigorous trade, aid, and cooperation. "The world of tomorrow is in our hands. . . . It can be a free world or it can be a world poisoned by statism and totalitarianism."[23]

Nothing worked, in part because of journalistic skepticism, in part because of the administration's seeming contradictions, but mostly because the dilemma posed by Sputnik was insoluble. How to outcompete the Soviets in "technological imperialism" (a *Life* phrase) without engaging in it oneself? Dulles told the press that Sputnik was useful because it "created a unity of purpose" among Americans and dispelled "a certain complacency." Was there complacency, then, in the government? No, said Dulles, but there had been a certain complacency in that it had been generally felt we were automatically ahead of the Soviets in every respect. "Mr. Secretary," confessed a reporter, "I'm confused."[24] Was Sputnik important or not? If important, was it for military or psychological reasons? If the latter, was it so because it purged American complacency, or because it impressed the underdeveloped world? Was the appropriate American response one of calm pursuit of current policies, or accelerated efforts befitting the end of "complacency"? If acceleration and competition, did this not mean imitation of the Soviets rather than rededication to

American liberties? When Sherman Adams announced a few days later that U.S. policy was not to win "high score in a celestial basketball game," confusion was compounded.[25] By the end of October even conservative newspapers admitted that Sputnik had shaken public confidence that Republicans could do a better job than Democrats in defense and foreign policy.[26] Yet Eisenhower found it hard to understand the national dismay and fear. He was startled that the American people were so psychologically vulnerable.[27]

The damage-control operation failed. Old verities about the virtues of free enterprise and balanced budgets, local education, and American know-how gradually lost their hold on much of the nation, while much of the press implied that a sensational defeat required sensational departures. This exchange between muckraker Mike Wallace and ex-Secretary of Defense "Engine Charlie" Wilson illustrated the clash between old and new symbolisms:

> Wallace: How worried are you about Russian military superiority?
> Wilson: I don't think they have military superiority.
> Wallace: Then why is the panic button being pushed?
> Wilson: So many people react that way. You know, that's the thing about people. They're so cracked loose in the Buck Rogers age that they're seeing things. Even the Texans.[28]

The *press* assumed Sputnik meant Soviet superiority, and the *press* pushed the panic button. But the administration, admittedly stuffy, was also blind to the symbolic power of a revolutionary technology. And in Texas, they weren't seeing things in outer space. They were seeing the political window opened by Sputnik.

As the public uproar grew, Lyndon Johnson ordered studies, gathered aides, and signaled colleagues. Capitol Hill came to life. In the words of an assistant to the Secretary of Defense: "No sooner had Sputnik's first beep-beep been heard—via the press—than the nation's legislators leaped forward like heavy drinkers hearing a cork pop."[29] In the third week of October a former aide to Senator Lister Hill (D., Ala.) boarded an airplane in Boston and made his way by stages to Austin, Texas. There he passed on to George Reedy, an LBJ confidante, a political memo fully warranting the troublesome journey. Reedy forwarded it to the senator at once with a cover note:

> The issue [Sputnik] is one which, if properly handled, would blast the Republicans out of the water, unify the Democratic party, and elect you President. . . .
> Eye [sic] do not pretend to estimate the effect that the satellite will have on the American people. But eye am convinced that on the basis of merit there is no more important issue. Eye am also convinced that you are the only man who can handle it. Symington practically puts the label "politics" on everything he

says about it. Jackson and Mansfield do not have the necessary force. Russell is so deeply involved in the segregation issue that he will be suspected of beating the President over the head because of Little Rock. You, on the other hand, have a reservoir of goodwill and an aura of statesmanlike handling of defense problems from the early Preparedness Committee days.

Eye think you should plan to plunge heavily into this one. As long as you stick to the facts and do not get partisan, you will not be out on any limb.[30]

The memo itself, drafted by Charles Brewton, made no bones about the Democrats' problem. The stumbling block to victory in 1958 or 1960 was segregation, an "issue that is not going to go away." Their only chance was "to find another issue which is even more potent. Otherwise the Democratic future is bleak." Sputnik was such an issue. To be sure, Americans would not consent to Russian-style deprivation just to have a ball in the air going "beep." But they also hated to be second-best, and when two or three Russian satellites were up, what was now curiosity might turn into panic.

It did not matter, Brewton continued, whether the satellite had any military value: "the important thing is that *the Russians have left the earth and the race for control of the universe has started.*" In previous ages the Romans controlled the world because of their roads, then England controlled the world because of its ships. When humanity moved to the air, the United States was supreme through aviation. *"Now the Russians have moved into outer space."* The first step was a calm, nonpartisan inquiry by Congress, and the questions to pose were obvious. Has the civilian economy been overly pampered? Have balanced budgets jeopardized world security? What is wrong with American education? *"If the issue has merit, the politics will take care of themselves."*[31]

LBJ needed no staffer to instruct him in politics, but the Reedy/Brewton memos reveal the thinking in the Democratic camp. As two Republicans noted at the time: "We live in a political world, and no greater opportunity will ever be presented for a Democratic Congress to harass a Republican Administration, and everyone involved on either side knows it"; "LBJ was eager to get out front in space because it was the new national toy. He was trying to get to become President of the United States. . . . You do like Robespierre—'There goes the crowd. I must get in front of them, I'm their leader.' So LBJ wanted to get out in front of the space rush."[32]

For a month Johnson had been preparing and observing the national reaction. *Sputnik II* was the clincher. The following day Russell and the Senate approved a special inquiry into the satellite and missile programs of the Eisenhower administration. A media riot, Republican confusion, and *Sputnik II* thus combined to make the Soviet space coup not a nine-day wonder but a durable permacrisis that broadened the ordinarily narrow margins for change in a complicated, pluralistic democracy.

SPUTNIK II

The Soviet Union celebrated the fortieth anniversary of the revolution with the launch of a second heavy satellite on November 3, 1957. Korolev's powerful test ICBM orbited a payload of 1,121 pounds, but since the satellite remained attached to the spent upper stage, the weight placed into orbit was on the order of six tons. The satellite contained geophysical equipment, a life-support system, and a live dog named Laika. The fate of the space dog became a matter of sentimental concern, and Americans even hoped for days that the Soviets would return the pup safely to earth even though such a capacity would indicate a much greater Soviet lead in space technology. When it became clear that the dog was to die in orbit, Communist "beastliness" was confirmed. The tremendous size of *Sputnik II* worried Americans, since the promised U.S. satellites would be baubles by comparison. The presence of Laika also suggested that the Soviets were pushing on at once toward manned spaceflight.

Johnson was free to call witnesses, define issues, sustain a mood of danger and humiliation, and pose as the initiator of measures necessary to the defense of the free world. He was aided after *Sputnik II* by a replay of the lamentations and jeremiads of the previous month. If the American people still felt inclined to trust old Ike, *Life* magazine corrected them again by "Arguing the Case for Being Panicky."[33]

The only way for Eisenhower to tranquilize the nation was to get the United States a satellite. After *Sputnik I,* Army Secretary Brucker leaned on the harried new Secretary of Defense on behalf of the ABMA, and on October 14 McElroy gave the go-ahead for refinement of satellite instrumentation for a Jupiter-C. But he refused a green light to prepare for launch. Medaris was stunned—anticipating immediate approval, he had already ordered alterations on Jupiter-C by "borrowing" funds from other departments.[34] In early November this flirtation with a court-marshal ended when approval came through. But McElroy still held back authorization to launch.[35]

Also in the wake of *Sputnik II* Eisenhower discarded the October tactic of playing down the importance of satellites. In a radio-TV address on November 7, he initiated the American people into the facts of life in the missile age. The country was strong, he said. American B-52s and H-bombs could annihilate the war-making capabilities of any nation on earth. The United States had tested its own long-range rockets and solved the reentry problem. Here the President held up a nose cone recovered from a Jupiter test launch. As for satellites, they had no direct military significance but did imply a technological capability that required a response. Therefore, the United States was engaged in building a continental defense system and ground-to-air missiles, and was seeking scientific cooperation with its allies. To ensure scientific expertise at the top levels of government, the President named James Killian to a new post, Special Assistant to the President for Science and Technology, and elevated the ODM science committee to the White House as the President's Science Advisory Committee (PSAC).[36]

What the President did not say was also significant. For way back in the spring—it must have seemed another eon—Eisenhower had commissioned another top-secret strategic review. Entrusted to H. Rowan Gaither, Jr., of the Ford Foundation, who fell ill, it was completed under Robert C. Sprague (a veteran consultant from the Killian panel) and such luminaries as William C. Foster, John J. McCloy, Frank Stanton of CBS, and Jerome Wiesner. They reported on November 7 that if current trends continued, the United States would face a critical threat, possibly as early as 1959 or 1960. The panel found no evidence "to refute the conclusion that USSR intentions are expansionist, and that her great efforts to build military power go far beyond any concept of Soviet defense." This controversial Gaither Report recommended acceleration of missile development, a crash program on R & D for defensive systems, a national fallout shelter program costing upward of $25 billion, and a frank campaign of public awareness of the dangers of the nuclear age.[37]

Ten days before, a CIA committee of experts submitted its view that the Soviets had been pushing a thoroughly thought out missile program for years. Their tests showed "unusually high reliability" and "extremely high proficiency in guidance." The CIA assumed that the USSR would have a dozen operational ICBMs by the end of 1958 and that the United States was lagging by two or three years. "Your consultant panel believes that the country is in a period of grave national emergency."[38]

Some of the Gaither Report recommendations found their way into Eisenhower's televised speech. But he rejected its call for a crash defense effort, especially in fallout shelters. Instead, Ike told the NSC that the "gloomy findings in the report would panic the American people into going off in all directions at once."[39] It was, in his view, a worst-case analysis that failed to take into account the United States' allies and overseas bases, and the political and economic factors on which security also rested. The NSC backed the President. It accelerated missile R & D, early warning and defensive systems, but did not approve vast supplemental appropriations.[40] Despite a series of invitations to panic, Eisenhower held to his course.

Lyndon Johnson, it was said, "scoops people like peanuts." When he spied a young man of talent, he did him a good turn or put him on the payroll in Texas or on the Hill, and then tapped his loyalty and ability when needed.[41] Now he quickly assembled a blue-chip staff for the most complicated and fertile investigation of his career: Edwin Weisl, Cyrus Vance, Gerry Siegel, Sol Horwitz, Homer Stewart, and the indefatigable Eilene Galloway to ferret out data. The staff viewed its task as one of exposing the truth about space and missiles that the Republicans were apparently keeping from the American people. Johnson billed the hearings as nonpartisan, but in the eyes of the administration they were clearly rigged to embarrass the President and elevate Johnson to a leadership role. Weisl and Vance had no trouble finding disaffected personages in the military; few generals and admirals were content with their funding, and it was a simple matter to give them a forum and make the administration seem careless with the national defense. The staff also hit on the ploy of requesting reports of extant programs meant to "close the gap" with the Soviets. Since these were classified, the committee could lift ideas contained in them and then release them as committee recommendations, creating the impression that the Senate was taking things in hand and the Cabinet doing as it was told.[42]

The monumental Inquiry into Satellite and Missile Programs began on November 25, on the third floor of the Senate Office Building.[43] Johnson welcomed the participants and informed eager reporters, photographers, and spectators that their purpose was to get the facts on the state of the nation's security (implying that the administration's "facts" were untrustworthy). "Our country is disturbed over the tremendous military and scientific achievement of Russia" (contradicting experts' claims that Sputnik per se was not militarily important). "Our people have believed that in the field of scientific weapons and in technology and science, that [sic] we were well ahead of Russia" (implying that public disillusionment prompted the inquiry). "With the launching of Sputniks . . . our supremacy and even our equality has [sic] been challenged" (implying that the President's reassurances were disingenuous). "We must meet this challenge quickly and effectively in all its aspects" (implying a prescriptive mandate for the committee). Johnson concluded that the Sputniks were a challenge greater even than that of 1941, they were a technological Pearl Harbor.[44]

Edward Teller, "father of the H-bomb," spoke first. He was a physicist not directly associated with missiles and space, but politically his selection was ideal, for he was the chief "hawk" in the scientific community. Teller attributed the American missile lag to delays pending development of low-weight thermonuclear warheads. Clearly this had been a mistake. Teller did not know whether the Soviets already had operational ICBMs, he could only hope not. They took bigger gambles in their R & D, and by dint of more spending and concentration progressed to the point where the United States no longer had a decisive lead in almost any field. The "captured German scientist" story was a myth; the Soviets had done it on their own. In prescribing remedies, Teller struck the fetching chord of technocracy in his own disarming style:

Shall I tell you why I want to go to the moon? . . . I don't really know. I am just curious. . . . If you asked me about ballistic missiles in 1945 or 1946, I would have said, "Let's do it and let's do it fast," and then you would have said, "In what particular way will you apply this in a possible war?" and I would have told you, "I don't know, but once we make it we will find some use." And I think going to the moon is in the same category.... It will have both amusing and amazing and practical and military consequences. This is how it always was in the world.[45]

Vannevar Bush followed Teller. He admitted having underestimated the pace of ICBM development and Soviet capabilities in general. By comparison, he said, Americans were complacent, egoistic, and spoiled. But Sputnik was "one of the finest things that Russia ever did for us. ... It has waked this country up." Bush urged greater support for science and education, and respect for the scientist "as a fellow worker for the good of the country" and not a "highbrow or egghead."[46] Jimmy Doolittle of NACA then warned that the USSR would soon surpass the United States in every field unless it gave immediate stimulus to military R & D. Senator Johnson replied with doggerel:

> I'd rather be bombed than be bankrupt.
> I'd rather be dead than be broke.
> Tis better by far to remain as we are
> And I'm a solvent if moribund bloke.

Doolittle allowed as how he had no sympathy with such an attitude.[47]

Day by day the witnesses rose to confirm the committee's suspicions and provide quotes for the next day's front pages. Administration figures had their chance. McElroy explained why Vanguard had been given a lower priority than the ICBM and IRBM programs. Quarles insisted that the United States was ahead in electronics, warhead design, aviation—everything except the highest-thrust rockets, which were too large for efficient ICBMs in any case.[48] But their accounts rang hollow as long as Sputniks circled alone overhead. When probed about satellites, in which the United States was undeniably behind, Quarles ducked: letting von Braun launch in 1956 would have delayed Jupiter by three months, he said. "We felt it wise not to buy the program at that kind of a price. ..."[49] The whole story of the Vanguard/Jupiter decision, with its connections to spy satellites and freedom of space, did not come out, at least in public session. When this first grueling day of testimony ended at a quarter hour before midnight, Johnson reminded the remaining onlookers that this was indeed another Pearl Harbor and that in such an atmosphere there were no Democrats or Republicans, only Americans.

When the Johnson committee reconvened in December, Vanguard's insufferable blow to American pride and a series of angry witnesses put the hearings back on the front page. The public was told that it demanded scapegoats and saviors, while the committee invited opinions and heard of bitter rivalries in the military, underfunding, and bureaucratic blindness. Such *specific* explanations should have served to cast doubt on the *general*

VANGUARD TV-3

The Sputniks put tremendous pressure on the Vanguard team to produce a satellite. NRL and Martin engineers rebounded from the summer's frustrations and readied their four-stage configuration, TV-3, in December. The White House insisted, however, that the scheduled test flight be billed as a full-fledged attempt to orbit a satellite. Reporters from around the world converged on Cape Canaveral and cranked up suspense during a two-day delay due to weather and holds. Shortly before noon on December 6, 1957, the countdown finally reached zero. Two seconds after ignition the Vanguard rose four feet off the pad—and suddenly exuded thunder and flame "as if the gates of Hell had opened up."[50] TV-3 settled back to earth and was consumed, but the nose cone fell clear, and the little ball of satellite came comfortably to rest nearby, chirping innocently.

While Martin and GE technicians pointed the finger at each other, newsmen and notables proclaimed their country's humiliation. Vanguard was Kaputnik, Stayputnik, or Flopnik, and Americans swilled the Sputnik Cocktail: two parts vodka, one part sour grapes. At the UN, Soviet delegates asked if the United States was interested in receiving aid to underdeveloped countries. And the Vanguard program, which after all was only in the test stage and would soon launch satellites for the IGY, suffered an ignominy from which it never escaped.

Worse yet for the navy, the army finally got the green light to prepare a launch with the Jupiter-C.

---

explanations including American complacency, self-indulgence, and poor education. Simon Ramo even denied that there was a shortage of engineers, and officials from engineering schools roundly resented implications that their graduates were inferior to Soviet technicians. But it did not matter—the general and specific accounts of American humiliation flowed through the press and public mind together, weakening faith in the administration and its values.

An army parade—Secretary Brucker, Generals Gavin and Maxwell Taylor included—then regaled the Johnson committee with attacks on the penury that led to strategic vulnerability even as the New Look stripped U.S. ground forces just as emerging nuclear stalemate restored their importance. Gavin, who announced his intention to resign (he had been passed over for his preferred command, but the committee implied he was a martyr to procrustean Republicans), thought it "absolutely vital" for the United States to occupy the moon. He also chronicled the frustrations of the ABMA in the satellite debacle. Medaris and von Braun elaborated, the former praising Soviet wisdom in placing missilery in the artillery instead of the air force, and promising LBJ a satellite and a giant space booster if the army were given the go-ahead. Such a rocket would not be an ICBM, explained von Braun, but the key to military control of outer space. Still, the best means of conquering space, he thought, was to centralize the effort in a national space agency. It would not handle military missions, but rather long-range space development including manned spaceflight and a space station.[51]

News of the Gaither Report leaked in December. Johnson requested its release, Eisenhower refused, and suspicions grew of a "cover up."[52] The hearings dragged on, as the navy testified, then USAF generals LeMay and Schriever recounted the history of the ICBM. When the committee adjourned for the holidays, Johnson again punctuated the proceedings: "I think that all of us remember the day after Pearl Harbor. There were no internationalists and no isolationists; no Republicans and no Democrats. . . ."[53]

January brought almost daily sessions. Contractors appeared from Lockheed, RCA, Aerojet General, General Dynamics, and more, defending their performance and subtly plumping for a technology race with the Soviets. Key witnesses had a second time in the dock. Schriever returned to explain the importance of WS-117L in public session and to praise the USAF contractor system over the army's arsenal system of R & D. He also opposed the idea of a single space agency. Medaris agreed with his USAF rival, for a change—none of the services wanted to lose its share of the action to a single space agency.[54]

By mid-month the hearings finally ceased to make news. On January 23 they came to a close. Johnson found the Soviets to be ahead in missile development and numbers of submarines, and closing the gap in aircraft, more efficient in R & D, leading in space, and turning out scientists and engineers at a faster pace. He released to the Senate, the White House, and the press a list of seventeen recommendations including the strengthening of SAC, acceleration of missile production and antimissile missile R & D, a program to build a rocket engine of one million pounds thrust, manned missile systems, and reorganization of the DoD. For the United States had lost not just a race to manufacture a weapon but the battle of organized brainpower.

We have reached a stage of history where defense involves the total effort of a nation. . . . There can be no adequate defense for the U.S. except in a reservoir of trained and educated minds. . . . The immediate objective is to defend ourselves. But the equally important objective is to reach the hearts and minds of men everywhere so that the day will come when the ballistic missile will be merely a rusty relic in the museums of mankind. . . .[55]

"What were the terrible 1960s and where did they come from?" asked longshoreman-philosopher Eric Hoffer twenty years later. "To begin with, the 1960s did not start in 1960. They started in 1957. . . . The Russians placed a medicine-ball sized satellite in orbit. . . . We reacted hysterically."[56] Indeed, in a matter of a few months the rhetoric, the symbology of American politics had left Eisenhower behind. Teller and Gavin, Lyndon Johnson or Henry Luce of *Life* magazine—their words no longer sounded like those of some future decade. Rather Eisenhower's suddenly sounded like those of a past.

CHAPTER 7

The Birth of NASA

Whatever his insistence on restricted federal spending, Eisenhower could not refuse to respond to the Sputniks. It might not be true that American science was slipping. (It was surely not true, as retired President Truman claimed, that the Russians led in "this satellite proposition" because of the "character assassinations of Oppenheimer and others.")[1] It might not be true that American education was inferior, or that high-school physics had anything to do with Vanguard's flop. It might not be true that the U.S. military posture was inferior, that the Pentagon spread money around in wasteful rivalry, or conversely did not spread enough money around, as in the single satellite program. Indeed, all the charges made in the wake of Sputnik may have been false, contradictory, or beside the point. Nevertheless, the new symbolic value of space, science, and education demanded action.

For the charges did spring from an apt intuition. A new age was dawning, in which organized brainpower for military and civilian science and technology was the dearest national asset. Eisenhower, however, rejected the demands of generals and congressional "hawks" for a crash buildup and opted instead for sufficiency. But "sufficiency" implied mutual deterrence, and that only meant that the Cold War would be expanded beyond nuclear weapons and espionage into a competition of entire systems, each claiming to be better at inventing the future. Hence Sputnik posed an insoluble dilemma for Eisenhower's United States: either it must race headlong for strategic superiority, compromising fiscal integrity and militarizing much of the private sector, or it must accept strategic parity, in which case *all* aspects of national endeavor, including conventional weaponry, economic growth, "social justice," and the hearts and minds of Third World peoples became yardsticks of Cold War competition. Either way Sputnik invited another American lurch toward technocracy. The Eisenhower response that addressed the Sputnik challenge head-on, but that also expressed his ambivalence to the new age, was the National Aeronautics and Space Administration (NASA), another federal agency devoted to the conduct of a specific technological revolution.

The 1958 State of the Union message was an echo of Khrushchev's 1956 foreign policy speech, and spelled out the new and subtle challenges to an audience still obsessed with the satellite problem (when do *we* get one?). "Honest men differ," Eisenhower began, "in their appraisal of America's material and intellectual strength. . . ." But, Sputniks notwithstanding, the American people "could make no more tragic mistake than merely to concentrate on military strength." Hence the paradox: the Soviet rockets actually blunted, rather than sharpened, the military component of the Cold War. Communist imperialism was still the threat, said Ike.

But what makes the Soviet threat unique in history is its all-inclusiveness. Every human activity is pressed into service as a weapon of expansion. Trade, economic development, military power, arts, science, education, the whole world of ideas—all are harnessed to this same chariot of expansion.
The Soviets are, in short, waging total cold war.[2]

American progress in strategic technology was extremely rapid, the President insisted, especially the navy's Polaris. But Communist regimes, frustrated in attempts to expand by force, were concentrating as well on an economic offensive, especially in developing countries, that could defeat the free world regardless of its military strength. Eisenhower confessed his failure to anticipate the psychological impact of the first satellite and warned against a repetition of this failure in the economic field. Hence aid, trade, and mutual security efforts were even more important than strategic arms, and the United States' major Cold War asset was its economic health, sustained by "tremendous potential resources" in education, science, research, and, not least, "the ideas and principles by which we live."[3]
So Eisenhower issued no call to arms. Rather he recognized the changed nature of the Cold War and the new themes and symbolism of the Space Age (all of which would find sharper and unrestrained expression in the inaugural address of the next President). But Ike still hoped to meet the demands of total Cold War with limited government. He called in his speech for (1) defense reorganization for unity in strategic planning and R & D; (2) acceleration of R & D; (3), (4), and (5) mutual aid, trade, and scientific cooperation with allies; (6) investment of a billion dollars over four years (a fivefold increase) in teaching and scholarships in fields vital to national security, and a doubling of research funds for the NSF; (7) supplemental appropriations for defense of $1.3 billion and another $4 billion for missiles, science, and R & D in FY 1959. But these increases would come from expected revenues and not unbalance the budget.
In the eyes of the President this was a decisive but prudent response, sufficient to show the world that "the future belongs, not to the concept

of the regimented, atheistic state, but to the people. . . ."[4] The first post-Sputnik budget, "adhering to those principles of governmental and fiscal soundness that have always guided this administration," amounted to a rise of only 1.5 percent over the previous year. Defense spending was still lower than in FY 1954. In his budget message, Eisenhower felt obliged to justify even this small increase by the need "to keep pace with the rapid strides in science and technology."[5] The dual nature of the administration's domestic response to Sputnik revealed itself in its four main initiatives: science and R & D, federal aid to education, defense reorganization, and the space program. In each case, the proposed changes were explicitly designed to be temporary in duration, limited in scope, or self-mitigating in execution: a nod, but not a bow, in the direction of technocracy. Let us see how this was so.

The federal role in R & D, as has been seen, was a headache dating back to World War II. Its complexities and contradictions were such that almost all federal funding of research fell, *faute de mieux*, to the AEC and the military. Throughout the 1950s, however, professors and administrators calling for direct government aid grew louder, more numerous, and less sensitive to dangers of politicization. The federal scientific community, such as it was prior to Sputnik, backed its colleagues in academe. Eisenhower declared in 1954 that the NSF should henceforth be responsible for all federally funded basic research, while other agencies stuck to applied research related to their missions.[6] But the effect of this executive order was to reduce DoD and AEC support for pure science, while the NSF lacked the funds to take up the slack! Alan Waterman protested, styling his appeal to the White House as a program for "Maintenance of Technological Superiority." In July 1957, before Sputnik, I. I. Rabi and the ODM Science Committee reported to the White House that "the welfare of the U.S., incomparably more than at any other time in its history, is dependent on new scientific knowledge for the welfare of its people, for the advancement of its economy, and for its military strength. . . . Research is a requisite for survival." Rabi pleaded for military and AEC support of basic research, since the military itself now pushed against the frontiers of knowledge. To be sure, government could encourage private investment in R & D, perhaps through tax policy, but the time had passed when national needs could be met from private sources.

The U.S. has reached a "point of no return" in Federally supported research. Our American society, our standards of health and living, our modern defense, all require large scale research. . . . [We] cannot take the risk of falling behind in our military technology which would almost certainly occur if the DoD depended on other agencies to plan and sponsor research. . . . *There is a need for a strong and wise protagonist of basic research in the DoD in the interest of maintaining our military superiority.*[7]

Here was a remarkable reversal! After 1945, scientists advised Truman that even military-related research ought to be directed and funded by a *civilian* agency; in 1957, scientists advised Eisenhower that even civilian basic research ought to be sponsored by the *military*! In August a classified Cabinet paper generated by the NSF and the BoB seconded the motion. After Sputnik the ODM scientists had little difficulty persuading the President to appoint a Presidential Assistant for Science and Technology and to release far greater sums for basic research through both the NSF and the DoD: $55 million for NSF grants (up from $38 million in FY 1958) and $53 million for science education (up from $17 million). Compared to the banquet of the 1960s, these sums were only hors d'oeuvres, but they quickened appetites in an age when a mass spectrometer costing $60,000 was rudimentary equipment, a serious chemistry lab went for $750,000, and a cyclotron or radio telescope many millions. Nor did Ike give a blank check to the military; he had enough trouble trying to rein them in on *applied* research. But these first increases proved to be a lever for many educational and research groups with "national" goals to pursue. In 1958 the vice president convened a panel on federal support for *social* science on the premise of countering Soviet psychological warfare and even drug-induced behavioral control, while in and out of government the proponents of job training, social welfare, mental health programs, and so on set new goals for the "national agenda" and comprised a vast academic/bureaucratic lobby demanding federal financing of the quest for new knowledge.

The great leap into federal support of local education, the National Defense Education Act of 1958, was another paradigm designed as a stopgap. And again Sputnik acted as catalyst in a volatile mixture that had bubbled up since World War II, when the GI Bill legitimized federal aid to education. Various sorts of reformers cashed in on the Cold War alarm to sell the notion that government money was a panacea for a variety of deficiencies.

By the late 1940s the reigning philosophy of American schools, John Dewey's "Progressive Education," came under attack. Built on a "new humanism" that stressed "life adjustment" rather than "the three Rs," Progressive Education encouraged two pernicious mentalities, according to later critics: "The almost frightening belief in education as a sovereign remedy for all our social problems" (James Killian) and "The naive egalitarianism which urged in the name of democracy the same amount and kind of education for all individuals. . . ." (Education Policy Commission, 1956).[8] The 1949 bestseller *And Madly Teach* excoriated an educational philosophy that discriminated against brighter students and enlarged the areas over which "the authority of the social whole is supreme." In this view, progressive education taught relativism and egalitarianism, thus undermining the moral confidence of young people and rewarding "grey conformity." But as the 1950s advanced, social "progressives" insisted that public education was not equal enough, given discrimination against children from poorer school districts and racial minorities, while Cold War pragmatists stressed that since education was the United States' first line of defense, excellence should be set apart and cultivated. Opposite emphases, but the same solution: more federal direction and subsidy. Admiral Hyman Rickover frankly urged Americans to imitate Russian education: the Cold War, he believed, was a race between "opposite systems of management," not ideologies. Von Braun denounced "life adjustment" curricula, considered egalitarian education a contradiction in terms, and ridiculed the notion that an intellectual elite was incompatible with democracy.[9] After Sputnik, these many threads intertwined as social liberals and Cold Warriors found common ground.

Eisenhower himself sponsored brick-and-mortar bills from 1955 to 1957 to help states cope with the baby boom, but attempts to channel federal dollars into curricula, teaching, and equipment repeatedly failed. Confusion among the reformers, resistance on principle to governmental meddling in the classroom, and thorny issues raised by parochial schools and desegregation all contributed to deadlock. Some Southerners advocated federal support but feared forced integration, while Catholics were loath to pay for programs from which their schools would be summarily excluded. But after Sputnik educational lobbies and their bureaucratic allies unabashedly exploited the panic and denounced U.S. schools as second rate. A National Education Association lobbyist admitted that "the [education] bill's best hope is that the Russians will shoot off something else,"[10] and the three Rs of educational legislation came to be known as "Race, Rome, and Russians."

The conflict of views was clearest, perhaps, in the pronouncements of the current and former presidents of Harvard University. Nathan Pusey said bluntly that Sputnik required a vast increase in the share of the national product devoted to education. But former prexy James Conant cautioned Eisenhower against crash programs that could damage schools, confuse school boards, and undermine confidence in what was generally an outstanding school system. "Those now in college will before long be living in the age of intercontinental ballistic missiles," said Conant. "What will be needed then is not more engineers and scientists, but a people who will not panic and political leaders of wisdom, courage, and devotion . . . not more Einsteins, but more Washingtons and Madisons."[11]

Eisenhower sided with Conant and, working closely with Killian and Health, Education and Welfare (HEW) official Eliot Richardson, designed a bill that served his rearguard view against the pretensions of technocracy. He granted the need for more scientists and engineers, but resisted the notions that this need was permanent and that technology alone could solve military and social problems. His program for aid to students in science, engineering, and foreign languages was meant explicitly to be temporary and not to imply control of local education by the bureaucracy.

"The federal role," Eisenhower insisted, "is to assist—not to control or supplant—[local] efforts." The program was to run for seven years only.[12]

Richardson joined with congressional leaders, especially Alabama Democrats Carl Elliot and Senator Lister Hill, to steer the bill "between the Scylla of race and the Charybdis of religion."[13] In the end, twenty-three Southerners and twenty-four Republicans who had previously opposed education bills shifted to support this carefully worded National Defense Education Act (NDEA), a bellwether bill of the young Space Age. For despite presidential warnings, the act still pointed in the direction more liberal Congresses would take. A Democratic rider on the bill earmarked $60 million in vocational grants for students *not* going to college. After all, if some youngsters were to be privileged on account of their scientific bent, did not equity demand that those less gifted or otherwise inclined also receive help? This may have seemed fair, but once federal responsibility for private opportunity was established, and the principle of equity applied, there was no stopping point at which government could resist claims upon the public purse. Each extension or increase of federal aid to one or another collectivity, defined by specialty, financial station, region, race, sex, or whatever, proportionally increased federal power over the recipient institutions. This was Eisenhower's premonition, hence his NDEA, another vanguard action forced by the Cold War, was drafted as a rearguard attempt to contain the domestic drift toward centralization.

The third Eisenhower initiative was reorganization of the DoD. Ever since the 1947 legislation was whittled down to win naval and congressional sufferance, civilian officials hungered for further reform. Sputnik and the Johnson hearings provided the opportunity. Even the testimony of disgruntled generals, admirals, and industrial contractors, each touting his own efforts and complaining of everyone else's, only strengthened Eisenhower's hand in his effort to push through DoD reorganization. The State of the Union message made it a major goal of 1958, the new defense secretary endorsed it, and the administration named blue-ribbon panels (including Rockefeller and the three most recent chairmen of the JCS) to design it.[14]

The bill sent to Congress in April invoked the technological revolution to explain the need for change. Thermonuclear weapons, missiles, and atomic submarines increased the destructiveness of war, reduced warning time, eliminated breathing space after the onset of hostilities, and placed a premium on efficient R & D. Hence Eisenhower asked Congress to unify operational commands and place them directly under the Secretary of Defense, enhance the power of the Secretary and enlarge his staff, allocate all military funds directly to the Secretary and not to the services, and centralize all R & D functions under a Director of Defense Research and Engineering (DDR & E). In addition, the JCS must cease to be a committee of rivals, but must act as a single corporate body with an integrated staff capable of directing all the armed forces of the United States in peace and in war.[15]

The bill drew stubborn resistance from quarters attached sentimentally or selfishly to the autonomous services: veterans' organizations, service advocates in Congress, contractors, and the navy. But a White House public relations campaign recruited distinguished advocates in the public and private sectors and dispelled the specter of "military dictatorships" that opponents claimed to see in centralization. "There will be," said the President, "no single chief of staff, no Prussian General Staff, no czar, no forty billion dollar blank check, no swallowing up of the traditional services, no undermining of the constitutional powers of Congress." Rather, the reorganization would meet the needs of the nation by streamlining operations and R & D in an expensive, technically dynamic age.[16]

With minor amendments the administration bill became law on August 6. At first glance it seems another innovation forced on Eisenhower by the outcry over Sputnik. In fact, it was as much another example of Ike's campaign to help civilian leadership hold the line on R & D and keep technology policy subservient to national strategy and economic prudence. Its significance was evident in the fracas over the defense budget for FY 1960, as the Cabinet squarely faced the problem of adjusting American strategy to the coming age of mutual deterrence. Was massive retaliation still valid, now that the USSR had an ICBM? Even Foster Dulles had his doubts: Europeans worried whether the United States would use its nuclear arsenal in case of Soviet conventional attack. Perhaps tactical nuclear weapons might suffice. But Secretary of Defense McElroy feared that tactical nuclear warfare would escalate. Generals Nathan Twining and Taylor and Admiral Burke all liked the flexibility offered by tactical weapons but observed that an inventory of small yield warheads did not yet exist. Deputy Secretary Quarles stuck with massive retaliation: the nuclear age was inevitably one of deterrence, not war-fighting. But what if deterrence failed? asked Navy Secretary Thomas Gates. In that case, said Twining, nuclear attacks would be directed at military targets, not population centers. But that in turn required more numerous and sophisticated delivery systems than a simple "city-busting" strategy.[17]

Here were the leaden questions of the missile age. The United States had to maintain a sufficient and technically current deterrent. But since the Soviets would, too, tactical weapons became important for the defense of Europe. Since crossing the nuclear threshold risked escalation, conventional forces must be beefed up to avoid that option. And if Khrushchev intended to foment brushfire wars in the decolonizing world, then counterinsurgency forces must be purchased as well. Thus there was a great temptation to increase one's options with an across-the-board buildup of military force. But buying the maximum of flexibility, like extending aid to more and more social groups, was ultimately ruinous.

There were two ways of budgeting for defense. One was to ask each service to declare its needs, urging restraint of course. When this was done, the total would come to something like $100 billion—everyone asked for everything. The second way was to impose a ceiling, $44 billion in FY 1960, allocate a share to each service, and let each set its own priorities. The latter method, initiated by Truman and revived by Ike after Korea, was arbitrary and annually assaulted. As soon as the budget was released the aggrieved services rallied every congressman, contractor, and columnist in their camp to protest the budget and strategy that produced it.[18] But with the Secretary of Defense and JCS enjoying centralized direction and control of all funds, the ceiling system could be reinforced and made less arbitrary. The Pentagon reform was a tool, therefore, of efficiency and economy, which helped the United States to adjust to the missile age and helped Eisenhower rein in the services.

In all these areas—science, education, and defense—the President hoped to restrain the growth of government even as he expanded federal activity into domestic arenas relevant to total Cold War. The most revolutionary issue of all, however, was space exploration. It had so many unique elements, organizational anomalies, and conflicting political implications that the administration's best efforts could not untie its tangles, but only reduce them to a few, tight knots. It was also the issue most closely connected with the new symbolism of politics and technology, and potentially the most expensive.

By the mid-1950s the venerable NACA was slumping. It was the best equipped aeronautical research organization in the world, but institutional conservatism and financial strictures rendered its very future dubious. Jet aircraft were becoming routine, the future lay in spaceflight, but since 1947 NACA's role in rocket research had been circumscribed by the military. To be sure, the NACA participated in the "X-series" of rocket planes that were carried to high altitudes above the California desert then shot upward on their own rockets to record heights. The Bell X-1 first broke the sound barrier in this way in 1947 and soared fourteen miles high. The planned X-15 would eventually reach fifty miles above the earth, the fringes of space. Nevertheless, as late as 1955 only a small portion of the committee's budget went for space-related research, and Chairman Jerome Hunsaker gladly relinquished "the Buck Rogers jobs" to the USAF and JPL. According to JPL luminary Theodor von Kármán, the NACA was "skeptical, conservative, and reticent."[19]

From its peak in World War II, the NACA budget shrank steadily until, in 1954, it received only half of what it asked for. This institutional weakness was due in part to Eisenhower's cost cutting, but NACA also lacked powerful allies. It channeled only 2 percent of its funds to private contractors, while the military services were pleased to assume tasks in which the committee showed no interest. NACA was an adjunct, not a rival, of the Pentagon and industry. As such it performed well, but if the USAF or army or NRL came to monopolize the next great stage of flight technology, NACA might lose its lease on life. This prospect inspired young NACA engineers, mostly from Edwards AFB and Langley Research Center, to organize a "frontier faction" and agitate for future-oriented programs. Meanwhile, the traditionalist Hunsaker was replaced by Jimmy Doolittle, who not only embraced "Buck Rogers" but commanded respect in Congress, industry, and the military. By October 1957, one-fifth of all NACA work was space-related.[20]

After Sputnik, the timid NACA leaders still held back, however, until internal protest (punctuated by the "young Turks dinner" of December 18) and talk of new space agencies forced them to choose between pushing NACA forward or floundering in the backwash of the Sputnik tide. By mid-January, NACA director Hugh L. Dryden, Doolittle, and chief counsel Paul Dembling had in hand a coherent space program based on NACA in cooperation with the DoD, NSF, NAS, universities, and industry. David challenged the Goliaths for the limitless and potentially richest fiefdom of all—outer space.[21]

In a liberal society government grows by accretion. A foreign threat or new political symbolism can bestow prerogatives on the state that it must exercise if it is to maintain its international status and domestic legitimacy. But once these are acknowledged, struggle ensues within government for control of the new tasks and the budgets and power they confer. Sometimes existing agencies win out, sometimes new ones—such as the AEC—are created. The victorious organization, finding its place in a pluralistic system, can then forge alliances inside and outside of government and sustain itself into the far future, outlasting even the threat or symbolism that first gave it life. Space was likely to be just such a "big ticket" enterprise, and Eisenhower accordingly pursued an apparatus for space R & D that was subservient to the White House, isolated from its most powerful claimants, but still adequate to discharge legitimate space missions for science and defense.

The management of public tasks, therefore, is both a function of policy and an influence on it. Who does something, and how, go far to determine what gets done. Was space technology a military problem rightfully devolving on the DoD? If so, how could space science receive the attention it deserved? If space was awarded to a civilian agency, how would legitimate military functions be performed? Was space inevitably tied to Cold War competition, or could it spawn global cooperation? If competition prevailed, the space program must be national and secret; if cooperation, then international and open. The same questions tormented Truman and the Congress at the time of the Atomic Energy Act: civil or military control, secret or open; stress on science or weaponry; in-house research by government or contracted research by universities and industry; control of patents by the state or encouragement of private development; international cooperation, regulation, or laissez-faire competition? Atomic energy policy retained these tensions. The main business of the civilian AEC was still warheads for the military; the main research was done at Los Alamos and Livermore, not Westinghouse; the International Atomic Energy Agency and the Atoms for Peace program fell far short of their promise. Could space be handled differently? Or were the nuclear arrangements the best that could be had? The dawn of the Space Age did differ from that of atomic energy in one happy way: the first satellites were peaceful contributions to the IGY, not weapons of war. Perhaps U.S. policy could help to prevent the extension of the Cold War into the serenity of space and head off a literally limitless technology race that would inevitably make the Space Age an age of technocracy for the United States and all the world.

Such reasoning made elevation of the innocuous NACA an attractive answer to the question of what to do about outer space. But competition was strong. As early as December 1957, Medaris and von Braun submitted a fifteen-year space program based on development of heavy boosters by the ABMA. It forecast lunar reconnaissance and two-man satellites by 1962, manned lunar circumnavigation by 1963, and a fifty-man moon base by 1971.[22] The army's ABMA/JPL team gave it the best in-house capacity for the space job—and space, after all, was just "high ground," the taking of which was the army's job.

The USAF meanwhile anxiously monitored the army-navy race to launch the first U.S. satellite and hoped to persuade Washington that space was its rightful domain.[23] USAF public relations specialists promptly invented the term "aerospace" to suggest that air and space were a continuum. The X-15 program meant that the USAF was already working toward manned spaceflight; it possessed the biggest boosters then under development, the Titan and Atlas; and it would soon test the Agena spacecraft for WS-117L. But the navy was also in the game. The NRL inaugurated American satellite research in 1945; it managed the official U.S. satellite project, Vanguard; it, too, had missions in space: satellites for navigation, weather, and fleet communications. And when science fiction wrote of space travel, it always spoke of voyages in *ships*.

The Johnson hearings gave voice to all sides, and each service had its tribunes on the Hill, but senators, too, were perplexed about what to do with space. Johnson's seventeen recommendations only mentioned improved control of space-related work "within the DoD or through the establishment of an independent agency." Backed by special pleaders, "each political participant sought to convince the administration of its own special capability in space by calling loudly for recognition of its skills and resources. It was a veritable 'Anvil Chorus.' "[24] Candidates

"Whew! At First I Thought It Was Sent Up by One of the Other Services"

From *Herblock's Special for Today* (Simon and Schuster, 1958). Originally appeared in the *Washington Post*, November 21, 1957.

included the three services, an independent, unified DoD office, the AEC, the NACA, the NAS and NSF in cooperation with any of the above, a brand-new space agency, or a Cabinet-level Department of Science. The last was an updating of the Kilgore notion, which smacked of socialism to some but was a pet project of Senator Mike Mansfield (D., Mont.).[25]

The military claim to space, on the basis of mission, priority, and capability, was too strong to ignore, while satellite programs currently underway needed at least a temporary home. So in mid-January 1958

EXPLORER 1

Army-navy competition for the honor of launching the first U.S. satellite built to an excruciating climax throughout January 1958. A political cartoonist caught the mood by depicting a Soviet rocket whizzing above a military base, with the brass below gasping in relief, "Whew! For a minute I thought it was launched by one of the other services!"[26] The presumptive reward of victory in the race was an inside track to future space missions. While the ABMA hurriedly prepared a Jupiter-C for launch on the 29th, the navy combed another Vanguard for bugs before its next chance on the 18th. But the NRL pushed the date back to the 23rd, then to the 26th because of rain and technical problems. Finally, a second-stage engine was deemed faulty, and Vanguard missed its chance entirely.

The ABMA, possessed of a thoroughly tested booster and experience at the Cape, geared up in very little time. General Medaris insisted on scanty publicity; he wanted no repetition of the Vanguard debacle in case the worst should happen. High winds in the jet stream stopped the countdown for twenty-four, then forty-eight hours. January 31 would be the army's last hope before Vanguard got another crack. Medaris resumed the count-down. At 10:48 P.M. the Jupiter ignited. The first U.S. spacecraft, like Sputnik four months before, rose like a Roman candle in the dark, lighting up the swamps of the Banana River instead of Asian steppes, free from the humbling competition of God's own sunshine. The guidance system functioned; the upper stages fired. Now there was nothing to do but wait, for perhaps an hour or more, for news from the tracking stations. Medaris fought with the press and his own nerves, Army Secretary Brucker complained from Washington of shortages of coffee and cigarettes. Like Korolev and his comrades, they all acted like expectant fathers. Finally someone shoved a slip of paper into the general's hand: "Goldstone has the bird."[27] *Explorer 1* was in orbit.

Hagerty phoned Eisenhower, who was standing by at the Augusta National Golf Club. "That's wonderful," said Ike. "I surely feel a lot better now." The country felt better, too. But Ike's next thought was characteristic: "Let's not make too great a hullabaloo over this."[28]

Explorer 1 weighed in at 10½ pounds and established a lasting American superiority in miniaturized electronics. The two micrometeoroid detectors, a Geiger counter, and telemetry returned more, and more useful, data than the giant Soviet Sputniks—and discovered the Van Allen radiation belts girdling the earth.

Jupiter and Vanguard each failed in February attempts, but the navy evened the score when the diminutive *Vanguard 1* reached orbit on March 17. Its Geiger counter sent back more data on the Van Allen belts, and its proton-precession magnetometer established beyond the doubt the geologists' suspicions that the earth is pear-shaped. If the Sputniks argued persuasively for the political/military importance of the space technological revolution, the American "moons" proved it to be a scientific leap of unparalleled promise.

Secretary McElroy created the Advanced Research Projects Agency (ARPA) within the DoD, headed by GE executive Roy Johnson and physicist Herbert York. ARPA would run U.S. space programs on an interim basis by authority of the Secretary.

After the welcome relief provided by *Explorer 1*, Killian appointed a PSAC panel to study the space problem, while the bustle and rhetoric on Capitol Hill gave the impression that the administration was indecisive. But Congress, too, had to endure a period of education before reaching conclusions. Senator Clinton Anderson's plea for help from the president of DuPont is indicative. A patron of atomic energy from New Mexico, he was introducing a bill to give the space mission to the AEC:

I had a professor in math—calculus, I think—who said I could solve most problems in math if I could state them correctly. If I could state my current problem to you, I would probably have it half-solved. My trouble is that I can't.

I went to see LBJ and pointed out that this problem was likely to be tossed into the lap of Congress. . . . I want the military to have the fullest opportunity to push satellites into outer space and to explore outer space for every military reason which now occurs to them.

But if that is the only thing we do, then the Russians, who are very adept at propaganda, will say that the President's program for peaceful uses of outer space is hypocrisy. . . .

I have not tried to foreclose the possibility that the conquest of outer space may be left to a completely separate civilian agency. . . . It may be NACA or NSF should take charge. In my bill I assigned it to the AEC. . . .

Now you can see what considerations of this kind do to an individual whose business life has been devoted to running a little insurance company in a small Western city. . . .[29]

In those same hectic days after *Explorer 1* the Congress organized itself for the Space Age. In so doing, it paid tribute to its extraordinary symbolism. There had not been a new standing committee in the House since 1946, yet the reconvened Congress moved quickly to create committees for space. An aide to Overton Brooks (D., La.) recalled:

We were staying at the *George Cinq* [Paris] and we came out of the hotel and bought an American language newspaper . . . and here on the front page is the headline—Russia had orbited a satellite. Well, Brooks about jumped out of his skin. He could talk of nothing else. As a matter of fact, we came home two days early. He said, "The first thing I'm going to do when Congress goes back into session is to drop in a bill to form a special committee because we have to catch up with them or surpass them."

Speaker Sam Rayburn agreed, and the committee formed under John W. McCormack (D., Mass.) in early March. But as usual Johnson was first out of the gate. The Senate named its Special Committee on Science and

Astronautics on February 6, with LBJ as chairman and a membership composed of other committee chairmen.[30]

A new congressional committee is no light undertaking. It invariably sparks jealous jurisdictional struggles. The prestigious membership of the space committees was also a testimony to the importance vested in space. Oversight committees for a federal activity guarantee visibility and support, since committees do not generally want to see their federal charges lose budgetary power and importance. Hence the space program, wherever it came to reside, was assured in advance of a strong political alliance. In addition, the special committees gave impetus to a civilian solution, for purely military space activities would remain under the aegis of the armed services committees.

What might Congress do to influence space policy? Here again Anderson's musings give a clue to congressional thinking. His long experience on the Joint Committee on Atomic Energy, Anderson lectured his colleagues, had taught him that "Committee members cannot compete with scientists on their own ground. So we stay in our field—the objective." What ought to be the objectives of a U.S. space program— propaganda, military power, science? "We should not," he continued, "encourage an all-out effort in all three fields. Let one man go and let the two others work as fast as they can." His own pet project was a nuclear rocket, but if immediate propaganda results were deemed the first priority, then the Congress should "turn von Braun loose" on his million-pound-thrust chemical rocket.[31] That, in turn, would suggest a civilian space agency independent of the AEC and the military. Such were the interconnections of politics, organization, and technology.

The PSAC, reporting in just two busy weeks, identified two distinct objectives in space: exploration and control. The PSAC discounted most of the Buck Rogers notions, but granted the military importance of surveillance, meteorology, and communications. Such uses, however, raised questions of international law such as where outer space began, how to allocate radio frequencies, the legality of overflight, and the regulation of space vehicles, since within the ten years orbital space might become a "celestial junkyard." "The problems involved are tremendous and the programs which must be undertaken will be lengthy and costly." All this suggested to the panel the wisdom of a civilian agency. But to be effective, it must have access to the necessary brainpower, which meant freedom from civil service pay scales and restrictions, freedom to draw on all talent inside and outside of government, and broad contractual powers in the private sector. The various civilian options all had their drawbacks. A new space agency would take time to organize and require extensive legislation and facilities. The AEC could be easily expanded, but at the expense of interference with its current function. NACA had the experience in flight technology, but its governing committee was cumbersome and it had only partial relief from civil service and contractual rules. The new ARPA could take on the whole job without retooling, but that would seem to make spaceflight solely a military enterprise. The United States had lost the prestige of being first; at least it should project an open, peaceful program in contrast to Soviet secrecy. The preliminary PSAC report, therefore, tended to favor the eventual creation of a new Space Exploration Agency by legislation.[32]

Even as the PSAC staff drafted these preliminary thoughts, the heavyweights were moving to a decision. Vice-chairman of the PSAC, James Fisk, and retired General James McCormack, a vice-president of MIT, favored the NACA. The Bureau of the Budget, always hesitant to create new agencies, also favored expansion of the NACA. Gradually, a consensus emerged. McElroy and Quarles, impressed by the history of NACA/DoD cooperaton, came on board. So did Rockefeller, who stressed the importance of a peaceful space program in world opinion, Don K. Price, an advocate for civilian science, and Milton Eisenhower, President of Johns Hopkins University and Ike's brother. The PSAC then concluded that, apart from reconnaissance satellites, the major goals of spaceflight in the near term were scientific and political. "The psychological impact of the Russian satellites suggests that the U.S. cannot afford to have a dangerous rival outdo it in a field which has so firmly caught, and is likely to continue to hold, the imagination of all mankind." An American space organization should leave military satellites in the Pentagon, but otherwise be lodged in an open, civilian agency. NACA was the preferred choice by dint of its experience, facilities, and, not least, "its long history of close and cordial cooperation with the military departments."[33]

As currently constituted, however, NACA was too small. The rocket and space engineers were all in the ABMA, NRL, USAF Ballistic Missile Division, JPL, and the aerospace firms. NACA's basic laws must be amended to tap these sources, to provide for a single director appointed by the President, to free it from civil service, to retain an in-house capacity but permit contracts with private industry, and to provide for coordination with the DoD.[34]

On March 5, Eisenhower approved a final memorandum ordering the BoB to draft a space bill based on NACA before Congress recessed for Easter. Three weeks later the draft was done and, as Senator Johnson sneered, "whizzed through the Pentagon on a motorcycle." Nevertheless, the BoB, ARPA, State, and even NACA's Doolittle had their chance to complain.[35] By and large, they rallied to the administration, but the proposed space agency was already stepping on toes and eliciting yelps that presaged the interagency skirmishes NASA would spark in years to come.

The PSAC moved to support the maturing space act through the release of its essay, "Introduction to Outer Space." It was, to Killian's delight, a best seller.[36] In it the PSAC explained the four reasons why space technology was important, urgent, and inevitable: (1) man's compelling urge to explore; (2) military security; (3) national prestige; and (4) science. It went on to instruct the public on why satellites "stay up," rocket thrust and staging, what satellites can do in orbit, and the potential for exploration of the moon and Mars. It noted the military value of reconnaissance satellites but denied the efficacy of such things as satellite bombs and moon bases. Finally, it offered a vague timetable for space exploration, beginning with satellites and moon fly-bys, leading "later" to manned flight and "still later" to manned landings on the moon. But the cost, noted PSAC, would not be small. Scientists and the general public must somehow decide if "the results possibly justify the cost" even though scientific research "has never been amenable to rigorous cost accounting in advance."[37]

The administration bill, the National Aeronautics and Space Act of 1958, entered Congress as S. 3609 in early April. The preamble argued that "the general welfare and security" required adequate provision for aeronautical and astronautical activities, and that they should be the responsibility of a civilian agency except where associated with weapons systems, military operations, and defense. The purposes of space activities were the expansion of human knowledge, improvement of aircraft and space vehicles, development of craft to carry instruments and living organisms through space, preservation of the United States as a leader in space science and applications, cooperation with other nations, and optimal utilization of American scientific and engineering resources. The bill established an independent office of government, the National Aeronautics and Space Agency, under a single director. Its Space Board would subsume the old NACA governing board and consist of seventeen members (nine from outside government).[38] The bill met PSAC guidelines and accounted for the various, sometimes conflicting, considerations of space policy. As such, it sketched a controversial structure that satisfied no one fully and placed a stamp of ambiguity on the enterprise that has never been erased. By splitting responsibility between the new NASA and DoD, the bill chartered two parallel space programs, one open, scientific, and devoted to research, the other closed and devoted to military applications. It was also a significant step toward state-directed mobilization of science and technology, but only to ensure that the United States remain *a* leader, not *the* leader in space. It did not commit the nation to an all-out race. It mentioned several goals for space R & D—science, prestige, and so on—but left open the priorities among them. Perhaps a fuzzy mandate was inevitable or even preferable in the unknown matter of spaceflight. But it ensured that the struggle over space policy began, not ended, with the space act.

Congress now had something to chew on. Indeed, the space act attracted more interest on the Hill than anything since atomic energy. While the space committees held hearings, Senator Johnson maneuvered behind the scenes and tidied up the messier provisions of the bill. Researcher Eilene Galloway ably seconded Senator Johnson and House Majority Leader McCormack in these months with penetrating memos on the issues, the most intractable being the division of responsibility between civilian and military agencies.[39] Johnson himself buttressed the Pentagon's claim to a share of space, but publically identified the United States—and himself—with the peaceful uses of space. "There are three kinds of records that can be made," wrote his staff. "(1) record of the U.S. as a leader in *international* space activity; (2) record of the Congress; (3) record of the Democrats since they control the Congress." LBJ was advised that he had received the most favorable publicity when speaking of the international aspects of outer space. Stressing this aspect in the fight over the space act would create an "opportunity for inspired leadership."[40]

Throughout April, congressional deliberations came to focus on these military-civilian and national-international problems. The administration downplayed military potential, yet the Congress learned from General Schriever and others that the military side of space technology, like pitching in baseball, was 75 to 90 percent of the game. The proposed bill was vague on the division of labor, while the language on the House side seemed to give NASA all responsibility for R & D. The USAF denounced this version and succeeded in rallying NACA, which had no desire to become a fourth armed service, to its position.[41]

After thousands of pages of testimony, the congressional melée resolved itself to one between the House Committee, which stressed civilian control against the presumptuous generals, and the Senate, which played up international cooperation but was anxious to protect the military space role. The House bill called for a liaison committee (modeled on the atomic energy act) to "feed" useful space technology to the Pentagon. The Senate drafted an article creating a National Aeronautics and Space Council composed of Cabinet officers and chaired by the vice president to plan space strategy. But Eisenhower believed such a mechanism endowed space with an unwarranted importance, while the House feared such a high-powered council would subordinate space policy to strategy and diplomacy and shut out the scientists. What institutional arrangement could prevent NASA from co-opting military functions, yet prevent the Pentagon from "swallowing" NASA?[42]

A Senate committee staff memo explained the military-civilian confusion by the fact that "some people are trying to divide things which cannot be divided. . . ." Scientists want to engage in scientific research. "The fact that one scientist wears a uniform while his co-worker wears a civilian suit does not mean that the uniformed scientist is an incipient Napoleon. . . ." Civilian control was a red herring—in a democracy all policy is guided by the elected representatives of the people. "The main reason why we must have a civilian agency," the memo suggested, "is because of the necessity of negotiating with other nations and the United Nations from some nonmilitary posture."[43] This truth sank in when Johnson and the House leaders sat down to draft a compromise version of the space act. McCormack confided to industrialist Victor Emanuel that "you know 60 percent of it is military, but I am sure the President . . . and the Department of State want to stress in language the civilian approach rather than the military approach for reasons I am sure you can guess." McCormack thought "he did a great deal when he put in the Bill that the [space] agency should cooperate with the military, instead of the military should cooperate with the agency." Management consultant Donald Wilkins admitted that it was "unanimously apparent to the knowledgeable members of the Space Committee, the Atomic Energy Joint Committee, the Foreign Affairs Committee, and the Leadership of both parties in the House that for the next decade it is extremely likely that the dominant agency of the U.S. Government [in space] will be the Department of Defense."[44]

How should such statements be interpreted? Eisenhower knew how vital spy satellites might be, but he purposely played down the general military importance of space in the near term. Johnson did the same. The House leadership was strongly opposed to military control of the space program. Both houses were preparing resolutions endorsing "space for peace" and "the benefit of all mankind." Yet even staunch civilians admitted under their breath the genetic dominance of the military in their new baby. There is no telling which of several explanations apply to any individual, but all the following have their place. First, there was confusion about what militarization of space entailed. Some had in mind ICBMs as well as spacecraft. Others thought of militarization in terms of "ray guns" and "orbital bombs," not passive satellites. Still others grasped that almost all space technology could be put to military as well as civilian use with no way of sorting it out. To ban the Pentagon from using space without an agreement with Moscow would amount to unilateral disarmament. Second, there was widespread concern, born of idealism and propaganda both, that the United States show the world an open space program. Third, perhaps most important, was the growing realization that separation of military and civilian activities was increasingly artificial in an age of scientific warfare and total Cold War. Even scientific programs, under a civilian agency, were tools of competition in so far as an image of technical dynamism was as important as actual weapons. The space program was a paramilitary operation in the Cold War, no matter who ran it. All aspects of national activity were becoming increasingly politicized, if not militarized.

The House passed its version of the space act on June 2; the Senate followed two weeks later. Among the novelties in the House bill was an upgrading of the proposed agency to an administration and its director to an administrator. This was "a mighty promotion in Washington

---

SPUTNIK III

On May 15, 1958, Korolev's big booster launched one and one-half tons into orbit. The payload included a geophysical laboratory but no animals. At a Soviet-Arab friendship meeting in the Kremlin, Khrushchev told his visitors that the United States would need "very many satellites the size of oranges in order to catch up with the Soviet Union." His country, it seemed, had outstripped the United States in science and technology.[45]

Although the Soviets still dominated the weightlifting category, the numerical score was even. Von Braun launched a second Explorer, the third American satellite, on the twenty-sixth of March.

---

bureaucratic terms."[46] Another was an article on patent policy borrowed from the atomic energy act. It gave the government sole rights to all inventions derived from NASA-sponsored research. The patent problem, as always, placed in jeopardy the incentives to American industry to help mobilize the nation's talent for the space effort and, ultimately, the principle of free enterprise. If NASA opted for the arsenal system of R & D, the army model recently vindicated by the ABMA satellites, a state monopoly of patents would pose no problem—but it would also make the government the senior partner in the performance as well as funding of R & D. If NASA opted for the contract system of R & D, the USAF model, a measure of private enterprise would remain—but the monopoly of patents would discourage private firms from wholehearted participation. Nor was private assignment of inventions financed by the state good capitalism. The Congress had to decide, therefore, whether the United States would tend toward an outright statist technocracy or a mixed contractor-state technocracy in which the private sector performed public chores.

On July 7 Eisenhower invited LBJ to the White House. The President disapproved of the Senate's space council idea, but Johnson would not sacrifice this assurance that space got the attention it deserved. Instead, he sought to satisfy Eisenhower by making the President himself the National Aeronautics and Space Council (NASC) chairman. Then he could do with it whatever he liked. "Yes, that might do it," said Ike.[47] The House-Senate conference then hammered out a common version. Both the Civil-Military Liaison Committee favored by the House and the space council favored by the Senate survived. The issue of patents, however, reached a deadlock when the House decided to place all NASA-derived inventions in the public domain for anyone's use. The makings of a horse trade emerged when both houses took steps to create permanent, standing space committees. There was talk of a joint committee such as that for atomic energy, but congressmen feared it would be dominated by the prestigious senators. And so, in Johnson's office, "where two larger than life paintings of him and Lady Bird dominated the room," Johnson surrendered on the joint committee (which he may not have wanted anyway) and McCormack yielded on patents. "That's the sign of a big man," said LBJ.[48] The patents section, longest in the act, conferred on the government all rights to inventions made in NASA programs, but gave the administrator the freedom to waive such rights at his discretion.

The conference bill passed both houses the next day and Eisenhower signed it two weeks later. On October 1, 1958, the NACA would disappear and reemerge as the National Aeronautics and Space Administration (NASA). And when the 1959 Congress reconvened, it would have two new standing committees, the Senate Committee on Aeronautical and Space Sciences and the House Committee on Science and Astronautics. They inherited the chores of trying to sort out, in conjunction with the administration, the unsortable issues of civil-military relations, cooperation versus competition with other nations, the appropriate spending levels for space R & D, and the role of the space program in determining the future relationship between the state and the creation of new knowledge in a capitalist democracy. In subsequent years veterans of PSAC and of the Congress both claimed the civilian space agency as their baby. Both played indispensable roles, as befit the American system. But why either was so eager to take credit for the space act is less clear. It was an extraordinary piece of legislation fashioned in very little time. But it sewed as many snarls as stitches in the fabric of American government.

In response to Sputnik and the national outcry that followed, Eisenhower took initiatives with which he was not wholly comfortable. He accelerated military R & D, approved unprecedented peacetime funding of civilian science, moved the federal government to fund and direct education, and created a new agency dedicated to state-financed and -directed R & D in a critical and "civilian" branch of technology. That he took these steps with misgiving rather than confidence is indicated by his prior attempt to remove the military from basic research, his watch over military spending, his reform measures to tighten control over military R & D, his insistence that the education act was not a precedent, and his decision to make space, as far as possible, a civilian mission under White House control. This is not to minimize the vanguard aspects of his initiatives. Still, Ike hoped to adjust to the apparent demands of the space and missile era, and of total Cold War, without giving over the government to a technocratic faith that he himself rejected. In any case, organization was only a third of the battle. If Eisenhower's delicate balance of vanguard activity checked by rearguard philosophy was to succeed, prudent management would have to be reinforced with unmistakable policy directives and stringent budgeting. Instead, Ike would learn how difficult it is to preserve one's equilibrium and sense of direction in the topsy-turvy canopy of outer space.

CHAPTER 8

A Space Strategy for the United States

Strategy is a form of economy, a function of scarcity: unlimited resources render strategy unnecessary. But according to Eisenhower, American resources were decidedly limited, not because the United States was poor but because it was rich through private enterprise. For the government to sequester too large a share of the national wealth meant to kill the goose that laid the golden eggs. Truman's government was too big, which was why Eisenhower relied more completely on high-technology nuclear deterrents. But high tech might no longer be a cheap option if, as Sputnik suggested, the United States must mobilize more and more to stay ahead of the Soviet technocracy. How could the United States escape this dilemma? What sort of strategy in space best served American national interest?

Space posed two of the overarching international problems of the twentieth century: how to contain expensive arms races despite bitter competition and distrust, and how to manage the use of nonterritorial regions like the sea, air, Antarctica, or outer space, within the system of sovereign, territorial states? The answers to both seemed to lie in treaties—for arms control and international law to fill the legal vacuum in outer space—and neither was really new. Missiles and military spacecraft merely extended the arms race dating from the atomic bomb, while legal questions raised by spaceflight merely extended the quarrel over verification of arms control, especially through "Open Skies." But space also presented some novelties, including the definition of where "air" ended and "space" began.

Scholars had anticipated the problems posed by satellites, and after Sputnik a spate of articles and books appeared on space law. Journalists and congressmen seized on such exercises, either through idealistic urge or the titillation attending questions such as "who owns the moon?" To the administration, abstract theorizing was sterile. For strategy must serve values, and practical steps, while promoting ideals, cannot be a function of them. That is, one cannot establish harmony and a united humanity simply by wishing them into existence, or eradicate armaments or greed simply by renouncing them oneself. In the tentative atmosphere after Sputnik, two earnest hopes conflicted in the West: hope that the United States might respond with even greater vigor to counter Soviet space power; hope that space could be made off-limits to weaponry altogether. Eisenhower had to allow for all possibilities by speaking of idealism and acting with realism. The dual space program and the space policy derived from it in the first years of the Space Age reflected this complexity. Hence U.S. space strategy aimed at the establishment of a legal regime in space that complemented the American propaganda line of openness and cooperation in space and held out hope of agreements to "put a lid on the arms race," and at the same time preserved American freedom to pursue such military missions in space as were needed to protect and perfect the nuclear deterrent. But the dual thrust of American space strategy also opened the United States to charges of hypocrisy from Moscow and Western critics, which only increased as hopeful rhetoric found little echo in deeds.

The RAND Corporation weighed in first with a study of the political implications of the Space Age. Despite the flights of fancy of some space law theorists, there was no "escape velocity" that took one beyond the political rivalries of this world. The Soviets had already made clear the uses they saw in space triumphs, that is, to support their claims that the USSR was the strongest power on earth, that the U.S. deterrent was obsolete, that smaller countries would do well to expel American bases. Meanwhile, Khrushchev made his usual offers of bilateral accords that would isolate the United States and make its allies feel abandoned. While Sputnik was not likely to smash NATO, "it would be folly to deny that the allies' estimates of the balance of power in the future are based in part on the expectation that Western science and technology will maintain a decisive lead over the Soviet bloc." Hence prestige and perceptions were as important as actual military force. The security of the United States might depend solely on the latter, but the health of the free-world alliance and the liberal values that cemented it depended on continued belief in American dynamism. Space strategy could not dispense with prestige no matter how silly a space race might seem. "From now on, the U.S. should recognize the need for restoring credibility in U.S. superiority, stress our peaceful intentions and their aggressive ones, and *disclose* and *publicize* U.S. outer space activities according, first and foremost, to the effect on the U.S. international position."[1]

A similar analysis emerged from the office of the Secretary of Defense. It insisted that national policy provide for the imminent use of satellites for reconnaissance, tracking, early warning, satellite interception, antimissile systems, communications, navigation, weather forecasting and perhaps control, as well as civilian uses. It stressed the importance of a positive

American position on proposals for space law at the UN and prior consultation with allies lest they make embarrassing proposals out of ignorance of U.S. requirements. But, while freedom of space should be upheld in principle, the right to interdict hostile spacecraft must also be reserved. "There is a real danger that we may harm ourselves by too early commitments before the full implications of space potentials are known. *Our policy and national interest* should be permitted to develop first: the law and commitments should follow, and be consonant with the former."[2]

Diplomatic thinking tended naturally to emphasize an American commitment to space cooperation and UN involvement in space law. But perceived commitment was more important than results. State Department counsel Loftus Becker testified that "any sound body of law is based on a system of facts that we just don't know at the present time with respect to outer space. . . . There is no magic in a rule. The very nature of international law is that it is consensual."[3]

Throughout the first half of 1958, while the space act was drafted and passed, the administration contemplated space law and policy. In the public domain, Eisenhower responded to American and world opinion, to his own hopes for control of technological competition, and to the needs of American propaganda, when he initiated exchanges with the USSR on outer space. In a letter of January 12, 1958, to Nikolai Bulganin, Eisenhower proposed "to solve what I consider to be the most important problem which faces the world today." He suggested that the United States and the USSR agree "at this decisive moment" to use outer space for peaceful purposes only. He recalled the failures of the previous decade regarding atomic power and urged a halt to the testing of missiles in outer space, as well as to their improvement and production. But "the capacity to verify the fulfillment of commitments is of the essence. . . ." Foster Dulles agreed that the time to control space development was now. In ten years it might be too late. Bulganin replied that the USSR was also prepared to discuss ICBMs and that the Soviets endorsed a multilateral petition to the UN including a ban on the military use of space, liquidation of foreign bases, and creation of "appropriate international control" and a UN agency to devise and supervise an international program for launching space rockets.[4]

As usual, however, the two sides divided over procedure. UN Ambassador Lodge called first for a technical study of controls for all missile testing, leading later to a ban on the use of missiles that plied outer space for aggressive purposes. But controls on missiles, as opposed to just spaceflight, would rob the USSR of its mighty ICBMs and offer nothing in return. Besides, wrote Bulganin, it was not the missiles that threatened the world but the warheads they could carry in place of "peaceful sputniks." Of course, the first argument—that banning ICBMs would only hurt the USSR—was the same argument the Soviets rejected in 1946 when the United States enjoyed a weapons monopoly; and the second argument served no purpose unless the USSR permitted on-site inspection to determine the presence of warheads or sputniks.[5] Throughout the summer of 1958, Khrushchev discussed a nuclear test ban treaty but never agreed to the technical study on means of controlling missiles and space.[6]

In the meantime, Eisenhower ordered the NSC to do its own study and to draft an American strategy for space. Following the 1950 RAND report and the space act, this was the third, and most comprehensive entry, in the documentary history of the U.S. space program. It necessarily involved some compromise among the agencies: the BoB wanted to suppress alarmist language lest space command too many funds; State and the DoD conflicted on the extent of international cooperation to seek in space.[7] But the draft paper was completed and approved by the President in mid-August 1958. It was NSC 5814/1, "Preliminary U.S. Policy on Outer Space."

"The USSR," the document began, "has . . . captured the imagination and admiration of the world." If it maintained superiority in space, it could undermine the prestige and security of the United States. The connection between long-range missiles and space boosters was intimate, but, the NSC declared, missile policy would be treated separately from space. This was a decision of great importance, for it meant that U.S. diplomacy, and thus UN controls, for space would be restricted to satellites. Even a UN agreement on "space for peace," therefore, would not mean a freeze on missile technology. NSC 5814/1 also explained that this policy statement was "preliminary" because the implications of space research were still largely unknown.[8]

What was outer space? The NSC noted that no definition existed, although the question bore on the legality of overflight. It would, however, "appear desirable" to promote a common understanding of the term "outer space as related to particular objects and activities therein."[9] In other words, the United States favored a functional definition of space (an object in orbit was ipso facto in space) rather than a schematic one (space starts fifty miles up). For while the United States did not want to forfeit its freedom to launch satellites of any sort, neither did it wish to give up the right to denounce hostile craft or develop aerospace craft that could fly in the atmosphere *and* orbit in space.

The NSC then underscored the scientific potential of spaceflight and its applicability to civilian and military missions alike. Imminent military systems included satellites for reconnaissance, communications, weather, electronic countermeasures, and navigation. Future missions included manned maintenance and resupply vehicles, manned antisatellite vehicles, bombardment satellites, and lunar stations. "Reconnaissance satellites are of critical importance to U.S. national security," the paper emphasized, and went on to describe the spy satellites then under development. They would serve missile targeting but also implement "Open Skies" policing of arms control. There were still potentially adverse implications, however, and "studies must be urgently undertaken in order to determine the most favorable framework in which such satellites would operate."[10]

Policy on manned spaceflight was also crucial. Present space research could be carried on with unmanned vehicles, but "the time will undoubtedly come when man's judgment and resourcefulness will be required. . . ." Furthermore no unmanned experiments could substitute for manned flight in psychological effect.[11]

International cooperation also appeared desirable from scientific, political, and psychological standpoints. The United States should cooperate in space so as to enhance its position as a leader in the peaceful uses of space, conserve American resources, speed up space progress by pooling talent, open up the Soviet bloc, and achieve international regulation. But genuine U.S./Soviet collaboration appeared unlikely. In March, at the time of Eisenhower's demarche to Bulganin, an NSC Ad Hoc Working Group on the Monitoring of Long Range Rocket Agreements found that much of the test data required for missile testing could be gleaned in the guise of "peaceful" space launches. It was American policy to try to prohibit the military use of space, but "contingent upon the establishment of effective inspection." Given continued Soviet secrecy, such a policy was probably barren. But since the UN would discuss space questions anyway, the United States ought to "take an imaginative position" in the General Assembly.[12]

The legal problems of space were already manifold, the NSC continued, and more were not even identifiable as yet. "The only foundation for a sound rule of law is a body of ascertained fact." Thus many legal questions could not now be settled. The United States ought to reserve its position on whether celestial bodies were open to national appropriation and declare an insufficient basis for drawing the boundary between air and space. Instead, the United States ought to make an analogy to the proposed treaty on the Antarctic and seek agreement on which activities in space would be permissible or prohibited. *"Generally speaking, rules will have to be evolved gradually* and pragmatically from experience. . . . The field is not suitable for abstract *a priori* codification."[13]

The NSC then lowered its gaze to the steppes where it all began. Conclusive evidence showed that the USSR placed a high priority on spaceflight but would not let it interfere with its ICBM program. The Soviet space program was believed to aim at manned spaceflight for military and/or scientific purposes. It would continue to lead in orbital payload for several years, but the American lead in miniaturization meant that the effectiveness of U.S. satellites was greater on a per-pound-in-orbit basis. The NSC assumed rapid American progress, and made the following prognosis:

Earliest Possible Time Periods of Various Soviet and
U.S. Accomplishments in Outer Space

	Soviet[a]	U.S.[b]
1. Scientific Earth Satellites (IGY Commitment)	1957–58	1958
2. Reconnaissance Satellites[c]	1958–59	1959–61
3. Recoverable Aeromedical Satellites	1958–59	1959
4. Exploratory Lunar Probes or Lunar Satellites	1958–59	1958–59
5. "Soft" Lunar Landing	1959–60	early 1960
6. Communications Satellites	—	1959–60
7. Manned Recoverable Vehicles		
a. Capsule-type Satellites	1959–60[d]	
b. Glide-type Vehicles	1960–61	1960–63
8. Mars Probe	Aug. 1958[e]	Oct. 1960
9. Venus Probe	June 1959[e]	Jan. 1961
10. 25,000 pound Satellite—manned	1961–62	after 1965
11. Manned Circumlunar Flight	1961–62	1962–64
12. Manned Lunar Landing	after 1965	1968

SOURCE: NSC-5814/1, "Preliminary U.S. Policy on Outer Space," 18 Aug. 1958, p. 16: DDE Library, Office of The Special Assistant for National Security Affairs.
[a] Estimate by the Guided Missile Intelligence Committee of the IAC as of June 3, 1958.
[b] Source: Department of Defense, June 4, 1958.
[c] Defense comment: The United States plans to launch a reconnaissance satellite of approximately 3,000 pounds in later 1959. . . .
[d] The Joint Staff member of GMIC reserves his position on the date 1959.
[e] The Soviets most likely would attempt probes when Venus and Mars are in their most favorable conjunction with the earth for such an undertaking.

These predictions were understood as the "earliest possible" dates. They were not presented as a function of given spending levels, nor could either country meet all the goals in any case. It is still remarkable how optimistic the NSC experts were about the rapidity of space technological development—especially on the Soviet side. If it was U.S. policy to win the space race, its chances seemed slim. The one clear prerequisite to any vigorous American space program, however, was rapid development of big boosters. NSC 5814/1 did not specify program recommendations, but did recommend basic and applied research and exploration to determine the military and nonmilitary potential of outer space, and planning for at least a decade in the future. Immediate action should include "projects which, while having scientific or military value, are designed to achieve a favorable world-wide psychological impact."[14]

In the international arena, the United States must "seek urgently a political framework which will place the uses of U.S. reconnaissance satellites in a political and psychological context more favorable to the U.S. intelligence effort." At the same time, the United States must maintain its position "as the leading advocate of the use of space for peaceful purposes. . . . Recognize UN interests in outer space cooperation, but do not encourage precipitous UN action to establish permanent organizational arrangements." A UN planning committee should be established, but not an international space agency. The United States should also reserve its position on legal issues, but study them urgently.[15]

In the aftermath of NSC 5814/1, Eisenhower's Operations Coordinating Board (OCB), responsible for executing NSC decisions, formed a Working Group on Outer Space. For its first meeting, OCB Vice Chairman Karl Harr drafted a briefing on the importance of space activities, the management of which, "particularly the emphasis on military or non-military aspects thereof," went far to define "the basic attitude and philosophy of all government programs."[16] Preparatory to the UN General Assembly session, the OCB put together coherent policy on international aspects of spaceflight. The State Department saw in this a double goal: the United States must maintain its image as a force for cooperation but also establish "an acceptable policy framework for the WS-117L program as a priority task."[17] But since nothing could be done at the UN without Soviet compliance, what, asked the OCB, was "the feasability of developing a cover for such reconnaissance satellites?"[18]

By the time the General Assembly convened in September, American officials had pondered the wisdom of various approaches to international control of space technology. A maximum solution—complete prohibition of military use of space—required nothing less than a comprehensive arms control treaty including on-site inspection or an operational UN agency to manage space activity. The NSC had already nixed the latter, romantic idea, while the former depended either on a complete change in Soviet policy or on the perfection of satellite reconnaissance, which must, in that case, be exempted from control! A minimum solution offered a better chance of meeting American desiderata, as the OCB concluded in October. The UN delegation should seek to: (1) create an informed and understanding national and world opinion identifying the United States with peaceful uses of space for the benefit of the whole

---

PIONEER 1

The next event in the space olympics, beyond the first satellite and weightlifting, was "shooting the moon." Smaller rockets made the United States an underdog again, but it made the first try in this round on October 11, 1958, when a Thor-Able (the IRBM plus a modified Vanguard) sent *Pioneer 1* on a trajectory for the moon. The media speculated whether the first country to achieve lunar impact, or plant a flag, or land a man, might "claim" the moon. But one sensitive guest at a Cocoa Beach party, gazing at the heavy half-moon on a languid Florida evening, told an air force officer: "If you try messing up anything as beautiful as that, I hope you miss it by a mile—by a thousand miles!"[19] It did miss, but reached a record distance from earth of 71,300 miles and discovered the radial extent of the Van Allen belts. *Pioneers* 2 and 3 (the latter an army spacecraft) failed in November and December, but returned more data on particle fields in cislunar space.

world; (2) create a worldwide understanding that the U.S. military space program helped to provide the free world with a deterrent against Soviet aggression or control over outer space; (3) promote free world progress in space; (4) establish a global climate of opinion that condoned operation of certain classified space programs. To these ends, the United States should cover its military program with a rhetorical blanket of "space for peace" and define it as vital to deterrence and therefore peaceful. Since the Eastern bloc and other states would oppose or misunderstand American intentions, a minimum of international control was desirable. The OCB foresaw a UN committee to pursue agreement on satellite orbits and radio frequencies, and bilateral cooperation in space science, but nothing more.[20]

The State Department, nevertheless, still hoped for direct U.S./Soviet cooperation in space. The difficulties in sharing strategic technology were obvious, but the gap between appearance and reality in the "space race" was what really stymied such cooperation. The Soviets seemed to be way ahead in space and did all they could to sustain that impression. In fact, they trailed in everything except big boosters and possibly space medicine preparatory to manned flight. Therefore, the United States would gain little from bilateral programs that "gave away" technology to the Soviets, especially since the world would assume it was the United States that sought help in rocketry from the Soviets! The USSR, in turn, had no desire to reveal how backward it really was in overall technology. Dulles, supposedly intractable where the Communists were concerned, was the only leading figure who still favored cooperation with the Soviets. The PSAC and OCB were both skeptical, except for sharing of scientific data "in matters on which we had equality with the USSR."[21]

Nevertheless, the U.S. delegation prepared to make a great display of its concern for international cooperation in space.[22] In September 1958, Dulles called on the UN to take immediate steps for an Ad Hoc Space Committee and study further "organizational arrangements": "As we reach beyond this planet, we should move as truly 'united nations'."[23] Ambassador Lodge renewed his request for Soviet participation in a technical discussion of inspection systems for space technology. In November, even Senator Johnson addressed the General Assembly to demonstrate the unanimity of American opinion behind "space for peace." He asked, among other things, that a UN space committee "consider the future form of internal organization in the UN which would best facilitate cooperation in this field."[24]

Such language could easily be interpreted as an invitation to the UN to assume strict management of all human activity in space. Certainly no enterprise fell more clearly under UN jurisdiction, but neither had any been so charged with the Cold War politics that made the UN ineffective. The Soviets' own resolution called for a ban on all military uses of space, elimination of foreign bases, international control of space, and a UN agency to include an international program for launching long-range rockets. Having made points with this offer, the Soviets hastily withdrew it and called instead for the same Ad Hoc Committee on space as the United States. But the USSR envisioned a committee made up of three Western, three neutral, and five East bloc countries. The Western proposal named a prospective membership of eighteen that more accurately reflected the physiognomy of the UN, but restricted Soviet-bloc participation to a small minority. On November 24, the General Assembly defeated the Soviet plan and opted, fifty-four to nine, with eighteen abstentions, in favor of the Western resolution.

This vote gave birth to the UN Ad Hoc Committee on the Peaceful Uses of Outer Space (COPUOS), the forum in which space law would be crafted in coming decades. Its first instructions were to survey the resources of the UN relating to space, report on areas of likely cooperation, organize exchange of information, and suggest future organizational and legal problems for UN consideration.[25] The rhetoric was uplifting; the mandate restricted. There would be no UN space agency, no discussion of space disarmament, no action of any kind without agreement between the two space powers. What was more, the USSR protested the "unbalanced" composition of the COPUOS and boycotted the committee's labors.

The circumstances in which space technology emerged, the military and political importance of it for the Superpowers, American policy as drafted by the NSC in 1958, and the deadlock at the UN all meant that there would be no "control at the outset" of space technology. U.S. and Soviet stances both made the outcome inevitable—but whether the outcome was vexatious is itself debatable. The United States surely won out in the short run, for its goals were fulfilled by passage of the Western resolution. "Space for peace" came to be associated primarily with the United States, but there was no danger of its being translated into perverse UN restrictions on national technology. The American formula of space for "peaceful" rather than for explicitly "nonmilitary" purposes also won out and served to guard the U.S. military space programs.

Few diplomatic issues seemed as urgent and loaded with implications for world peace as the law of outer space. Here were a new complex of frightening technologies *and* a virtually limitless medium, opened up simultaneously to human exploitation. And just as the voyages of the Age of Discovery stimulated inquiry into the law of the sea that advanced international law generally through the work of Hugo Grotius and others, so the launching of the Space Age inspired a burst of inquiry on the fundamental principles that ought to guide *all* the deeds of nation-states. The most beguiling legal problems were those tied to sovereignty: could nations claim space; divide it into zones according to some scientific, political, or technical principle; make it off-limits to weaponry;

extend the cooperative framework of the IGY? What legislative and enforcement mechanisms were preferable for space law? What arrangements could be made for advance notice of launches, exchange of data, assessment of liability for damage caused by space vehicles? Who owned the moon or the electromagnetic spectrum? How could space boosters be distinguished from military missiles? Was space development best served by an international effort or by national programs operating under ground rules?[26]

A handful of visionaries tackled such puzzles even before Sputnik. John Cobb Cooper, air law expert and fellow of Princeton's Institute for Advanced Study, took up the question of sovereignty in a 1951 article, reviewing the history of air law from the Romans (who said land ownership extended *"usque ad coelum"*) to the great jurisprudential theorists of the seventeenth and eighteenth centuries (Samuel von Pufendorf limited sovereignty in the air to the ability for "effective control"), to the Chicago Convention of 1944 (which recognized complete and exclusive national sovereignty over air space). But how far up did air extend? Sounding rockets revealed that the atmosphere did not just stop, but gradually dissipated. Cooper opted for "effective control" (also the formula chosen by the 1885 Berlin Conference, which set rules for the colonization of Africa). "The territory of each state extends upward into space as far as the scientific progress of any state . . . permits such state to control it."[27]

After Sputnik, numerous proposals were advanced for defining outer space. The so-called von Kármán line set the boundary at the point at which a vehicle traveling seven kilometers per second loses aerodynamic lift and becomes a "spacecraft." Such an event would occur about fifty-three miles up. Cooper and common law (post–October 4, 1957) indicated that space simply stopped at that point below which an orbit could not be sustained. But such "lines" were a function of velocity and therefore of technology, and were in no way innate. Everyone knew where land ended and the ocean began, but now man had entered a realm that, in a real sense, did not exist except as a function of man's own tools. Any definition of outer space was a solipsism.

The critical variable in the definition of space was perceived military interest. The higher the boundary of national sovereignty, the greater the protection against unfriendly overflight, but the lesser the ability to ply the lower reaches of space for any purpose. It was guesswork in 1958 as to which would best suit American or Soviet interests. Similarly, whether a low limit was good or bad depended on the international regime that would obtain in space. If a rigid system of international control was instituted, then national freedom was best served by a high boundary. If a laissez-faire regime arose in space, then national freedom would be greatest by lowering "outer space" as close to the earth as possible: "Open Skies."[28]

These ambiguities gave spacefaring nations no incentive to solve the riddle. State Department counsel Becker explained that the United States, while not recognizing any top limit to its airspace, nevertheless granted that existing space activities conferred the right to ply space wherever it was. In short, the United States believed in "freedom of space," but reserved its position on what that freedom entailed or where it took effect. "Moreover," he continued, "there are very great risks in attempting to transmute a body of law based on one determined set of facts (e.g., air or sea law) into a body of law with respect to which the basic facts have not been determined." The State Department was "inclined to view with great reserve any such suggestions as that the principles of the law of space should be codified. . . ."[29]

The principal concern of American policy was always the protection of spy satellites. But the right to launch satellites over the territory of other states was already established during the IGY. In this connection George J. Feldman, counsel to the Senate Space Committee, declared that security considerations alone would preserve the principle of sovereign air space and work just as powerfully against a definition of where that air space ended. Satellites had already been launched without protest, implying that formal consent to satellite overflight was either unnecessary or implicitly given. "It is tempting to accept the first explanation—which would mean, for example, that President Eisenhower's Open Skies proposal is an accomplished fact. However, any such assumption would be premature and unjustified." Limited agreements on space might be made, but none should be sought "which are more comprehensive or explicit than our present knowledge warrants."[30]

The same caution obtained in debate over sovereignty on heavenly bodies. As early as 1952 a UN lawyer, Oscar Schachter, asked "Who owns the universe?" and worried that we might someday read of colonial rivalries in space, of "lunar Washingtons and New Yorks, perhaps of King George mountains and Stalin craters." He suggested that space and celestial bodies belong, like the high seas, to all mankind. States should be allowed to develop settlements and mineral deposits, but in such a way as not to cause waste and destruction "against the general interest of mankind."[31] The fear of a "scramble for colonies" in space, more rapacious even than the nineteenth century's scramble in Africa, also motivated space law theorists after Sputnik. But if space was not subject to sovereignty, what was its legal status? Was it *res nullius*—space as belonging to no one, but presumably subject to claims? Or *res communis omnium*—space as "the heritage of all mankind" with an implied right for all powers to regulate and reap the benefits of spaceflight? Or *res extra commercium*—with sovereignty and jurisdiction vested in the UN? The first threatened to stampede the powers, but the others implied an international control over national technology that the US and USSR alike were unlikely to accept.

Early discussions of such problems fell roughly into two categories, a fact acknowledged by the leaders of the schools themselves, Andrew Haley and Myres McDougal. The former, an amateur rocketeer turned lawyer, counsel to the ARS and president of the International Astronautical Federation, was the major exponent of the "natural law school." According to Haley, law rested on universal moral principles derived from the nature of man: moral precepts such as the Golden Rule that found expression in all the great religions. Codified natural law theory arose, significantly, in response to the problems posed by European discovery of the New World. But the law of nations, as the moral law of individuals writ large, did not constrain the states of early modern Europe, with unfortunate results. Now the world's governments again faced virgin territory. This time states must join in advance of the conquest of space to set standards and principles of conduct, and so avoid the old pattern of abuse and competition.[32]

The "positivist school" of space law, associated with McDougal of Yale, argued that law emerged from patterns of common usage and could not be invented in advance of knowledge of the facts and emerging national interest. The difficulty in separating military and civilian activities rendered prohibition of the latter all but impossible, and space law in any case would always be a function, not a determinant, of international politics. High-blown principles and futile attempts to shackle the space powers would only make the ideals that inspired the principles appear ridiculous. Instead, the patterns of usage of space must be allowed to establish themselves before codification.[33]

The two schools could aptly be termed the idealist and the realist. The most striking vindication of the realistic positivists was the fact that the secret NSC decisions had already rendered the space law debate academic. The reasons for the Superpowers' aloofness included the one offered in disparagement by the natural law idealists—that nations were obsessed by power and flouted the ethical imperatives imbedded in every human being—*and* the one offered in sweet reason by the positivists—that it would be folly to make artificial rules for a vast area of human activity before the facts were known. Hence the USSR boycotted the Ad Hoc COPUOS entirely, while the United States sharply circumscribed its agenda.[34] The upshot was that discussion would proceed on such things as spacecraft registration and liability, sharing of the radio spectrum and scientific data, but not on restrictions on the development and use of space technology by competing national states. Many space law theorists expressed their disgust with this narrow nationalism and hypocrisy, but their cries of "space for peace" and "space for all mankind" carried no further than if they had been shouted in the vacuum of space itself. The irony is that those enthusiastic about the human adventure in space should have been rejoicing. Competition was the engine of spaceflight. Had space exploration been truly internationalized or demilitarized, the

Superpowers would have had little incentive to make huge investments for its realization. Space programs would have been stunted with malnutrition.

Congress and the press came only gradually to understand. Throughout 1958, "space for peace"—implying demilitarization—seemed an unassailable proposition. A Library of Congress study in February 1958 even sketched out a UN space agency to conduct all exploration—though its authors doubted that the United States would propose it or the Soviets agree to it.[35] But the leaders of both houses of Congress carefully guarded the clauses in the space act that committed the United States to peaceful space exploration for all mankind. In June John McCormack introduced a resolution to "ban the use of outer space for military aggrandizement" and pursue space exploration for "the good of all mankind rather than for the benefit of one nation or group of nations." The purpose of the resolution, which was reported out by the Committee on Foreign Affairs and passed unanimously, was to make clear to the world the repudiation by the American Congress of "narrow nationalism."[36] In July the Senate passed a similar resolution.[37]

This summer of the space act, hearings and resolutions on space law and cooperation, and preparation for the UN General Assembly session marked the zenith of American sentiment for the demilitarization of space. By late autumn the fears of a Soviet Damocles' sword in space had receded (the Sputniks being apparently harmless for the moment), the USSR had declined to participate in initial UN studies, and the U.S. government showed no interest in UN space agencies. By the time Congress reconvened in 1959, its leaders had also presumably been briefed on the importance of distinguishing "peaceful" and "nonmilitary" uses of space. U.S. military space programs, especially spy satellites, did serve peaceful purposes in that they promised to strengthen the deterrent, keep watch on the Soviets, and prevent a Soviet hegemony in space. Demilitarization, therefore, would not serve the cause of peace. As for the Soviet response to U.S. military programs, Sol Horwitz advised LBJ, "The Russians will scream on any occasion they think it desirable to scream." The only way to avoid denunciation was to have no satellite programs at all.[38] In subsequent years, critics on the Left would intermittently denounce American "militarization" of space, but the congressional mainstream never again took "space for peace" to mean closing down the Pentagon space programs.

While American diplomats maneuvered to establish the virtue of military spacecraft, ARPA projects bloomed like Mao's hundred flowers. To be sure, ARPA was given direction of all military space programs precisely to prevent interservice rivalry and runaway R & D programs. But space was unknown, and even skeptics like Roy Johnson and York had to grant that its military potential would never be known except at the cost of chasing up some blind allies.[39] Two philosophies of R & D

SCORE, LUNIKS, AND DISCOVERERS

The Soviet advantage in weightlifting could last only until the American ICBM entered the testing stage. Hastened along by Schriever's "concurrency" tactics, Atlas was ready for an orbital mission by the end of 1958. On December 18 world opinion was stunned by the news that the Americans had placed a four-ton satellite into orbit. This constituted the weight of the entire upper stage, of course—the payload was about 150 pounds—but the United States had learned from the Soviets how to manipulate data. Project Score was also the first communications satellite, a primitive relay device that broadcast Christmas greetings from President Eisenhower to the peoples below. It suited well the NSC requirement for otherwise useful projects designed for propaganda impact.

In January the Soviets entered the moon derby with *Luna I*. The rocket missed the moon by 3,000 to 4,000 miles, but it sped past into a solar orbit, the first manmade object ever to escape the gravity of the earth.

On the last day of February 1959, a more substantive mission blasted off from the scrub and dunes of Vandenberg AFB, California. A Thor-Able A launched *Discoverer 1*, the first test satellite of the WS-117L program. Lockheed's Agena spacecraft, a cylindrical upper stage measuring about five by twenty feet, carried instrumentation in the front and command, guidance, and propulsion systems in the rear. Once lodged in its polar orbit, the Agena could circle the earth every ninety minutes while the globe rotated beneath it. The first Discoverers carried no film packs, but ultimately they would discharge their photographic intelligence for reentry and recovery in the ocean or by an airborne "snatch."

Discoverer 1 tumbled wildly while in orbit due to malfunction in the stabilization system. *Discoverer 2* (launched April 13, 1959) carried a biomedical capsule. It performed well, but human error resulted in a botched reentry. The capsule landed somewhere in northern Norway and was lost. *Discoverer 3* and *4* failed to orbit, and the next failed to reenter when improper orientation caused it to lurch into a higher orbit when retrorockets fired.[40] Spy satellites proved as tricky as a carnival shooting gallery—but the prize was worth waiting for.

inevitably clashed: the one that saw wisdom in spreading seed money liberally on the expectation that the few winners would soon become evident, and the other suggesting that no poker player ever won over the long haul without folding a few winning hands. The secret of efficient exploratory research was to cancel unpromising programs before they reached the expensive hardware stage. But R & D programs, like federal agencies, tend to acquire lives of their own. Big-ticket items of dubious promise but durable political backing included Project Rover, a nuclear rocket under study by the AEC, and the USAF follow-on to the X-15, called the X-20 Dyna-Soar (for "dynamic soaring"), a Sänger-type spaceplane expected to provide the USAF with a manned military space program.

Applications satellites had more promise. The USAF and CIA cooperated, then clashed, over control of Discoverer, and the USAF instigated two more programs, the observation satellite Samos, and the infrared early-warning satellite Midas. They also pushed ahead on designs for communications, navigation, maintenance and repair, weather, and geodesy satellites. These last were especially vital components of the ICBM effort, since precise measurement of the shape of the earth and its gravitational and magnetic fields was a prerequisite to improved missile accuracy. Strange as it may seem, traditional survey methods had never established the exact relationship between the American and Eurasian land masses. Scientific and observation satellites not only located precise targets halfway around the world but increased one's chances of hitting them.[41]

Military space technology suggested other, more alarming novelties. Bombs in orbit had to be studied, if only to demonstrate their impracticality, as well as fractional orbital bombardment systems that traveled the long way around the earth before diving to their target. Since the Soviets would presumably develop their own military space systems, the USAF also researched antisatellite and antimissile weapons. All told, at the very moment when the President signed the space act with its commitment to a civilian program, Budget Director Stans was authorizing $294 million for ARPA and only $242 million for the new NASA.[42] The figures were small, and the balance soon shifted in NASA's favor, but the military space program had a huskier stature than its low profile suggested. In December 1958 the OCB space working group adopted a public information policy on U.S. space activities, and the administration imposed increasingly rigorous ground rules throughout 1959 to reduce publicity of DoD space launches.[43]

Indeed, the military space program caused increasing frustration. In private, civilian officials (not to mention the military) felt no shame about pursuing military advantage in space. Spy satellites in particular promised to be a tremendous boon to free-world defense and the prospects for arms control. Yet the subtleties were lost on most people, especially overseas, and the United States had to preserve its peaceful image. How to protect the coming spy satellites? The Itek Corporation, a contributor to Agena and consultant on space law, reported that "information from overflights of the USSR is now vital for U.S. security. . . . The problem is *not* a problem of technology. It is *not* a problem of vulnerability to Soviet military measures. The problem is one of the political vulnerability of current reconnaissance satellite programs." The Soviets would take powerful countermeasures, just as they had when the United States tried balloons and aircraft. "Satellites are our last chance. Should recon sats be 'politically shot down,' no scientific or technological opportunity can be foreseen to obtain this security information during the forthcoming critical years. What is needed is a program to put recon sats 'in the white' through early and vigorous political action. . . ."[44] Indeed, the new NASC approved a strong position at the UN opposing "any activities which put unacceptable limits on U.S. freedom of action" in space.[45]

The UN Ad Hoc COPUOS completed its survey in July 1959. Its report waxed enthusiastic on the human benefits promised by satellites: scientific advances of all kinds, better weather forecasting, communications, mapping, navigation, and manned exploration. It pointed up the need for allocation of radio frequencies, registration of spacecraft, and other managerial functions. It made no mention of demilitarization or internationalization of spaceflight. On legal problems, the report endorsed the "freedom of space," stating its belief that, given universal acceptance of IGY satellites, "there may have been initiated the recognition or establishment of a generally accepted rule to the effect that, in principle, outer space is, on conditions of equality, freely available for exploration and use by all. . . ." The COPUOS reasserted the sovereignty of states over air space, but admitted no consensus on where outer space began and did not regard it a priority consideration.[46] In all these matters the American position triumphed.

What had become of Eisenhower's bold invitation to ban or control "outer space missiles"? It was not simply eyewash. Eisenhower put his PSAC on the task of studying the technical potential for a verifiable nuclear and space missile test ban as soon as the committee formed, and it remained one of its most time-consuming activities until the end of his term. But its findings were discouraging. A working group chaired by George Kistiakowsky reported in March 1958 that detection of Soviet rocket tests could be made reliable through expansion of intelligence systems then in place (in Turkey and Iran) and by new techniques under development (spy satellites). But the complications that would arise for space programs were consequential. "A complete prohibition of the launching of all large rockets leaving the atmosphere . . . would freeze the development of ballistic missiles and space vehicles near their present status and would prevent their use for 'peaceful purposes.' " Agreement to permit space launches under a U.S./Soviet or international agency was a possibility, but that would not prevent the USSR from going ahead with an operational ICBM force, if it was ready to go into production at that time. The only way to stop an expanded Soviet missile force was to ban manufacture of warheads and missiles, which posed a far more difficult problem of verification.[47]

Foster Dulles concluded from this evidence that a ban on long-range missile tests must come within the next six to eight months if it was to prevent an operational Soviet ICBM force and permit adequate inspection. After that time, "the only sure method of preventing such a capability would lie in controls on production and deployment which would be very difficult to inspect." He did think an immediate freeze that prevented

Soviet ICBM deployment while U.S. IRBMs were in place might be to American advantage.[48] Thus the United States could retain its foreign-based bombers and intermediate missiles, while the Soviets would have to give up their best means of reaching the United States. Such logic, of course, ensured that the Soviets would ignore such a proposal, which they did when Lodge called for a study of missile test verification in the fall of 1958.

Despite the technical problems, the Soviet snub, and the contradiction embedded in the need for secret reconnaissance satellites to verify a ban on secret rocket programs (!), the notion of a missile freeze persisted. Jerome Wiesner, PSAC member, urged immediate action. If missiles were frozen now, he wrote in November 1959, each side would possess a barely adequate deterrent inhibited by the cost, size, unreliability, and inaccuracy of first-generation ICBMs. A freeze would slow down the missile race, and if it prohibited space shots "as it must to be effective, it would also get the U.S. out of the space race, which otherwise will continue to be a serious source of embarrassment and frustration."[49] But others thought such ideas unrealistic. It was true that the passage of time would make arms control increasingly difficult, but the realistic goal for the next five years, according to arms expert George Rathjens, was not a freeze but an increase in "stability." The United States should court a situation in which the deterrents of both sides were *more* secure so that no one would have an incentive to strike first or retaliate hastily. "Any proposed changes such as a cessation in testing must be examined with regard to whether they increase or decrease stability."[50] In short, the United States could not afford a freeze until its own missile deterrent was assured. In December 1959 the panel again thought a freeze on missiles in the primitive stage had "favorable implications," but warned that it would mean controlling space activities more tightly ("Is this realistic now?") and giving up the pursuit of a more stable deterrent through smaller, mobile missiles.[51]

Thus the two arguments that came to dominate American missile and space policies over the next decades had already surfaced by the end of 1959. The first was that stability, not disarmament, was the key to security in the missile age.[52] Once mutual deterrence was in place, both sides could pursue arms control with preservation of "stability" the determining factor. The second was that a missile test ban would shoot down the space programs of the world, a regrettable development for secular reasons but a tragic strategic contradiction, since ever more sophisticated spy satellites promised a technological end run around Soviet secrecy, itself the greatest barrier to arms control!

Space technology, like atomic power, was not to be controlled at the outset. Instead it would develop according to national interest in an international environment of distrust and competition. Each Superpower blamed the other for the loss of these critical years after Sputnik when neither the COPUOS nor the UN Ten Nation Disarmament Committee made progress toward agreements on missiles and space technology. Khrushchev spoke of U.S. militarism, Eisenhower of "fleeting opportunities." But the fact was that neither was in a rush to engage even the narrow range of questions within the competence of the COPUOS. U.S. space strategy developed on a line from its initial consideration by RAND in 1950. First and foremost, space was about spying, not because the United States was aggressive but because the USSR was secretive. Whether arms competition or arms control obtained in the future, American space strategy must spin off from its first space program—reconnaissance satellites. This dictated a policy subtle in conception and delicate in execution. The United States must become the champion of "freedom of space," which sounded virtuous (and, in American eyes, was), but translated into a laissez-faire regime for space that other UN members, who tended to identify virtue with "controls," might well take amiss. But Eisenhower, with overwhelming congressional support, also identified the United States with "space for peace" and "space for all mankind," a thread in American policy that stemmed from traditional idealism and respect for the rule of law on the one hand and from Cold War competition for prestige on the other.

The same impulse that gave birth to NASA also produced the line that the U.S. space program was open, peaceful, and cooperative, in contrast to the Soviets. They had been first in space, and were likely to pile up more "firsts" for some time. The United States, at least, could rally its allies and neutrals alike with the promise of a vigorous but salutary space technology in the interests of humanity. All this made sense, even if it meant an abiding awkwardness in U.S. international space policy. But the lack of controls, the impossibility of cooperation, and the continued symbolic importance of space policy and achievement in the eyes of the world also meant that space technology would continue to evolve as a race. Eisenhower accepted, regretfully, the need to keep ahead of the enemy in military technology. He also feared that the technocratic method might come to be applied to civilian pursuits as well. But the peaceful, open image that he wanted to convey for the U.S. space program required precisely that a space race be civilian, not military. Unless Eisenhower and his successors junked the attempt to restore American prestige in space, or chose to ignore world opinion and pursue a heavily military program, then the space program would have to become just what Eisenhower hoped to avoid: a model for the application of the technocratic method to civilian goals.

CHAPTER 9

Sparrow in the Falcon's Nest

If strategic considerations were of surpassing importance in U.S. space policy, what was NASA all about? Was the main reason for a civilian agency, as Johnson's staffer wrote, the need for some nonmilitary body to present to the outside world? Or just for the propaganda value of a civilian space program? Or to conduct basic R & D and space science not immediately of interest to the services? All three played a role, but none of these necessarily implied a large and vigorous space program. Indeed, Eisenhower was skeptical of large-scale prestige programs in space, and a weak NASA fit his "rearguard" predilections concerning the role of government in technological change. But a weak NASA might also fall into the same relationship to the military as its parent NACA, and thus fail even as a showpiece for the civilian space program. If administration policy required the creation of NASA, it also required a willingness to shelter the agency from the military, sustain its image, and nurture it to maturity as a sparrow in a nest of jealous falcons.

The threat to NASA from the DoD was no delusion. For the army, USAF, OSD, and ARPA had all favored either no space agency at all or one patterned on the pliant NACA. The ABMA/JPL team itched for the primary role in space and even after the space act retained a near monopoly over the talents and facilities needed for big space R & D. The tension in Ike's policy for NASA, therefore, stemmed from the need to fashion a strong, competent civilian agency while still restraining the overall space effort. For a strong NASA, buttressed by congressional and industrial friends and feeding on the new symbolism, might itself promote the spread of command technology to wider spheres of civilian government. That was the danger of a space race and hence of placing inordinate value on prestige. The trouble for Eisenhower was, how many more Soviet triumphs could he, and his policies, stand?

The first step in building and controlling the new agency was to choose a suitable administrator. Throughout the spring of 1958 the frontrunner for the job was NACA Director Hugh Dryden. He was a renowned aerodynamicist and manager of research, but his reputation for professional conservatism troubled congressional leaders who wanted a daring space program designed to "leap frog" the Soviets. Dryden's integrity—he was also a lifelong lay preacher in the Methodist church—served him poorly when he told the House Space Committee that sending a man into orbit inside a Redstone nose cone "has about the same technical value as the circus stunt of shooting a young lady from a cannon." This, congressmen sneered, was not the man to command a space race.[1] Killian thought this animus against Dryden "another unhappy result of the exaggerated notion abroad in Congress that space would revolutionize everything." He himself then inherited the task of choosing an administrator. After Doolittle refused, Killian settled on the president of Case Institute of Technology, T. Keith Glennan. Eisenhower told him that he wanted a space program "sensibly paced and prosecuted vigorously." After some soul-searching, Glennan took the job on condition that Dryden remain as his deputy.[2] They were sworn in on August 19, 1958.

Glennan's philosophy of government must have gratified his chief. In his diary Glennan recorded the convictions he brought to the new NASA: (1) Government was getting too big, hence the bulk of NASA work should not be done in-house, but channeled to industry and universities under contract; (2) NASA should transcend the "missile mess" and build an orderly program for large launch vehicles; (3) the propaganda value of spaceflight was not of primary importance, but neither could it be ignored—NASA had a unique mission in this area, and one likely to confer a high profile on the young agency; (4) programs must nonetheless be structured according to long-range goals, and not simply as propaganda ploys; (5) NASA must take over much of the ARPA program, but forge its own broadly based plans for increasing *capabilities and options* for future spaceflight, rather than moving prematurely to specific *goals*.[3]

To begin with, NASA inherited the existing facilities of the NACA: Langley and Ames Aeronautical Labs, the Lewis Flight Propulsion Lab, the High Speed Flight Station at Edwards AFB, and the Wallops Island rocket range. But these were not the core of a space effort. The NRL readily relinquished the Vanguard program, and Congress authorized a third major NASA center, appropriately christened the Goddard Space Flight Center when it opened in May 1959 at Beltsville, Maryland. The ARPA lunar probes also moved over to NASA, as well as the satellite projects underway at Huntsville. But most important for NASA's leadership in space was the ABMA program for a single-chamber rocket engine of one million pounds of thrust. The von Braun team and Schriever's Ballistic Missile Division, in league with North American Aviation, both coveted it. But the ARPA and Secretary of Defense remained unconvinced of the military requirements for such a giant booster. Either this F-1 engine, and the Saturn rocket based on it, would be built by NASA, or it would not be built at all.

How could NASA discharge such responsibilities without the personnel and facilities residing in the military services? Glennan and Dryden understood that NASA needed more than paper jurisdiction over projects; it had to raid the services for the wherewithal. But transfer of military facilities would do more than make NASA a going concern—it would also determine which armed service would win control over military spaceflight. Thus three struggles raged simultaneously: one within the USAF between the "big bomber boys" and the missile boys over the long-range future of the USAF; one among the services for major shares of the military space mission; and one between the services and NASA for the basic R & D missions in space.

The race for space among the services in the post-Sputnik months was intense but inconclusive. By the time NASA emerged, the ABMA, USAF Ballistic Missile Division, and ARPA all had long-range space programs on the table. Medaris and von Braun hoped to capitalize on their Explorer victory with a ten-year program based on the Saturn.[4] But the navy's Vanguard finally succeeded as well, while the USAF pushed ahead with ICBMs, the Agena spacecraft, and manned space research. The latter touched on the biggest prize of all in the sweepstakes: manned spaceflight. The preferred mode of USAF test pilots was the X-series of winged rocket planes climbing higher and faster until they crossed the boundary into space.[5] But if competition with the Soviets demanded a "quick and dirty" manned space program—blasting astronauts into space inside nose cones ("Spam in a can")—then the USAF must corral that assignment as well or risk losing the whole mission. In March 1958 the Air R & D Command requested $133 million from ARPA for a project called "Man-In-Space-Soonest." Using Thors, "Super-Titans," and the Agena spacecraft, the USAF would proceed from simple manned orbital flights to eventual landings on the moon, all for the bargain price of $1.5 billion.[6]

ARPA itself studied the long-range military role in space, including space weapons, bases, satellites, and the moon.[7] Roy Johnson, while not bucking the administration's policy of preserving a large role for NASA, professed openly that *all* space technology would find military applications, hence the DoD was fully prepared "to sponsor pure research against the judgment that it might lead to valid military applications without proving the judgment in advance."[8] ARPA was not an operational agency, but it served as clearing house for Pentagon space work. Its "Long Range Plan for Advanced Research" listed no less than nineteen military space requirements for the army, ten for the navy, and fourteen for the USAF, including the Defender antisatellite system and Mrs. V, an early inquiry into orbital weaponry.[9] Such energetic canvassing of space missions within the DoD and the prevailing suspicion that virtually all space technology had military application made NASA vulnerable to the charge of redundancy. But NASA had its purposes. The nation needed two space programs—but no more than two. To build up NASA, Glennan and the NSC resolved to reallocate American space resources. Their target—and the loser—was the army.

In October 1958 Glennan bid for the JPL, the Caltech lab funded largely by the army, and half of the ABMA Development Operations Division, the von Braun team. Glennan visited Huntsville and came away with the impression that the Germans were less than sincere in insisting on their military importance: "They were working on the Pershing (a battlefield missile), but they were really interested in the moon."[10] Yet von Braun was loyal to Medaris and probably thought that a feeble, civilian agency would never command big money. Army Secretary Brucker also "read Glennan the riot act" about his "crazy plan to split up the rocket team,"[11] and the NASC chose to leave the ABMA intact. But bereft of the JPL, it was a rocket team without spacecraft. Then in September 1959 the DoD assigned military space operations to the USAF, and the army's Saturn promptly ran into funding difficulties. Medaris now faced a Hobson's choice—fight to keep von Braun but have nothing for him to do, or haul down the army's space shingle for good. In any case, ARPA, the USAF, and the Secretary of Defense were more than a match. When ARPA's York approached Glennan in October and asked if NASA were still interested in the ABMA team, the NASA chief demurely assented. Medaris telegraphed Brucker to express his chagrin and soon retired from the Army.[12] Congress ratified the decision, which went into effect on July 1, 1960. For the first time since he was visited by Reichswehr Captain Dornberger in 1932, Wernher von Braun was out of the army and on his way to the moon.[13] The NASA chunk of Huntsville became the George C. Marshall Space Flight Center, Eisenhower presided over its inauguration, and the name itself symbolically "Americanized" the von Braun team for its new, highly public mission.

Stripping the army put NASA in business and simultaneously solidified the USAF hold on military spaceflight. But boundaries between the two realms were still disputed, and nowhere so much as in the juiciest province of all—manned spaceflight. As bluntly stated in its own chronicles, the USAF was "the most logical agency to achieve this military [space] power." NASA was "made by taking up room previously occupied by the military departments," "peaceful purposes," in USAF eyes, were a front for prestige, and a divided space program was irrational. "Lacking changes in top level policy attitudes, the armed services would have to await their 'day in the sun' until science and technology or the onrush of international affairs made military space travel a recognized, permanent, national necessity."[14]

Are "militarism" and "institutional politics" sufficient to explain the USAF lust for control of the space program? No, they are not. USAF apologists had been upheld, though *sub rosa*, by leaders of Congress and the administration in their judgment that most space R & D had military

"And That One Was General Medaris!" October 1959. Courtesy of C. Werner, the *Indianapolis Star*.

potential. They were right in arguing that a divided space program was problematical—even NASA leaders admitted as much. They would soon be right, if not at once, that the Soviet space program was largely military. Schriever understood the new age well when he said, "Today, as never before, our military and civilian aims and actions are inseparable. . . . The challenge is total. Our response must therefore be total."[15] That very politicization of national technology suggested it did not matter whether NASA or USAF did the job, as long as glaring inefficiencies were avoided. But the prestige motive cancelled out all USAF logic. In the eyes of the world, NASA astronauts (even if drawn from the services) would be "envoys of all mankind," and not secret soldiers in space. Finally, even if space technology did have military implications, the USAF failed to demonstrate immediate military missions for manned spaceflight that required that they do the basic R & D. Instead, USAF reveries of rocketing pilots in "aerospace planes" to "orbital bases" for purposes that could be better fulfilled with instrumented satellites only convinced Eisenhower and his lieutenants that the USAF had to be reined in, not encouraged.

The manned spaceflight issue reached the White House in August 1958, where Eisenhower came down in NASA's favor. There were technical reasons—NACA's frontier faction had drafted a credible, low-cost plan for manned capsules. But the decision was also political. The United States' image required that such a high-profile program be civilian. There was no evidence that the military could do the job more quickly, while there were signs that the USAF hoped to grab manned spaceflight and run with it, which tripped the signal flags in the budget-conscious administration.[16] In September a Joint Manned Satellite Panel declared its objective "to achieve at the earliest practicable date orbital flight and successful recovery of a manned satellite. . . ."[17] Glennan and Dryden, on November 26, gave it a name: Project Mercury. Its unstated task was more precise: to beat the Soviets to the first man in space. NSC 5814/1 declared that a longshot, but whether the United States had a chance to avoid "another Sputnik," as Killian put it in February 1959,[18] depended in part on the outcome of another great debate, a Hamlet-like meditation on the question "To Race, or Not to Race?"

One purpose of Eisenhower's strategic posture was to restrain those elements in government and society willing to jettison limited government and financial restraint in order to prove American superiority. Racing with the Soviets for space spectaculars ran against his grain.[19] Yet Sputniks were helping the USSR to remake world politics into a total competition in which prestige was as important as power, and the apparent ability of societies to force "progress" as important as their ability to nurture what is enduring in human culture. Hence the dilemma— freedom and stability served the deepest desires of human nature, but organization and technical revolution seemed increasingly necessary to maintain freedom and stability. The Marxist claim that capitalist societies were based on contradictions was right, but the contradictions emerged only as a result of the existence of a Marxist competitor. A traditional conservative might refuse to compete on technocratic terms, reject modernism wholesale, and take the consequences. A modern conservative sits still for an inoculation of social regimentation, be it in defense or collectivist welfare, in hopes of preserving a measure of freedom.

Eisenhower was a modern conservative: if a vigorous civilian space program served to blunt the appeal of Moscow, it might save the United States far greater sums in future military spending. A space race would not be the moral equivalent of war, but it would be a less expensive and more benign struggle within the total Cold War of which Sputnik was the trumpet.

The administration enjoyed unusual freedom when it came to budgeting for space. As Glennan noted in his diary, "Congress always wanted to give us more money. . . . Only a blundering fool could go up to the Hill and come back with a result detrimental to the agency."[20] But Glennan and the PSAC thought a crash program unnecessary and dangerous, and Eisenhower hoped to hold the NASA budget for FY 1960 to about $485 million. ARPA requests would push total space spending to $830 million, an imposing figure for 1959. Senator Johnson had already boasted that he would add substantially to the administration's space budget no matter what it was, and Ike feared that Congress would "break loose under the pressure. World psychology on this matter has proven tremendously important. . . . people are demanding miracles." The President granted that big boosters were "the visible element in affecting world psychology," but he balked at Glennan's estimate that Saturn costs might reach $2 billion. Ike concluded: "we must balance measures of fiscal soundness against extra measures in this particular field. . . . At the same time, the relationship of the program to the Soviet rate of advance must be clearly recognized."[21]

In May 1959 Glennan bore witness to this tension in calling for a broadly based review of all space activities. The enormous technical challenges ahead were now apparent—a "business as usual" basis was impossible. But, he added, "it is not clear that a crash program is warranted, or indeed would be substantially more productive" than an orderly approach. He lauded ARPA for refusing to treat space as a special field but simply as another environment in which space activities competed with other means of achieving the same objectives. But NASA could not do that. It had no objective. He thought it imperative that the administration develop a clear and supportable position for nonmilitary space.[22]

Glennan's discomfiture grew when he was called upon to justify his FY 1960 budget according to urgent goals that he privately admitted the government did not have. Consequently, his testimony invoked arguments that became mainstays of NASA self-justification throughout its history: the Soviet challenge demanded a vigorous response; uncommitted nations were influenced by space achievement; space investment paid for itself many times over in economic benefits "that will dramatically affect the lives of all of us."[23] He got his money, but patterns were set that ran counter to Ike's own proclivities: NASA clearly needed the Soviets; its *raison d'être* was not science but competition with the technocratic USSR

LUNA II AND III

The Soviets tried again for the moon on September 12, 1959. Their *Luna II* spacecraft sped on a collision course for the earth's natural satellite and smashed into its surface just 270 miles from the moon's visible center and eighty-four seconds after the predicted moment of impact. This astonishing exercise in guidance had sober implications for the accuracy of Soviet military missiles.

Luna II carried the hammer-and-sickle flag of the Soviet Union, and a wag suggested that space would soon be *res communist*, not *res communis*. Another wondered if the Soviets planned to splash the moon with red paint, after which the United States might follow with blue paint to make the moon red, white, and blue. Khrushchev observed that capitalism was what made U.S. moon rockets fall into the ocean.[24]

Three weeks later, on the second anniversary of *Sputnik I*, *Luna III* went to the moon. The sophistication and potential of Soviet technology silenced all doubters when the spacecraft circled its objective and returned the first photographs of the far side of the moon. The *New York Times* still managed to exaggerate, however, declaring *Luna III* the first "Space Station in Orbit Around Moon."

on its own terms; the Soviets were implicitly correct that state-managed R & D made for economic vitality; the free market was increasingly obsolete in the most important new industries. As for Glennan's request for explicit goals, Eisenhower met it in October 1959, when he broke space spending into three elements: first, be sure the military got what it really needed; second, see that real advances are made so that "the U.S. does not have to be ashamed no matter what other countries do; this is where the super-booster is needed"; third, see to an orderly scientific program.[25] Prestige had moved up another notch in the national space priorities.

Glennan's plea for a review of the civilian space program was heard at the top. In the autumn of 1959 representatives from NASA, ARPA, and the private sector met in various forums to debate the goals of space activity, while the NSC reviewed 5814/1 pursuant to a new directive on space. At issue in both was the question "To Race, or Not?" The temptation must have been great to throw open the Treasury, unleash American know-how, and whip the upstart Soviets once and for all in this technocratic tournament. But to do so might kick off an orgy of state-directed technological showmanship that would be hard to stop, might spill over into other policy arenas, and would relinquish to the Soviets the initiative in defining the fields of battle for the hearts and minds of the world.

Meetings were numerous, debates often circular as befit a liberal society confronted by a technocracy. For instance, at a September 1959 meeting of scientific advisers, York insisted that the Soviets were not really ahead in rocketry, they had merely designed their ICBM for heavy, outdated bombs. There was no military requirement for big boosters, but they should be built anyway. Glennan agreed, but was cooperation with the USSR a viable path? Kistiakowsky, who had succeeded Killian as Presidential Science Adviser, thought not. Space was too involved with the military for cooperation. Robert Murphy expressed the view from Foggy Bottom: "The Russians' use of their space advantage has had a tremendous impact around the world. We cannot afford to discuss whether or not to compete—we must compete." How much? asked Glennan, and answered himself: the competition was not essentially military, but in prestige, and that dictated the pace. Karl Harr agreed, but was gloomy: "We cannot undo the Sputniks. . . . The situation we have now is that of two world powers. We cannot permit an image to exist that this is the end of the U.S. Golden Age . . . and the advent of a new, progressive USSR era." Only Kistiakowsky dissented, pleading that the United States not let itself be forced onto a battlefield chosen by others. The importance of space, he warned, might diminish just as the United States was heavily committed.[26]

A week later NASA and PSAC leaders met in the White House. They agreed that the United States should develop the F-1 engine with a million or more pounds of thrust, but that still larger boosters made by clustering F-1s await a decision on "all-out competition with the USSR."[27] Glennan and Dryden then formed a committee of outsiders "to examine into the significance of competition with the USSR for space leadership as a determinant of the magnitude, scope, and urgency of U.S. nonmilitary space efforts."[28] Its chairman was Crawford Greenewalt, president of DuPont, and its members a balanced hand of five businessmen, five scientists, and two "jokers" (in the words of Paul Nitze)—Walt Rostow and Nitze himself, both academic strategists of the first rank.[29]

After much study, briefings, and a preliminary meeting, the Greenewalt Committee held a day-long conference at NASA headquarters on December 10, followed by dinner at the White House. Greenewalt reviewed some facts. The United States could not compete in moon shots or large-scale manned spaceflight until the maturation of the Saturn rocket in 1964–65. Competition was impossible in any case when you do not know what your competitor is doing. But the impact of Soviet successes was real and damaging, and space "firsts" would continue to go to the other side for some time. The panel decided that the United States must respond in other ways: a presidential address making clear the irrelevance of space to the military balance, a clear outline of U.S. space objectives, maximum use of instrumented satellites, and concentration on impressive missions that did not require big boosters, like communications or solar energy.

Two factions emerged in the committee: the "science group" and the "space race group." The former petitioned for practical experimentation, fearing a prestige race that might drain all the money from scientific missions into hardware and spectactulars. The "space racers" called this short-sighted and stressed the psychological thrust of space technology: "If a third country interprets Soviet space leadership to mean the triumph of socialist over capitalist science and industry, the interpretation will color that country's expectations about the outcome of the Cold War . . . and lead it to act in a way as to help validate them. . . ."[30] The "science group" responded that it was unsound to think of overtaking the USSR at an early date, and thought in terms of an annual budget of $100 million. The "racers" spoke of $2 billion per year. Greenewalt himself leaned to the antirace position but refused to force his views on the committee.[31]

The debate reached a climax after dinner in the basement of the White House. Vice President Nixon presided. He had studied and listened carefully, and revealed a technical knowledge greater than that of some of the panelists. Speaking without notes, Nixon rambled on for forty-five minutes, the august audience listening in confusion, boredom, or admiration to a man who grasped, rightly or wrongly, the political symbolism of the Space Age. Politics, thought Nixon, had to rank higher than science. Congress would seek to make the U.S. program seem a failure and try to vote more money whatever the budgetary consequences. The real motive in space was prestige, but the excuse for action would be the presumed military implications. Sputniks had a tremendous impact in the uncommitted world because of the example of "a backward country coming up from nowhere." The key time period would be 1963 to 1966, "when the USSR will have moved out from under our major counter-deterrent. The eyes of the world will be directed toward the competition between the U.S. and USSR. . . . The question will be how many more Soviet successes can we stand." Khrushchev had made it a race, and

combined with the missile problem, with the exploding problems in the under-developed countries, the two or three years which could be gained in the space field [are] not just any three years, but vital, important years. . . . Space and the new world concept captures [sic] the imagination. It indicates power; the people do not downgrade the military potentiality of space. I would hope otherwise, but I do not think this is the case.

There might be more desirable crusades, but space had it all over them from the point of view of appeal. "If I thought Congress would support increased expenditures for medical programs, for foreign aid—dramatically larger—I would trade space for this, but they will not buy it."[32]

Nixon's political sagacity looked beyond the Eisenhower years to the mid-1960s. He iterated the need for balanced budgets and thought them possible. But the new symbolism dictated a push in space regardless of its merits. Outside the White House basement, meanwhile, the nation received less-informed judgments from peddlers of anxiety. But their thrust was in the same direction. "How to Lose the Space Race!" screamed *Newsweek.* "(1) Start Late; (2) Downgrade Russian Feats; (3) Fragment Authority; (4) Pinch Pennies; (5) Think Small; (6) Shirk Decisions." The *New York Times,* with little interest in accurate reportage, judged "U.S. Space Program Far Behind Soviets." "Johnson Ready to Open Fight—Demo Chief Believed Set to Blast Nation's Space Age Leadership," said the *Washington Star,* and "Economy Curbing U.S. in Space Race" announced the *Times* after the Soviet moonshot. "After Ike, the Deluge," wrote Joseph Alsop in a paroxysm of gall.[33]

The "To Race, or Not?" debate played itself out in the NSC, entering the final stages of its review when the Greenewalt Committee adjourned. The splits between agencies revealed the tensions in a democracy waging total Cold War. The BoB fought to delete the stated NSC objective of "overall U.S. superiority in outer space" and the commitment to achieve goals "at the earliest practicable time." The State Department spoke strongly for a commitment to *surpass* the Soviets in order to improve American stature abroad. The DoD and State split on the use of space for peaceful purposes, the JCS opposing any international agreements that resulted in a net military disadvantage for the United States. PSAC even challenged the assumption that "the U.S. is behind the USSR in total space achievement," and Kistiakowsky made a last-ditch effort to block language making competition for superiority a basic objective of space policy.[34] But NSC-5918, "U.S. Policy on Outer Space," approved by the President on January 12, 1960, granted the important scientific, civilian, military, and political implications of space technology, including the psychological impact of broad significance to national prestige. Thanks to their space accomplishments, the Soviets' "baldest propaganda claims are now apt to be accepted at face value." Failure to satisfy expectations that the United States would catch up might give rise to the belief that the United States was now "second best." American science and even space technology might in fact be superior, but to the layman the true conquest of space would be represented by manned spaceflight. It was therefore the American objective, among others, "to achieve and demonstrate an overall U.S. superiority in outer space without necessarily requiring U.S. superiority in every phase of space activities." To minimize Soviet psychological advantages, the United States should select and stress projects that offer the promise of obtaining a demonstrably effective advantage, and proceed with manned spaceflight "at the earliest practicable time [the BoB objected to this phrase]." This was not yet a green light for an "all-out" race, but a clear recognition that the United States must compete vigorously for space spectaculars.[35]

The outcome of the 1959 debate had immediate budgetary consequences. The NSC policy paper recommended a 60 percent increase in the NASA budget for FY 1961, and estimated that it would reach $2.1 billion by FY 1964. The BoB relented and even chipped in another $113 million for acceleration of Saturn and $108 million for Mercury. The budget went to Congress in February 1960, where avid committees anted up $50 million more for a total award of $964 million. The U.S. space program had only moved into second gear by the standards of the next decade, but accelerating through the lower gears always requires more torque. In only two and a half years, Eisenhower came to commit the nation to a vast enterprise in civilian command technology, while the Saturn, Mercury, and ARPA programs, to shift metaphors, bequeathed the foundations of a far loftier structure to the freemasons of technology in the next administration.

The only hope of checking the slide into a space race lay in arms control or cooperation in space. These two were also explicit elements of U.S. space policy.   What became of the commitment to cooperation? To begin with, negotiations at the UN were still constrained by U.S. half-heartedness and Soviet insistence on "general and complete disarmament" and refusal to permit inspection. Similarly, the military importance of space technology, technical asymmetry, and mutual distrust prevented bilateral cooperation. Indeed, NASA's rising budgets depended on the assumption of competition with the USSR. Consequently, NASA itself, the civilian agency meant to express American policy on cooperation, became an institutional skeptic toward large-scale cooperation with the Soviets. Glennan urged Eisenhower, preparatory to his planned trip to the USSR in 1960, to refrain from proposing cooperation in space beyond safe and useful sharing of meteorological data.[36]

The clause in the space act enjoining NASA to seek international cooperation was a departure in the history of state technology policy. Its origin was twofold: propaganda value and the hope of preventing an arms race, where the Soviets were concerned; propaganda and the hope of tapping the science and resources of the allies, where the Europeans were concerned. Sputnik initially caused a widespread feeling that the United States was incapable of going it alone and should not try to carry the whole strategic burden on its back. But R & D is a complicated enough affair without trying to do it in an international setting. In any case, the French and British had little to contribute to a free-world missile and space effort. As Lord Hailsham observed in Britain: "International cooperation is no substitute for national excellence."[37]

Still, NASA had to confront its dual birthright: it was in a space race yet being told to be cooperative. Having won the facilities, missions, and mandates for a prestige race, NASA turned abroad and defined the relationship of the U.S. space program to the world.

The IGY with its international Committee on Space Research (COSPAR) was a powerful precedent. Certainly scientific data could be shared despite the bitterest competition. But hands-on collaboration with the Soviets or allies presented innumerable problems. Both Glennan and his hard-headed appointee for international affairs, Arnold Frutkin, had experience in international atomic energy programs, and both were keenly disappointed in Atoms for Peace. Many countries, it turned out, had cheaper sources of energy or insufficient demand to warrant nuclear plants. Few had the universities and industry to support a domestic nuclear program. In these cases, vigorous "cooperation" simply meant bestowing U.S.-made reactors at U.S. expense, with U.S. technicians, in the hope that U.S.-trained nationals could someday take them over. Then there were the questions of proliferation and congressional suspicion of "give-away" programs.

Frutkin therefore opted for a new realism when addressing space cooperation. U.S. policy would be based on literal cooperation, not aid or support. Underdeveloped countries frankly had no role to play in space technology, and raising their hopes did no service to them or the United States. Each nation should decide for itself whether it wished to spend money on some aspect of space technology. If it did so, the United States would welcome mutually beneficial proposals. But there must be no courting of such proposals or U.S.-inspired boondoggles, just for the sake of "cooperation."[38]

NASA did have some mandatory foreign chores to perform in order to set up a global tracking network. Seeking diplomatic aid, Glennan found few State Department officials versed in scientific matters. What was more, the State Department's negotiating procedures were cumbersome and their demarches to target countries tended to get bound up with other diplomatic issues. Glennan then encouraged Frutkin to set up his own "little state department" within NASA. With commendable skill and energy, Frutkin acquired tracking stations in two dozen countries over the years, including some, like Mexico, that had snubbed American requests for military bases. Frutkin also expanded cooperation to such activities as launching by the United States of other nation's satellites, integration of foreign experiments on U.S. spacecraft, joint high-altitude sounding rocket projects, foreign data analysis, and extensive training arrangements in NASA and American universities for foreign technicians and students.

These pragmatic programs paid off in goodwill and a positive image for NASA abroad, at least in the first six or eight years. But they were not what idealists had in mind when they spoke of a humanity united in space.[39] Nor did they do much to pool the resources of free-world science and technology. But, as Frutkin emphasized, each nation must decide for itself, and in the wake of Sputnik, the other states with serious rocket programs, Britain and France, defined their needs in fatefully different ways, each of which constrained further the possibilities of cooperation with the United States.

The new French Fifth Republic, established after the return to power of Charles de Gaulle in 1958, committed itself to an ambitious program of technological independence that included nuclear and missile forces. Britain, possessed already of the world's third nuclear deterrent, judged the costs of competing in the missile age beyond its means, cancelled its own IRBM program, and accepted dependence on the United States for advanced strategic technology. The United States could not, therefore, cooperate extensively with France in rocketry and spaceflight without violating its policy on nonproliferation, while the British had little to offer at all. In coming years the Europeans formed their own space research organizations, but these would be designed precisely to promote European competitive independence from the United States in advanced technology. With the exception of some scientific programs and military space systems for NATO, U.S.-European cooperation in space was limited. This fact, combined with the free regime developing at the UN, suggested that space technology, like atomic power, was destined to develop in a context of competing national states.

Rapid expansion in manpower and facilities, clear and large-scale missions for the next decade, security from military rivals, and a well-defined political stance vis-à-vis the outside world—all this made NASA by 1960 a "line agency" with little in common with the old NACA. It won primary responsibility for space R & D and execution of the most expensive, prestigious programs. To meet these tasks, it absorbed not only the NACA centers, Goddard, the Huntsville rocket team, and JPL, but a rocket test facility and launch complex carved out of the USAF Eastern Test Range at Cape Canaveral. NASA's national complex had taken shape already in 1960, even if it had not reached its ultimate girth. Born as a civilian sparrow in a nest of warbirds, NASA grew up and flew.

Conventional wisdom portrays Eisenhower as skeptical and tight-fisted regarding space, in contrast to his enthusiastic successors. This is part of the picture, to be sure, the "rearguard" aspect. But it obscures the fact that Eisenhower also secured NASA's place as a growing technocratic enterprise. Ike founded the civilian agency, nurtured it, gave it the major missions and the tools it needed, and linked it to national prestige. Once the critical judgment had been made that the United States should promote its space program as open, peaceful, and scientific—but still in competition with the Soviet—the future of NASA was assured.

Ironies behind the decisions nevertheless caused American space propaganda to diverge from reality. The United States spoke of civilian spaceflight, yet pushed a broad, secret military program in NASA's shadow. The United States made much of cooperation, but worked to prevent UN limitations on national prerogatives while displaying conservatism even in cooperation with friends. Did these indicate bad faith? Only to the most ungenerous. For the space race posed root questions of principle for the Eisenhower administration, and the dualities of space policy followed less from a desire to dissimulate than to uphold traditional values. Eisenhower compromised but did not surrender to a technocratic cynicism that relegated all creative enterprise to state control or all "peaceful pursuits" to politicization. An "honest" space program might have been one single, coordinated effort run by the DoD and pursued in outspoken competition against an "inferior, flawed Communist rival." This would have been candid, but would also have been a mirror image of the Soviet posture. U.S. space institutions at least reflected the values of free, open, international inquiry and discovery for the elevation of the human spirit, even if deeds did not always measure up. The United States under Eisenhower traveled far on the road to technocracy, but it still sheltered the memory of goals loftier than those of the power-state.

CHAPTER 10

The Shape of Things to Come

Jeffersonians, and their heirs in both parties, suspected that the state and society were natural adversaries. Unless the people were vigilant, the government would exploit its powers to tax, regulate, and coerce to overwhelm private life. Even the Progressive Era and the New Deal, the latest in a series of reform movements enhancing federal power, were aimed at the goal of preserving in the main both individual liberties and the free market by compensating for human corruption in the prior case and natural disaster and the business cycle in the latter. As late as 1955 Eisenhower Republicans still regarded most state intervention as a necessary evil. But contrary intellectual trends, supported by but not born of Sputnik, had grown so powerful by 1960 as to capture many leaders of journalism, academe, business, and politics who would set the agenda for the next decade of American history, while the 1958 elections returned the most liberal Congress since the 1930s.

Indeed, a curious reversal had occurred: the spokesmen for "society," including many Republicans, urged sharply increased government activity regardless of cost or principle, while the executive branch alone championed curtailment of its own scale and scope! This anomaly could last only as long as Eisenhower, the "tired old man," remained in office. Another president, in step with the trends, would surely exploit the calls in the land for the government to set and fulfill, on behalf of the people, a bold agenda of national change.

Can one really speak of dominant trends among elites when the many critics of Ike's America held different and sometimes contradictory views? The Democratic Congress, for instance, was cloven by Stevenson doves and Symington hawks (to borrow the parlance of the next decade), Hubert Humphrey integrationists and segregationist Dixiecrats, incipient Keynesians like Paul Douglas and traditional liberals wanting to balance higher budgets through higher taxes. But almost all agreed on the need for sharply increased federal action in one or another arena of public policy. The post-Sputnik critiques, beginning with defense policy and quickly encompassing American science, education, materialism, social inequalities, and general "softness," helped make possible a strange alliance between the social activists and the military activists, each lamenting the country's drift and malaise and each extolling vigorous and self-confident action to "get the country moving again." The alliance showed signs of life in 1958, coalesced in the Kennedy/Johnson ticket, and lasted for another eight years until it was shattered violently by "Great Society warfare" in Vietnam.

Attacks on Eisenhower multiplied after the installation of the Eighty-sixth Congress in January 1959. Recession, Khrushchev's rocket rattling in Berlin, Castro's triumph, insurgencies in Laos and the Congo underscored Communist virulence and American indolence. At home the President vetoed two large housing bills, two large public works bills, and other liberal initiatives. In 1960 a second civil rights bill passed, while acts to increase the minimum wage and aid public schools, housing, and "medicare" were hotly debated. Defense-minded senators pressed for large increments in military spending. By the end of the term it was clear to all Democratic constituencies that fulfilling their agendas meant cracking traditional beliefs on the proper role of government.

In national defense, the marriage of Cold War vigor and incipient technocracy was most clearly illustrated by the new prominence of civilian strategists. Composed mostly of political and physical scientists, the "strategic community" sold itself to the politicians, public, and even military chiefs as the locus of expertise on arcane questions of strategy in the age of high tech. The assumption (which they themselves promoted) that nuclear weapons changed forever the nature of warfare and the usually unspoken but traditional American assumption that generals were bellicose, undemocratic, and not very bright supported the civilian claim to co-opt the strategic function.[1] Warfare had now been politicized and democratized such that it involved the manipulation of entire societies—science, industry, and economics. Hence military work lost its autonomy, and the officer corps became only one body of specialists in a pervasive and functionally differentiated national security apparatus. The opinion spread that the task of overall planning fell naturally to those political scientists who also knew something of weapons technology and to those physicists who also knew something of politics and war— the strategic community.[2]

RAND epitomized the sort of think tanks—born of World War II "operational research" and weapons R & D—that spread across the country on the assumption that strategy was susceptible to the scientific method.[3] And the proliferating strategists, moving from institute to government and back again, writing for everything from dense tomes to Sunday supplements, achieved a growing prestige from their willingness to treat matters both mysterious and loathsome to the public at large. These were the men who invented, applied, and debated the concepts and variables of the nuclear age: deterrence, first strike/second strike,

city busting/war-fighting, countervalue/counterforce, stabilizing/desta-
bilizing, tactical and theater weapons, throw-weight, megatonnage, circular
error probability, systems and cost-benefit analysis—the arcane language
of the age. Generals, senators, journalists, and assistant undersecretaries
might all learn the lingo, but their choices were limited to what the
language could express. Whatever eddy they chose to bathe in, the
aqueous element was the same throughout the sea of academic strategy.
It was John von Neumann who invented "mutual assured destruction,"
and, together with Oskar Morgenstern, pioneered the use of "game
theory," Thomas Schelling and Morton Kaplan who applied game theory
to nuclear strategy, Bernard Brodie who established assured destruction
as the "strategy for the missile age," Henry Kissinger who popularized
"limited nuclear war," Herman Kahn who made nuclear war "thinkable"
and, he argued "winnable."

The middle 1950s were a fruitful time for the strategists. Massive
retaliation had been under attack since its inception,[4] the young Kissinger,
under the auspices of the Rockefeller fund, called for across-the-board
rearmament in pursuit of flexibility and proportionality,[5] while the dean
of civilian strategists, Brodie of RAND, concurrently composed the
"bible" of deterrence. His *Strategy in the Missile Age*, first published in
1959, bluntly preached mutual deterrence: short of preventive war, there
was no alternative to doing whatever was necessary to erase the perceived
advantage of a first strike. This meant reducing the vulnerability of one's
own retaliatory force and promoting "stability" by every means. Stalemate
was the best that could be hoped for. "As far as limited wars are
concerned, they can have little more than the function of keeping the
world from getting worse."[6]

Brodie warned against the "academic vice" of assuming that atomic
weapons made war unthinkable. But a second school, led by Herman
Kahn of the Hudson Institute, looked into the nuclear abyss without
growing dizzy. His calculations of postholocaust recovery rates, depending
on how many tens of millions of deaths resulted from a nuclear exchange,
offended readers but established the case that nuclear destruction was
both possible and finite. To deny such facts only disarmed those who
hoped to prevent it, by "deterring the deterrers." To avoid the choice of
war or surrender, "we must have an alternative to peace." But under
current programs the United States might find itself unwilling to accept
a Soviet retaliatory blow and be unable to uphold its commitments to
allies. Where Brodie insisted that "strategy wears a dollar sign," Kahn
called for a minimum increase in the national budget of 10 to 20 percent.[7]

Massive retaliation and Eisenhower's penury did not lack for critics
within the military itself. What is significant about the complaints of the
"brass," however, is that what they protested, without putting their
finger on it, was just this assumption of the strategic assignment by
civilian theorists and managers, who thereby stole the responsibility,

dignity, and autonomy from the military profession. This by-product of scientific warfare dated from World War I in Europe, but only after 1945 did the U.S. officer corps lose its grip on the fundamentals of war-planning, for only then did the United States not revert to "normalcy." The roles and missions controversy left all the services with permanent grudges against civilian-imposed strategies, but Sputnik legitimized their dissent as nothing before.

To the intelligentsia, Maxwell Taylor seemed that rare thing—a "good general." He was urbane, erudite, and articulate. After four frustrating years as Army Chief of Staff, he retired in 1959 and turned public advocate for "flexible response." He traced the trouble in U.S. defense to the fact that military strategy was now a function of civilian constraints, not military judgments. This in turn politicized the military. The wholesale turnover of the Joint Chiefs at the end of Truman's term (on the recommendation of Senator Robert Taft [R., Ohio]), was a damaging precedent, for it suggested that the Chiefs were no longer professionals but creatures of the political party line.[8] Of course, Taylor himself came to practice what he condemned, for he made himself the most acclaimed of "political generals" and prepared his return to power under the Democrats. His notions suited well the demands of the new consensus. The army in particular needed air support capability as well as IRBMs. All the services had legitimate roles in outer space. But the principal chore of the military in the coming decade was to gear up for limited wars and counterinsurgency in the underdeveloped world. This, plus antimissile defense (another army mission) and fallout shelters, would cast for the United States a "new and certain trumpet" in the age of mutual deterrence.[9]

The frustrated ABMA chief, Bruce Medaris, also lamented the shift in military policy away from the professionals and into the hands of civilian bureaucrats. Strategy was co-opted by accountants, officers won promotion through conformity, budgetary ceilings obliged the services to argue for plenty when need was absent and suffer poverty when need was urgent. Medaris's remedies included reunion of the army and air force into a single service, a single space program, a single Chief of Staff, reduction by *90 percent* of the civilian staff of the Secretary of Defense, retraction of the power of the BoB to reallocate funds passed by Congress, and breaking the power of pressure groups (especially the aviation industry). Medaris did not hide the fact that he saw the defense problems of the United States in moral terms. Current trends encouraged greed, deception, and toadying in the once-proud officer corps. If Americans quit paying the price in discipline and sacrifice, they would lose their lease on freedom, and the great forces of science and technology would fall to the purveyors of Communist ideology.[10]

What were the new frontiers of strategy in the age of mutual deterrence? Surely they were in the jungles of the new nations on the one hand and in the vastness of outer space on the other. Even before the éclat of his retirement during the Johnson hearings, ex-chief of Army R & D James Gavin recorded his thoughts on *War and Peace in the Space Age*. Sputnik and the space threat were a challenge without equal, wrote Gavin, for beneath their new technological canopy the Soviets would exploit the yearnings of dissident nationalists and agrarian upheavals in the developing world: "Space is the theatre of strategy of tomorrow—space and the human mind."[11] Soon military satellites of all kinds would become routine, as would manned orbiting stations and perhaps even control of the weather and outposts on the moon. The United States must organize for the new age with a unified Space Command. Once space was secured, the UN could be brought in to establish, once and for all, a lasting peace.

The Gavin plan for space required a well-conceived technological strategy for which scientists, industrialists, and soldiers must coordinate their activities. But currently, he said, U.S. strategy was stymied by politics. Liberals called for big budgets to provide flexible response to any sort of conflict; conservatives said, "Pick the best weapons system and put your money on it. Reduce all others to a minimum." Such conflicts became embedded in the procurement process, with industrial contractors creating disputes that appeared to the public as interservice rivalry. Such disputes had to be transcended. "[We must] not wait until science gives us the hardware to decide what is to be done." A space strategy must emerge first, assuring "that the demands made upon science will meet our actual needs."[12]

These and other military critiques in the post-Sputnik years differed in perspective and emphasis but had telling points in common. All sought to take the missile revolution to its logical conclusions. All cried for the liberation of the military from bureaucratic and budgetary constraints. All promoted across-the-board defense in place of massive retaliation. All relied on intensified exploitation of command technology. The contradiction in this, suggested by Medaris, was that reliance on command technology was precisely what led to civilian preemption of military decision making, corruption of the arms industry, the politicization of strategy, and the blurring of roles and values among previously autonomous realms of social work. The age of perpetual technological revolution and total Cold War was inevitably the age of politicization of the military and the replacement of intuition, honor, and battlefield courage by the exploits of the machine. In such an age, what training or virtue made the soldier more qualified to judge matters of national defense?

The trend toward overlapping authority in strategic planning promised a plethora of problems for relations among the state, the military, corporations, and universities. But for the moment there was a campaign to conduct, and many Democrats eagerly seized on the *cris de coeur* of

Looks Like We May Have To Buy Them Both

March 3, 1960. Courtesy of D. Dowling, N. Y. *Herald Tribune.*

dissident generals to bolster the attack begun by civilian strategists and social critics against the Republicans. They called for reinvigoration and sacrifice, which translated into sharply increased spending for arms, foreign aid, internal mobilization, and forced economic growth. The consensus of the late 1950s conflated the growing strains of protest against limited government. All that was necessary was a political synthesis of the Cold War posture and the "progressive" social posture and leaders willing and able to sell the synthesis with the aid of sympathetic media. The first task was performed elegantly by Walt Rostow and Nelson Rockefeller; the second by Kennedy and Johnson.

Rostow was to political economy what Kissinger was to strategy—a brilliant, synthetic intellect encased in a personality presumptuous enough to offer solutions to the problems of an age. He understood the terrible reality of the Soviet challenge and the handicaps of democratic capitalism in the struggle to woo developing countries. Western values and institutions were subtle, gradualist, and based on faith in the invisible mechanisms of enterprise—process rather than prescription. How could the West hope to compete with Marxist promises, even if one discounted the heritage of imperialism and climatic and cultural factors that seemed to make liberal capitalism irrelevant to peoples of Asia and Africa? The Eisenhower administration was not as blind to these problems as often supposed—Dulles had made eloquent speeches about the challenges of decolonization—but it had little in the way of a plan for coping with the political dangers stemming either from stagnation or from rapid change in the new nations. Rostow's *Stages of Economic Growth,* unabashedly subtitled *A Non-Communist Manifesto,* purported to offer an alternative to the Marxist model.[13]

Rostow boldly stood Lenin on his head. Where the Marxists claimed bourgeois capitalism to be a temporary stage within a larger historical tableau tending toward universal socialism, Rostow called communism an aberration of immature economies within a larger tableau tending toward liberal institutions and individual emancipation. Western values were the ones in accord with human nature; the West had no need to be defensive. Rather the United States should help new nations resist the socialist temptation during the unsettling stages leading to economic "take-off" by helping to build the infrastructure of public services ("social overhead") necessary to industrialization and then by channeling in enough investment to push countries beyond the critical level needed for "take-off" (estimated at 5 percent of national income).[14]

Rostow's model, oversimplified here, was nonetheless simple. One is even tempted to conclude that he reasoned in reverse, beginning with the question "What can the United States do?" and working backward from the answer "Invest" to a model endowing that answer with teleological force. Adlai Stevenson certainly reasoned this way when he said that the United States could best meet the Sputnik challenge by helping the "have-nots" to get "20th century technology and the high living standards it inevitably produces."[15] Whatever the merits of the "non-communist manifesto" (its economic and historical soundness was debated for two decades), it had a profound influence on American foreign and domestic policy. For if the United States was to peddle an investment model to the underdeveloped world, then it must surely demonstrate that the model worked at home. If high levels of state-directed investment in infrastructure, plant, and technology were to work miracles in regions where capitalism suffered handicaps, then it surely must work in spectacular fashion in the industrial democracies. If the U.S. message was to be heard by the poor, colored masses of the world, then the United States must show that its affluence reached the poor and colored at home.

Rostow also had strong views on space policy. As a member of the Greenewalt Committee, he considered technological competition to be critical and had "a bias toward hope rather than skepticism." But the United States' "inferior position in space exploration" was only one area in which he wanted "to see American energy, talent, and resources allocated." The others were the missile gap, aid to underdeveloped areas, the disarray of NATO, and the weakening dollar. "It follows that I would advocate an increase in the federal budget along a broad front, rather than in the space field alone."[16] Rostow then embarked on a three-volume lambasting of Eisenhower's fears of federal growth for the Subcommittee on Economic Statistics. Democracies, he wrote, could not afford to duck important challenges in hopes of saving democracy.[17]

Reinvigoration of American life was the *leitmotif* in the symphony of protest from 1958 to 1961. The United States seemed lifeless in a dynamic world. Yet Americans were idealistic, striving, "can do" people, or so went the venerable myth. They hungered for a sense of mission beyond consumerism and yearned for sacrifice. In retrospect, one may ask why, if the myth was true, Americans were so needy of communal missions, why the search for meaning and charity in one's own life was not enough. Yet the public was told that it needed goals and that the agenda must be set, and fulfilled, by elites. No clichés were more abroad in the land than those about "goals" and "excellence," and nothing so invested them with meaning as the humiliations in space. Chief among the many studies of national goals were the Rockefeller Panel reports, begun in 1956 after Eisenhower's refusal to boost military and foreign aid spending, and published over the years 1958 to 1960. They began with defense and moved on to foreign policy, economics, education, social policy, and "the future of American democracy." The panels brought together hundreds of "leading citizens" from business, labor, education, journalism, religion, and elsewhere, an agglomeration of elites that laid down *Goals for Americans.* They recommended greatly enlarged outlays on strategic and conventional arms, increased investment to accelerate growth in the developing world, and "radically increased public expenditures" at home for urban renewal, health, social security, and education. These programs would be financed thanks to a real growth rate "sharply increased and steadily held at 4 percent or higher." This was possible thanks to the new economics plumped by Paul Samuelson, Galbraith, and others that asserted the ease of "managing growth" through public investment, deficit spending, exploitation of new technology and "human resources."[18]

Intensified exploitation of brainpower was the special concern of a PSAC committee appointed in 1960 to study science education. Its report, named for chairman Glenn Seaborg, chancellor of the University of California, Berkeley, confirmed Ike's fears of snowballing federal intervention in education. Starting with the assumptions that "both the security and the general welfare of the American people urgently require continued, rapid, and sustained growth in the strength of American science" and that "the defense and advancement of freedom require excellence in science and technology," the Seaborg Report justified a revolution in science and education policy. "The right word [for state funding of science] is *investment*," it said, for science had extraordinary economic power; its returns were "literally incalculable." Simple self-interest dictated that "we increase our investment on science just as fast as we can, to a limit not yet in sight."[19]

Especially important to the Seaborg Committee was the need to erase distinctions between pure and applied science (in Marxist terms, science should be considered a "direct productive force") and for the federal government to accept its responsibility for all science. *"From this responsibility the Federal Government has no escape."* A government/university partnership must form that does "much more good than harm." But since "mediocre research was generally worse than useless," the government must reinforce outstanding work wherever it was found. Recommendations ran to form: vast new sums for facilities, faculties, and students, scientific guidance left in university hands, federal policies to govern support of research and education drawn up in consultation with academic administrators.[20]

Academic strategists, disgruntled generals, ambitious economists and scientists, social reformers, and politicians—all contributed to the new consensus in an enlightened United States. Common to all was the perception of need for vastly increased federal spending and power, a perception singularly favorable to an alliance between defense and space "hawks" and social "doves" built on the presumed realization that state investment in science, technology, and education would refill the coffers of a generous Treasury. Ike alone did not buy the new orthodoxy and was hooted down. His successors bought it.

On January 2, 1960, Walt Rostow, on board with the Kennedy campaign, drafted an electoral strategy to defeat the presumptive rival, Vice President Nixon. The Democrats should hit the Republican record, "which has so substituted rhetoric for action." The United States had been living off capital for seven years, wrote Rostow. The decisive issue was the military one, because "nothing would swing votes like the conviction that the Republicans have endangered the nation's safety.

. . ." But the military issue should be expanded by identifying the Republican party as primarily interested in restraining the federal budget, thus the cause of the military problem also accounted for the inadequate programs in education, transport, cities, and so on. "The missile gap can be used as the Charles Van Doren of the Republican Administration: the most flagrant symptom of a generally bad situation."[21] The Democrats should hit hard on the lack of evidence of Soviet commitment to arms control and the fact that American weakness only reduced chances of agreement. The urgent correction of the missile gap should be presented as a precondition for peace.[22]

Nixon would be torn, thought Rostow, between presenting himself as Eisenhower's man or as a young man of independence. Democrats should exploit this. Nixon might break ranks with Eisenhower, but that would only seem "the ultimate zag of Tricky Dick" and "a confirmation of the phoniness of the peace and prosperity slogan." Rostow continued:

Whether or not Nixon cracks, the fact should be borne in mind that every responsible politician and civil servant—of both parties—with minor exceptions, and virtually every responsible journalist knows the facts to be much as I have assumed them here; they were waiting eagerly for Rockefeller to get the mush out of his mouth and say it, which he never did; and my guess is that there will be a great lift of support for the first Democratic candidate to take this bold line.[23]

Rostow was right about the breadth of agreement with his "facts." But Democrats needed no prompting to play up the missile gap. George Reedy told LBJ in 1957 that he could ride it to the White House. The Sputnik horror show played to crowds everywhere during the 1958 campaign. Now Kennedy adopted it for his own. The best that can be said for the Democrats' alarmism is that no one knew for sure how many ICBMs the USSR had in place. Some CIA estimates were as high as 100 Soviet ICBMs deployed by 1960, 400 in 1961. But these "worst-case analyses" assumed that Korolev's test rocket went into immediate production and that maximum industrial capacity was assigned to it. Khrushchev did all he could to encourage such estimates, and when U.S. generals and aerospace journals echoed Soviet claims, they drowned out the expressions of calm emanating from a President whom the media depicted as feckless and complacent.

Hard evidence came from U-2 flights. While irregular and spotty in their coverage, they seemed to indicate no deployed ICBMs at all! Secretary Gates informed the Senate in January 1960 that intelligence had downgraded the Soviet threat. But that only prompted Symington to charge that Eisenhower had "juggled intelligence data so that the budget books may be balanced" and LBJ to call such estimates "incredibly dangerous." *Newsweek* reported the "most stinging expression of disbelief ever hurled at this Administration," and Joseph Alsop accused Ike of playing Russian roulette.[24]

On May 1, 1960, a Soviet surface-to-air missile shot down the U-2 piloted by Francis Gary Powers. The real circumstances surrounding the flight are still secret,[25] but as Oliver Gale noted in his diary, a happy consequence of the episode was that "now we can back up our 'intelligence estimates' as something other than the 'guesses' we have been accused of. It was frustrating to have hard evidence and not be able to use it in the face of vigorous attacks on our credibility."[26] Still, the U-2 incident was a disaster in other respects. The administration first issued the story that it was a NASA weather plane, which NASA officials were obliged to parrot until the Soviets unexpectedly produced the pilot and wreckage. Next, Eisenhower refused to repudiate the surreptitious mission or save face for all parties by blaming subordinates, whereupon Khrushchev exploded the Paris summit and Ike's last hope for a nuclear test ban.

The congressional inquiry that followed was thorough, suspicious, and secret. Allen Dulles testified for five and a half hours on the U-2 and displayed photographs of the USSR. The stereographic tapes were shredded, save one copy for the committee safe. There is little doubt that Democratic leaders learned whatever the CIA knew about Soviet missile deployment, if indeed they had not known all along, and that no basis existed for panic.[27] The Killian report still held true; the SAC B-52s more than outweighed whatever handful of ICBMs the Soviets might possess. Jupiters had been placed in Britain in 1958, the first battery of twelve Atlas ICBMs was commissioned in May 1960, and the first Polaris submarine, with sixteen nuclear missiles, would go on station in November. The campaigning Democrats and much of the media, however, preferred to believe the news releases of *Tass*.

In the battle on space policy, Eisenhower did not help matters. He never could get across with enough force that Soviet space shots did not bear on the military balance, and he continued to suggest that the United States was above the battle. In a news conference a week after his lackluster budget message, the President was asked why, in view of the prestige at stake, the United States did not move with more urgency to catch up in space. Eisenhower questioned whether prestige was at stake. "Well, sir, do you not feel that it is?" the reporter asked.

"Not particularly, no," said Eisenhower, and he defended the U.S. space program as one to be proud of.[28]

Another week passed and Eisenhower countered another query about his sense of urgency: "I am always a bit amazed about the business of catching up. What you want is enough. . . . A deterrent has no added power once it has become completely adequate."[29]

Yet another week, and a reporter asked, "Mr. Khrushchev is quoted as saying, 'Our flag is flying on the moon. This means something. Is this

TIROS, TRANSIT, KORABL SPUTNIK, AND ECHO

American satellite programs, while not as spectacular as the Luniks, did far more to advance applications of space technology. *Tiros 1* and *Transit 1B* were space "firsts" of far more meaning than hurling a flag to the moon. *Tiros 1* was the first weather satellite, launched by a NASA Thor-Able on April 1, 1960. It orbited 450 miles above the earth, whence the spin-stabilized spacecraft televised awe-inspiring pictures of the earth's cloud cover. *Tiros 2* through *10* followed over the next five years, all successful, and set the stage for a second generation of more sophisticated weather satellites. The United States shared its meteorological data with the world, making the first major and lasting contribution to cooperation in space for the benefit of all mankind. Tiros also impressed on thoughtful citizens the potential of military reconnaissance from outer space.

On April 13, 1960, the first nagivation satellite, the navy's *Transit 1B*, went into orbit on a Thor-Delta. Ships below could home in on its transmissions and calculate their own position with astounding precision. This first test model promised that soon operational systems would permit atomic submarines to cross an ocean without surfacing and still thread a needle like the Strait of Gibraltar and merchantmen under many flags to ply the seas with efficiency and safety.

But the Soviets always seemed to get the jump on "important" things. In May 1960 they launched the five-ton *Korabl Sputnik I*, a test vehicle for manned spaceflight. Recovery failed after sixty-four orbits due to incorrect attitude at retrofire, but three months later *Korabl Sputnik II* successfully parachuted to earth, returning two live dogs, Strelka and Belka. The Soviet promise of men in space drew nearer to fulfillment.

Eisenhower's approval of low-cost prestige shots resulted at the same time—mid-August—in a passive reflecting communications satellite called *Echo*. It was a balloon thirty meters in diameter that inflated upon reaching orbit. *Echo* outshone everything in the sky save the sun and moon, and was an impressive sight to the peoples of the earth.

not enough to prove the superiority of communism over capitalism?" What do you think of such remarks?"

Eisenhower: "I think it's crazy. . . . We should not be hysterical when dictatorships do these things."[30]

The 1960 campaign was not fought on the space and missile gap alone. But no issue better symbolized the New Frontier that the Kennedy/Johnson ticket asked the American people to explore. Indeed, the new symbolism was just as real to Nixon, who had invoked it in his San Francisco speech of 1957 and again to the Greenewalt Committee. The space race was a metaphor for the varied challenges facing the United States at home and abroad. As Kennedy explained:

The people of the world respect achievement. For most of the twentieth century they admired American science and American education, which was second to none. But they are not at all certain about which way the future lies. The first vehicle in outer space was called Sputnik, not Vanguard. The first country to place its national emblem on the moon was the Soviet Union, not the United States. The first canine passengers in space who safely returned were named Strelka and Belka, not Rover or Fido, or even Checkers.[31]

Eisenhower came to Nixon's aid in August with a statement on U.S. achievements in space. It catalogued American successes, especially those "that promise very real and useful results for all mankind." Nixon applauded this record and rebuked Kennedy for attempting to "hitch his political wagon to the Soviet Sputnik." It was "irresponsibility of the worst sort for an American presidential candidate to obscure the truth about America's magnificent achievement in space in an attempt to win votes." Eisenhower, said Nixon, had just about closed the missile gap inherited from the Democrats. At present thirty-six satellites had been orbited; twenty-eight were American. In practical research and benefits the United States was still farther ahead. Even Soviet booster power would soon be matched by the Saturn, which would make possible, during the next president's tenure, manned space stations and circumlunar flight. Nixon pledged that under his administration, the United States would be second to none in space.[32]

The candidates placed their space planks on record in response to an invitation from *Missiles and Rockets* magazine. The editor's open letter displayed the bias of the trade press against "the ambiguities of the Eisenhower era," assumed that the USSR was "forging ahead," and called on the candidates to promise a space station by 1965, lunar landing by 1967–68, and reusable space plane by 1968–69. It advocated a dominant role in space for the military and a defense budget set by need and not ceilings.[33]

Kennedy agreed with *Missiles and Rockets* that the United States was losing the space race. Control of space would be decided in the next decade, and the nation that controlled space would control the earth. Kennedy pledged immediate acceleration of all missile programs and basic research, but he backed off from the demand that space be turned over to the military or that objectives be reached by specific dates. Nixon pointed to the record, insisting that "if the Eisenhower Administration had not long ago recognized that we were in a strategic space race with Russia, our space record would not be as creditable as it is today. . . . Today we are ahead of the USSR." Nixon promised continued progress. Moon landings, he said, were scheduled for 1970–71, but "it is entirely possible that this target date will be advanced."[34]

Moon landings scheduled for 1970? Nixon was exaggerating, but studies of such advanced programs were indeed underway. NASA's charter and NSC directives made it incumbent upon the agency to plan for the long term. Glennan, too, thought this prudent, to avoid being yanked to and fro by every Soviet surprise. The Mercury program, placed under Robert Gilruth's Space Task Group at Langley, was frankly a stopgap aimed at getting a manned capsule into space as quickly and safely as possible. But in mid-1959, when the first seven astronauts were introduced to the public, a Research Steering Committee on manned spaceflight, led by Henry Goett of Ames, looked beyond Mercury to identify the problems NASA must tackle in the future. It designated a manned lunar landing as an appropriate and "self-justifying" goal. With these first studies the NASA twig was bent again, because alternative goals considered and set aside included a manned orbiting laboratory, or space station.[35]

NASA headquarters created an Office of Program Planning and Evaluation to draft an official long-range plan. This ten-year prospectus was complete by the end of 1959. It projected steady increases in funding, peaking at $1.6 billion per year—sufficient, it said, to fund manned circumlunar flight by 1966–68 and a moon landing sometime after 1970. NASA conceded that the USSR would continue to win some "firsts," especially in the early 1960s, but that this plan sufficed for the United States ultimately to dominate the "space Olympics."[36] It must have seemed an age ago when Sherman Adams disparaged "celestial basketball games."

Any follow-on to Mercury relied on the F-1 engine and its clustering into tremendous rockets to lift great weights into orbit, or to the moon. These blueprint behemoths, the Huntsville equivalents of Peenemünde's A-9/A-11, were designated *Saturn 4* and *Nova*. Since acquiring the von Braun team, NASA's Space Flight Development Office under Abe Silverstein moved quickly on specifications for the big boosters and settled on the tricky, but powerful, LH_2 fuel for the upper stages and the kerosene-based RP-1 and LOX for the first stage. These decisions rested in part on prior assumptions about the kind of missions the Saturn would be called upon to do: first priority, lunar and deep-space missions with escape payloads of a ton or more. Once made, such decisions introduced an element of technological determinism to the subsequent political commitments in space. Eisenhower continued to support the Saturn and even invited Glennan to request additional funding to accelerate it in January 1960. The same week Saturn won a DX (top) priority.[37]

NASA entered the FY 1962 cycle in May. Its $1.25 billion request agitated the BoB and alerted the White House to the full financial implications of manned spaceflight. Eisenhower was reconciled to Mercury but began to balk at this new beast called Project Apollo, and asked Kistiakowsky for a PSAC review. In public, however, the President ceased to make disparaging remarks about space races and did what he could to breathe wind into the sails of Nixon's campaign.

As the 1960 campaign drew to its close, it was hard to tell the

MIDAS, DISCOVERER, AND SAMOS

Three weeks after the U-2 went down the next generation of technological spooks made its appearance. An Atlas-Agena lifted *Midas* 2 (for *missile defense alarm system*) into orbit. *Midas* 1 had blown up during stage separation in February. No firm decision had been reached in the administration about whether to impose a news blackout on military launches. Killian thought such satellites ought *not* to be played up—if the United States seemed to threaten or embarrass the Soviets, they might feel compelled to take reprisals. On the other hand, news that the United States had secure means of watching the USSR might reassure allies and dampen missile gap demagogy. In the absence of firm directives, the OCB advised minimal publicity, but the *Washington Post* touted Midas in bold face, and the State Department positively advertised U.S. ability (thanks to infrared sensors from ITT) to detect all missile launches as a step toward arms control.

Meanwhile, Discoverer engineers soldiered on. Stabilizers failed again on *Discoverer* 7, the parachute failed on number 8, then 9 and 10, in February 1960, failed to achieve orbit. *Discoverer* 11 appeared to function, but tracking stations lost it. Number 12 flopped off the pad. Finally, on August 10, 1960, *Discoverer* 13 sailed through seventeen orbits, reentered smoothly, and was tracked all the way down by Hawaiian radar. When navy frogmen fished it from the ocean, they completed the first recovery of a man-made object from space. It happened a mere week before the Soviets did the same for Strelka and Belka. *Discoverer* 14 followed at once. This time the USAF did not claim, as it had on 13, that no sensor equipment was aboard. It was another success, hence it is likely that Eisenhower was mulling over the first space photographs of the USSR as early as August 1960. Five more Discoverer tests followed before 1961; two succeeded, one of which may have had a complement of cameras.[38]

Spy satellites proved successful beyond the most sanguine expectations of laymen (what Edwin Land and the technicians expected is unknown). Satellites were invulnerable, could cover more ground than a whole fleet of airplanes, and could return to the same spot in a short time. But the principle that photographic resolution deteriorates with distance would seem to make space-based spying a folly. The U-2 sailed thirteen miles high; satellites were at least eight times higher. To be sure, the absence of engine vibration and atmospheric disturbance worked to a satellite's advantage, but would these make up for the extreme altitude? Von Braun lectured in 1956:

So far as photo reconnaissance is concerned, I believe a very high quality of photos can be obtained. . . . Taking photos with a similar optical system outside the atmosphere, from outer space, the disturbances and turbulence are far away. The atmosphere is much more transparent from without than from underneath. Pick up a piece of wax paper. Hold it close in front of your face and you see only a blur. Hold it on a piece of newsprint, and it is perfectly transparent. You have much the same situation here.[39]

candidates apart on the space program, except in their assessment of its current health. Both granted the importance of prestige and both promised visionary exploits. But the missile gap myth and the tired image of the Republicans made Kennedy's vague call to arms more effective than Nixon's specific survey of past and future deeds. Kennedy raised the "space gap" over twenty times in the campaign; Nixon responded half as often.[40] And the "experts"—Scientists for Kennedy, academics, and aerospace executives—sided with the challenger. The strategy of linking Nixon to the "ineffectual" Eisenhower also remained in force to the end.[41] A week before the election, LBJ released a "space gap" broadside that began by charging that "the Administration had committed the Republican Party to such a slow motion policy that the Republican Party could not now reverse it," and ended by insisting it was "not an overestimate to say that space has become for many people the primary symbol of world leadership. . . ."[42]

Would a Nixon victory have yielded a significantly different set of space and technology policies from what occurred? On the basis of Nixon's expressed understanding of the symbolism of space exploits and his own campaign promises, one is tempted to conclude no, not much. The demands of total Cold War created an imperative that even Eisenhower, after all, had to accommodate. Nixon may not have issued a dramatic call for moon landings in the 1960s, but he assuredly would have accelerated a high-profile space program to match the Soviets during his "critical years" of the mid-1960s. What might not have happened under Nixon was the triumph of a Cold Warrior/domestic liberal alliance and the extension of the technocratic method from defense and space to social and political arenas ranging from American ghettos to the Third World.

After the election, Eisenhower made two decisions that expressed, one last time, his instincts on the relative importance of space activities. The first involved the Apollo moon program, the second a USAF program for satellite inspection and destruction. In November the Ad-Hoc Panel on Man-In-Space, chaired by Donald Hornig, reported that a circumlunar voyage would cost an additional $8 billion and a manned lunar landing an extra $26 to $58 billion.[43] With the danger of saddling Nixon with an apparent retreat no longer present, Eisenhower reverted to his "rearguard" posture. Such costs were outlandish. Supporters compared lunar exploration to the voyages of Columbus, but Ike insisted that he was "not about to hock his jewels." Glennan agreed. "If we fail to place a man on the moon before twenty years from now," he told the President in October, "there is nothing lost." He was "screwing up his courage to state publically that this should not be done."[44] Ike took that burden on himself and vetoed Apollo.

In the same budget cycle, the USAF space men pushed hard for Saint (*satellite interceptor*). If some favored diplomatic efforts to protect spy satellites, and others cloaking them in secrecy, the USAF argued with clumsy honesty that military assets must be defended with military means, Soviet propaganda and world opinion be damned. The PSAC, BoB, and State Department all objected at first, but a secret OCB report in November asked what the United States would do should *Soviet* "killer satellites" appear in orbit? This fear of technological surprise obliged Eisenhower to approve development and testing of Saint.[45]

In his last weeks in office, Eisenhower recoiled from the cost of escalating the prestige race in space but acquiesced in the necessity for provocative military uses of space. His "hock my jewels" remark helped to place a stamp of backward conservatism on his space policy that remains to this day. These lame-duck actions were of little consequence anyway, since they could be reversed by the incoming administration. But the last six weeks of the Eisenhower era were also the ones in which the elder statesman reflected in the White House on a suitable Farewell Address. During the same weeks the Kennedy transition team eagerly polled scientists, academics, and military and civilian strategists for their views on the shape of things to come and the optimal battle plan for assaulting the New Frontier.

Conclusion

According to Rostow, Eisenhower substituted rhetoric for action and had no business taking credit for "peace and prosperity." In one sense Rostow was palpably wrong. The eight Eisenhower years were the only period of sustained peace the American people had known since 1941. The GNP grew 44.8 percent from 1952 to 1960, over three times faster than the population, with no appreciable inflation. In another sense Rostow was right. Eisenhower would agree that he was not responsible for these achievements. The citizens were, by dint of imagination, work, and risk taking. Government could help to create an environment in which private virtue was rewarded, but it could not "create" peace and prosperity. This was signal humility for the man who organized D-Day.

Eisenhower's misfortune was that Sputnik, in his own words, made the Cold War total and validated attacks on the principles most Americans had taken for granted since his boyhood in Kansas. Before 1957, the domestic effects of Cold War had been contained. East Asia and Europe were the battlelines, subversion and the Red Army the threats, the CIA and SAC the requisite deterrents. But Sputnik signaled imminent strategic parity and a new credibility for Soviet propaganda, especially in the postcolonial world. Technology, education, even race relations became points of comparison between communism and democracy. The blows to U.S. prestige also helped unite Democratic Cold Warriors and social liberals beneath the banner of vastly increased federal activity in all areas, not just to close the missile gap but to construct an American society that matched its preferred image of affluence and justice.

At first the impact of Sputnik confused Eisenhower. But in time it sank in that a new political symbolism had arisen to discredit the old verities about limited government, local initiative, balanced budgets, and individualism. The United States had to respond in kind to Soviet technocracy. Of course, the United States already had a missile and space program geared to defense, and at first the administration formulated policy in line with this priority. Space policy must shield spy satellites and other military systems to support the deterrent, provide accurate intelligence, prevent the military from going "hog-wild," and monitor hoped-for arms control accords. Second, the U.S. space program ought to appear open and cooperative in contrast to the secret, rocket-rattling

Soviets. A civilian space agency (and thus a divided program) served these goals. NASA emerged in part as eyewash, in part as expedient.

But NASA had to be sheltered and fed if it was to resist the predations of the USAF. So Ike transferred JPL, the von Braun rocket team, and the Saturn. NASA grew and laid foundations for still more growth in the future. But Eisenhower's legacy was paradox. Born of competition, the space program was dedicated to cooperation. So dedicated, it still relied on "racing" to prosper. Hailed as civilian, much NASA R & D was judged militarily relevant, while even civilian technology was "militarized" insofar as it became the most visible symbol of national vitality. A godson of the IGY, the space program was the true child of the missile age, and its future depended on the *un*willingness of the powers to submit to bilateral or UN controls.

Ike hoped that once U.S. satellites were up, the panic would subside. Instead, the NSC and the President himself gradually promoted prestige to the top goal of the space program and accelerated development of the militarily useless Saturn. By 1960 the infrastructure was present for a massive manned space enterprise.

Then Ike balked, though only at the end and only on the biggest stunt of all, a moon landing. This gesture should not obscure the real saltation in American policy since Sputnik. In 1955 the United States spent $6.2 billion on R & D, 53 percent of it by the federal government. By FY 1961, R & D jumped to $14.3 billion—131 percent in five years—and the government's share was 65 percent. In 1955 the United States trailed the USSR in R & D spending as a percent of government budget 4.8 to 5.3 In FY 1961 the United States share almost doubled and led the Soviets 9.5 to the same 5.3. Compared to the redoubled push of the 1960s, Eisenhower's seemed halfhearted. In fact he presided over a vast expansion of peacetime federal involvement in technology.

Eisenhower also laid federal foundations, especially in education, for the gingerbread architecture of social legislation to come. In retrospect, therefore, one could criticize him from either side: he was neither true to his conservatism nor truly committed to the new progressivism. He was a pragmatist, compromising where real emergencies could be said to exist, but reminding Congress and the people that these were only inoculations, that they took the nation farther from the values and ideals that made the United States worth defending. But his appeals became wooden by the end. When did he realize that he was a Quixote? That "the long haul"—what his successor would baptize a "long twilight struggle"—was just the sort to transform American government into an engine of forced, planned change?

Dedicating the Marshall Space Flight Center in September 1960, Ike insisted that everything Americans had and would accomplish was the product of "unrestrained human talent and energy relentlessly probing for the betterment of humanity" rather than "the outgrowth of a soulless, barren technology, nor of a grasping state imperialism." This proved, he said, "that hard work, toughness of spirit, and self-reliant enterprise are not mere catchwords of an era dead and gone."[1]

Who could doubt the industry and toughness of spirit of the Huntsville engineers? Nor was big technology innately un-American. They were not the cause of Eisenhower's distress. Nor were the experts such as Killian and Kistikowsky, whom Ike praised as "My scientists, . . . one of the few groups I encountered while in Washington who seemed to be there to help the country and not help themselves."[2] His fears fell rather on the growing complex of public and private interest groups whose "main chance" lay in promotion of ever greater subsidies for weapons, technology, education, and regulation of private life. The services were never satisfied, he wrote in his memoirs, the arms makers formed powerful lobbies, their factories locked in local voters and congressmen, and soon a whole array thrived on fear of the enemy and disparagement of oneself.[3] As big spending moved into civilian areas, he might have added, the process repeated itself.

Out of this gloom came Eisenhower's startling conclusion that the trend toward technocracy was the most important theme he could raise in his Farewell Address. It reads like prophecy now, its phrases sagging with future memories. Like Medaris, Ike cried forth the economic, political, even spiritual dangers posed by the growth of a "military-industrial complex" and a "scientific-technological elite." What most Americans do not recall is that Eisenhower attributed the unhappy trends to the march of technology, and that he himself did not know what to do about them:

Akin to, and largely responsible for the sweeping changes in our military-industrial posture, has been the technological revolution during recent decades.

In this revolution, research has become critical; it also becomes more formalized, complex, and costly. A steadily increasing share is conducted by, for, or at the direction of, the Federal government.

Today, the solitary inventor, tinkering in his shop, has been overshadowed by task forces of scientists in laboratories and testing fields. In the same fashion the free university, historically the fountainhead of free ideas and scientific discovery, has experienced a revolution in the conduct of research. Partly because of the huge costs involved, a government contract becomes virtually a substitute for intellectual curiosity. For every old blackboard there are now hundreds of new electronic computers.

The prospect of domination of the nation's scholars by Federal employment, project allocations, and the power of money is ever present—and is gravely to be regarded.

Yet in holding scientific research and discovery in respect, as we should, we must also be alert to the equal and opposite danger that public policy could itself become the captive of a scientific-technological elite.[4]

The next day a science reporter asked the President what steps he recommended to prevent the capture of policy by a scientific-technological elite.

I know of nothing here that is possible, or useful, except the performance of the duties of responsible citizenship. . . . When you see almost every one of your magazines no matter what they are advertising, has a picture of the Titan missile or the Atlas . . . there is . . . almost an insidious penetration of our own minds that the only thing this country is engaged in is weaponry and missiles. And, I'll tell you we just can't afford to do that. The reason we have them is to protect the great values in which we believe, and they are far deeper even than our lives and our own property, as I see it.[5]

The dangers posed by concomitant Cold War and technical revolution were moral and insuperable. How could "good citizens" decide when, as the Farewell Address warned, military-industrial influence had become "unwarranted" or power "misplaced" or public policy a "captive"? Formerly autonomous scientific, military, industrial, and academic institutions themselves were increasingly enmeshed in a national technocracy—which, as Ike also admitted, was an "imperative need."

An air force writer dismissed the Farewell Address: "President Eisenhower . . . had his eye on a place in history as a military hero who revolted against war."[6] Edwin Land took a broader view in private council with Eisenhower:

While in the past the earning of a living was tied to individualism and afforded a sense of personality and achievement, we are now coming to a time in which it is very hard to maintain private initiative and private property because so much of what we do—including science—is done by large groups or by the state itself. More and more we tend to resemble the Soviets, however much we disclaim this.[7]

PART IV

Parabolic Ballad:

Khrushchev and the

Setting of Soviet

Space Policy

Rockets are weapons and science.   —S. P. KOROLEV

We are not concerned so much with the fate of an individual as with the
successes of our science.   —A. ALEXANDROV, 1962

Harebrained scheming, hasty conclusions, rash decisions, and actions based
on wishful thinking, boasting and empty words, bureaucratism, the refusal to
take account of all the achievement of science and practical experience—all
these defects are alien to the party.   —*Pravda*, 1964

We still have a lot to learn from the capitalists. There are many things we
still don't do as well as they do. It's been more than fifty years since the
working class of the Soviet Union carried out its Revolution under the
leadership of the Great Lenin, yet, to my great disappointment and irritation,
we still haven't been able to catch up with the capitalists. Sometimes we
jokingly say that capitalism is rotten to the core. Yet those "rotten" capitalists
keep coming up with things which make our jaws drop in surprise. I would
dearly love to surprise them with our achievements as often. Particularly in
the field of technology and organization, "rotten" capitalism has borne some
fruits which we would do well to transplant into our own socialist soil.
                               —N. S. KHRUSHCHEV (in retirement)

[Gauguin] sped away like a roaring rocket . . .
And he entered the Louvre,
Not through stately portals,
But like a wrathful parabola
Piercing the roof.   —A. VOZNESENSKY, "Parabolic Ballad," 1960

THE LAST American President to stress the distinction between military
and civil order was, naturally, the general Eisenhower. Military work is
differentiated and hierarchical. Each man, each unit is trained for a
specific task and told in essence: "Don't worry about the big picture; if
each unit does its job, the whole will take care of itself." But even under
the best of circumstances, victory is not a sum. The many local successes
of an army, or a militarized society, can add up to one big failure. Civil
order, by contrast, minimizes centralized direction in order to permit
spontaneous individual and group activity. In such a society businesses
flop and chaos appears to reign. Yet all the local failures can add up to
one big success.

The clue to this paradox is the presumption of those who would militarize and direct. Action requires an assumption of effects, which in turn requires simplification. The larger the stage one is trying to direct—a factory, a central bank, a whole economy—the greater the simplification, the greater the assumptions, the farther from reality the action taken. This presumption was the root sin Metternich spied in the French Revolution and the one against which the American colonies in part rebelled. The Soviet Union, on the other hand, was founded on such simplification: "You have nothing to lose but your chains"; "Communism equals Soviet power plus electrification"; "In the era of reconstruction technology decides everything"; "Sputniks and Luniks prove the superiority of socialism." Such slogans do little damage when exhortatory or propagandistic; they are disastrous when taken as true. Under the inspiration and goad of Sputnik, Nikita Khrushchev launched the USSR from the more "solid" ground of bluff and promise into the transparent air of reality. But the militarized society could not achieve the orbit he planned for it and instead arced back to earth in an ever angrier parabola.

Khrushchev, it is said, was exceptional among Soviet leaders for being a true believer.[1] He was certainly a revolutionary. In his brief decade of primacy, Khrushchev denounced Stalinism and launched major reforms of the Party, the bureaucracy, industry, agriculture, the military, and R & D. He revised Communist ideology in foreign affairs and predicted nothing less than the final attainment of communism at home. All of Khrushchev's "harebrained schemes," though tied to his own campaign to consolidate his power, made little sense except in terms of genuine Leninist faith. But the success of a militarized society depends not only on the soundness of its directors' plans for inventing the future at five- or seven-year intervals, but also on the cooperation of all people, at home and abroad, who are in a position to muck up the plan by capricious action. The need to control all variables tempts the militarized state to exert greater coercion and direction and to blame shortfalls on "wreckers" and foreign enemies, which in turn reinforces the need for terror against the whole world.

Khrushchev had his share of uncontrolled variables: the Western imperialists, the scurrilous Chinese rivaling Moscow for Communist leadership, the holdover Stalinists, the interest groups in the military and industry that stood to lose from reform. But a true Communist also believes that he has objective forces on his side, and such forces are lodged, in the first instance, in technology. Khrushchev had the same task as Eisenhower: to usher his country into the age of missiles, space, and "scientific-technical revolution." But while Eisenhower sought to do this with a minimum of violence to American traditions, Khrushchev seized on the "new circumstances created by technology" to justify sweeping reforms and even more sweeping promises. He discovered, in the latter stanzas of his parabolic ballad, that the Soviet Union was not yet equal to his goals, and—more shocking—that ruling circles in the Party and economy did not want to attain them. Believing himself the revolutionary who would pick up where Lenin left off and turn decades of lies and promises into glorious reality, Khrushchev ended by relying more on bluff than any Soviet autocrat before or since. In nothing was this sad progression from aspiration to dissimulation more evident than in the Soviet space and missile programs; and nowhere more important, for Sputnik was both starting pistol and symbol of the Khrushchevian revolution.

In 1956 Khrushchev made his startling attack on Stalinism. In 1957 he sacked Marshal Zhukov and began to recast the armed forces according to the needs of the missile age. In 1958 he pushed agricultural reforms and set aside august procedure to call an extraordinary party congress for January 1959. Its sole agenda item was a new economic plan, a Seven Year Plan, in which Khrushchev, assuming the mantle of the master Lenin, declared himself the theoretician of a new Party program, based on "many brilliant and new principles," that would carry the USSR in short order from the lower socialist stage to the ultimate Communist stage of history. Soon the USSR will have "provided a complete abundance of everything needed to satisfy the requirements of all the people. Communism is impossible without this." Thanks to Soviet missiles, the imperialists had no choice but to reconcile themselves to peaceful coexistence. Communism would continue to expand as more and more peoples threw off exploitative masters by peaceful means or wars of liberation. Capitalist encirclement was broken and the correlation of forces had shifted irreversibly in favor of socialism.[2]

This was a unique vision, for the transition to pure communism could only happen once. No more would the USSR be "behind, but catching up"; "inevitably triumphant, but for the moment encircled." The USSR was coming of age. It was a daring vision, because if imperialist and counterrevolutionary demons were indeed exorcised, then failure to achieve the goal could only be blamed on the leadership itself; daring as well because the whole Communist power structure, based as it was on the endless "transition period," stood to lose its *raison d'être*. One by one Khrushchev's initiatives alienated every established interest group. Either his plans would miscarry, endangering the security and stability of the regime, or they would succeed, in which case the existing elites might lose their power and privilege. Either way, Party, military, and economic leaders could not stand Khrushchev's communism. The ultimate wreckers could only be the Communists themselves.

The final weakness in Khrushchev's grand design was that the technological base for the whole edifice, rocketry, was itself more scaffolding than concrete. The ICBM program quickly fell behind the American, and the space program was even more in arrears: the missile gap was a myth. In later years, when defectors and Western experts insisted on the

"Potemkin village" character of Soviet space exploits, few believed them, such was the force of propaganda and secrecy. But the stakes in projecting the image of a mighty space program were high, for Khrushchev's foreign and domestic programs rested on the proposition that nuclear and rocket technology had made the USSR safe from attack. If that were known to be false, his programs and his own power would be placed in grave danger. But even if it were thought to be true, then the burden of initiative in the world struggle between "camps" fell, for the first time, on the USSR. With war unthinkable, that meant the "inevitable march" of socialism must proceed in other ways, through intimidation, proxy battles, or competition for prestige in real values. Either way, Khrushchev's domestic program was linked to the space and missile bluff.

For seven years the Soviet space program crackled and hissed like the medieval Chinese rockets that panicked Mongol ponies, distracting attention from the confusion inside the fortress walls. By the time the bluff was called and Khrushchev dumped, the Soviets found they must return to the old formula, "We are behind now, but . . ." even in the arena of their greatest triumph, outer space.

CHAPTER 11

Party Line

Sputnik was a press agent's dream come true. But it was apotheosis to a totalitarian state boasting of technological progressivism in a decade when the exploits of jet aircraft circling the world nonstop or polar icebreakers and submarines daring the Arctic titillated the public. The Russian people, too, must have felt justifiable pride. Despite Stalinist claims that everything from the telephone to baseball had been invented by Russians, the West tended to denigrate their achievements even in fields, like nuclear physics, where they had contributed.[1] Now the Soviets were leaders in an awe-inspiring technology that was arguably a Russian preserve since the days of Tsiolkovsky. It would have been peculiar had the Soviet regime *not* celebrated its feat in hyperbolic terms, especially since Sputnik seemed strikingly relevant to every major plank of the 1956 Party platform.

True, Khrushchev may have been no more aware than Eisenhower of the impact the first satellite would have. But once evident, it must have seemed so ideal a conjuncture as to validate the premier's reading of the material laws at work in the history of his time. In a matter of days and weeks after Sputnik every propaganda theme of the next seven years was coined and in circulation. First, sputniks verified the Soviet claim to an ICBM, hence the new vulnerability of the United States to nuclear assaults. Second, the space achievement irrefutably demonstrated the scientific, and ultimately economic, superiority of the USSR. Third, the Soviet space program was dedicated solely to peaceful conquest of the cosmos in the interest of all mankind. Fourth, the leading inspiration and guide of the Soviet space effort was none other than Premier Khrushchev himself, and it illustrated his dynamic leadership.

"When we announced the successful testing of an intercontinental rocket," Khrushchev told James Reston a week after *Sputnik I*, "some American statesmen did not believe us. The Soviet Union, they claimed, was saying it had something it did not really have. Now that we have successfully launched an earth satellite, only technically ignorant people can doubt this...." The Soviet ICBM was "fully perfected," and the USSR already possessed "all the rockets it needs of all ranges."[2] The

United States, on the other hand, clearly did not have intercontinental rockets or it would have launched its own sputnik. Instead, the United States was planning only a tiny satellite weighing eleven kilograms. *Sputnik II* weighed 508 kilograms and "if necessary, we can double the weight of the satellite."[3] The military implications were made clear by the end of November: "The fact that the Soviet Union was the first to launch an artificial earth satellite, which within a month was followed by another, says a lot. If necessary, tomorrow we can launch ten, twenty satellites. All that is required for this is to replace the warhead of an intercontinental ballistic rocket with the necessary instruments. There is a satellite for you."[4]

During the months when Eisenhower was trying to reassure Americans, Soviet disinformation sought to persuade them that the balance of power had suddenly and irreversibly shifted. *International Affairs*, a Soviet journal published in English, proclaimed the U.S. bomber force obsolete. What was more, U.S. overseas bases were now subject to Soviet rocket attacks and served only to involve other countries in the destruction sure to follow U.S. adventures.[5] Khrushchev repeatedly stressed the risks attached to continued alliance with the United States and spoke of states being "wiped from the face of the earth" in the event of war.[6]

The second theme—Sputnik as proof of the superiority of socialism—emerged in the very first press release from *Tass*, and was elegantly expanded by *International Affairs* the following spring. History, it proclaimed, knew few examples of a scientific discovery giving its name to a whole epoch, but this was surely one: "We can say that the era of the conquest of the Cosmos has dawned." But Sputnik was a "many-sided phenomenon." Could so amazing an advance in scientific and technical thought be separated from its social roots? Could it arise independent of industrial, scientific, and technical development in the sputniks' mother country? And if the sputniks were indeed a social phenomenon, did this not mark the dawn of a new era in international relations? *International Affairs* answered its own questions: yes, the sputniks proved the social maturity of the USSR, and yes, such a breakthrough did revolutionize the social and political development of other peoples. Just as the steam engine had signaled the maturity of capitalism, the sputniks were the mark of an emerging socialist system. As the French Communist physicist Frédéric Joliot-Curie observed, "It is no accident that the Soviet Union was the first to launch a sputnik. In this fact lies the law of development of the new society. This outstripping of Western science will each year become more frequent." Hence, the most telling fact about Sputnik was its parentage: "scientists, engineers, and workers reared by our Communist Party in accordance with an advanced ideology. The whole world saw yet one more extraordinary, important demonstration of the Socialist system's superiority to the capitalist system."[7]

Of course, the West would eventually have satellites, too, granted

International Affairs. American militarists were already scheming to place weapons in space and on the moon. "Marxist science long ago established the truth . . . that only in a socialist society do scientific discoveries and technical achievements serve human progress and the welfare of mankind." Socialist priority in space was a potent force for peace. Furthermore, the sputniks reaffirmed the correctness of Marxist materialism by refuting "idealistic theories that the world is 'unknowable,' that the objective world does not exist but is only a product of the human consciousness. . . . The appearance in outer space of man-made satellites also proved the stupidity of attempts to combine science with clericalism, with religious theories regarding the divine origin of the world." The sputniks also enhanced the moral and political influence of the USSR and have "taken the ground from under the feet of malicious propaganda regarding the alleged economic and cultural backwardness of our country. It showed the world the Soviet Union as it really is."[8]

Finally, there were the military implications of the sputniks. All American strategies of terror—"situations of strength," Cold War, Truman Doctrine, containment, liberation of "captive peoples," brinksmanship, massive retaliation—all these varieties of "atomic blackmail" were bankrupt. Imagine the threat to the world, the author shuddered, if the United States had retained its atomic monopoly and become sole possessor of ICBMs as well! But the innate superiority of socialism ensured that that nightmare did not come to pass. Instead, "The world has entered a new stage of co-existence . . . *when any attempt by the imperialists to launch a new world war will inevitably boomerang against the entire capitalist system and lead to its complete downfall."*[9]

The Communist Party line depicted the space program, unencumbered by any artificial division between civilian and military programs, as entirely peaceful. Despite Khrushchev's talk of the interchangeability of H-bombs and sputniks and the utter secrecy surrounding the Soviet program, Marxist science was deemed progressive and in the interest of all. Candid U.S. statements about the military uses of satellites, on the other hand, were carefully gathered, polished, and flung back by the Soviet agitprop slingshot. Khrushchev accused the United States of seeking to militarize the cosmos, warned of devastating Soviet ripostes, and demanded the dismantling of all (now "useless") foreign bases. American refusal to comply then permitted Khrushchev to ignore American requests for technical study of orbital disarmament.

Throughout the period of the missile gap (1958 to 1961) Khrushchev punctuated his ultimata against the Western presence in Berlin with "rocket rattling," made credible by the space shots. On November 10, 1958, Khrushchev unilaterally renounced the Potsdam agreement on four-power control in Berlin and threatened to recognize East Germany's claims to sovereignty. Four days later he boasted that Soviet ICBMs were rolling off the assembly line.[10] The Luniks were "irrefutable proof" that the USSR could land a hydrogen bomb on a kopeck anywhere in the world. *Luna III,* launched on the second anniversary of *Sputnik I,* signaled Khrushchev's arrival in the United States in October 1959. Nevertheless, the peaceful nature of all Soviet space missions continued to be stressed. *Korabl Sputnik I,* according to *Pravda,* was "a peaceful explorer of the secrets of nature. . . . in the brilliance and glory of our science, the light of its noble aims, the bandit flights of the American spy plane [U-2] look pitiful and lowly before the whole world."[11] *Korabl Sputnik II* "delighted the whole world," but the "bourgeois scribblers" wasted no time concocting evil fabrications about Soviet intentions. "It is not on Soviet soil that one should look for madmen outlining plans for military bases in space. They are entrenched on the other side of the ocean, in the madhouse of the Pentagon, which has become the main asylum of spies and provocateurs."[12]

The ICBM was another matter. The Soviets did all they could in 1959 and 1960 to feed American panic over the "missile gap." Marshal Malinovsky thanked Khrushchev at the Twenty-first Party Congress for "equipp[ing] the armed forces with a whole series of military ballistic missiles [including] intercontinental." Marshal Grechko announced that the Red Army had "received" the ICBM, and Malinovsky praised its "pinpoint accuracy."[13] The "serial production" of ICBMs, coming off the line "like sausages," Khrushchev told the Party Congress, meant that "Socialism has triumphed not only fully, but irreversibly."[14] In May 1959 a delegation of West German editors was told that the USSR had enough bombs and rockets "to wipe from the face of the earth all of our probable opponents."[15]

The single most important audience for Soviet chest-thumping was Western Europe. The threats against foreign bases and allies of the United States cut directly at the bonds holding together the NATO alliance. Even American determination to resist, to go to the brink over Berlin, could be made to seem "insane" in light of Soviet prowess. While if the United States could be made to relent, and make even minor concessions, NATO might begin to unravel. Of course, Eisenhower's confidence in the continued value of U.S. bombers, the solidarity and nerve of his, Adenauer's, de Gaulle's, and Macmillan's governments, frustrated Khrushchev in Berlin.[16] But there is no question that the space campaign hurt American prestige. In June 1955 only 6 percent of West Europeans thought the West weaker than the USSR in military might. In November 1957, after two sputniks, 21 percent in Britain thought the West weaker, 20 percent in France, 12 in Italy, and 10 in Germany. But the United States, taken alone against the USSR, was deemed inferior by 50 percent of the British and a quarter of the French.[17] By 1960, after two years of missile propaganda, the Soviets were judged superior in military strength to the United States by 59 to 15 percent in Britain, 37 to 16 in France, 45 to 15 percent in Norway, and 47 to 22 percent even in loyal West Germany (the remainders judged the powers equal or were undecided).[18] More troubling still was European opinion of global trends. An April 1960 poll revealed that a plurality of Europeans in every country expected the USSR to be stronger than the United States after twenty years of "competition without war." Only a fourth of the British and one Frenchman in fourteen thought the United States could prevail over that long haul![19] Soviet propaganda was so demoralizing that the impression continued to prevail in Britain (and even in the United States) that the Soviets had launched more (as well as bigger) satellites than the Americans, long after it ceased to be true.

Secrecy, the most potent weapon of the totalitarian state, permitted the deception and fed Western imaginations. The origins of the Soviet rocket program, its chief designers, the uses to which they planned to put their heavy sputniks, imminent launch plans, even the location of their test sites, design bureaus, and missile plants were hidden. Launches were not announced beforehand, failures not at all. Official spokesmen predicted space stations and lunar flights, but provided no details or schedules, or estimates of Soviet spending. Soviet scientists at international conferences were often third-raters and presented papers based on published material or data from foreign sources. The configuration of the great booster, designated by NATO the SS-6 "Sapwood," was itself unknown for some time. All this fed the anxious expectations of Westerners, waiting and wondering what startling new capability would be revealed by the Soviets' next "surprise."

The burden of concocting the next surprise fell to the unacclaimed hero of Soviet engineering, Chief Designer Korolev. He was awarded additional bureaus, factories, and laboratories, which he fashioned into a national space complex. But he won no honors or fame, and had to divert his talent toward "spectaculars" for which Khrushchev took credit and used to back the party line and rocket-rattling. But the "ol' number seven" rocket had demonstrated its maximum payload. The addition of an upper stage permitted the moonshots of 1958 and 1959. The next spectacular, by which the relative standings in space would surely be measured, was launching a man into orbit. Korolev was expected to hurry no less than NASA's Mercury executives.

The lead time necessary to any venture involving new hardware, especially life-support systems, suggests that research on manned space-flight began even before Sputnik. High-altitude flights of animals dated from the earlier 1950s, *Sputnik II* carried the dog Laika, and in 1958 an obsolete aircraft factory near Moscow was turned over to Korolev and converted into a "manned spaceflight center." Certainly Project Mercury, underway by the end of 1958, lit a fire under the Soviet man-in-space program in more ways than one. According to defecting journalist Vladimirov, a special translation bureau existed to gather and distribute all data from U.S. sources on spaceflight. For two years Soviet staffs pored over American articles on manned rocket flight and life support.

The Vostok program, though built on indigenous foundations, thus grew in parallel with Mercury. Cosmonaut recruitment began in 1959, as did design and fabrication of a man-rated upper stage for the SS-6 and the spherical space capsule itself. Given the rudimentary state of Soviet space facilities in these years, the devolution of responsibility for all space missions on the same teams, and the political pressure to schedule launches around anniversaries and political events, one can imagine the pressure building on Korolev and his lieutenants. Khrushchev complicated matters by insisting, apparently for reasons of security, that capsules be brought down on Soviet ground instead of in the sea as the Americans intended. This required a far heavier and sturdier craft, not to mention a prodigious parachute system, all adding considerable weight. The makeshift solution was to arrange for the pilot to eject after reentry and descend on his own parachute, allowing the capsule to land at a much higher velocity.[20]

The first test of this Korabl Sputnik system was in May 1960; the second was when Strelka and Belka were recovered in August. Predictions appeared in the Western press of a Soviet manned flight. But behind the veneer of boastful self-confidence, the harried Soviets charged with maintaining appearances suffered several crippling blows that put the United States back in the chase. And if the Americans should succeed in getting the first man into space, much of Sputnik's impact might be undone. It was a very near thing.

On September 19, 1960, the Soviet liner *Baltika* docked in New York harbor. On board was Premier Khrushchev and in his luggage were some "miniature spaceships," presumably to be trotted out during his stay in celebration of some space triumph. A moon shot heralded his visit the previous year; this journey coincided with a "launch window" for Mars. Three rockets were apparently prepared, but nothing happened. Khrushchev even extended his time in New York, grew more bellicose with passing days, and finally pounded his shoe on the table at the UN. He left for home on October 13, and several days later, Soviet tracking ships were observed to turn homeward as well. Still nothing happened.

Over the years accounts of launch failures and even a catastrophe trickled to the West. Intelligence leaks, Soviet defectors, the double-agent Oleg Penkovsky, and Soviet historian Zhores Medvedev all testified to the worst space-related disaster in history, attributable in part, according to Medvedev, to Khrushchev's "misuse of space research to boost Soviet political prestige." Apparently, two Mars-bound rockets fizzled after launch on October 10 and 14. By the evening of the twenty-third, with the "window" about to close, the third rocket failed to ignite. Field Marshal Mitrovan I. Nedelin, commander of the new Strategic Rocket Forces, ordered technicians to leave their blockhouses and examine the rocket close-up. Suddenly the propellants in the great rocket exploded.

MERCURY-REDSTONE 2 AND VOSTOK I

In mid-January 1961 both contestants in the race for manned spaceflight seemed to be having their troubles. The Soviets aimed directly at orbital flight, but their Korabl Sputniks had had uneven success. The U.S. Mercury program aimed first at the more modest goal of "shooting a man from a cannon" up into space and back into the sea 300 miles downrange. Still, an American first man in "space," even if not in orbit, might steal the acclaim from a subsequent cosmonaut's orbital flight. It was well within the realm of possibility. The smaller, more thoroughly tested Redstone could be man-rated far more quickly than the giant, pressurized Atlas booster that would serve for later orbital missions. The army's Jupiter had carried two monkeys, Able and Baker, on a suborbital trajectory as early as May 1959. But the first Mercury/Redstone mating (November 1960) had flopped.

On December 19, the *MR (Mercury/Redstone)-1A* performed well, reaching an altitude of 131 miles. *MR-2* went up on January 31, 1961, and inside it was another "astronaut"—a chimpanzee named Ham. Several malfunctions resulted in the chimp spending four hours on the ground strapped inside the Mercury, blasting up to a velocity a thousand feet per second more than expected, enduring a load of 17 g's and a cabin pressure that dropped, thanks to a faulty valve, from 5.5 to only one pound per square inch, six and a half minutes of weightlessness, and a landing 130 miles beyond the target, and thus having to bob in the Atlantic for two hours forty minutes before retrieval by a helicopter. Free at last and aboard ship, Ham sampled an apple and an orange, and generally behaved like a chimp. The engineers were unhappy with the eccentric behavior of *MR-2*. With the sturdiness of primate physiology they could only be pleased.

The tentative schedule pointed to launch of *MR-3*, with a man aboard, in late March. Quick fixes had been made at Huntsville in response to *MR-2*, and the seven Mercury astronauts were eager to go. But safety was at least as important to NASA's political and budgetary position as rapid results, and the seven planned modifications suggested another test flight. On February 13 at Huntsville, the decision was made to keep *MR-3* and astronaut Alan Shepard waiting until late April at the earliest. The extra test, *MR-BD*, went off without a hitch on March 24 and got the MR system "man-rated." But it could have carried a man.

Korolev and the Vostok team must have applauded the doughty Ham. How much they learned from the data made available by the open NASA program is debatable. But a healthy chimp, combined with two Korabl Sputnik recoveries in March, must have boosted Korolev's confidence considerably. On April 12, in the midst of the period between the extra Mercury test and Alan Shepard's moment, Korolev lit the engines of his modified ICBM and flung Yuri Alekseyevich Gagarin into orbit about the earth. He circled the globe in ninety minutes, completed his assigned tasks, and was brought back into the atmosphere by ground control, landing from his parachute in a pasture in central Russia. The Vostok capsule— minus the cosmonaut—came to rest in a "plowed field" near the Leninsky Put collective farm not far from the towns of Marx and Engels on the banks of the Volga. A very appropriate spot.

Many (estimates range from scores to hundreds) of the Soviet Union's most skilled space personnel perished in a thunderous fireball on the pad in Kazakhstan. Korolev and Yangel, now his deputy, were indoors and survived. Marshal Nedelin, who gave his name to the disaster, did not. Two days afterward Moscow reported that Nedelin had died—in a plane crash. Only in 1979 did an official biography substitute for this a vague account that he had died "tragically in the performance of his official duties." Khrushchev's memoirs admit that Nedelin was killed in a missile "malfunction."[21]

The loss of men and machines, though extensive, could be made up. The effects on morale, on Korolev's nerves and conscience, and on his relationship with Khrushchev were probably more damaging and irreversible. But the pressure for continued space triumphs only increased: NASA was now projecting suborbital manned flights by the spring of 1961. Korolev wanted to fly tests with primates, like the Americans, but was not given the time. Then on December 2, 1960, *Korabl Sputnik III*, with dogs Pcholka and Mushka, reentered the atmosphere at too sharp an angle and the capsule burned up. The next day Korolev had a heart attack. Doctors also diagnosed a kidney disorder, a common ailment in survivors of the *gulag*, and prescribed a lengthy rest.[22]

Whether on his own drive, the knowledge that continued funding hinged on beating the Americans, or on callous orders from Moscow, Korolev went back to work. Two more test flights failed in February. Redesign and preparation of two more proceeded at a breakneck pace. *Korabl Sputniks IV* and *V* were recovered successfully in March. If his luck held, Korolev would soon give Khrushchev a great gift, another four or five years of life to the myth of Soviet space leadership, and the party line based on it.

Yuri Gagarin's flight was a second Sputnik (or third if one counts the 1959 Lunik that helped jolt Ike into competing for prestige). Three and a half years of American protestations and accomplishments could not stand before this latest "proof" of Soviet technological might. When the first American blasted off three weeks later, his suborbital trajectory and "sloppy splashdown" seemed a poor imitation of the Soviet feat. The space gap, in the eyes of the world, had widened.

Vostok I was a measure of the genius of Korolev, the competence of Soviet engineers, and the courage of Gagarin. But it was said to "testify to the scientific, social, economic, and moral superiority of the Socialist system," while "increas[ing] immeasurably the strength and authority of the Socialist world in its struggle to insure a peaceful future for the inhabitants of our planet."[23] To Khrushchev, Gagarin's flight marked

a stage of the scientific/technical development of our country and reveals its mighty upward flights from backwardness to progress. In this flight are reflected the heroic accomplishments of the Soviet people, the working class, collective

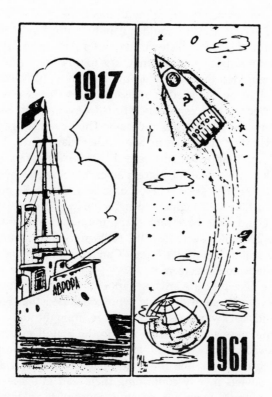

"The Dawn (Aurora) Always Heralds a New Day." The cartoon plays on the names of the naval ship *Aurora* (the dawn), a cradle of Russian Revolutionary agitation, and *Vostok* (the East), the first manned spacecraft, implying that the future is always made in the Soviet Union. From *The Morning of the Cosmic Era* (Moscow, 1961).

farm peasantry, working intelligentsia, and our wonderful scientists, who have created a technical present-day miracle, the cosmic ship Vostok, this greatest triumph of the immortal Lenin's ideas.[24]

The very name "Vostok," a word like "Orient" that meant "upward flow" and may have been chosen by Korolev to signify the "upward flow" of humanity, was translated for foreigners as "The East," to suggest the rising sun of communism.[25]

To *International Affairs*, the real secret of Soviet spaceflight was

rooted in the specific features of Socialist society, in its social structure, its planned economy, the abolition of exploitation of man by man, the absence of racial discrimination, in free labor and the released creative energies of peoples. Our achievements in the field of technology in general and in rocketry in particular are only a result of the Socialist nature of Soviet society.[26]

Gagarin himself, it was said, sent greetings during his hasty trip over

Africa to the peoples below struggling to break the chains of imperialism.

Success and world acclaim could not induce the Soviet leadership to lower its curtain of secrecy. Instead, they set a pattern of official lies and coverups about the space program that lasts to this day. Korolev remained a nonperson, guarded by security agents and given his medals in secret lest, it was said, he be abducted or assassinated by the CIA. Academicians Keldysh and Sedov, and especially Khrushchev himself, were paraded as the masterminds of the space program. They even fudged the details of Vostok I, claiming that Gagarin had ridden his craft to the ground, a stipulation of international flight records. The mysterious Soviet space launch complex was—like all place names on Soviet maps—given an erroneous location. The great cosmodrome was said to lie near the town of Baikonur. In fact, U-2 flights and ingenious private spacewatchers in Japan and England, who backtracked sputnik trajectories, established the location some 200 miles away near the railhead of Tyuratam, whereupon this bustling Kazakh "Huntsville" and "Canaveral" all in one disappeared from Soviet maps and gazeteers.[27]

Propaganda, like any advertising, appeals to unconscious fears and desires. For every Western observer who concluded that Soviet secrecy hid falsity and backwardness, many more thought the Soviet space program all the more sinister. According to the State Department, Vostok I earned "extraordinarily heavy" media coverage in Europe, perhaps greater than Sputnik I. While the note of astonishment was no longer present—the Soviets merely confirming their lead in space—the overall impact was as great. Even conservative journalists in Europe granted Soviet leadership and speculated uneasily on the use the Kremlin might make of its leverage. A poll taken after Shepard's flight in May revealed that West Europeans believed the USSR ahead in total military strength by 41 to 19 percent, and in overall scientific achievement by 39 to 31 percent.[28]

Reactions in Latin America, Africa, South Asia, and the Middle East varied with the political slant of nations and journals, but none disputed the Soviet claim to scientific leadership. An independent paper in Manila saw the real importance of Soviet cosmonautics in "its effect on the people of the uncommitted countries, who see in all this the supposed superiority of the Communist way of life, economic system, and materialistic philosophy." A conservative paper in Kuala Lumpur: "It is evident that the U.S. is losing the space race with Russia. . . ." Sukarno of Indonesia was "confident that the Soviet feat would eventually contribute to the progress and prosperity of mankind as well as to world peace." Nehru of India thought that it revealed the narrowminded folly of preparing for war and considered Vostok a victory for peace. The official journal of Iran thought Gagarin's flight more important than the discovery of the new world; that of Tunisia, more important than the invention of the printing press; that of Kenya, more important than the discovery of

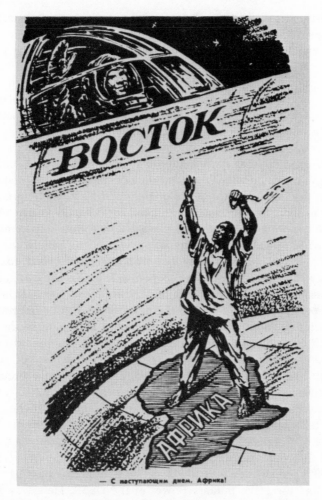

"In Tune with the Times . . . Africa!" The cartoon depicts Yuri Gagarin saluting the African people from space, implying that each is engaged in the same, mutually supporting struggle against imperialism. From *The Morning of the Cosmic Era* (Moscow, 1961).

the wheel. Egypt's Nasser praised the "gigantic scientific capabilities" of the Soviet people and had "no doubt that the launching of man into space will turn upside down not only many scientific views, but also many political and military trends."[29]

The global impact of Soviet space shots was, of course, intangible. No uncommitted nation can be shown to have "chosen" socialism or allegiance with Moscow on the basis of the standings in the space race. But Soviet successes must have made the socialist model far more respectable than it was ten years earlier, when U.S. hegemony was assumed, and provided Third World intelligentsias with excuses and encouragement to lean toward a national socialism, neutralism, or anti-Americanism to which they tended for other reasons.[30] More interesting, perhaps, was the reverse phenomenon: the effect of Cold War rivalry in the Third World on the subsequent history of spaceflight. For just as Sputnik sparked great changes in U.S. domestic policy, so the narrow Soviet victory in inaugurating manned spaceflight helped to complete the Space Age revolution in the United States and, in the process, send men to the moon.

Space propaganda was equally important in Soviet domestic politics. The party line held that space triumphs proved that socialist science and technology must inevitably advance more rapidly, and had in fact overtaken the West. But the proof of the pudding in Communist theory was overall industrial production and technical sophistication. After all, it was the *imperialists* who tended to place new technology at the service of war, while socialist inventions were released to benefit the people. Hence the space shots supposedly symbolized the broad advance of technology throughout the Soviet economy. This was the case, according to Khrushchev, who boasted that the USSR would surpass the United States in per-capita production in the near future. But Soviet rocketry in fact reflected isolated, force-fed development in a few sectors of the militarized economy. In other words, space travel *was* the triumph of a command economy, but its fruits were bitter, not sweet. For, as Soviet nontraditionalists had always contended, broad economic progress depended on massive shifts of resources away from the military sectors to agriculture and light manufacturing. Thus, even as Khrushchev pointed to space technology as proof that communism would soon provide "all the needs of the people," the space and missile expenditures may have inhibited that very achievement.

The space propaganda was also designed to impress on foreign opinion the new and awesome military power of the USSR. Here was a second contradiction. For despite its "firsts," the USSR was in fact behind the United States in every meaningful category of missile technology. Even the size of the Soviet rocket was a measure of backwardness, for it made a poor ICBM. Third, Khrushchev perceived the Space Age as the age of the attainment of communism. As such, it meant a profound change in domestic Soviet organization. But insofar as these reforms relied on the expertise of the technical intelligentsia, they accentuated the old friction between technicians and the Party and between interest groups and Khrushchev himself. Desirous of consolidating his power and promoting his policies, Khrushchev set up the space program, much as Stalin had aviation, as a personal trophy. Hence he depended on the technicians, Korolev most of all, to provide him with glory and leverage, even as he frustrated them and twisted their efforts to serve his personal rule. From the Red Terror to Stalin's attack on "technocrats" and the aviation designers, to Khrushchev's clumsy interference with the space program, such was the pattern of Soviet-style technocracy.

Khrushchev claimed to be the inspiration and leader of the space program. He played "father" to the cosmonauts after the fashion of Stalin. And yet he showed little interest in the scientific benefits of spaceflight and apparently never attended a launch. But he obliged Korolev to set aside followup missions and move on at once to new "breakthroughs" more likely to stun the world. Hence the fitful quality of the early space programs in the USSR. The first Sputniks soon gave way to the Luniks, then to the manned program. If Korolev and his associates had begun to work on the various applications of space that the Americans undertook from the start—and later programs suggest that they had—they must have done so in their spare time. Korolev had to bow to Khrushchev's wishes or lose his budget and perhaps his position to envious rivals like Yangel or Chalomei. His health was broken, many of his comrades dead, and his resources and technology inferior to those available to his American competitors. How long could he maintain the fiction of Soviet leadership, and at what cost to the real progress of Soviet space technology? He must have yearned to attend foreign conferences, accept the acclaim due him, and consult with foreign colleagues about techniques for the conquest of space. But he could not. By the summer of 1961, he must also have sketched out a rational plan for future development. If so, he could not implement it. In mid-July Khrushchev summoned Korolev to his Black Sea dacha and informed him that the next manned flight must surpass the achievements of Gagarin, and it must take place within a month—to punctuate, as it turned out, the erection of the Berlin Wall on August 13.[31]

In his memoirs Khrushchev confessed, "Of course, we tried to derive the maximum political advantage from the fact that we were first to launch our rockets into space. We wanted to exert pressure on the American militarists—and also influence the minds of more reasonable politicians—so that the United States would start treating us better."[32] But the quest for "maximum political advantage" led to the espousal of a "party line" that not only hindered rapid and rational development of space technology but encouraged a dangerous deception in military policy as well.

CHAPTER 12

The Missile Bluff

For it was all a bluff. At the very time Khrushchev boasted of the obsolescence of American "massive retaliation," the U.S. deterrent was at the height of its effectiveness and the USSR had yet to deploy a single ICBM. It is not known whether the design of the first Soviet ICBM was approved just before or just after the first Soviet H-bomb test of August 1953.[1] Possibly the conventional wisdom to the effect that the ICBM was so big because it was built to accommodate atomic bombs is an oversimplification. The first Soviet H-bomb was itself a cumbersome affair, since it contained a much larger detonator of fissionable material than the U.S. bomb and its yield was estimated to be only half a megaton. Thus the Soviets may have assumed that mightier H-bombs would still require a large rocket.[2] The critical determinant of ICBM size, however, may instead have been the relative inefficiency of Soviet rocketry. Apparently unable to construct a jet nozzle capable of withstanding the high temperatures generated by a larger engine, Soviet engineers made do with clusters of the smaller RD-107 engine. This meant a lower thrust-to-weight ratio than in the first U.S. ICBMs and a shorter range. For all its size, the SS-6 had a range of only about 4,340 miles.[3] Thus giganticism in Soviet rocketry was a sign of primitivism, and the very clusters of clusters system that made the Soviets appear to be ahead in the 1950s inhibited their efforts to keep up in the 1960s.

What was worse, the SS-6 relied on an unstorable kerosene and LOX combination and could not be launched quickly. Nor could the unwieldy rocket be easily fitted in silos. Finally, as the Soviets had not mastered inertial guidance, the SS-6 had to be radio-guided at intervals in its boost phase. All told it was a decade behind the solid-fueled Minuteman already under design. Yet world opinion, including American, was quick to tab Sputnik a sign of Soviet superiority! What was the Kremlin to do? Throw a wet blanket on the otherwise delightful chorus of panic and praise by admitting that the USSR was years away from an operational missile? Or encourage the impression of Soviet might? Deploy the SS-6 despite its faults? Or wait wisely for a better product and finesse the widening gap between talk and truth?

Subsequent developments reveal what decisions were made. The first confirmed total of deployed Soviet ICBMs is 150—in 1965! These were SS-7s, still relatively inaccurate, but with a longer range, a five-megaton warhead, and a silo basing mode. Backward extrapolation suggests that the SS-7 began to be deployed about 1961, starting perhaps with thirty. If the SS-6 ever went into "serial production," as Khrushchev claimed, it was only for testing and spaceflight. By the time of the Cuban missile crisis, the USSR had made twenty-eight orbital launches with the SS-6. Assuming the same 20 percent failure rate as in the United States, about thirty-five units were committed to spaceflight during the years of the missile bluff. This was a highly economical use of the hardware—the Soviet space show had far more publicity value than an embryonic ICBM force would have had military value.[4]

The United States also held off on large-scale deployment of the first ICBMs. Eighteen Atlas missiles were ready by the end of 1960, perhaps 126 by the Cuban crisis, after which they were decommissioned. The Titan 1 was not deployed until 1962, then quickly replaced by 54 Titan 2s, which remained in the inventory until the 1980s. More important, about 150 Minutemen occupied silos in 1962, 700 by 1964, and 1,000 by 1967.[5] Only after Khrushchev's demise did the second-generation Soviet missiles, SS-9, SS-11, and SS-13 (the last a solid-fuel rocket), come on line and pull the USSR even in deployed ICBMs by 1969. During the entire period of Khrushchev's missile bluff, the USSR was markedly inferior to the United States in strategic forces.

Of course, one cannot compare nuclear forces simply by counting missiles like dreadnoughts before World War I. Warhead yield, accuracy, throw-weight, and the K-factor (or probability that a missile will destroy a given enemy silo) must all be considered.[6] But given that the United States already led in ICBMs by 1961, outnumbered its enemy by about 194 to 72 in late 1962, still held a large lead in manned bombers, and had begun to deploy submarine-launched missiles as well, it is clear that the USSR was passing through a gaping "window of vulnerability."

In his memoirs, Khrushchev admitted that

launching Sputniks into space didn't solve the problem of how to defend the country. First and foremost we had to develop an electronic guidance system. It always sounded good to say in public speeches that we could hit a fly at any distance with our missiles. Despite the wide radius of destruction caused by our nuclear warheads, pinpoint accuracy was still necessary—and it was difficult to achieve.[7]

The "ol' number seven," he admitted, was not a reliable weapon, but "only a symbolic counterthreat to the United States. That left us only France, West Germany, and other European countries in striking distance. . . ."[8] By 1958 the Kremlin opted against deployment of the SS-6 and committed resources instead to intense R & D on military rockets worthy of production. This meant a delay of five years or more before meaningful ICBM deployment. Did this indicate a public posture of caution and restraint? Would not premature bravado stimulate a panicky American buildup or embolden the radical Chinese to launch crusades that the USSR dared not support? Perhaps, but bravado might also help to restrain the West during this period of Soviet vulnerability, buy time for the USSR to make good its claims, and perhaps earn some diplomatic victories along the way. It would also be a grave embarrassment for the Soviets to undercut their own prestige by downplaying Sputnik. Indeed, Sputnik offered them a chance to retire once and for all the Soviet reputation for backwardness and replace it with an image of dangerous and inestimable strength. If your opponent always thinks you weaker than you are, you can profit from this misconception only by surprising him in a clash of arms. But if the enemy always thinks you stronger than you are, you can exploit his error through a diplomacy of intimidation. What was more, the risks involved in a bluff were reduced by secrecy and by the American unwillingness, evident since 1945, to use its nuclear superiority in preemption. A strategy of deception, therefore, capitalized on Soviet strengths and cloaked Soviet weaknesses.

The missile bluff played to gullible audiences in the West, but Peking was so credulous as to be vexatious. After 1949 the two Red giants cooperated in foreign policy and technology. The Chinese parroted Stalin in disparaging nuclear weapons and stressed the decisiveness of people's war. In 1954 the Chinese began to integrate nuclear weapons into their strategic thought and embraced Soviet technicians. But the advent of a mature nuclear deterrent in the USSR inevitably drove the Communist powers apart. Either the Chinese would themselves affirm Khrushchev's theories on the changed nature of warfare, expecting the Soviets to share their technology in socialist brotherhood, or the Chinese would continue to reject nuclear deterrence and "peaceful coexistence" as antipopulist revisionism.[9] In any event, Chinese reaction to Sputnik was as euphoric as the American was splenetic. Editorials called the Soviet ICBM "epoch-making" and "the ultimate weapon." It all meant that "any imperialist warmonger cannot but take stock of the political, economic, and moral strength of the Soviet Union. . . ." Marshal Ho Lung wrote that the USSR had surpassed the United States in science and technology; Chang Wen-tien and Mao Tse-tung himself wrote that a third world war must now spell the end of world capitalism.[10]

Unfortunately, Mao refused to accept Khrushchev's corollary to the effect that war was now unthinkable. He went to Moscow after *Sputnik II* with a revised assessment of the global correlation of forces—"the East Wind was prevailing over the West Wind"—and rather than whimpering about peaceful coexistence, exhorted Communists to overthrow the bourgeoisie by armed force if it sought to oppress the peoples.

"Leninism teaches, and experience confirms," said Mao, "that the ruling classes never relinquish power voluntarily. . . . [If] the worst came to the worst and half of mankind died, the other half would remain, while imperialism would be razed to the ground and the whole world become socialist. . . ."[11] At the November summit Mao paid obeisance to the Kremlin as the sole, natural leader of the revolution, and in December won a pact of assistance to include atomic energy. But the Soviets soon informed their eager allies that (1) they intended to retain control over all nuclear warheads, (2) they would not back up Chinese adventures with their nuclear force, and (3) they would not share missile technology. Peking then flip-flopped again with equal suddenness. By April 1958 the Chinese declared it "despicable to rely on foreign countries," denounced "slavish dependence on the Soviet Union,"[12] and reverted to a military doctrine that belittled high technology even as they promised they would soon develop their own nuclear weapons with or without Soviet help.[13] As the Sino-Soviet rift widened to a gulf, they vilified the Soviets for negotiating with the United States on a nuclear test ban, for timorous behavior in light of its missile prowess, for refusing to back China against Taiwan on the islands of Quemoy and Matsu, for adopting neutrality in the 1959 Sino-Indian War, and for a host of ideological errors stemming from the Twentieth Party Congress, including a misreading of the place of technology in Marxist dialectics.[14]

In 1959, Khrushchev recounts, the Politburo

knew that if we failed to send the bomb to China, the Chinese would accuse us of reneging on an agreement, breaking a treaty, and so forth. On the other hand, they had already begun their smear campaign against us and were beginning to make all sorts of incredible territorial claims as well. We didn't want them to get the idea that we were their obedient slaves who would give them whatever they wanted, no matter how much they insulted us. In the end we decided against sending them the prototype.[15]

At the same time, neither Maoist ideology nor Chinese *amour-propre* (like Gaullism and French pride) could abide Superpower exclusivity in advanced technology. Sputnik and the missile bluff blew the Communist schism into the open and gave way within months to indigenous Chinese nuclear, missile, and space programs.

To sum up, soon after *Sputnik I* the USSR chose to delay deployment of an ICBM force until better missiles came along, but to use the space program as a deception. According to RAND, the initial motives for the bluff were (1) to conceal the decision not to deploy, while reaping diplomatic rewards as if the missile gap were real; (2) to help deter an aggressive NATO stance on critical issues lest the USSR have to retreat; (3) to put the United States on the defensive and to create a climate for Communist initiatives in Europe and the Third World.[16] Its dangers lay in the chance that Soviet adversaries might respond perversely. Indeed, the United States did not blink in the Berlin crises and opponents of Eisenhower called for a rapid strategic buildup. The Chinese were so impressed by Soviet might that any restraint on Khrushchev's part exposed him to Maoist accusations of bourgeois peace-mongering. Finally, there was the danger that the Soviets would be caught out. The Soviet bluff was ably seconded by space successes and by the willingness of many in the West (out of paranoia, self-interest, or funk) to accept the most alarming Soviet claims. But it was threatened by American intelligence operations designed to reveal the true state of Soviet missile technology and by the possibility that Soviet-sparked crises might get out of hand and force a crossing of swords. The latter consideration shaped the character of the Berlin crises: the maddening succession of ultimata and threats, protracted and sterile negotiations, ending in every case with a Soviet excuse for relaxing tensions "in the interest of world peace." The former consideration—Western intelligence operations—in turn determined Soviet policy on law and controls for outer space. For the greatest threat to the Soviet missile bluff was the spy satellite.

At the Twentieth Party Congress Khrushchev pronounced nuclear weapons and rockets so terrible that even the imperialists were cowed and competition confined to peaceful pursuits. The Hungarian uprising in 1956 punctured Soviet buoyancy, but Sputnik reinflated it and corroborated Khrushchev's prophecies as nothing else could. "In peaceful competition, we will work to win out," Khrushchev told William Randolph Hearst, Jr., after *Sputnik II.* "Here, if I may say so, the Soviet people will be on the offensive . . . and are confident of their victory."[17] These fetching slogans compared favorably with American Cold War rhetoric, warning as it did of Soviet militarism and duplicity and urging the peoples of the earth to be vigilant, eschew neutralism, and risk terrible destruction in defense of freedom. Similarly, the USSR called piously for the renunciation of force through general and complete disarmament, while the United States responded with complicated plans to ban this or that technology and refusals to dismantle overseas bases. In an arms race, said the Soviets, their superior system would win out; in a disarmed world, in peaceful competition, again they would win out. Only in a militarized world, however, could the capitalists hope to frighten their minions into subservience.

This elegant propaganda line accorded the USSR a lasting moral advantage. This was so because disarmament, partial or complete, never happened. And it did not, in part, because the Soviets refused to permit inspection to ensure their own compliance. They claimed that the *imperialists* were the untrustworthy ones, yet if that were so, they (the Soviets) should have been the ones to insist on strict controls. They called for abolition of all modern weapons, yet claimed that advanced technology had made possible the reign of peaceful coexistence. They insisted on the peaceful nature of their space program, yet conducted it in secrecy and used it to boast of country-busting and "fantastic new weapons."[18] They proclaimed their support for keeping space free of all military activities, yet in the absence of general and complete disarmament declared themselves free to develop both ICBMs and military spacecraft. Finally, the American efforts to use space technology (spy satellites) to relieve the Soviets of the onus of on-site inspection were denounced by Moscow as evidence of U.S. militarization of space. It is frankly impossible to construct a rational hypothesis that explains the Soviet position on disarmament in terms of a genuine desire to promote arms control.

Foremost among the circumstances that encouraged obstructionism in the Kremlin was the missile bluff. Khrushchev recounts:

I believe at that time the U.S. might have been willing to cooperate with us, but we weren't willing to cooperate with them. Why? Because while we might have been ahead of the Americans in space exploration, we were still behind them in nuclear weaponry. . . . Our missiles were still imperfect in performance and insignificant in number. Taken by themselves, they didn't represent much of a threat to the United States. Essentially, we had only one good missile at the time: it was the Semyorka, developed by the late Korolyov. Had we decided to cooperate with the Americans in space research, we would have to reveal to them the design of the booster for the Semyorka. . . . In addition to being able to copy our rocket, they would have learned its limitations; and from a military standpoint, it did have serious limitations. In short, by showing the Americans our Semyorka, we would have been both giving away our strength and revealing our weakness.[19]

Recalling the discussions in the Eisenhower administration on the same topic, one must comprehend, with a weary shake of the head, the perverse logic of technocratic competition. U.S. experts knew that they led in military missilery, yet to seek cooperation might convince the world that the United States was hoping to poach Soviet know-how. The Soviets could not permit cooperation because it would reveal how backward they were. In arms control most U.S. advisers favored postponing serious discussions until the United States deployed a mature, dependable missile force, while the Soviets likewise had to forestall controls until they had made good their bluff and had a missile force in the first place. Khrushchev again:

I must say that the Americans proposed certain arms control measures to which we could not agree. I'm thinking about their insistence that a treaty include a provision for on-site inspection anywhere in our country. In general, the idea of arms control was acceptable to us. Zhukov, who was our defense minister at the time, and I agreed in principle to on-site inspection [for a limited nuclear test ban, 1957] of the border regions and to airborne reconnaissance of our territory up to a certain distance inside our borders, but we couldn't allow the U.S. and its allies to send their inspectors criss-crossing around the Soviet

Union. They would have discovered that we were in a relatively weak position, and that realization might have encouraged them to attack us.[20]

Whether or not the Soviets really feared an American attack, the missile bluff reinforced their opposition to inspection. In January 1958 Khrushchev explicitly required the West to ban all nuclear weapons and evacuate overseas bases before the USSR would even discuss the peaceful uses of outer space.[21] And why not? As Khrushchev explained to Bertrand Russell:

We agree to discuss the control of cosmic space, which is in fact the question of intercontinental ballistic rockets. But it must be examined as part of the general disarmament problem. . . . We are tying them together in the same way that they are tied together in real life; for if we did otherwise, instead of an end to the arms drive, this drive could develop speeds such as the world has never known and lead to a holocaust brought on at the behest of the imperialistic circles.[22]

The United States had a great bomber force, "but the ballistic rocket is, of course, an improved weapon. This is why we can understand the U.S. interest in the problem of outer space. It demands the prohibition of the intercontinental ballistic rocket in order to put itself in a more advantageous position, should war break out."[23] In *International Affairs*, space law expert Ye. Korovin elaborated: ". . . the meaning of the U.S. proposal to neutralize the cosmos practically comes down to forbidding the Soviet [missile]. . . . It follows that Eisenhower's proposals correspond to the security interests solely of the U.S., but envisage no measures lifting the threat created by the presence of U.S. military bases on foreign soil. . . ."[24]

This made military sense, but its implications extended to space technology as a whole. By linking the peaceful uses of outer space to missiles and then dumping the whole package into the context of the UN Disarmament Committee, the Soviets helped to block global consideration of a legal regime for space technology. Their propaganda, of course, stressed their special dedication to the peaceful uses of space and to the "unlimited" possibilities for cooperation.[25] But once the U.S. delegation succeeded in garnering overwhelming support in the General Assembly for an Ad Hoc COPUOS, the Soviets switched their tactics from emphasis on disarmament to emphasis on "equal representation" in the COPUOS. Korovin justified the Soviet boycott:

The U.S. delegation . . . used its "voting machine" to carry a resolution on the creation of such a rigged preparatory group (two-thirds of whose members are tied to U.S. military blocs) as would give the U.S. complete control of it. Thus, the U.S. Government has again showed its intention to subordinate cosmic research to its expansionist and aggressive plans. . . . This is why the Soviet Union naturally refused to participate in this body.[26]

ОБРАЗУМЬТЕСЬ!

"American Imperialism and Bonn's Revanchism try to play games with the Soviet Union, but they forget about the might of the Soviet Government—Come Back to Your Senses!" Some missile bluff propaganda from 1962, depicting the wagging tongues of Uncle Sam (the rich policeman) and the West Germans (often shown as neo-Nazis in Soviet cartoons) being silenced by a giant Soviet missile. From M. A. Abramov, *Byl' nacheku* (Moscow, 1962).

Despite the absence of the USSR from the Ad Hoc COPUOS in 1959 and 1960, Soviet views on space law were no mystery. Rather, the Academy of Science formed a new Committee on the Legal Problems of Outer Space, and a spate of legal scholars, by dint either of personal interest or official assignment, published a sizable corpus on space law. Before 1957 the USSR had vigorously upheld the extension of national sovereignty *usque ad coelum,* and on that basis condemned U.S. attempts to spy from balloons and aircraft.[27] This was the attitude that caused Washington to worry about Soviet reaction to the first satellite. But after Sputnik, the "principles of socialist humanism" proved not to be etched in stone. The USSR had no choice but to uphold "freedom of space" and renounce its belief in unlimited "vertical sovereignty." (Sputnik, it was said, did not violate sovereignty because it did not fly *over* countries below, rather the countries themselves rotated beneath the Sputnik!) Space was now considered analogous to the high seas, beyond the "effective control" of governments.[28]

Having argued necessarily for the legality of their satellites, the Soviets then had to deal with the hidden American agenda, the use of satellites for espionage and military support. Soviet writers asserted that ". . . there are no definite norms of international law with respect to cosmic space. . . . any state can freely use interplanetary space and can launch its satellites and rockets therein without requesting permission for this from other states." However, the "highest principle of international law" is national sovereignty and the right of states to self-defense, the principles of peaceful coexistence were universal, and space could not be used to prepare attacks and threats on others.[29] Articles of these early years followed the Soviet line at the UN to the effect that controls for space would not suffice in the absence of terrestrial disarmament. But the looming threat of satellite reconnaissance nudged the Soviets into trying to separate the debates on disarmament and space law and then to mitigate the principle of "freedom of space."

The only arena in which the USSR could promote these new formulations against the imminent success of the U.S. Discoverer program was in the UN COPUOS. So the Soviets reconsidered their boycott, and in the fall 1959 UN session acquiesced suddenly in a new standing COPUOS, consisting of twelve Western states, seven Communist states, and five neutrals. The COPUOS then constituted itself as two subcommittees of the whole, one legal, one technical, and took up the charge of recommending action to the General Assembly. But the agreement on membership was a false dawn, for the committee fell into a two-year lassitude during which Soviet and Western delegations clashed over the designation of officers, procedural issues (especially majority voting versus unanimous, or consensus, decision making), and the details of a world scientific conference planned for 1961.[30] Meanwhile, the Soviets used the platform to fulminate against the "illegal" efforts of the

American militarists to use space for espionage against other countries.

The U-2 incident the following summer was a mixed blessing. It sorely embarrassed the Americans, advertised Soviet air defenses, and gave weight to Soviet accusations about American intentions for space. But it vexed the Kremlin to admit that such flights occurred at all, and it threatened to cast doubt on their claim to a great missile force. Two days after the destruction of the spy plane, Khrushchev announced the appointment of Nedelin to command the Rocket Troops and iterated that the armed forces were "being converted to rocket weapons."[31] He spoke complacently about the "undoubted" superiority of the USSR in delivery systems.[32] To allay suspicions that the U-2 might have been able to locate Soviet rocket bases, Khrushchev displayed film allegedly from Powers's flight that showed airfields and factories but no ICBM bases. Of course, there were none to photograph, but Khrushchev claimed the photos proved "that spy flights were carried out precisely over regions which have no rocket bases."[33] At any rate, the downing of the U-2 proved that U.S. bombers could no longer knock out Soviet bases: "they would be shot down before reaching their targets."[34]

The U-2 affair was only a prelude to a game with higher stakes. Quoting a New York *Times* article, Georgi Zhukov, the dean of Soviet space law theorists, warned in October 1960 that since the USSR had proved it could shoot American spy planes out of the sky, the United States would rush development of a new method via satellites in space. The kind of information provided by spy satellites "can be of importance . . . solely for a state which contemplates aggression and intends to strike the first blow. . . ." The USSR, decreed Zhukov, had the right and the ability to defend its sovereignty in space just as it had in the air: "If other espionage methods are used, they too will be paralyzed and rebuffed." And so he sharply qualified the original Soviet position on freedom of space: rather than space being different from the air, now the same "considerations of state security" applied to both. Espionage was unlawful wherever it was conducted, and the USSR was prepared to cooperate with any state to prevent the militarization of space.[35]

The Soviets' discomfiture over American spy satellites presented an ironic picture since they were telling the truth about their own space program, which was designed primarily for prestige. The Americans *were* the first ones to place military systems in space—not because they were more aggressive, but because they were actually *ahead* in satellite technology, more practical in their allocation of funds, and more in need of intelligence about the doings of the adversary. Otherwise, on specific issues of space law, the Soviets had little to dispute with the Americans.

For instance, the USSR joined the United States in skirting the issue of "where space begins." The Soviets also took the functional approach in declaring that whatever is in orbit is in space. They also ridiculed those who feared that the USSR would claim the moon in "imperialistic fashion" and agreed that celestial bodies were off-limits to national appropriation. They, too, preferred a legal status for space of *res communis omnium*. Furthermore, once the Soviets' early call for a UN space agency was withdrawn, they showed no more desire than the United States to be constrained in their technology by a gaggle of lesser beasts in the UN barnyard. Indeed, the only immediate and meaningful clash between East and West over space law concerned reconnaissance. The United States insisted on language banning all "aggressive" use of space, the USSR all "military" use of space.[36]

In November 1961, when the two-year term of the original COPUOS members was about to expire, Valerian A. Zorin notified U.S. Ambassador Adlai Stevenson that the USSR was prepared to end the deadlock in the committee. They tacitly agreed on the principle of consensus, the United States giving up on majority voting and all members therefore retaining an informal veto. This in turn reduced the importance of the membership issue. The resulting resolution voted by the General Assembly increased the size of the COPUOS to twenty-eight, adding one Communist and three neutral states. Otherwise, Resolution 1721 (XVI) of December 20, 1961, held that exploration of space be used only for the betterment of mankind, that international law including the UN Charter extended to outer space and celestial bodies, that outer space was free for exploration and use by all countries, and that spacefaring states report all launches to the Secretary-General.[37] This first agreement on principles and procedure fulfilled the prerequisites for further progress on space law. But again the price to be paid was the granting of a soapbox to Soviet diplomats. Even as Zorin and Stevenson bargained behind the scenes, the Byelorussian delegate struck the chord of Soviet rhetoric:

> The U.S. representative has said that space was as yet free from earthly conflicts and prejudices. Unfortunately ... military projects play an important part in the U.S. space program, and any new success by the Soviet Union is always viewed by the U.S. from a military standpoint. The U.S. press, complying with the wishes of monopolies which profit from the arms race, seeks to present the Soviet Union's peaceful achievements in space as an attempt at world domination. While publishing such slanders, the U.S. attempts to make up for the failure of the U-2 policy by launching satellites designed to spy on the territory of the socialist countries. Such actions do not accord with its pious statements.[38]

Rather than signaling rapid progress in space negotiations, full Soviet participation meant eighteen *more* months of deadlock in which militarization was the principal issue. In 1962 the Soviet Academy of Sciences sponsored a major symposium on space law chaired by Korovin. The Space Age, it was declared, was one of struggle between competing systems, of socialist wars and wars of national liberation, and of the downfall of imperialism. One aspect of this universal competition was the struggle for outer space between the policies of peaceful coexistence and of war. Citing U.S. budget figures and military leaders on the strategic importance of space, Korovin argued that the doctrine of "space war" dominated U.S. policy. At the same time, the United States slandered the peaceful Soviet conquest of space and sought to strip the socialist camp of its ICBM deterrent to American aggression. The USSR rightly insisted on complete disarmament, but until that was achieved, space, like the sea and air, would inevitably be a theater of military operations. The right of overflight was indisputable, but so was the illegality of espionage, hence the right of a nation to destroy a spy satellite. "Freedom of space" assumed respect for forms and methods established by the mutual consent of all. This was the meaning of *res communis omnium* rather than *res nullius*, in which "anything goes." The American plan for "freedom of space" in the broadest sense, combined with U.S. military space doctrine, meant that "the American plan for control in space is a plan for American world domination."[39] The United States shamelessly admitted the true aims of such programs as Midas and Samos, information from which could only be of use for preparing an aggressive nuclear rocket war. The USSR could not remain indifferent to such espionage, and its right "to destroy a spy satellite, and in general any space ship encroaching on the security of this nation, is indisputable."[40]

Sputnik was almost too much of a good thing for the Soviet Union. Khrushchev can perhaps be forgiven for taking so seriously this "objective proof" of his 1956 theories. But he also chose to make the Soviet space program a cover for a brazen military bluff. Perhaps he thought there was no other choice but to run the bluff. Retreat and retrenchment, after Sputnik, was too bitter to contemplate. But it bought him a pack of troubles. The Americans accelerated their own missile programs and the Chinese demanded to know why Moscow was so timid if its might was as great as it claimed. Most of all, the Soviets faced the danger of being found out. At this point the original American space program, spy satellites, became a serious threat.

Having balked at first from joining UN deliberations dominated by the West and aimed, presumably, at cooperation and inspection that Khrushchev could not permit, the Soviets finally took up the COPUOS as a forum for denunciation of American militarization of space—that is, reconnaissance. From late 1959 through 1963 the Soviet chorus against spy satellites droned on with growing volume in the UN. It demanded the prohibition of the use of space for "propagating war, national or racial hatred, or enmity between nations" and prior agreement concerning any use of space that might hinder exploration for peaceful purposes, and insisted that "collection of intelligence information by satellites is incompatible with mankind's space objectives."[41] The United States just as doggedly refused to take up such demands, and the General Assembly could only "note with regret" that the COPUOS was unable to make further recommendations regarding the peaceful uses of outer space.[42]

The boycott, then the deadlock, in the COPUOS only meant that the infancy of the Space Age passed without Superpower agreement on the optimal use of mankind's latest, most spectacular complex of technologies. The U.S. government, as has been seen, was not terribly distressed. As Soviet UN delegate Platon D. Morozov sighed: "Let us say that in science we can cooperate, but in law we cannot."[43] Only after American satellite photography was a fait accompli, only after the Soviet missile bluff was exploded, only after the Soviets, too, began to realize military applications in their space program, and only after a "young Turk" faction in the Soviet military agitated for Soviet space weapons and doctrine could the two Superpowers resume their negotiations on a legal regime for outer space. For then the early Space Age pattern would be broken: the pattern of missile bluff, made credible by prestige-oriented spectaculars, threatened by U.S. spy planes and satellites, which in turn dictated a Soviet hard line at the UN. The pattern held from Sputnik to the aftermath of the Cuban missile crisis. By then the technological revolution proclaimed by Khrushchev in 1956 was finally setting in, though only after an interval of seven years, the years of bluff.

CHAPTER 13

Hammers or Sickles in Space?

Some Americans consoled themselves with the thought that the Soviet system was not better at promoting technology, but only specific technologies, none of which (short of a Dr. Strangelove "doomsday weapon") was liable to be ultimate. Soviet doctrine itself held that no single weapon could ever be decisive in war. Yet Stalin had cracked the whip on missiles in the hope of acquiring, all at once, an equalizer to the U.S. threat against him. When the day came, Khrushchev talked and acted as if the hope were fulfilled. Instead, Sputnik laid on the Soviets all the same burdens it dumped on Americans: they must make their crash program permanent, increase R & D still more, while maintaining their advantage in conventional arms. What was more, as the Soviets moved to redeem their bluff in strategic arms, they had to shift emphasis in their space program from prestige stunts to the ancillary satellite systems attending a missile force—to choose between hammers or sickles in space. For once the effects of Sputnik took hold in the United States, the Soviets, too, were confronted by a rival technocracy, and one with superior resources. If Sputnik was a "technological Pearl Harbor," then the Soviets, like the Japanese, found they had awakened a sleeping giant.

Soviet strategy is a dim realm. Indigenous writings on the subject are often meant for foreign consumption, or else reflect debate, not decision—for once an issue is settled at the top, it tends to disappear from the literature. Published budgetary figures likewise are unreliable, incomplete, and undifferentiated. Space budgets in particular are hidden among several "scientific" and "military" categories. The most reliable means of reconstructing Soviet planning is simply to examine what came off the test or production line after an estimated lead time. But observing what the Soviets did cannot always settle why they did it. To be sure, Soviet strategy can be presumed to derive in part from geography, economy, and technology. But all three are in flux (for the third can alter the other two). Soviet strategy is also a function of ideology, which helps to determine the perception of the enemy and the definition of national interest. Similarly, "bureaucratic rivalry" also exists in the USSR, but is itself subsumed in the larger context of military-party relations.

In all these arenas the opening of the Space Age would not seem the shock to the USSR that it was to the United States. For Americans, Sputnik brought down the curtain on the unique isolation that had been their happy birthright. But Russians had lived under direct foreign threat for centuries, suffered terribly just fifteen years before, and lain in the U.S. nuclear shadow since then. Similarly, Sputnik alerted the United States to the economic sacrifices and contingent security of the nuclear age. But the Soviets had been geared to a warfare economy since the 1930s—there was no need to compromise age-old civic values in the USSR. Sputnik also shocked the United States into facing up to a total Cold War—B-47s and the CIA alone were no longer enough. But the Soviets had existed in a state of "cold war" against their neighbors since 1917. Peacetime arms races and wars of nerves were not new to Khrushchev's generation. Indeed, if the Americans had been first with ICBMs and satellites, as everyone expected, it is hard to imagine any crisis resulting in Soviet government. Rather, the difficult transition that did result in the USSR was a crisis of success. The Soviets finally got a leg up on the imperialists, the system had worked—and that, more than anything, threw Soviet strategy into confusion. To be sure, debate over military autonomy, spending priorities, and R & D were all continuations of old struggles within the regime. But the crisis of the 1960s, the one that toppled Khrushchev, was the product of the new adventurism and millenarian promises inspired by the glow of Sputnik.

Sovietologist Roman Kolkowicz has described Communist Party-military relations as cyclical. The Party's goal is to subordinate all institutions to itself, the military's goal is to preserve autonomy in professional realms like strategy, doctrine, and procurement. When the Party is united it can purge and intimidate the military, but the latter's indispensability will always provide the officer corps with a chance to recoup, especially during periods of Party strife.[1] Similar cycles described the relationship between the Party and the technical intelligentsia. In the formative years of the missile age, the two cycles overlapped like sine curves on a graph. After 1945 the victorious Red Army went into eclipse: Stalinist science distilled strategy from Marxist theory, not "bourgeois expertise," defined war as a "social enterprise" and victory as the result of "permanently operating factors" rather than strategic surprise and command decision. This formula exonerated Stalin for the debacle of 1941, but the result was stagnation in Soviet doctrine lasting until Stalin's death.[2]

During the "Malenkov interregnum," 1953–55, the army regained its autonomy. *Military Thought*, the general staff organ, declared that "It is necessary to say frankly that in connection with the cult of the individual, no science sinned so much as did military science."[3] Rather, "the military art of the Soviet Army must take account of a whole series of new phenomena which have arisen in the postwar period." Chief among these was the atomic bomb. The army paper *Red Star*, silent on the subject from 1947 to 1953, published over fifty pieces on the implications of nuclear weapons in the two years after Stalin's death. Did the advent of such weapons mean that strategy must now fall to politicians, scientists, and ideologists? On the contrary, the laws of war applied to capitalist and socialist societies alike: "to include in military science questions beyond its competence means to ruin military science as a specific branch of knowledge."[4]

Zhukov's brief period as defense minister under the rising Khrushchev brought increased authority and social prestige for the military, as well as the first Bison long-range bombers and H-bombs. But far from embracing the bomb as a kind of military philosopher's stone, the general staff ridiculed talk of "one-weapon strategies" as adventurous, bankrupt, and false, the "arbitrary fabrication of bourgeois military theorists."[5] Historical experience taught that with the appearance of new, more destructive technologies, *"the significance of men on the battlefield not only does not decrease, but increases all the more. . . ."* Zhukov wrote that "one cannot win a war with atomic bombs alone," and the editor of *Military Thought* that "Atomic and thermonuclear weapons at their present stage of development *only supplement the firepower of the old forms of armament."*[6] Malenkov's notion that thermonuclear war "would mean the destruction of civilization"[7] became heretical; Bulganin instead warned that the "impossibility of war" doctrine only bred complacency.[8]

The upshot of this military renaissance was renewed stress on a combined arms strategy favorable to heavy industry and the army. But the renaissance was brief. Neither Khrushchev nor any party leader could long tolerate the influence of heroic military rivals with potentially "bonapartist" tendencies. On October 4, 1957, Zhukov was sent abroad on a trivial tour of the Balkans—then sacked. Khrushchev loyalists, the so-called Stalingrad Group, ascended, including A. R. Malinovsky (Minister of Defense), A. A. Grechko (Commander of Ground Forces), and K. S. Moskalenko (eventual successor to Nedelin as commander of Strategic Rocket Forces).[9] These Khrushchevian appointees tried to limit the return of party control, but personal ambition checked their institutional loyalty.

Khrushchev's greatest ally in the struggle against the army, however, was the ICBM. Soviet H-bombs and missiles enabled him to resurrect the doctrine of noninevitability of war, reduce the size of the army, declare war on the "steel-eaters," and plot the future road of Communist expansion through political means under the banner of peaceful coexistence. They allowed him to pressure Berlin without fear of reprisal. They stood behind his drafting of the new Seven Year Plan with its stress on chemicals and agriculture. Finally, in January 1960, they permitted him to pronounce a new military strategy for the missile age that cut at the personal and professional vitals of the officer corps. Justifying a one-third cut in the standing army, Khrushchev declared to the Supreme Soviet:

Our state now has at its disposal powerful missiles. The air force and the navy have lost their former importance in view of the contemporary development of military technology. This type of armament is not being reduced but replaced. Almost the entire air force is being replaced with rockets. We have now cut sharply, and will continue to cut sharply, even perhaps discontinue production of bombers and other obsolete equipment. In the navy, the submarine fleet assumes great importance, while surface ships can no longer play the part they once did. In our country, the armed forces have to a considerable extent been transformed into rocket forces.[10]

A brazen statement! Not only because it was false—part of the bluff, partly aimed at foreign audiences—but because it ensured the hostility of virtually every branch of the Soviet military. Malinovsky and others fought back as stridently as was prudent, stressing the old line that only combined arms brought victory. But deterrence, not combat, was the main goal of such one-sided dependence on nuclear missiles. Indeed, the goals of Soviet foreign policy could now be met without war, since the West was stymied from using its nuclear club to frustrate the march of history. The USSR, by definition the expanding power and able to extend its influence in other ways, stood to gain more from deterrence than the United States.[11]

Whatever the political logic of deterrence, Soviet officers must have bitterly resented Khrushchev's new Strategic Rocket Forces. First, they knew better than anyone that it was a bluff. Second, the severe reduction of the army justified by the missiles even weakened the conventional deterrent (i.e., ability to overrun Western Europe). Third, the building of a genuine second-strike capability paired with Khrushchev's attack on heavy industry could only mean less money for the old services. Fourth, nuclear deterrence was abhorrent to military professionals of any culture, for it was not based on trained, courageous, and patriotic soldiers engaging the enemy in something resembling honorable combat but rather on holding hostage the enemy's civilian population, on a technical means of terror.

This last sentiment, rarely stated, surely ate away at the morale of American soldiers as well. It completed the relegation of men to machines, rendered leadership, discipline, and *esprit de corps* obsolete, and rewarded the skills of the engineer and civilian strategist. In the United States, the air force embraced this decidedly unmilitary role in part through interservice rivalry and in part because of the residual glamour of manned bombers, which still called for human character. But there was no bluesky faction in the USSR, no intermediate "bomber phase" in the Soviet transition to deterrence. If half of what Khrushchev predicted came to pass, much of the officer corps would be redundant and much honor removed from service for those who remained. Who stood to gain from a massive shift to Strategic Rocket Forces? Only the technicians in the artillery corps. Even in the other services, accelerated introduction of new technology redounded to the benefit of young technical officers, whose values and habits were as alien to those of World War II veterans as those of "Red Experts" were to holdover industrial managers in the 1930s. By pressing the technological transition so loudly and quickly, Khrushchev exhibited a civilian control that Eisenhower or McNamara could only have envied, but he also stirred resentments that presaged his own demise.

So why did he do it? Why did Khrushchev seize on Sputnik to overturn the military structure of the USSR? In his memoirs, he refers offhandedly to traditionalism that had to be overcome in order to "build a missile army." He had "to do away with this sort of old-fashioned thinking," implying that he had a better understanding of military needs than the military. For instance, wrote Khrushchev, he himself conceived of placing missiles in underground silos only to hear the experts scoff. (In fact, as has been seen, the Soviets had first to develop a missile compatible with silo deployment.) But "the new cannot live side by side with the old in military policy. We had to hasten the process of replacing the old with the new."[12]

The macroeconomic context of these years also suggested a shift in strategy. The early 1950s were years of phenomenal growth, averaging 11 or 12 percent, which helped to support military budgets that rose from 8.0 to 11.2 billion new rubles, twice the peacetime level of 1940. After Khrushchev double-crossed his military supporters in 1956, the official defense budget held steady at about 9.6 billion rubles for the remainder of the decade. Yet economic growth also leveled off, excepting the boom year of 1958, to an average of about 6 percent. If Khrushchev were to make a serious shift to light industry and agriculture, he had to make good on his missile and space pretensions without big increases in defense spending. The logical conclusion would appear to be increased effort in missile R & D and deployment combined with severe cuts in conventional forces and bombers.[13]

All these features—deterrent strategy, increases in R & D, cutbacks in conventional arms—invite comparison with Eisenhower's New Look. The latter stemmed from the assumption that nuclear weapons gave "more bang for the buck" and thus were cheaper than large armies and navies. Indeed, American pay scales and support costs made conventional forces expensive, while the sophistication of American science and industry made high-tech weapons relatively cheap. But in the USSR the reverse was probably true. Soldiers came cheap in the Soviet Union, while high-tech systems absorbed relatively more resources than in the United States.[14] Why then might the Soviets expect a deterrent strategy to reduce economic strain? Perhaps the Politburo, scarcely up to date in its accounting methods and long-range systems analysis, simply underestimated the costs of missile competition.[15] Or perhaps the reduction in conventional manpower found its source not in macroeconomics but in demographics. The Stalin era dealt an axe blow to the Russian population curve. Not only did total population decline by 20 million during the war, and perhaps by an equal number during the collectivization and purges, but the low birthrate of those years meant a considerable drop in the number of adolescents entering the workforce in the 1960s. The CIA projected that Soviet population would rise at an annual rate of only 1.5 percent during the 1960s, with the increase in those of working age even less.[16] If the country were to sustain its growth rate, then productivity must increase at an even faster pace—meaning large doses of new technology—and/or more workers must be found—meaning a reduction in the army.

Thus the Soviet New Look differed from that of Eisenhower. The Soviets entered deterrence running a bluff; the American New Look was an effort to capitalize on superiority. Soviet deterrence was offensive in that the Kremlin could expect to make political gains under its cover; the American New Look was designed precisely to freeze the status quo. Soviet deterrence strengthened political control over the military and released men for civilian labor, but, unlike the American deterrent, it was not cheap. Within a year Khrushchev retreated from his bold pronouncements of 1960 about the obsolescence of the army and navy and cuts in military spending. The defense budget shot up 29 percent in a single year, and army cuts were restored. These moves were attributed to American belligerence over Berlin. In fact, the Kremlin probably absorbed the fact that the Red Army remained its trump card, especially after spy satellites exploded the missile gap, and that deployment of a secure nuclear deterrent would take longer and cost more than expected.[17]

The irony in the Soviet defense posture over the years of Khrushchev's reform, therefore, is that it led to sharp increases in R & D *and* expensive deployment of ICBMs *and* restoration and modernization of conventional units. The primacy of "massed armies" seemed to have won out after all, albeit armies fortified with the newest technology.[18] At best, Khrushchev's rhetoric reflected undue optimism about the benefits of deterrence; at worst, he was running another bluff, this time for domestic consumption, granting the recalcitrant military much of what he claimed to take away, while promising "goulash communism" to the masses.

Whatever the motives for the Soviet "New Look," Khrushchev launched a great technical enterprise evident both in the new systems that came on line in the mid-to-late 1960s and in profound strategic debate. Some officers remained loyal to his shifting policies, while others, including Marshals Malinovsky, Grechko, and G. I. Voronov, began to form an opposition. The old disputes surfaced once more over the competence of the military to formulate strategy and the role of all the armed services in future warfare.[19] In 1961 Malinovsky presented a "new military strategy" to the Twenty-second Party Congress affirming deterrence but also the importance of the "multi-million man army." All the old questions reemerged: the proper size of the armed forces, the decisiveness of the early stage of modern war, the length of such a war, the inevitability of escalation, and thus the relative importance of military planning versus technical and political-economic calculation.[20] The product of debate over these questions was nothing less than the first major inquiry into Soviet strategy since 1926. Assumed by Western observers to be an official statement, the volume was later judged to be a measured, thumping blow in an ongoing debate. On the one hand, it accepted the importance of the new rocket forces; on the other, it sidestepped vulgar deterrence and laid the groundwork for a dynamic Soviet strategy aimed at winning a nuclear war should it break out. And central to that war-winning effort would be command of outer space.

Voennaia Strategiia (Military Strategy), was edited by Marshal V. D. Sokolovsky and appeared in the summer of 1962.[21] Its implications for space policy, though understated in the first edition, were clear. By 1962 the Soviets had to face up to the bitter irony that their space program had been used as indicative of a missile force that did not exist. To promote the bluff, Khrushchev had given priority to spectaculars that did little to advance Soviet competence in the real military uses of space. By 1962 the Soviets were finally prepared to deploy ICBMs, whereupon they woke up to the fact that now it was the space program that represented a bluff! For all the glory of the Luniks and Vostoks, the Soviets were at least five years behind the United States in the military satellite systems—for reconnaissance, meteorology, communications, geodesy—indispensable to a mature ICBM force. Hence, even as traditionalists fought for a continued role for large armies and navies, the avantgarde technical officers tried to alert the military and political elites to the fact that technological revolution was not proceeding quickly *enough*.

The Sokolovsky volume approached the space problem delicately, given official Soviet dedication to the peaceful uses of space. The 1962 edition exploited the technique of quoting American sources on military uses of space, presumably in a condemnatory mode, then not refuting the American arguments. Hence the U.S. "doctrine" on military uses of space was left to stand. American "militaristic circles," it was said, saw mastery of space as the path toward world hegemony; and President Kennedy said that "space supremacy is the aim of the next decade, and the nation which controls space can control the earth." American "scientific" launches were held to be a cover for "far-reaching military plans."[22] U.S. reconnaissance satellites were already in operation, and

used for detecting and determining the coordinates of military-industrial objectives, the launching sites of ICBMs, military bases, airfields, and other objectives. . . .

Due to the fact that reconnaissance satellites moving in known orbits could be destroyed, creation of maneuverable manned space ships with various reconnaissance apparatus is planned [by the United States—a reference to the Dyna-Soar]. . . .

Great attention has been devoted to navigational satellites . . . which will be used to facilitate aerial and fleet navigational support, particularly for submarines, the compilation of navigational charts, study of the shape of the earth, etc.[23]

But these were

only a minor part of the U.S. program of mastery of space for military purposes. The main part of the program is the creation of aircraft-satellites or other aerospace vehicles carrying nuclear warheads. . . .

Finally, a considerable part of the U.S. program of the mastery of space for military purposes is the creation of antispace weapons. . . . These satellites will presumably be used to destroy, on command from the ground, satellites and other space vehicles as well as ICBMs.[24]

Such militarization of space was clearly aggressive, in contrast to the

inflexible tendency of the entire Soviet nation toward enduring world peace.

However, the Soviet Union cannot disregard the fact that the U.S. imperialists have subordinated space exploration to military aims and that they intend to use space to accomplish their aggressive projects—a surprise nuclear attack on the Soviet Union and the other socialist countries.

. . . It would be a mistake to allow the imperialist camp to achieve superiority in this field. We must oppose the imperialists with more effective means and methods for the use of space for defense purposes.[25]

The lesson for Soviet readers was unmistakable: nowhere was the U.S. effort condemned as erroneous; nowhere was it said that such efforts were wasteful; if anything, the potential of space-based weaponry was overstressed.

The passage on space weaponry in the Sokolovsky volume proved to be part of a growing campaign to alert civilian leadership to the strategic possibilities of orbital space.[26] Of course, the vast majority of published commentaries stressed the inevitability of Soviet primacy in space and the economic rewards to flow from spaceflight. *Pravda* even compared the rocket to a horn of plenty.[27] But some dissenters apparently existed: Academician Sedov found it necessary to argue as late as 1963 for the "usefulness of spending great resources and creative efforts on space research" and regretted that even some "major scholars and writers" questioned the space program.[28] The dominant line of criticism, however, and that most likely to win the ear of the Kremlin, assailed the *civilian* emphasis in the space program. Among the most prominent publicists were two chiefs of the Red Air Force—Marshals S. I. Rudenko and S. A. Krasnovsky—and two generals in the Engineering-Technical Branch of the military—I. I. Anrueev and G. I. Pokrovsky. Coming to their aid were a number of lesser technical officers, who must have grasped, like their American counterparts, the strategic potential of spacecraft. But it was not until 1962, in the midst of strategic controversy, the beginnings of ICBM deployment, and continuing investment in space spectaculars, that this "military space lobby" was moved—or allowed—to express its grievances. The campaign opened in March with two articles in *Red Star*. The first attacked American military space programs. The second argued the importance of space for Soviet strategy, both in terms of support of land, sea, and air operations and in the strategic nuclear realm. Indeed,

the creation and employment of various space systems and apparatus can lead immediately to major strategic results. The working out of effective means of striking from space and of combat with space weapons in combination with nuclear weapons places in the hands of the strategic leadership a new, powerful means of affecting the military-economic potential and the military might of the enemy.[29]

The years 1961 to 1963 were also those of the loudest Soviet protests against U.S. spy satellites. Many articles denounced American militarization of space only to publicize at the same time the importance of satellite systems for the USSR.[30] Other articles in line with Soviet diplomacy at the UN accused the United States of designing orbital bombardment systems and antisatellite weapons,[31] only to suggest that the USSR must retaliate with space weapons of its own. Malinovsky welcomed the task of countering the aggressors' "attempt to reconnoiter our country from air and from space,"[32] while another high-ranking officer warned that space-based selection of targets could decide the outcome of a nuclear battle.[33] The "military space lobby" made these same points explicitly or implicitly over and over again.[34]

In October 1962 Khrushchev, for whatever motives, tried to close the reverse missile gap by insinuating medium-range ballistic missiles (MRBMs) into Cuba.[35] The eventual settlement of the crisis was not unfavorable, but the Soviets' apparent admission of strategic inferiority raised military "self-criticism" in the USSR to a higher, shriller plane. A second edition of *Voennaia Strategiia* was hastily prepared and published in August 1963. It differed little from the first in generalities, but among the specific changes was a more explicit and expanded acknowledgment of the decisiveness of the space theater. It stressed the need for antisatellite and antimissile defense, and promised a wide range of chilling space weapons: "Various radiation, antigravity, and antimatter systems, plasma (ball lightning), etc., are also being studied as a means of destroying rockets. Special attention is devoted to lasers ("death rays"); it is considered that in the future any missile and satellite can be destroyed with powerful lasers."[36]

The campaign on behalf of military space programs was as loud as one could expect in the muted military forum of the USSR, and thus bears some resemblance to the "missile gap" mania in the United States. By 1962 the Soviets, though first with an ICBM and satellite, found themselves outclassed both in missiles and military space systems, and determined to build a missile force equal to or greater than that of the United States, while giving high priority as well to passive military satellites and active space weaponry. The lead times of test and operational systems suggest that while some systems were already under research, the big jump in Soviet military space investment occurred in 1962–63, the years of agitation by the "military space lobby."

At the start of 1962 the Space Age was but four years old. The United States had launched sixty-three payloads into orbit or beyond: thirty-four, over half, were military. The USSR could boast only fifteen launches; none were presumptively military. But R & D had surely begun on reconnaissance satellites and probably other military applications by 1959–60. Of course, the Soviets did not need spy satellites as badly as the United States, but they, too, required pinpoint targeting and had a special interest in monitoring the movements of the U.S. Navy and Chinese Army. Soviet perfection of recovery techniques, demonstrated with Strelka and Belka in 1959, required only that they replace life-support apparatus with cameras in order to achieve film-pack reconnaissance. The first such satellite was launched April 6, 1962, from Tyuratam at the standard inclination of 65 degrees. It orbited for three days to a perigee of about 190 miles, then parachuted to earth inside the USSR. It was labeled, blankly, *Kosmos IV.*

In the same spring of 1962 the United States began to black out its military satellite programs as well, releasing only the fact of a launch and the orbital data, as required for UN registration. When the Soviets began their military applications programs, they chose the cover name "Kosmos" for all, saying only that they involved various scientific types and would originate from "different cosmodromes" in the USSR.[37] Indeed, *Kosmos I,* which flew on March 16, 1962, inaugurated a new space launch capacity from the old IRBM test range at Kapustin Yar near the Caspian Sea.

Recoverable spy satellites, *Kosmos VII, IX,* and *X,* followed in the summer and fall of 1962. Six more flew in 1963, avoiding the winter months, perhaps because inaccurate reentry might lose the spacecraft in the snow cover. By late 1964 nine more had been launched, as well as the first three tests of a second-generation spy satellite in a lower orbit that probably produced pictures of a higher resolution.[38] By 1965 the USSR was launching on average more than twice the number of spy satellites as the United States. Pressure on the Tyuratam facility began to build—military launches, for instance, appeared to cease whenever a manned space mission was in the offing. Furthermore, its geographical location at 45 degrees north latitude was not ideal for space spying. The standard due-east trajectory, which made maximum use of the earth's rotation, yielded orbits that failed to cover the northern half of North America. To achieve higher inclinations, the Soviets had to launch east-north-east, lose part of the boost from the earth's rotation, and thus shed payload weight. By 1965 construction was underway of a new, secret military launch site at Plesetsk, 600 miles north of Moscow at 65 degrees north latitude. The first launch from Plesetsk was *Kosmos CXII*, a recoverable observation satellite, in March 1966. The decision to build the military space center was probably made in the last year or two of the Khrushchev regime.[39] The Kremlin never publicly acknowledged the existence of the new complex.

By 1964 the Kosmos series accounted for the great majority of all Soviet spacecraft and for the relatively sudden jump in overall Soviet space activity. The Soviets launched only six spacecraft in 1961, twenty in 1962, and seventeen, thirty, and forty-eight in the following years. Most were on observation missions of one sort or another, but new military space R & D also hid under the Kosmos rubric. In August 1964 a triple launch—three satellites on one booster—marked the first test of a navigation satellite system. The exercise was repeated in February 1965 and led to three quintuple launches that summer. The first Molniya communications satellites flew in April and October 1965; tests of a fractional orbital bombardment system and maneuverable satellites for inspection and destruction of hostile spacecraft, in 1967.[40] Indeed, virtually every category of military applications attributed to the Pentagon "lunatics" must have entered the R & D stage by 1964 in order for these test flights to have occurred in the subsequent years.

The original, prestige-oriented Soviet space program continued. But the shift in spending and overall increase in resources devoted to space technology were extraordinary. Together with the ICBM deployment of the mid-1960s, it meant that the promise of Sputnik was beginning to come true.

Now there were two space races. The first, for prestige, dated from Sputnik and involved a highly publicized effort by the United States to catch up. The second, for military exploitation, dated from the first Discoverer and involved a highly secretive effort by the USSR to catch up. Not surprisingly, official American pronouncements in the early years denounced Soviet secrecy, while insisting that Soviet space shots had no real bearing on the balance of power; the Soviet line, in turn, denounced American militarization of space, while insisting that Soviet feats, though mostly for prestige, proved military superiority. During these years the two Superpowers sparred at the UN about the principles and organizations that ought to guide human activities in space. But the onset of the most intense period of competition, in both space races, proved to be precisely the moment for reconciliation and progress in space *law*. This, too, made sense, for the space powers now had little to argue about: their space programs were both dualistic and roughly symmetrical.

The Cuban missile crisis blew away the fog of Soviet missile bluffs, and in that clarified atmosphere Premier Khrushchev saw the utility of détente. On August 5, 1963, Soviet, American, and British representatives initialed the Limited Test Ban Treaty banning nuclear explosions in air, sea, and outer space. When the COPUOS reconvened in September, the Soviet delegation promised "new measures in order to bring together the different points of view" on principles for space law.[41] These two arenas overlapped—for the tacit acceptance of space-based reconnaissance (the American goal) required not only that space law be debated without further reference to the "illegality" of military uses of space, but also that no active steps be taken toward a capability to shoot down such satellites. Early antisatellite plans involved detonation of a nuclear warhead in the vicinity of a hostile satellite, hence the banning of nuclear tests in space contributed to the tacit legitimation of spy satellites.

In October the Soviets and Americans jointly resolved to refrain from placing nuclear weapons in orbit. In December the Soviet delegation ceased its three-year-long tirade against U.S. spy satellites.[42] The result was UN Resolution 1962 (XVIII) of December 13, 1963, a Declaration of Legal Principles Governing Activities of States in the Exploration and Use of Outer Space. It declared outer space free for exploration by all and out of bounds to national sovereignty; space activities to be carried on for the benefit and in the interest of all mankind in accordance with the UN Charter and international law; states to bear responsibility for all their national space activities, whether carried on by government or nongovernmental agencies; states to be guided by principles of cooperation and mutual assistance, with "appropriate international consultations" to precede any activity potentially harmful to peaceful uses of space; spacecraft to remain under the jurisdiction of the launching state, with the latter accepting liability for any damage caused to foreign property by accidents; astronauts to be regarded as "envoys of all mankind" and rendered every assistance in case of peril.[43]

This resolution signaled a breakthrough in the evolution of space law. It ratified the role of the COPUOS as the formative body for space law, showed that progress could be achieved by consensus, and anticipated the terms of the later Outer Space Treaty. But while the specific principles were sensible enough, the declaration was still a minimum program and the "breakthrough" amounted essentially to a retreat by the USSR to the positions advocated by the more "mature" space power, the United States. Expressly "military" uses of space were not banned, while the prohibition on nuclear weapons in orbit, though reassuring to a nervous world, meant little. Land- and sea-based missiles were more economical, effective, and concealable than bombs in orbit anyway. In sum, the principles accepted by both sides in the first flush of détente represented no self-abnegation, but rather a recognition by the USSR that it had the same interest as the United States in developing a panoply of military satellite support systems without interference from third parties. The Soviet campaign against U.S. espionage abruptly ceased, observation from space became tacitly legitimate, and the hope first voiced by RAND long before Sputnik came to pass: "Open Skies" were a reality.

In strategic theory and practice, the first half of the Space Age brought new hopes, risks, and turmoil to the USSR. Facing challenges to his leadership and resistance to his economic reforms, a continuing Cold War with the United States plus new discord with China, Khrushchev seized on to Sputnik as an all-purpose technological fix. Instead he found that the space and missile age brought as many problems as solutions. Rocket-rattling deterrence seemed to promise foreign policy triumphs and opportunities to trim and reform the military at home. Unfortunately, Soviet technology was not as advanced as the space program made it appear, and therefore in the latter stages of the "missile bluff" he had to reaccelerate R & D, restore cuts in conventional forces, and approve a massive reorientation of the space program toward military applications. The first fruits of the latter decision were an end to the diplomatic campaign against American spy satellites and the beginnings of a UN code for outer space. "Freedom of space" triumphed—which pleased the United States—but only because the Soviets determined to race the Americans in ICBM deployment and the military applications of space— which scarcely pleased the United States.

The second fruits of Khrushchev's reversals included turbulence in the Soviet officer corps. For if Khrushchev's initial overselling of the technical revolution and his military reforms drove the "old school" generals into opposition, so the delays in "real" exploitation of Space Age technology irritated the new technical-minded cadre as well. Nor were the trends confined to the military sphere; they operated in many sectors of Soviet life, as Khrushchev labored to build, on his own authority and the strength of the sputniks, a Space Age communism.

CHAPTER 14

Space Age Communism: The Khrushchevian Synthesis

"We stand at the threshold of a new scientific and technical revolution, the significance of which far surpasses the industrial revolution associated with the appearance of steam and electricity."[1] Nikolai Bulganin, speaking to the Central Committee in 1955, had in mind mostly atomic energy. But in the decade after Sputnik, the "scientific-technical revolution" became the basis for a new Soviet reading of the current stage of history. The new Party Program in 1961 declared: "Humanity is entering a period of scientific and technical revolution connected with the mastering of nuclear power, the conquest of space, the development of chemistry, the automation of production, and other achievements. . . ."[2] A special section of the Academy of Sciences Institute of the History of Natural Sciences and Technique took up the study of the Scientific-Technical (or S-T) Revolution. Longstanding theories of capitalism, international relations, and Soviet administration were set aside in favor of ones that took account of the S-T Revolution. Khrushchev made it his keynote, rose on the strength of it, and eventually fell on the many conflicts engendered by it. But the S-T Revolution would survive him and leave both the Soviet reality and Communist theory profoundly changed.

As in military science, so in natural science the passing of Stalin reopened debate. Not since the silencing of Bukharin in the early 1930s did such a plethora of new formulations enter the Communist oeuvre. Was science in the foundation or the superstructure of society? Was it objectively neutral, or did "bourgeois" and "proletarian" science exist? Did technology dictate social relations, or did class rule dictate technology? Could unity of theory and practice in science be achieved under socialism? What indeed was technology? Such were the questions, left over from the 1920s, that took center-stage in the wave of post-Sputnik enthusiasm over the S-T Revolution.

The answer to the last question—what is technology?—changed during the 1960s when *tekhnika* came to be broadened from "the totality of means of labor" to "the artificially created means of activity of people."[3] This opened up to the realm of technique all the methods of research, data processing, and management that went hand in hand with the new hardware, and thus made *technologiya* a subset of *technika*. But what was its place in Communist theory? Lenin had silenced the prophets of "proletarian science," but the claim that socialism "freed up" creative energies and pushed technology ahead more quickly than capitalism implied that proper social relations triggered new technology, not vice versa. This was Bukharin's theory, in which ideological revolution fostered political revolution, which yielded economic revolution through reshaping social relations, which finally stimulated technical revolution, "not in the relations between people, but in the relations between the human collective and external nature."[4] In the 1920s Bogdanov and Trotsky declared that the transition from socialism to communism depended on "the scientific pursuit of science."[5] To be sure, even capitalists could adopt planned R & D, research institutes, and so on, but only under socialism, as Modest Rubinshtein explained, could technology "break out of the shell" and become the basis for true communism.[6]

In the 1960s the Academy of Sciences found the key to interpretation of the S-T Revolution in this stress on the role of social relations in carrying to fruition a "mere" technical advance. Science was declared a "direct productive force" but was subordinated, like labor itself, to the pattern of social relations defined by the Party.[7] It also explained how specific breakthroughs could occur without supplanting the role of class conflict as the engine of history. To illustrate this, the academy's team of scholars showed that the industrial revolution in England was only partly technological, but encompassed as well a "production revolution" (the factory system), new forms of energy (coal and steam), and a social revolution (bourgeois capitalism).[8] Similarly, technical advances such as atomic energy, radioelectronics, computers, automation, and space research were not "changing the world" by themselves, but comprised the technical *side* of a new phase in *social* revolution. The increasing importance of intellectual work was indicative of this larger revolution. Just as the industrial revolution had replaced muscle with machines, so now labor was being supplanted in the logical and control functions that ran the machines. In this new age, man would stand above machines and nature both, and all work would become increasingly scientific and creative.[9]

The problem for Soviet theorists was not in "communizing" the technological explosion of the postwar world but in explaining how it was that capitalism (presumably in its third or fourth dotage) managed to participate so mightily in the explosion. It was this question that led Soviet theorists to a new view not only of the place of science but of the character of (disturbingly dynamic) capitalism itself.

Soviet discussion over the relative timing of the S-T Revolution in various countries is reminiscent of Western debates about the timing of economic "take-off" under Rostow's stages of growth. Soviet scholars dated the U.S. S-T Revolution from 1953, noting that the Korean War "created favorable conditions for the adoption of advanced technique." What "advanced technique"? The Korean War did not produce the atomic bomb, or the ICBM, or the computer—but it had locked into the American system the techniques of large-scale government R & D, integration of science, industry, and government, and widespread economic controls that—in the Soviet view—were of the essence, not the effervescence, of the S-T Revolution.[10] What the United States had done, in the Soviet view, was to assimilate the superior methods of planned science and R & D that socialist countries adopted as a matter of course. And the United States had done so not in response to the workings of capitalist society—that was inconceivable—but in response to pressures from the socialist camp itself!

This opened a second, vast field for Soviet revisionism under Khrushchev. After Evgeny Varga's disgrace (see chapter 2) notions of a dynamic capitalism were anathema. But more or less objective study of American and international politics revived in the 1950s as a sibling of military science.[11] For if, as Khrushchev told the "fanatical" Chinese, "the atomic bomb does not observe the class principle" and general war was not inevitable, then (1) there must be independent variables such as weapons technology that shape the policies of capitalist and socialist states alike and (2) there must be differences among the foreign policies even of socialist states, and some means of accounting for those differences. In short, "the role of international relations [as opposed to the primacy of domestic politics] in the life of human society has grown sharply." Yet almost nothing was being published in the USSR worthy of the name political science or diplomatic history, and the only "area study" of note was the same ideologically safe field that the tsars had permitted: oriental studies. At the Twentieth Party Congress in 1956 Mikoyan deplored this neglect, and the dean of Moscow State University lamented that training in international affairs had ceased "at a time when we are in dire need of research workers in historiography, international relations, problems of scientific socialism, and other vital branches of history."[12]

The Twentieth Congress reversed this trend. In April 1956 the Institute of World Economy and International Relations was resurrected with a mandate to study present-day capitalism and its relations with the socialist world. It was encouraged to tackle the delicate issues and stress *aktual'nost* (topicality or "relevance"). In the pages of the institute's journal and a flood of dissertations and collections on postwar politics, Soviet scholars began to engage such root problems as whether "the State" was a monolithic actor in politics or an amalgam of competing bureaucracies; whether smaller countries were autonomous or merely subordinate to the Cold War blocs; whether American officials were independent actors or tools of monopoly capitalists; whether the Western economy was dynamic or stagnant; whether dynamism or stagnation meant belligerence or quiescence in foreign policy—in sum, the very questions that Western analysts dance about in circles while trying to make sense of the Communist world.[13]

Soviet writers began, naturally, with the objective forces of history as outlined by Khrushchev. Atomic weapons and missiles in Soviet hands ensured peaceful coexistence, made U.S. supremacy "short-lived and gone with the wind," and unleashed the "third crisis of capitalism." The first two such crises had been associated with world wars and each brought gains for socialism. The third, held to begin in the mid-1950s, would not encompass a world war but would still lead to socialist gains thanks to Soviet technology and the Third World revolution.[14] The central importance of Sputnik as a symbol and proof of the opening of the third crisis was repeatedly stressed.[15] Indeed, peaceful atomic power and earth satellites were "the *standard bearers of socialist civilization*" and as symbolic of socialism as the steam engine had been of nineteenth-century capitalist civilization.[16]

Looking west, the Institute of World Economy also fashioned a new and flexible model of the bourgeois state that allowed for a certain independence from class interests. Thus the U.S. government was no longer a creature of Wall Street or the armaments manufacturers, but was influenced by exogenous forces. The presidency, in particular, had been strengthened by past wars, so that the White House could sometimes override big business. This more subtle portrait permitted Khrushchev to justify détente with the Democrats after Cuba and speak of "progressive forces" at work in the United States. On the other hand, the same tendencies toward autonomous central power made this imperialist rival more dangerous than ever, for increased government regulation and management of science meant that the United States could, within limits, assimilate the fruits of the S-T Revolution. Under Stalin, Soviet writers had facilely described the relationship of the bourgeois state vis-à-vis the capitalists as *podchinenie* (subordination). The new political science allowed for evolution in the West under the impact of mobilization for war, Keynesianism, and state-supported R & D of a "state-monopoly capitalism," while the relationship of the state to the capitalist elites evolved into *srashchivanie* (interdependence).[17]

The implications of such change were enormous. On the one hand, the fact that "the bourgeois state sometimes even acts against the will of the majority of the ruling class" prepared the ground for détente. On the other hand, "state-monopoly capitalism" made the United States a dynamic adversary that could become terrible if the fruits of the S-T Revolution should fall into the hands of war-crazed reactionaries. Rather than predicting depressions and collapse in the West, as Soviet theory had done for decades, the new orthodoxy predicted high growth rates under capitalism.[18] As a 1969 conference later made explicit, "The S-T Revolution . . . has become one of the main sectors in the current phase of the historic competition between capitalism and socialism."[19]

The new scholarship created some difficulties for Khrushchev, but they were not so antagonistic to his party line as, for instance, Varga's were to Stalin's. The new view of the United States at least underscored Khrushchev's contention that a new age was born and that the S-T Revolution made some Leninist tenets inoperable. It also suggested the inevitability of socialist expansion during the "third crisis of capitalism" and of competition on battlefields of technology and growth, rather than war. Eventually, as a later theorist argued in 1967, capitalism would either fail to overcome its contradictions through technical progress, or it would gradually adopt Soviet methods, whereupon the S-T Revolution would prove a force for "convergence" between East and West.[20] In either case, the Communist Party's responsibility was to push the S-T Revolution as far and as fast as possible.

To talk of booms in Soviet R & D can be misleading, since "science" budgets have expanded rapidly in every stage of Soviet history save the war. But by any standard the first decade of the Space Age was explosive, with R & D growing at 15 to 16 percent annually from 1953 to 1956, and 17 to 18 percent annually through 1960. Thereafter, with a less robust economy, sharply higher defense procurement, and agricultural crises squeezing the budget, R & D still grew 13 percent per year. Overall, the USSR from 1955 to 1964 multiplied its annual investment in new science and technology four and one-half times, while R & D as a percent of GNP rose from 1.2 to 2.8 percent. The 5.1 billion new rubles allocated in 1964 comprised 5.5 percent of all state spending, while the number of workers employed in "science"—about 600,000 in 1955— doubled by 1959 and more than tripled by 1963.[21] Nor do these figures include the money and personnel engaged in R & D hidden in the undifferentiated military budget.

The Academy of Sciences seized this opportunity to renew its call for more theoretical research and a loosening of its ties to the central economic Gosplan and industry. But the president of the academy, A. N. Nesmeianov, ran afoul of the Technical Division, whose secretary, "space man" Blagonravov, called distinctions between pure and applied science artificial. In April 1961 the Central Committee and Council of Ministers reorganized the academy for the S-T Revolution, endorsing theoretical research, removing applied technology divisions—but sacking Nesmeianov. His replacement was Mstislav Keldysh, another space scientist "irrevocably committed to our concept of what needed to be done in the development of nuclear missiles. . . ." Despite a noisy campaign for chemicals and fertilizers, the Gosplan devoted the bulk of R & D resources to missiles and space, nuclear warheads and reactors.[22]

Khrushchev himself, the "peasant's son," probably had little under-standing of science and technology. Indeed, he is most infamous for his continued support of the bogus genetics of T. D. Lysenko. But his willingness to believe in the limitless prospects of command technology made possible a flowering of scientific exchange such as the USSR had not known since the 1920s. Khrushchev reopened channels of international communication, invited Eisenhower to end trade restrictions, and ap-proached Britain and even West Germany to purchase chemical plants for his drive to cultivate "virgin lands." The Seven Year Plan projected a doubling of East-West trade.[23] Khrushchev's interest in good ideas "wherever they might be found" was best symbolized by his visit to an American friend's Iowa farm in 1959, during which they exchanged tips on corn and hogs.

Soviet scientists, meanwhile, appeared again at international conferences and foreign journals again flooded the libraries of the academy.[24] Khrushchev rehabilitated *sharaga* inmates, and when these were added to the one million new workers in R & D, an immense "brick and mortar" program was dictated. It involved nothing less than construction of ten entirely new "scientific cities" like Akademgorodok in Central Asia,[25] the doubling of university graduate programs in science and engineering (48,000 students in 1965 vs. 21,400 in 1955), the construction of new cyclotrons, a manned spaceflight center, and two new launch complexes. Cybernetics, too, was rehabilitated after its denunciation by Stalin, and the Soviets bridged the gap between their first "electronic brain" (1955) and first "second-generation" computer (1962), although they continued to lag far behind in quality.[26] Television was introduced, and automation which, Khrushchev promised, would soon free mankind from assembly-line labor.[27] All in all, it spelled an R & D boom unprecedented even for a relentless, forty-year-old scientistic state.

Years of growth, years of sensational statistics, years when many in the West traded their image of Ivan as a brutish *muzhik* for one that made him ten feet tall and a daring engineer to boot. The truth was, however, that Space Age communism failed even by its own standards. Soviet growth rates leveled off at a time when the United States began rapid expansion—clearly the USSR would not "overtake America in per capita GNP" for decades, if at all. Instead of outproducing the United States in eggs, milk, and meat, as promised, the USSR even faltered in cereals and began to import large quantities of grain. Most telling of all the failures of Khrushchev's "transition to communism" was political. Imagine the changes that have occurred in the United States since the 1950s—and then imagine the problems and responsibilities devolving on a Soviet dictator as a result of similar change in the rigid structure of his country. Introducing new, revolutionary technologies was a big enough problem; adjusting Soviet institutions and reconciling elites to such change was still more difficult. Thus, without denying any of Khrushchev's blunders, one can suggest that he was also a victim of progress as well.

One by one, Khrushchev alienated elites. He abolished the Stalinist police of L. P. Beria, subordinated its replacement, the KGB, to the Council of Ministers, and removed its empire of slave labor camps to the Ministry of Internal Affairs. At first a relief to the Party, the reform also meant a certain revival of dissent, alarming Party officials and frustrating the KGB. The industrial reforms of 1957 alienated the "steel-eaters"— heavy industry and the army—by shifting resources to consumer goods and light industry, especially electronics, vital to the S-T Revolution. Khrushchev favored the managerial rather than the ideological wing of the Party and the provincial Party cadres over the central planning organs. In Stalinist terms, these were indicative of "technocratic tendencies," and they antagonized not only the ideologues but also the local Party committees, which were suddenly handed more responsibility than they wanted. In hopes of fostering specialization, Khrushchev divided the Party and government at every level into industrial and agricultural sectors, and transferred Moscow bureaus to the *sovnarkhozy* (economic councils) in the boondocks, an unwelcome change for the displaced and for the inundated provincials alike. He persecuted village churches, helping to drive more peasants off the land at a time of hoped-for agricultural expansion. His chemicals program was poorly adapted to Soviet soil and crop requirements and subtracted resources from other sectors. In all these ways and more, Khrushchev found little constituency for his reforms based on the perceived needs of new technology, but found many old constituencies ready to protest.[28]

The policies associated with the S-T Revolution especially jolted the military. When one army chief of staff boasted "without exaggeration" that the future of the armed forces belonged to technical officer-specialists,[29] and his successor that the "native intuition" of past commanders was "very risky" in an age of nuclear warfare,[30] the careers and values of traditionalists were clearly threatened. Young technical officers began to outnumber the graduates of military academies and espoused professional values obnoxious to their seniors. The editor of *Red Star* upheld against them the concept of military honor: "Even in the old Russian army there were good traditions: bravery, selfless dedication, and military skill were revered." But, as the chief of armor feared, "Disproportionate stress on theoretical training may lead to the separation of officers from life, may transform them into scholastics who do not understand life at all but are capable only of citing the book."[31]

Where was the bravery and honor in a force of engineers preparing to atomize an enemy thousands of miles away with textbook calculations? This was the military dilemma of the nuclear age. As Malinovsky observed, it was just as important for *political* officers to have a "thorough familiarization with science and technology" as vice versa, while technical officers predicted that "The time is coming when the technician and engineer will assume one of the central places in war. . . . If military leaders fail to understand the changes that have taken place in military technology and continue to adhere to obsolete methods, [they will bring about] disasters. . . ."[32] Khrushchev admonished traditional officers not to discriminate against technicians, but the more he modernized the army, the more he conjured up the same problem Stalin faced: how to ensure the political reliability of "indispensable" technical experts? When he instituted "elaborate and cumbersome indoctrination" for technicians, they protested: "We are engineers, our element is technology."[33]

In addition to the three-way tension among the old guard, new technicians, and the Party, Khrushchev's apparent irrationality during the missile bluff, the Cuban crisis, and the seeming mismanagement of the space program must also have irritated the officer corps, traditionalists and "young Turks" alike. In the end, Malinovsky and the Soviet high command were either directly involved in the coup that overturned Khrushchev (Roman Kolkowicz expects they were), or at least conspicuously neutral.

Scientists and engineers themselves could only applaud Khrushchev's dedication to the S-T Revolution, but they, too, came to resent his handling of it. Results were never commensurate with investment, and Soviet technology remained, in policy and in fact, dependent on duplication (or "assimilative repetition") of Western products acquired through purchase, literature, or industrial espionage. Senior scientists and young graduates alike saw their experience, training, and imagination spent in large part replicating Western contraptions. Even in space and nuclear technology, the Soviets relied heavily on American publications and duplication of foreign components, while overall reliance on foreign technology, by some accounts, increased during the 1960s.[34] Sputnik proved an exception, not the rule, in the age of the S-T Revolution.[35]

Even Khrushchev's rapid expansion of the R & D sector was in some ways unsettling. The 1961 reform of the academy was achieved by spinning off the applied technology divisions to provincial industrial sectors. Some researchers had to accept lesser positions to avoid transfer, while those who "went east" to the new Siberian scientific cities had their promised salary increases summarily cancelled, probably for budgetary reasons. Khrushchev's manipulation of academic ranks also served to hinder promotions, provoke rancor and conformity, and reinforce gerontocracy despite his stated desire for "new blood."[36] Such aggravations formed the background of scientific dissent, which in turn reminded Party leaders and the clipped KGB why Stalin had suppressed the technical intelligentsia in the first place. Hundreds of scientists in all fields joined in protests against Lysenko and Khrushchev's patronage of him. Physicist Peter Kapitsa refused to do defense work and was forbidden to travel abroad. I. V. Kurchatov and Andrei Sakharov ("father of the Soviet H-bomb") publicized the dangers of nuclear testing and pleaded with Khrushchev to cancel the fifty-megaton tests of 1961.[37] A devastating 1957 incident in a nuclear waste dump in the South Urals and the Nedelin rocket catastrophe contributed to growing disgust with Party leadership in matters of science and technology. By 1962–64 young dissidents joined forces with Sakharov's generation, distributed *samizdat* literature, and traded nightmares of official incompetence.[38] Khrushchev's survival in power was damaging, it seemed, not only to the scientific effort but to the hold of the Party over the technical intelligentsia.

By October 1964, according to Medvedev, "there was not a single power group of any size—whether scientists, doctors, writers, educators, artists, business or industrial executives, factory or office workers, military officers, young people or old—that would have been willing to offer Khrushchev its backing and support."[39] The reasons were manifold, and not all connected with technical policies. But the phenomenon was special in Soviet history in this way: for the first time the regime, committed as always to pushing new technology, did not simultaneously seek to isolate the power structure from the second-order consequences of such technology. Rather, Khrushchev embraced the social and political consequences of the S-T Revolution, even glorified them, and called it all the fulfillment of communism. Finally, his tinkering reached Party organization itself, including mandatory turnover of committee member-ships and culminating in rumors of a special session of the Central Committee at which there were to be "many changes at the top." That may indeed have been "the last straw."[40]

Khrushchev did not invent the S-T Revolution, but he identified himself with it, tried to take credit for its progress, sought to revolutionize administration and industry according to its new realities. In so doing, he reaccelerated the drive for introduction of new technology across the board, changed the veneer of Soviet urban life, and put several more stories on the "house" that Russians symbolically build, floor by floor, with toasts of vodka. In the brief span of seven or eight years, Khrushchev introduced the USSR to the age of space travel, television, and computers. But his futile effort to build "true communism," Space Age or otherwise, frightened old elites in the army, the bureaucracy, and the Party, while his uneven management of new technologies even exasperated the new technical elites. Expecting specific breakthroughs to augur an entirely new age, Khrushchev only found that for some it did not arrive quickly enough, while others were uncertain they wanted it to arrive at all.

So Space Age communism fizzled like a capitalist rocket. During the last years of his rule, as pet projects miscarried, Khrushchev fell back on the tactics of his mentor Stalin, founding his own cult of personality and linking himself personally to the glorious achievements of the space program. Even after the missile bluff exploded, his push for space spectaculars continued, and the premier came to depend more heavily than ever on the man who had given him his first great triumph and enduring symbol of the S-T Revolution: Sergei Korolev. Irrational though it may have been, Khrushchev began by staking out a forward position with his ebulliant propaganda about Sputnik, tried to advance the entire Soviet technological front in order to bolster that forward position, failed to do so, and then, rather than call retreat, relied on the space technological rampart to hold the front alone. Khrushchev's swan song would be one last campaign of cosmic bluff.

To Western observers the United States seemed to be the hare in the space race, sleeping complacently then racing to catch up. But the Russians were no tortoises, crawling methodically along. Rather, after months of mysterious inactivity, they would suddenly bound forward with unlikely quickness on the strength of their outsized rockets, bulky instrumentation, and awkward procedures—more like a lumbering bear than a tortoise. American congressmen shuddered to contemplate what might come next, while NASA and DoD officials privately rejoiced at the fillip each ursine leap administered to their budgets. How could one read this riddle?

The early events of Soviet manned spaceflight seem to have unfolded under four influences: constraints in facilities and hardware, dictating the infrequency of launches (relative to the United States) and the nonrepetitive missions; constraints in launch technology and satellite instrumentation, obliging the search for new ways to make progress with the same booster and delaying the advent of space applications (weather, communications, etc.); political interference, inhibiting long-range planning in favor of quick propaganda successes; and Korolev's genius, permitting the successes the program did generate. This is not to say that the early "firsts" were devoid of scientific and technical value. But it is to say that the lesser resources commanded by the USSR and Khrushchev's insistence that the Soviets maintain their apparent lead for as long as possible probably damaged Soviet chances of grabbing a real lead in space.

A critical period in the Soviet space program, as in the American, was the weeks and months after the flight of Yuri Gagarin. Where did one go from here? The military space lobby was only beginning to press its case, and Korolev himself is said to have drafted a plan for longer and longer single-man missions with orbital maneuvers and experiments, leading eventually to two-man crews and orbital docking with a new spacecraft on the drawing board to be called the Soyuz. But Khrushchev demanded a command performance that would upstage Mercury, and that it be done by August 1961.

In June a front-page story in *Izvestia* credited Khrushchev with the inspiration and direction for the space program. He received the Order of Lenin in honor of his "leadership in creating the rocket industry and the successful achievement of the Gagarin mission." Other honorees

VOSTOK II

On August 6, 1961, Gherman Titov, test pilot and "cosmonaut-poet," rode the A-1 booster (Korolev's "ol' number seven" plus the Vostok upper stage) into the standard 65-degree orbit, circled the earth for twenty-four hours, and reentered. Like Gagarin, he ejected from the capsule at 23,000 feet and parachuted to earth. Titov had no control over his ship, apparently suffered throughout the flight from a nauseating disequilibrium in the inner ear, and was unable to receive Khrushchev's bear-hug until the day following his return. The premier then awarded Titov with instant membership in the Communist Party and presented him to the Twenty-second Party Congress then in session. He and Gagarin, announced Titov, "are very proud that Nikita Sergeevich has called us 'celestial brothers.' I must say with confidence that among us cosmonauts we call Nikita Sergeevich our 'space father' [stormy applause]. We constantly feel the concern of the party, its Central Committee, and Nikita Sergeevich Khrushchev personally for us cosmonauts and for the conquest of space [stormy applause]." Titov then took the opportunity to testify to the great firepower of his own Red Air Force, thanks to new types of guided missiles.[41] A week later, August 13, 1961, Soviet troops sealed off East Berlin.

included the Secretary of the Central Committee, the deputy prime minister . . . and Leonid Brezhnev. Academician Keldysh drew a lesser award.[42] After the Titov flight, *Pravda* boasted that the premier "directs the development of the major technical projects in the country, and determines the basic directions of planned growth in cosmic science and technology. In his able proposals there is evidence again and again of his great conviction in the triumph of Soviet rocket technology." It was said that he visited all the factories and test stands, knew the leading space scientists by name, and "participates in the discussion of all the most vital experiments."[43] Gagarin's official memoirs claimed that Korolev spoke often of his meetings with Khrushchev and that the latter "devot[ed] a great deal of care and attention to this new sphere of activity."[44] The political ghostwriter of Gagarin's book must have penned these words with a wry smile. Korolev doubtless did speak often of his meetings with the boss, and probably felt that he devoted rather too much "care and attention" to spaceflight.

How much direct control over the space program did Khrushchev exert? In the early years the evidence suggests that it was still an ad hoc enterprise, with budgeting and policy falling to the Central Committee, and spacecraft design, systems integration, and flight operations to Korolev's team at Tyuratam. In the post-1961 expansion, however, a Soviet space structure evolved that provided policy niches for the Academy of Sciences, the Gosplan, the Medium Machine Building Industry (which ran the rocket plants), the State Committee for Scientific

Research (founded 1961), as well as Korolev's arsenal, Voskresensky's manned spaceflight center, and Glushko's propulsion laboratory. The Interdepartmental Commission on Interplanetary Communications (later "on the Exploration and Use of Space") probably influenced the choice of experiments and payloads, while Blagonravov and Keldysh served as front men for the Soviet space program in international gatherings. The most pervasive influence below that of the Party, however, was military. The boosters (except the coming Proton) were adaptations of military missiles manufactured by the same division responsible for nuclear warheads. All launches from the three cosmodromes were conducted by the Strategic Rocket Forces. The tracking system was military, the cosmonauts trained in the military, even the leading personnel at the Academy of Sciences had military rank and specialties.[45] After 1962 most of the satellites themselves involved military applications.

Of course, military institutions, personnel, and hardware played almost as great a role in the U.S. space program. But there was no Soviet equivalent of NASA, or congressional and public opinion to weigh in on the side of science and "peaceful uses of space." Khrushchev, his successors, and their lieutenants might approve programs for prestige or science, and Korolev and others might extol space exploration for its own sake, but the actual organization appears to have been unitary and pervaded with military influence from the early 1960s.[46] Major General G. I. Pokrovsky, a scientist and military man who took it upon himself to educate Soviet leaders about the implications of the technological revolution, attested to this:

Soviet military science teaches that, under contemporary conditions, it is possible to assure the high combat qualities of our armed forces only with the harmonious development of all forms of military technology and all fields of military science, with comprehensive mutual relations between military science and all other sciences.[47]

Hence the new age demanded complete mobilization and integration of the national scientific and military efforts. Distinctions between military and civilian spheres were simply passé. Khrushchev obliged by pouring large sums into missiles and space and creating a powerful technical complex allied to the military, but then diverted much of the effort to self-serving, political space missions. Once the complex was in place and the importance of the S-T Revolution accepted by the Party as a whole, both space scientists and the military must have considered the frivolous Khrushchev dispensable.

In 1962 the United States geared up for the space race: Mercury pressed on with three manned orbital flights, the giant Saturn booster underwent live testing, the Apollo program was being planned. Korolev lobbied again for longer flights—up to three days—in order to expand

VOSTOK III–IV AND V–VI

On August 11, 1962, another Vostok trod the same path through the Central Asian skies into the same orbit as its predecessors (with an interior TV camera added for effect). It circled for the same seventeen orbits, bringing it back to the same point 113 miles above Tyuratam. It was now August 12. Suddenly the world learned that a second capsule, *Vostok IV*, had roared off the pad to join its sister ship in space. Korolev had turned around his launch facilities in twenty-four hours and boosted two ships in close proximity. Apparently the USSR was two years ahead of the United States, which would not attempt a similar feat until the follow-on Gemini program. But the two craft were not equipped for docking or maneuver. It all amounted to a neat exercise in pinpoint artillery and tracking. Both cosmonauts returned to earth on the fifteenth.

In June 1963 Korolev and the manned spaceflight team received permission for a week-long mission, necessary to test human endurance and physiology under sustained weightlessness. Otherwise it was another iteration: same ships, same orbit, same indigent passengers controlled from the ground. The difference this time was that one of them was a woman. Valentina Tereshkova, a sport parachutist of impeccable proletarian heritage, was another Soviet "first." Having no astronautical training, she was presented as proof of the routineness of spacefaring in the USSR—in fact, her lack of expertise only proved the superfluousness of the test pilots on the other flights. Unlike American astronauts, the cosmonauts did not share the piloting of their spacecraft. But *Vostok V* and *VI*, another group flight, set longevity records of five and three days in space, passing each orbit within 5 kilometers of each other. The medical data were doubtless valuable, but no advance was made toward rendezvous.

When Tereshkova returned to earth (just in time for an International Congress of Women to cheer her in the Kremlin), she became a symbol of emancipated Soviet womanhood. Khrushchev admonished American bourgeois society for referring to women as "the weaker sex." Tereshkova toured the world proclaiming the equality of sexes in the USSR and "the ever-growing superiority of the socialist order of society over capitalism altogether."[48] Americans bought the Soviet line and berated NASA for "sexism." But no other women would fly in space until the 1980s. Tereshkova was wedded to a fellow cosmonaut amid official theatrics, and Soviet cosmonauts subsequently declared spaceflight too demanding for females, especially potential mothers. "In such conditions we just had no moral right to subject the 'better half' of mankind to such loads."[49]

biomedical experiments and orbital maneuvers. He received approval on the condition that he add gimmicks to win the USSR more "firsts" in space.

The Vostok program ended in June 1963, after perhaps five years of existence, with the first man in space and five other manned capsules to its credit. But Vostok showed little innovation from its first mission to the last, and instead of giving way to a second-generation spacecraft, it perforce hung on, for another twenty-one months under the rubric Voskhod, thereby retarding Soviet progress.[50] For by late 1963 the Soviets had no choice but to take seriously the American claim that the real finish line in the prestige race was the moon. Could the Soviets resist the temptation to squeeze more "firsts" out of its primitive Vostok system and tackle the long-range planning necessary to reach the moon? Or, if that was deemed beyond their means, could they cease to talk of space technology as proof of socialist superiority and swear off "racing" entirely? Either decision would have made sense—Khrushchev made neither.

The trouble that a manned moon mission posed for the Soviets was that the world expected, on the basis of Moscow's own propaganda, that the USSR would be first. If the Soviets chose to take up Kennedy's challenge, they would, in the eyes of the world, be starting with the advantage. Khrushchev's reaction to American moon plans is imponderable. He did say in retirement: "I'm only sorry that we didn't manage to send a man to the moon during Korolev's lifetime."[51] But there is enough technical evidence, compiled by James Oberg, to suggest that the Soviets were in the race for the moon despite their disclaimers after the fact. When, therefore, did the Soviet moon program begin, and how did they intend to proceed? The defector Vladimirov reports that Khrushchev requested a briefing from Glushko, and was told of the (then) current von Braun plan to reach the moon. This limits the conversation to the period between May 1961 and July 1962. The von Braun plan, Glushko explained, involved construction of a platform in earth orbit, where the moon ship could be assembled beyond most of the pull of terrestrial gravity. Such a procedure, called earth-orbit rendezvous (EOR), involved many orbital missions, delicate maneuvers, extravehicular labor, and life-support systems that did not then exist. A second plan, direct flight to the moon from the surface of the earth, was less complicated, but required a first-stage booster with as much as 16 million pounds of thrust, or about fifteen times the force of the Vostok rocket! Vladimirov's hearsay holds that Khrushchev "calmed down" after this report and dismissed Kennedy's moon talk as propaganda.

Nevertheless, at least one Soviet space scientist suggested a third way to the moon: lunar-orbit rendezvous (LOR). The Russian Yuri Kondratyuk had surmised as early as 1929 that the best way to the moon was to fire a rocket into lunar orbit and then drop a space "dinghy" down to the surface. This would reduce the amount of fuel needed to escape lunar gravity, hence reduce the size of the ship and the initial rocket for escape from the earth. Soviet engineer Yuri S. Khlebtsevich, already engaged in plotting trajectories to Venus, stumped the halls of the Academy of Sciences and the Kremlin trying to sell LOR. According to Vladimirov,

"Even His Compass Won't Help Him. Which Way is West?" The cartoon depicts two "ten-foot-tall" cosmonauts riding Vostoks III and IV to glory, while an American on his hobby-horse, intimidated by Soviet technical superiority, can no longer tell West from East. From *Izvestia*, August 1962.

Glushko and others told him to keep his ideas to himself, either out of professional jealousy or political prudence.[52]

Doubtless there is truth in this story, but it cannot be the whole truth. By 1963 the Soviets were aware that the United States was committed to precisely this mode of moonflight. Perhaps the Kremlin scuttled the idea because of its cost. But this, too, breaks down if Western estimates of Soviet space spending are anywhere "in the ballpark." Analyses of Soviet budgets and hardware yield estimates in the range of 1.5 to 2.0 percent of GNP devoted to space by the mid-1960s, or about 3 to 4 billion rubles.[53] Given that the Saturn booster absorbed 21.2 percent of NASA R & D from 1962 to 1968 (and a far higher percentage in the peak years) and that the vast majority of Soviet space shots used the same small stable of derivative boosters involving little new expense, it is difficult to account for Soviet space spending without assuming a large launch vehicle development program.[54] More convincing still is the evidence from intelligence sources and occasional Soviet predictions that the USSR was at work on a *Saturn 5*–class booster in the mid-to-late 1960s.[55] But Soviet engineers never perfected this giant vehicle (the "G" booster in NATO parlance). It was apparently not flight-tested until 1971, whereupon it disintegrated in the air. There is evidence as well of disasters on the launch pad in 1969. In the meantime, Korolev (and presumably Glushko) fashioned another rocket, the Proton, that first flew in 1965 and seemed to involve still greater clustering of the initial RD-107 engines. No details of any kind on the Proton were released, but it appeared to have an overall thrust some three times that of the Vostok booster and came to be used in unmanned Soviet moon missions in the late 1960s.

Soviet testimony about the moon race in the Khrushchev era is mixed. Sedov and Keldysh implied more than once that a manned moon landing was a difficult chore to which they gave a low priority. Khrushchev himself "wished the Americans luck" in their Apollo enterprise and said that the Soviets had no deadline for such a mission. But he quickly retreated after U.S. headlines cried "Russ Drop Out of Moon Race!" "Gentlemen," he said, "give up such hopes once and for all and just throw them away. When we have the technical possibilities of doing this and when we have complete confidence that whoever is sent to the moon can safely be sent back, then it is quite feasible. We never said we are giving up our lunar project. You are the ones who said that."[56]

Soviet engineers must have wanted to give it a shot. In an anonymous interview in June 1963, Korolev himself confessed his belief that "prolonged interplanetary flights of man are not so far off." He admitted to "having in mind" a manned moon flight, which he described as "an extraordinarily tempting but a very difficult problem. . . . Soviet scientists are working on the solution of the problems involved here. I'm sure that the time is not far off when the journey of man to the moon will become a reality, although more than one year will be necessary. . . ."[57] Yet in a pseudonymous article in *Pravda* on January 1, 1964, Korolev spoke only of earth orbital operations as a near-term achievement.[58]

Despite their dutiful flights of optimism, therefore, neither Khrushchev nor the engineers publicly committed the USSR to a race for the moon. But the Soviets apparently continued to study the problem, perhaps in hopes of finding a cheaper way to accomplish a moon landing (or a less delicate mission, like manned circumlunar flight). After three years of inactivity in the lunar realm, the Soviets launched a second-generation moon program in 1963. A platform, called *Tyzazhely Sputnik*, was placed in earth orbit. Near the end of its first revolution, the platform jettisoned a rocket probe, or high-energy deep space stage, which fired for the moon. A similar technique had already been used for probes of Venus and Mars. On their third try, the Soviets crash-landed *Luna V* on the moon in May 1965. The R & D for this system was assuredly begun under Khrushchev, perhaps as early as 1961.[59]

Still, Korolev was not permitted to concentrate on the moon problem or on steady progress in earth orbital operations. For the USSR, theoretically free to defer gratification and command sacrifice on behalf of future achievements, was still victimized by an impetuous leader who linked his own legitimacy to immediate, spectacular results. In 1963 and 1964, when Korolev should have been devoting his talents to bigger boosters, the Soyuz, and/or the moon, he was obliged to play ringmaster in another Khrushchevian circus.

Over eight years had passed since the new Soviet premier closed the book on the hideous "aberrations" of Stalin and declared a new era of peaceful coexistence, S-T Revolution, and, ultimately, the arrival in secular heaven: true communism. Over seven years had passed since *Sputnik I* circled the earth to underscore the message. But now the most

---

VOSKHOD I

By the end of 1963 the United States had sketched out its intermediate manned program, Gemini. Beginning in April unmanned test flights of the new capsule would commence, followed by missions with two-man crews, docking, and spacewalks. Khrushchev was told that the Soyuz was still some years away, so in the interim the American two-seater would seem to indicate leadership. His instructions: launch a three-man capsule, and do it before the next anniversary of the revolution. But the only available hardware was the one-man Vostok! Korolev is said to have sunk into a cynical determination borne of a life in various *gulags:* physical, professional, psychological. Voskresensky had a nervous breakdown. There was nothing for it but to strip the Vostok of all equipment save life-support, cram in three seats, and pretend that this was a new-generation spacecraft, the Voskhod. The cosmonauts could not wear space suits—there was no room. This was touted as more proof of the "routineness" of Soviet spaceflight. Safety systems were discarded—too much weight. This was brushed aside as evidence of the reliability of Soviet rockets. The standard cosmonaut ejection/parachute system was impossible—men required suits and oxygen at 23,000 feet. So Korolev hastily designed a larger parachute to bring down the entire capsule. But test monkeys died in the less-than-gentle impact, so a still heavier parachute was installed, sending the cosmonauts themselves on special diets, like wrestlers trying to "make weight."[60]

Voskhod I orbited on October 12, 1964. It carried three men but weighed only 1,300 pounds more than Vostok. *Pravda* headlined "Sorry Apollo!" and claimed that the "space gap" was increasing: ". . . the so-called system of free enterprise is turning out to be powerless in competition with socialism in such a complex and modern area as space research."[61] And yet—the proud Soviets who briefed the press with such pride seemed unusually hesitant to say anything about their "new spacecraft."[62]

"glorious" of all Soviet space feats, the three-man flight of *Voskhod I*, was brought back to earth mysteriously, after a single day in space. The sardine-cosmonauts, despite their discomfort, radioed their protest to early termination of their mission. Korolev reputedly replied: "There are more things in heaven and earth, Horatio, than are dreamt of in your philosophies."

The date was October 13, 1964, and Nikita Khrushchev had just been relieved of all his party posts and responsibilities.

Conclusion

"The idea behind every creative art is the creation of another way of life," wrote Nicholas Berdiaev in 1906. "The breaking through from 'this world' ... the chaos-laden, distorted world, to the free and beautiful cosmos."[1] He was criticizing the "piecemeal reformism" of the new Russian Duma, or parliament, but he also sounded a note of the Communist harmony to come. For the psychological appeal of a secular religion lies not only in the perfection but in the mortification of this "chaotic, distorted world." Revolution is a political program; to be a creed it must aim higher. The Promethean promise of Bolshevism is what stirred hearts; the conquest of Nature, not only of men and classes, what inspired sacrifice. Like Eugene Zamyatin's fictional spaceship *Integral*, the Soviet purpose was to "integrate the indefinite equation of the Cosmos" and seed the planets with "the grateful yoke of reason ... a mathematically faultness happiness."[2] Zamyatin's *We*, of course, was a satire of totalitarian reality, but in the decade of the revolution, before reality was apparent, Russian artists and revolutionaries alike dreamed of storming the cosmos and expected of it an apotheosis—as if the gods were false, but their heaven real.

The Soviet technological push, especially into outer space, was not just a drive for power and affluence in a materialistic universe. For Tsiolkovsky, for Zamyatin's scientist-hero, perhaps even for Khrushchev, perfecting earthly society was a prosaic and intermediate goal. The "New Soviet Man" did not dream of new refrigerators for his children's children, nor of an abundance of tea and tobacco. Rather he was a Titan shouldering his way to the vanguard of the human army in its campaign to subdue Nature. Sputnik and Vostok meant more than Soviet leadership in big rocketry, more yet than evidence of the superiority of socialism. They reminded those Soviet citizens who retained at least an agnostic stance of the visceral appeal of communism. In only four decades it had become as distorted and meaningless as the society and polity it overthrew. But the assault on the cosmos, for a brief flicker, called back to mind the Communist vision: yes, *this* was the *mechta*, the dream (and the name given the first successful Lunik).

The link between rocketry and revolution was reforged in the years after Sputnik. It was fitting that Khrushchev should, on the strength of

Sputnik, "immanentize the eschaton," proclaim the S-T Revolution, and hope to end, once and for all, the discrepancy between promise and performance that had smeared the USSR since 1917. Now it would truly be the global leader in science and technology; it would truly surpass its enemies in military strength; it would even surpass the United States in economic power. As a true believer, he marked the new correlation of forces in the world and the objective forces of his age. Buoyed by triumphs in space, he went so far as to espy the Promised Land on the horizon and declare himself not another Communist Moses but Joshua. In time it became clear that Khrushchev had not the wit, nor the USSR the resources, to fulfill the hopes, given the many demerits of militarized social management. Nor did Khrushchev have the iron grip of Stalin to survive failure. That Khrushchev blundered repeatedly is beyond doubt; that he was a victim of the disaffections attending an era of explosive technological change is also true. But his radical error was hyperbole, for he plotted the Soviet curve in the Space Age as hyperbolic, when it in fact was parabolic. After straining upward on a dizzy slope, Space Age communism slowed, then arced downward, like *Voskhod I,* to a premature end.

John Foster Dulles suspected that Sputnik might prove to be "Mr. Khrushchev's boomerang." Indeed it was. It jolted the United States into technological end runs that left the Soviets craning their necks. It excited the Chinese into ambitions and demands the Soviets dared not honor. It tempted Khrushchev into commitments that alienated his own peers. Where Eisenhower underestimated the importance of Sputnik, Khrushchev overestimated it. Where Ike hoped to contain the social impact of technological competition, Khrushchev tried to catalyze it in a system far more resistant to change than the American. Against bureaucratic opposition a dictator has only three weapons: the power of the Party, the power of personality, and terror. Khrushchev undercut the second and third when he denounced Stalin and personality cults. As for the first, Khrushchev sacrificed it when he judged the Party itself to be in need of reform. He promised the dawn of true communism, peaceful coexistence, the recession of terror, the liberation of labor, all thanks to correct exploitation of the S-T Revolution. To the Party, it was an intolerable program.

The Khrushchev era nonetheless spawned irreversible change. He could be denounced for bungling the transition, but the march of technology was still the essence of Sovietism. The new regime, another ephemeral "collective leadership," might abolish the *sovnarkhozy* and state committees and return to the ministerial form of government. It might recall agricultural ministries to Moscow, reduce the virgin lands program, and rescind the mandatory turnover in Party committee membership, restoring job security to old cadres. It might favor the army, embark on unprecedented naval expansion, and restore power and autonomy to the KGB.[3] But the ideology and practice of the S-T Revolution would not be rescinded. How could they be? The march of technology under communism was an article of faith. Nor could the missile revolution be undone, whatever the nostalgia of veterans for close-order drill. In February 1965 the chief of the general staff wrote:

> A revolution has occurred in military affairs which is unheard-of both in extent and in consequences, a revolution that has produced a truly profound change in the organization, training, and education of the armed forces, and in the views, manners, and forms of armed struggle. . . . With the emergence of rocket-nuclear arms, cybernetics, electronics, and computer equipment, any subjective approach to military problems, harebrained plans, and superficiality can be very expensive and can cause irreparable damage. Only the thorough scientific foundation of decisions and actions . . . will guarantee the successful completion of tasks. . . .[4]

The revolution "has occurred." It "has produced" profound change: these *faits accomplis* were Khrushchev's legacy. The S-T Revolution, in the parlance of Merlin's magical anthill, was a "done thing"—its only word for "good."

In the first years of the new order, Brezhnev, Alexei N. Kosygin, and Nikolai I. Podgorny dispensed with boasts that the USSR would overtake the United States in everything—but they inflated still more the value of technology. Space "helps advance our entire economy," said Brezhnev; it had become a "key element of the contemporary technological revolution," said Sedov.[5] As soon as he took power, Brezhnev declared:

> We Soviet people do not look upon our space exploration as an end in itself, as some sort of "race." The spirit of gamblers is profoundly alien to us in the great and serious business of exploring and conquering outer space. We regard this enterprise as a component of the tremendous, creative work in which the Soviet people is engaged, consistent with the general line of our party in all areas of the economy, science, and culture, in the name of man and for the good of man.[6]

In short, the S-T Revolution was real and would continue, but would be drafted by the Party into its own service and not that of personal politics or fantasies. As years passed, and evidence appeared both of the Potemkin village nature of the early space shots and current stress on practical applications in orbit, it was tempting to conclude that the new leadership had abruptly changed Soviet space policy. In fact, the new regime apparently continued work on manned lunar flight in hopes of preempting Apollo and continued to parade the cosmonauts as heroes. But there would be no more "phony" missions, or claims of superiority, or talk of imminent apotheoses. Instead the Party leadership, increasingly dominated by men of technical training (thanks in part to Khrushchev's policy of providing it with new blood and promoting technicians),[7] returned with relief to the old formula: we are behind but will someday be ahead; we are the underdog, they the oppressors; we proceed scientifically, they according to mad ambitions.

In a 1967 history of Soviet cosmonautics, Korolev was described as being "merciless toward unfounded fantasizing [*prozhektorstvo*],"[8] the same word used to denounce Khrushchev. Now Korolev had the chance, at long last, to present his own plans; he proposed deemphasis of the moon and concentration on space stations in earth orbit with routine resupply by the new Soyuz craft.[9] For thirty-five years he had withstood misdirection, persecution, distraction, and interference. At last, it seemed, he would enjoy consistent and sensible support. But soon after *Voskhod I*, his nerve-wracked comrade Voskresensky was dead at fifty-two. Then, in January 1966, Korolev himself entered a hospital for removal of hemorrhoids. The surgeon found cancer of the colon and chose to attack it without proper equipment or preparation. The man who launched the Space Age died on the table at fifty-eight years.[10]

Korolev's achievement, Khrushchev's use of it, and the response to both by the Americans defined the politics of technology through two decades. And even though the Soviet bear ran in circles for a time, the scientists of Korolev's generation, struggling within, while working for, the Soviet system, laid the foundation for a postindustrial communist Superpower second to none. New ICBMs would come on line, new conventional arms, an intense and purposeful space program—and the Americans could always be counted on to take naps. Indeed, Khrushchev, for all his errors, bequeathed to Brezhnev the precise environment that gave the USSR its chance to catch up, at least in missiles and space, with the capitalist rival. That environment was détente.

PART V

Kennedy, Johnson,

and the Technocratic

Temptation

The quest for glory is the basest thing in man; but it is just this which is also the greatest mark of his excellence. —PASCAL, *Pensées*

If the newspapers printed a despatch that the Soviet Union planned sending the first man to Hell, our federal agencies would appear the next day, crying, "We can't let them beat us to it!" —HYMAN RICKOVER, 1959

The exploration of space will go ahead whether we join in it or not. . . . We choose to go to the Moon in this decade, and do all the other things, not because they are easy, but because they are hard. —JOHN F. KENNEDY, 1962

Science and technology are making the problems of today irrelevant in the long run, because our economy can grow to meet each new charge placed upon it. . . . This is the basic miracle of modern technology. . . . It is a magic wand that gives us what we desire. —ADLAI E. STEVENSON II, 1965

Oh, you may leave here for four days in space/ But when you return it's the same old place/ . . . Hate your next door neighbor, but don't forget to say grace. —STEVE BARRY and P. F. SLOAN, "Eve of Destruction," 1965

THE maître d' of Locke-Ober's, a Boston institution, was a longtime rocketry buff. On slow evenings Freddy would place an empty whiskey bottle on the bar, stick a pin through a straw crosswise, set the straw afire and gently lower the flame into the bottle. The pin rested on the lip while the alcoholic fumes expanded inside. Freddy then counted down dramatically to the "pop" that sent the straw shooting toward the ceiling.

Brothers Jack and Bobby Kennedy closed the Men's Clam and Oyster Bar at Locke-Ober's more than once during the 1950s. One night Freddy introduced them to another habitué, Charles Stark Draper of MIT, hoping to persuade the young politicos that rocketry was anything but frivolous. Instead, the Kennedys heatedly dismissed the whole business. Even after *Sputnik I*, Senator Kennedy "could not be convinced that all rockets were not a waste of money, and space navigation even worse."[1]

Kennedy was not defending a considered opinion; more likely, he just enjoyed the role of hard-headed skeptic. If so, he was no different from millions of intelligent Americans who supported science and discovery but as late as 1957 still shied from anything as outrageous as men on the moon. In 1960 Kennedy was pleased to campaign on the missile gap, but of all the issues he would face as President, he "probably knew and understood least about space."[2] Just months later, Kennedy stood before a joint session of Congress to ask Americans to bear the burdens and costs of rocketing men to the moon within six to eight years. This extraordinary commitment, proposed on May 25, 1961, was his most historic act—but the confession of faith in spaceflight comprised only a fifth of that Special Message to the Congress on Urgent National Needs. The moon appeal, like Ike's Farewell Address, must be studied in context.

Kennedy called it a second State of the Union Address. The tradition of annual reports had been broken in extraordinary times, he said, and these were such times. Since 1941 threats to freedom had been primarily military, but now "the great battlefield for the defense and expansion of freedom today is the whole southern half of the globe—Asia, Latin America, Africa, and the Middle East—the lands of the rising peoples." Their revolution was the greatest in history, but the adversaries of freedom were "seeking to ride the crest of its wave—to capture it for themselves." They fired no missiles but sent arms, agitators, aid, technicians, and propaganda to every troubled area. He cited Vietnam, where 4,000 civil servants had been murdered the previous year. The struggle to preserve freedom in these nations was a "contest of will and purpose as well as force and violence—a battle for minds and souls as well as lives and territories. And, in that contest, we cannot stand aside."[3]

Such was the preamble to varied proposals the President had come to set before the Congress. The first was stimulation of the economy with an "affirmative anti-recession program" lest "we handicap our effort to compete abroad and to achieve full recovery at home." Second, the United States must foster global progress, for "the most skillful counter-guerrilla efforts cannot succeed where the local population is too caught up in its own misery to be concerned about the advance of communism." This justified Kennedy's Act for International Development (AID) and a quarter-billion dollar contingency fund for foreign aid.

"All that I have said," the President continued, "makes it clear that we are engaged in a world-wide struggle in which we bear a heavy burden to preserve and promote the ideas that we share with all mankind, or have alien ideas forced upon them [sic]." This justified expansion of the U.S. Information Agency (USIA). The United States must also give all necessary aid to local forces with the will and capacity to cope with attack, subversion, insurrection, or guerrilla warfare—the $1.6 billion already requested for military assistance would not suffice. American military strength was also insufficient. Fourth, therefore, in the list of initiatives was reinforcement of the army and marines to provide "flexibility" below the threshold of nuclear war. Fifth on the list was civil defense, in case the nuclear-armed adversary took leave of his senses, and a tripling of the budget for fallout shelters and other measures. This in turn led to hopes for disarmament and Kennedy's call for an Arms Control and Disarmament Agency (ACDA).

"Finally," announced the President,

if we are to win the battle that is now going on around the world between freedom and tyranny, the dramatic achievements in space which occurred in recent weeks [Gagarin and Shepard] should have made clear to us all, as did the Sputnik in 1957, the impact of this adventure on the minds of men everywhere, who are attempting to make a determination on which road they should take.... Now it is time to take longer strides—time for a great new American enterprise—time for this nation to take a clearly leading role in space achievement, which in many ways may hold the key to our future on earth.

The Soviet lead in boosters promised them

still more impressive successes, but we nevertheless are required to make new efforts on our own. For while we cannot guarantee we shall one day be first, we can guarantee that any failure to make this effort will make us last. We take an additional risk by making it in full view of the world, but as shown by the feat of astronaut Shepard, this very risk enhances our stature when we are successful.

This was not just a race: "whatever mankind must undertake, free men must fully share."

Kennedy then stated his belief that the United States "should commit itself to achieving the goal, before this decade is out, of landing a man on the moon and returning him safely to earth. No single space project in this period will be more impressive to mankind, or more important for the long-range exploration of space; and none will be so difficult or expensive to accomplish." A moon landing would demand sacrifice, discipline, and organization: the nation could no longer afford work stoppages, inflated costs, wasteful interagency rivalries, or high turnover of key personnel. "New objectives and new money cannot solve these problems. They could, in fact, aggravate them further unless every scientist, every engineer, every technician, contractor, and civil servant gives his personal pledge that this nation will move forward, with the full speed of freedom, in the exciting adventure of space."

Then, suddenly, Kennedy stopped trumpeting the charge and prepared a retreat. He only "believed" that the nation should go to the moon; he warned that "this is a judgment which the members of Congress must finally make"; then "this is a choice which this country must make" and "a decision that we make as a nation"; then "I think every citizen of this country as well as the members of the Congress should consider the matter carefully"; then "you must decide yourselves" and "whether you finally decide in the way that I have decided or not, that your judgment—

as my judgment—is reached on what is in the best interests of our country."

Here were echoes of 1946 and the atomic energy debate. How could the Congress and people judge the worth of an exercise in technological futurism? How was this an "American" enterprise—the fixing of five year plans and national mobilization under federal bureaucracies? But Kennedy did not ask for judgment on the means, only on the goal. The context of the exhortation made the goal clear: to win a battle of image making in a total Cold War. In conclusion, the President praised the willingness of the American people to pay the price, to share resources, to join the Peace Corps or armed forces, to keep physically fit, to pay higher postal rates, to show friendship with foreign students that they might return home "with an image of America—and I want that image, and I know you do, to be affirmative and positive. . . ."

Daniel Boorstin, paraphrasing Oscar Wilde, wrote in 1962 that when the gods wish to punish us, they make us believe our own advertising.[4] The French pioneers of semiotics said the same, observing how the human use of symbols—words—can supplant the very things the symbols signify, to the point where the message conveyed by a symbol can shift 180 degrees over time. When a person, company, or nation sets out to sell its image rather than reality, the image making gradually absorbs the reality: values become pseudo-values and events staged to convey the image become self-fulfilling prophecies that impress, in the end, only those whose consciousness has been conditioned to respond to the false and not to the original.[5] The selling of the United States grew up with the Cold War. As early as 1955 press magnate William Randolph Hearst telegraphed the following to his editors, with a copy to the White House:

Kruschev [sic] who unquestionably the boss at present, made clear they still hope achieve communist domination world, but they want confine struggle measures short of war. That is meaning of competitive coexistence. It is that battle we must prepare for now. I think we should prepare for it with program of initiative and enterprise. It means convincing the people of Russia, China, India, as well as Europe, that our system is the best.[6]

The concomitant arrival of Sputnik and the Third World generalized the problem of the American image. The Soviet challenge and European colonial heritage made it vital for the United States to present an image of progressive anticolonialism. Perhaps, too, Americans just want to be loved. In any case, the policy or prideful want became a high priority in the context of total Cold War. But this meant the extension to foreign policy of a decadence in the United States that was the subject of Boorstin's book. Thanks to the spread of mass media—the graphic revolution—images and illusions replaced ideals in our sales-oriented society, newsmaking and public relations replaced news gathering, celebrities replaced genuine heroes, tourists replaced travelers, the imitation of reality eclipsed reality itself. In foreign policy, the very effort to publicize the wealth and style of American life was self-defeating: "We suffer abroad simply because people know America through images. While our enemies profit from the fact that they are known only, or primarily, through their ideals."[7] Asians and Africans were more likely to grasp Communist ideals (especially when tailored, like Jesuitical Catholicism, to local conditions) than American images of dishwashers and voting booths.

Prestige as we now understand it is a recent usage. The word originated in the Latin *praestigium*, an illusion or delusion, usually rendered in the plural to denote "juggler's tricks." In French and then English the word meant deceit: a "prestigious man" was a fraud. Only recently, especially in the United States, did "prestige" acquire a favorable connotation. In electoral politics the importance of image expanded with the advent of television, and U.S. politicians understandably transferred the techniques that won votes at home to the pursuit of goodwill abroad. Boorstin cited the example of a speechwriter who asked his client what he thought about a given issue only to hear him repeat the very phrases the agent had written for him weeks before. "It was disturbing," the speechwriter said, "to hear yourself quoted to yourself by somebody else who thought it was himself speaking."[8] Nowhere was this phenomenon so prevalent as in outer space policy. By the end of 1958 Lyndon Johnson knew a dozen catch-phrases by heart, all coined by aides seeking to create an image for their boss and a symbolism for the Space Age. Such phrases became the ammunition dump of the space lobby, then by the 1960s, conventional political wisdom.

The brief Kennedy years were those in which American space policy fell captive to the image makers. Looking for a hook on which to hang space policy, Kennedy tossed it into the same closet with all the other policies bearing on the "extraordinary challenge" to freedom. The Kennedy call to arms amounted to a plea that Americans, while retaining their free institutions, bow to a far more pervasive mobilization by government, in the name of progress. The Apollo moon program was at that time the greatest open-ended peacetime commitment by Congress in history, the Kennedy missile program was the greatest peacetime military buildup, and McNamara's imposition of stringent management on the Pentagon and the Kennedy economic program of Keynesian fiscal policy, "pump-priming" and "fine tuning" of the economy on the assumption that private behavior was susceptible to political control, all expressed a growing technocratic mentality. The justification, at least at first, was the need to compete with the Communist bloc, in nuclear arms, conventional arms, foreign aid, economic growth, space, and propaganda. The result was an American-style mobilization that was one step away from *srashchivanie* (interdependence) between the public and private spheres. That last step would come when the New Frontier gave way to the Great Society.

For the commitment to go to the moon did more than accelerate existing trends in space. It served as a bridge over which technocratic methods passed from the military to the civilian realm in the United States, to political problems at home as well as abroad. Sharp disagreements arose over the goals that government ought to pursue, but by 1964 little dissent remained over the methods. Under the impact of total Cold War, with the space program serving as lever, Left and Right, dove and hawk succumbed to the technocratic temptation.

CHAPTER 15

Destination Moon

"The generation that fought the war"—these were the Kennedy men. Convinced of their brilliance in comparison to the men who surrounded Ike, they extolled vigor, intellection, and movement. Behind the clichés about "company commanders" replacing the generals was the truth that World War II was the formative experience of their lives. They remembered the bitter fruits of appeasement, but above all the way war had galvanized science, industry, and government, and showed what Americans could do with technology, the proper leadership, and the inspiration of a mighty cause.[1] Kennedy commanded a PT boat, John Kenneth Galbraith helped to draft the strategic bombing survey, Walt Rostow picked targets for armadas of B-17s and B-25s, Robert McNamara and his "whiz kids" supervised development of the huge B-29. Except for the latter, they knew little of design and production, but that was the point. Scientists and engineers, while they welcome the financial rewards flowing from political promotion of technology, are less likely to oversell it as a cure for all ills; they know their limits. Rather, it is the lawyer, economist, journalist, or politician who is most susceptible to technocratic temptation.

Or the company commander. Eisenhower had little faith in centralized management of power outside the military arena. But of these Best and Brightest, David Halberstam wrote, "if there was anything that bound the men, their followers, and their subordinates together, it was the belief that sheer intelligence and rationality could answer and solve anything."[2] To set goals for the nation and devise methods for their achievement under state direction: this was the approach to public policy that captured university faculties and foundations in the late 1950s. Political scientists like James MacGregor Burns despaired of Eisenhower's passivity and wanted an activist presidency. FDR was the favored model. It seemed obvious that the United States could do better, that official reticence only perpetuated the ills of society, that power was not corrupting but a tool to be used for good. To set goals for the people, to assume command as the most intelligent and inspired citizens, this was simply leadership, no less.

The environment changed as well as personnel. Technological revolution was abroad in the world, and limits to action retreated beyond the horizon. In such a historical conjuncture Eisenhower's philosophy seemed not only obsolete but immoral, while a mobilized United States knew no limits. Kennedy said as much in his inaugural address: "The world is different now. For man holds in his mortal hands the power to abolish all forms of human poverty and all forms of human life. . . . Let the word go forth from this time and place, to friend and foe alike, that the torch has passed to a new generation of Americans . . . we shall pay any price, bear any burden. . . ."[3]

How different from Ike's words eight years before, when he hoped to liquidate the Korean War, slash defense spending, end regulation. Yet what expectations lay behind the eloquence? Within months Kennedy fired toward Capitol Hill a salvo of new spending measures and within a year the largest tax cut in recent memory. Only two assumptions could underpin such actions. Either a great surplus of wealth had built up in the 1950s (giving the lie to Democratic claims that the country had been "standing still" and "living off capital") or else explosive growth was expected in the coming decade sufficient to cover "any price, any burden." How could this be? Two to 3 percent growth would not yield "the revenues required for the welfare goals he had articulated, for the expanded infrastructure the cities required, or for the national security goals [Kennedy] had set," wrote Rostow. In other words, where traditional economics dictated the setting of state spending according to the ability of the economy to bear it, the new economics dictated stimulation of the economy to the point where it could sustain the desired level of spending. Kennedy and Walter Heller, chairman of the Council of Economic Advisors, forged a consensus in favor of "lifting the level of employment and the rate of growth by unbalancing the federal budget, grossly if necessary." Investment would be encouraged, wages and prices restrained by "jawboning" and a new "social contract," and business convinced that "a large, purposeful deficit" was sound policy.[4]

Technology did not emerge from the start as a primary tool for enforced growth. But the new dogma that federal spending was beneficial to the economy and the "pay any price" mentality conditioned the Kennedy team to think of space exploration in terms of ends (were they desirable?) rather than means (can we afford it?). When a new Soviet spectacular, Third World setbacks, and the energetic advocacy of Vice President Johnson combined to force a decision on ends in space, the outcome was assured. It was Destination Moon.

For all their "space gap" talk the Kennedy men had little notion of what to do with the space program after election day. Twice in December 1960 the President-elect met with one teammate who did, and Kennedy gave to LBJ the responsibility for space in the new administration. His vehicle for doing so would be the National Aeronautics and Space Council, created by Johnson in the space act but hardly active since. Its first meeting had taken up important matters such as the transfer of JPL to NASA. But in its second meeting Eisenhower dozed off during a discussion of the NASA logo. The council met seven more times, with the President usually in attendance. But he and Glennan resolved to abolish it as early as 1959, only to be blocked by the Senate.[5] Now Kennedy and his aide Theodore Sorensen decided to vest the chairmanship of the body in the vice presidency and provide for an executive secretariat. That post would be filled by Edward C. Welsh, economist, former aide to Symington in the 1956 air power hearings and contributor to JFK's speeches on space during the campaign.[6]

Johnson also grasped the threads of space policy in the Senate, where he chose his successor as chairman of the Space Committee. It was Robert Kerr (D., Okla.), an oil millionaire who knew little about space but was a cagy ally. (Kerr once boasted, "I represent myself first, the state of Oklahoma second, and the people of the United States third—and don't you forget it!")[7] With these institutional pieces in place, Johnson set out, just as in 1957, to marshal the information and influence needed to push through an accelerated space program.

What was to be done? Eisenhower had reluctantly granted the importance of prestige in space against the judgment of his scientists. Kennedy's scientific advisers also felt that prestige was overemphasized. Jerome Wiesner of MIT headed Kennedy's Ad Hoc Committee for Space and concluded, with support from the likes of Trevor Gardner and Edwin Land, that science was the only portion of the U.S. space effort free of severe defects. Their report denounced Project Mercury, which only "strengthened the popular belief that man in space is the most important aim of our non-military space effort," and held that *"a crash program aimed at placing a man into orbit at the earliest possible time cannot be justified solely on scientific or technical grounds."* The committee urged Kennedy to stop advertising Mercury lest he associate himself with a possible failure or even death of an astronaut. Instead, the U.S. government should concentrate on scientific and commercial applications such as communications satellites.[8] This Wiesner Report comprised a scientists' critique that would echo until the moon landing and beyond.

Kennedy found the report "highly informative" and promptly named Wiesner his Special Assistant for Science and Technology—then he set the report aside. "I don't think anyone is suggesting that their views are necessarily in every case the right views."[9] The admonition that seemed to affect the new President the most was that concerning Mercury—exploding rockets, dead astronauts, lost races—that is, not that manned spaceflight was wasteful or misguided, but that it might be a public relations failure. In his news conference of February 8, 1961, Kennedy

"They Went Thataway"

From *Straight Herblock* (Simon & Schuster, 1964). Originally appeared in the *Washington Post*, December 30, 1960.

demurred on the race for man in space, placing safety above the desire to "gain some additional prestige."[10]

Hence the first months of the new administration showed hesitancy about space rather than bold forays into this new frontier. Kennedy was learning, Johnson preparing his ground. The PSAC opinion was already on the table, but ran counter to the visceral enthusiasm of Johnson and the Congress. The only actor missing from the scene was the new NASA administrator.

Glennan resigned in December, and Dryden, the apolitical expert, was asked to stay on as acting administrator through the transition. But what should the new man be like: a low-profile technician, a businessman, ex-general, university president, political wheeler-dealer? The choice would be a function of what the NASA chief would be asked to do.

Reflecting its early confusion on this score, the transition team interviewed two dozen candidates, including James Gavin. He was an attractive choice, since he understood the R & D cycle as well as anyone, supported NASA despite his views on the military importance of the "space theater," criticized Eisenhower, and expected space technology to spark an economic revolution.[11] But Gavin either turned down the job or was scratched as a military man. Frustrated, Kennedy tapped Johnson to fill the vacancy, Johnson consulted Kerr, and the latter touted his business partner James E. Webb—the same man who had served Truman as Director of the Budget and axed the early ICBM and satellite programs!

Webb was fifty-four years away from the rural North Carolina of his birth when he took control of the civilian space program. Trained as a lawyer in the capital, Webb became a reserve pilot in the marines and an officer in Sperry Gyroscope Company in the 1930s. He joined the Truman administration in 1946. During the Republican ascendancy, Webb had made his fortune with Kerr-McGee Oil, sat on the board of McDonnell Aviation, and given considerable time to public service. This included leading roles in the Municipal Manpower Commission devoted to urban problems, the Meridan House Foundation, a center for foreigners in the United States, and Educational Services, Inc., in which Webb collaborated on a high school physics text to meet the needs of the Space Age. In sum, he was steeped in the post-Sputnik ethic of government activism, prestige, and scientific mobilization. But when Webb arrived in Washington on a weekend late in January 1961, he told Dryden, "Hugh, I don't think this job is for me." Dryden replied, "I agree with you. I don't think it is either." Webb sent friend Frank Pace to appeal to LBJ, but he got "chased out of the office." He then saw Philip Graham, publisher of the Washington *Post*: "Phil, I've got to get out of this, can't you help me?" No, said Graham, the only man who could was Clark Clifford of the Kennedy transition team. But Clifford had also recommended Webb: "I'm not going to help you get out of it." So on Monday morning Webb reported to the Oval Office. Kennedy explained that he did not want a scientist at NASA but "someone who understands policy . . . great issues of national and international policy." Slightly mollified, Webb accepted: "I've never said no to any President who asked me to do things."[12]

In the words of Abe Zarem, president of Electro-Optical Systems, a NASA administrator had to be

. . . a missionary, an evangelist, with a keen sense of our national rendezvous with destiny . . . an efficient manager . . . suave, a man of exceptional social manners, particularly for briefing Congress . . . able to understand human beings to keep in effective operation people of extremely diverse personalities . . . understand the relationship between scientific knowledge and industrial might . . . know generals and admirals . . . know the "spaghetti bowl," the Pentagon, how it works and how to get around it ... understand the workings of the Budget Bureau.[13]

James Webb was such an extraordinary man. But there would have been no point in placing him in charge of a space program limited to small-scale science or mortgaged to the military. In fact, Webb took office on February 14 anxious "to make unmistakably clear our support for the manned spaceflight program. . . ."[14]

The new team was in place. Lawyers, politicians, businessmen, academics, they were confident of their ability to manage a vastly expanded program of civilian command technology. But before the new team could even put the issue of NASA's future before the new President, they had to fight off another challenge to the *raison d'être* of NASA itself. The outcome of this skirmish, like the philosophy of the new administration and the choices of Johnson, Kerr, Welsh, and Webb, narrowed the possible futures of the American space program and pointed it, incredibly, toward the moon.

The melée over control of space R & D and operations after Sputnik left only two standards flying, those of NASA and the USAF. Military, and some civilian, critics still questioned the wisdom of a divided program. If military control was obnoxious, then let NASA do everything, but unify the program somehow![15] USAF space managers considered the verdict of 1958–59 irrational, unjust, and possibly dangerous. American rocketry grew up in the services. Farsighted officers had pleaded for years for the funds to launch the Space Age. But as soon as Sputnik had vindicated them, the government said, "OK, you were right. Now take all that you have done and hand it over to this new, civilian group." So the USAF space cadets waited, assuming the battle lost but the war still on, until the day when NASA might fade back into the status of the old NACA. This did not mean that the USAF did not cooperate; rather it must help NASA push space technology forward against the day when it might share in the spoils.[16] In the meantime, its skillful PR apparatus advertised USAF experience and prowess in space, kept the problems of a divided program before the public, and declared that "peaceful uses of space" were best ensured by a strong U.S. military presence in orbit.[17]

The presidential campaign, with its promise of change in the midst of "missile and space gap" mania, seemed an opportunity for the USAF to recoup. In October 1960, General Schriever established an Air Force Space Study Committee under Trevor Gardner, the man who had championed the crash program for an ICBM.[18] Meanwhile, the Air Force Secretary's office and aerospace trade press publicized the military shortcomings of the current space program.[19] The USAF Space Study Committee met five times over the winter and issued its top-secret Gardner Report on March 20, 1961. The first sentences revealed its

position: "The military implications of the frequency and payload size of
the Soviet space launches are a major cause of alarm for all members of
the Committee. Under existing U.S. schedules, and with the present
organization, it will be *three to five years* before we can duplicate the
recent Soviet performance." Soviet men in space, orbital rendezvous,
and lunar exploration posed an "impending military space threat" that
could not be met by current space organization. Among the hurdles was
"the insistence on classifying space activities as either 'military' or
'peaceful.' " The Soviets made no such distinctions, while American
niceties only exposed the United States to political attacks. Thus the
divided program was the worst of both worlds. The panel recommended
that a new Air Force Systems Command be given the task of developing
manned spaceflight, space weapons, reconnaissance systems, large boosters, space stations, and even a lunar landing by 1967–70. The U.S.
military, after all, had a long history of leadership in exploration, and in
any case the inhibitions against military spaceflight approached a unilateral
arms moratorium. "The U.S.," lamented the report, "has a consistent
record of under-reacting to the rate of Soviet technological and military
progress . . . in the military space field, we have continued to under-
imagine the possibilities of the future and are not yet organized to exploit
them." NASA, deemed superfluous, was scarcely mentioned in the sixty-
four-page report.[20]

It probably never occurred to USAF petitioners that the new administration would embrace their ambitions, grant their military importance,
and still weigh in on the side of the civilian agency. Yet the aftermath
of the USAF space gambit was precisely that. Even before Gardner
reported, the chairman of the House Space Committee, Overton Brooks
(D., La.) sniffed the winds and preempted the assault. He told a White
House conference in February that "any step-up in the [space] program
must be designed to accelerate a *civilian* program of peaceful space
exploration and use. . . . This is very important from the standpoint of
international relations." The military had a legitimate role, but "NASA
and the civilian space program badly needed a shot in the arm." A big
space program would have a "pronounced and beneficial effect on
America's civilian economy" as well.[21] Three weeks later Brooks expressed
to the President his serious concern about persistent rumors to the effect
that radical change was about to take place in space policy in the
direction of military uses.[22]

While awaiting Kennedy's answer, Brooks sponsored hearings on DoD
involvement in space. Undersecretary Roswell Gilpatric assured the
committee that the DoD did not want to control NASA: "We have plenty
of problems today. We don't need any more." When USAF General
Thomas White took the stand, he deftly retreated. To be sure, he had
spoken of NASA "combining with the military," but that was only a
statement of possible fact, not of advocacy, and was meant to encourage

USAF commanders to cooperate, not compete, with NASA.[23] Chairman Brooks was pleased with this assurance but even more pleased with President Kennedy's reply to his letter: "It is not now, nor has it ever been, my intention to subordinate the activities in space of NASA to those of the DoD."[24]

The USAF gambit was checked: if a major expansion in space should occur, NASA would be the beneficiary. But had the USAF been frustrated as thoroughly as it appeared? Did Generals White and Schriever really hope to persuade the politicians that a single, military-run space program best met all desiderata? No one knew more about management of space R & D than Schriever, yet even as the Gardner Committee slammed the civilian program, Schriever lectured to an engineering convention in Pittsburgh on the *divergent* needs of civil and military spaceflight. They were complementary, he said, and both must be pursued with imagination and vigor. But "the military and civilian missions . . . do not merge into a single image." The technology was essentially the same, but even this was a temporary condition. First, the military would need many more space vehicles (for surveillance, communications, etc.) than NASA, since the latter's would be exploratory in nature. Second, military spacecraft would have relatively longer lives and highly repetitive missions, while scientific spacecraft changed payloads almost with every shot. Third, military vehicles must be simple, reliable, and easy to maintain, while scientific ones would be complicated. Fourth, military missions were time-critical, while NASA, certain launch windows excepted, could choose when to fund and execute projects. Finally, military space technology required close coordination between developer and user, while in NASA programs the same team of technicians served as designers and users. All this meant that different management challenges faced the two programs. Schriever predicted that NASA and USAF efforts would diverge over time, implying that a dual space program was indeed appropriate.[25]

If these were the professional insights of the leading USAF space executive, then what are we to make of the Gardner Report he commissioned? It appears likely that the USAF deliberately overstated its case in order to educate the new administration into USAF assumptions about the Soviet military space threat. It hoped for greater, if not total, support for a "race posture," experimentation with military applications, and military participation in manned spaceflight and big boosters. General White might have been sincere in predicting large operational missions in space for the USAF, but that would come later, as the programs diverged, not at once, when R & D was still to be done. Schriever's own contribution to the policy process bears out this interpretation. He earnestly supported an accelerated NASA program, in order to stack the "building blocks" of spaceflight.

USAF background noise did make an impression. Webb's first priority at NASA was to cement ties with the DoD, while Secretary of Defense McNamara initialed Eisenhower's planned increase in military astronautics for FY 1962. Abraham Hyatt of NASA then proposed a division of tasks between the two agencies that granted the DoD primacy for military missiles, reconnaissance, military communications satellites, navigation, geodesy, satellite inspection and interception, and a joint role in launch vehicles and manned spaceflight.[26] On February 23 McNamara and Webb agreed that neither agency would initiate development of new launch vehicles without the other's consent, while large solid-fueled rockets were to be a USAF show. The DoD also retained a stake in manned flight with its X-20 Dyna-Soar and was promised the opportunity to observe and learn from Mercury.

By the end of March 1961, when Kennedy finally turned to space policy, not only the Gardner Committee but also the Space Science Board of the NAS had rebutted Wiesner and come out for a vastly expanded space program. Heretofore scientific views on the space program had been hostile to "big engineering" as opposed to research satellites. But the NAS Space Science Board, chaired by Lloyd Berkner, a close friend of Webb, recommended that *"scientific exploration of the moon and planets should be clearly stated as the ultimate objective of the U.S. space program for the forseeable future."* It considered that "[f]rom a scientific standpoint, there seems little room for dissent that man's participation in the exploration of the Moon and planets will be essential."[27] The board also held (a bit beyond its competence) that "the sense of national leadership emergent from bold and imaginative U.S. space activity" pointed toward a large manned program, and that "man's exploration of the Moon and planets [is] potentially the greatest inspirational venture of this century and one in which the whole world can share; inherent here are great and fundamental philosophical and spiritual values which find a response in man's questing spirit and his intellectual self-realization."[28]

Here was language to stoke the visionary, intellectual President! The scientifically sound but uninspiring caveats of the Wiesner Report fell flat by comparison. More important, the Space Science Board altered the terms of debate. Beforehand, the main conflict had been one of politicians and engineers pushing manned spaceflight for prestige, security, or big budgets, *versus* scientists and treasurers favoring unmanned flight because of greater scientific returns and much lower costs. But now a body of scientists had come out for a manned moon program, asserted its scientific value, and appealed to something more than "knowledge gained per dollar spent." Manned spaceflight could now be viewed as something over which "good scientists disagree"; the weight of purely political judgments was accordingly enhanced.

The minutes of the Space Science Board meetings, however, tell a different story. It seems that Berkner himself proposed that the board offer an opinion on manned spaceflight, and several members who spoke in favor of it did so in hopes that it would turn nations away from weapons and war, *not* because they held it to be an efficient scientific investment. Other board members bluntly criticized manned spaceflight as misdirected. Berkner then tried to close the meeting with the observation that "this sort of negativism always appears" and that a clear-cut national decision was needed at once. Several members then disputed the wisdom of the board making any policy recommendations, but the committee finally let Berkner "pull together a statement of the Board's position." That statement was hardly representative, for Berkner assured Webb in late February that his report would support the NASA chief's recommendations on manned spaceflight.[29]

The support was timely. Webb confronted the BoB in mid-February only to learn that "we are still pretty much in the dark as to what position the administration desires to take in the space field."[30] But Budget Director David Bell urged Webb to do a quick review (the delay in Webb's appointment had caused NASA's budget process to slip) and recommend any changes before supplemental requests went to Congress in late March. When Webb canvassed NASA opinion, he learned that staff recommendations included acceleration of the big Saturn booster, an even bigger Nova, and the start to Project Apollo that Ike had denied. The Manned Lunar Landing Task Force under George Low even thought the NASA plan for a lunar landing "after 1970" too conservative: it could and should be done before 1970.[31] Technical considerations were less important in selling the program, however, than political ones. "It is our responsibility," said Webb on March 17, ". . . to assess the worthwhile social objectives of our space program and to study our space effort in the context of our broad national and international goals."[32]

The same day NASA made its first pitch to the BoB, asking for a 30 percent increase in the last Eisenhower budget. The case to accelerate, wrote Bell, "was well presented by Mr. Webb and his associates." But budget directors are professional skeptics. Aide Willis Shapley and Bell questioned whether the United States should run races it might lose anyway, whether there were not better (and cheaper) ways of enhancing prestige, and whether "the total magnitude of present and projected expenditures in the space area may be way out of line with the real values of the benefits. . . ."[33] Bell wrote this to the President, then told Dryden not to expect rapid action as Kennedy had other problems to worry about. Dryden retorted: "You may not feel he has the time, but whether he likes it or not he is going to have to consider it. Events will force this."[34]

Webb got his first crack at the President on March 20. His presentation was a prototype of the technocratic argument he would make over and over again in years to come. He began by reminding the President of the effects of Soviet "firsts." To be sure, the Republicans had funded extensive scientific research in space (NASA's Robert Jastrow had just reported that the United States *led* the Soviets in every area of space science),[35] but they left the United States no room for initiative, the key to which was big boosters. Furthermore, the DoD benefited from NASA programs, while in foreign policy, the civilian space program was a positive force: "We feel there is no better means to reinforce our old alliances and build new ones. . . ." But future prospects were even greater, when

it will be possible through new technology to bring about whole new areas of international cooperation in meteorology and communications. . . . The extent to which we are leaders in space science and technology will in large measure determine the extent to which we, as a nation, pioneering on a new frontier, will be in a position to develop the emerging world forces and make it the basis for new concepts and applications in education, communications, and transportation, looking toward more viable political, social, and economic systems for nations willing to work with us in the years ahead.[36]

Prestige, cooperation, emerging world force, viable socioeconomic systems, a new frontier—Kennedy may not have known much about space, but he knew appealing slogans. The President agreed to include $125.7 million for rockets in his defense message to Congress of March 28. Moving Saturn ahead also kept his options open; it bought time.[37]

Two weeks later the time ran out. Yuri Gagarin orbited the earth, and American newspapers again echoed the Kremlin's judgment of it: "a psychological victory of the first magnitude"; "new evidence of Soviet superiority"; "cost the nation heavily in prestige"; "marred the political and psychological image of the country abroad"; "neutral nations may come to believe the wave of the future is Russian."[38] In Congress, Space Committee members insisted that the administration, committed as it was to vigor, determine once and for all whether the United States was going to be first in space. James Fulton (R., Pa.) demanded public acknowledgment that "we are in a competitive race with Russia. . . ."[39] To James Webb he said, "Tell me how much money you need and this committee will authorize all you need"; to the press corps: "I am tired of coming in second best all the time." Victor Anfuso (D., N.Y.) threatened a congressional investigation: "I want to see our country mobilized to a wartime basis, because we are at war."[40] Chairman Brooks demanded that the White House do whatever was necessary to gain unequivocal leadership in space. Webb observed: "The committee is clearly in a runaway mood."[41]

Kennedy's initial reaction to Gagarin was not unlike that of Eisenhower after Sputnik. His congratulatory letter to Khrushchev spoke of cooperation, and he told a press conference that "while no one is more tired than I am" of the United States being second best, he hoped "to go into other areas where we can be first and which will bring more long-range benefits to mankind."[42] Webb, uncertain of where the President stood, lauded the broad-based scientific U.S. space program and sounded like Glennan: "The solid, onward step-by-step pace of our program is what we are more interested in than being first."[43]

Had Kennedy and his people been genuinely contemptuous of manufactured prestige, they might have weathered this storm without major shifts of policy. But just two days after these initial utterances, Kennedy summoned Webb, Dryden, Wiesner, Sorensen, and Bell to the White House and invited journalist Hugh Sidey to observe as he played the leader intent on getting to the bottom of a crisis while others lost their heads. "Is there any place we can catch them? What can we do? Can we go around the moon before them? Can we put a man on the moon before them? What about Nova and Rover? When will Saturn be ready? Can we leapfrog them?" Webb assured him that NASA was moving ahead rapidly. Bell warned of the costs and Wiesner that "now is not the time to make mistakes." Kennedy assumed the burden: "When we know more, I can decide if it's worth it or not. If somebody can just tell me how to catch up. . . . There's nothing more important."[44]

Important for what? National defense, party politics, prestige, national morale? According to historian John Logsdon, Kennedy placed space within a domestic as well as foreign context. He had suffered embarrassments in Laos and the Congo, then Gagarin and the Bay of Pigs humiliation in Cuba. Somehow the trend must be reversed. But Kennedy also had a broad domestic agenda. Would an expensive space program help or hurt him in Congress? Webb believed that a big space initiative would help Kennedy with congressional power brokers and build a basis of support for all his plans.[45] Since the space message ended up as the climax of a lengthy appeal touching on foreign and domestic programs all tied together with the Cold War ribbon, it is likely that Kennedy shared Webb's analysis.

Yet Webb himself now displayed caution. He certainly favored a big push in space, but it was he who would be responsible for it. More money for more rapid progress was one thing, but to declare a specific goal, such as those mentioned by JFK in the Oval Office, was risky. What if a moon voyage proved impossible? Or accidents should happen? This was no Lewis and Clark expedition undertaken with discretionary monies—a moon mission would mean the partial transformation of the national economy! "My own feeling," wrote Webb to Glennan on Gagarin day, "in this and many other matters facing the country at this time is that our two major organizational concepts through which the power of the Nation had been developed—the business corporation and the government agency—are going to have to be re-examined and perhaps some new invention made." Ongoing Soviet competition required the United States to "utilize every resource we have in education, communication, and transportation to build a more viable economic, political, and social structure for the free world. . . ."[46] This was technocracy, and not to be undertaken lightly. Webb was not going to get out front on it without support from the highest quarters.

Kennedy was hardly thinking in terms of national restructuring. He worried about prestige. But his advisers shared Webb's belief in the growing obsolescence of free markets, balanced budgets, or limited government. Foremost among them was LBJ. Summoned to the White House on April 19, he requested a presidential mandate to make recommendations for space. Kennedy complied:

I would like for you as Chairman of the Space Council to be in charge of making an overall survey on where we stand in space. . . . Do we have a chance of beating the Soviets by putting a laboratory into space, or by a trip around the moon, or by a rocket to land on the moon, or by a rocket to go to the moon and back with a man? Is there any other space program which promises dramatic results in which we could win . . . ?[47]

The next day Kennedy announced the study to the press. The hinge was not cost—he admitted that "billions" were involved—but "whether there is any program now, regardless of its cost, which offers us hope of being pioneers in a project." When asked if the United States should beat the Russians to the moon, he replied, "If we can get to the moon before the Russians, we should."[48]

Johnson now had carte blanche to set a goal. Rarely had a great political issue been so clear cut, but rarely had the variables been so obscure. Could giant rockets be built? Could men stand long periods of weightlessness? Could orbital rendezvous be mastered? Was the lunar surface suitable for soft landing? What were the Soviets up to? Johnson consulted NASA, which saw a chance of beating the Soviets to manned circumlunar flight or a lunar landing, but at the cost of at least $11.4 billion extra dollars over ten years.[49] Then he asked McNamara, three business cronies, von Braun, Schriever, and Vice Admiral John T. Hayward. They all supported a moon landing, the last two stipulating only that it not detract from military missions. Von Braun even spoke of putting all other elements in the space program "on the back burner." One of the businessmen, Donald Cook, believed that an action must be "based on the fundamental premise that achievements in space are equated by other nations in the world with technical proficiency and industrial strength . . . and will be of fundamental importance as to which group, the East or the West, they will cast their lot. . . ."[50]

Johnson sent the President a report so loaded down with assumptions that a moon landing was the inescapable conclusion: (1) the Soviets led the United States in prestige; (2) the United States had failed to marshal its superior technical resources; (3) the United States should recognize that countries tend to line up with the country they believe to be the leader; (4) if the United States did not act, the Soviet "margin of control" would get beyond our ability to catch up; (5) even in areas where the Soviets led, the United States had to make aggressive efforts; (6) manned exploration of the moon was of great propaganda value but was essential whether or not the United States was first.[51] In another context, LBJ put it more pithily: "One can predict with confidence that failure to master space means being second-best in the crucial arena of our Cold War world. In the eyes of the world, first in space means first, period; second in space is second in everything."[52]

The persuasive vice president then worked on Webb in a "consultative meeting" packed with lunar zealots: Dryden, Welsh, the three businessmen (including Frank Stanton of CBS), Senators Kerr and Bridges. All sensed that the moment for NASA had arrived. Why did Webb, of all people, hang back? He said he wanted to be sure that NASA had enough support and really knew what it was getting into: if NASA appeared to be the initiator of an expensive, questionable project, it might be left twisting in the wind should the national mood change or delays and failures ensue. NASA must be *given* the task, asked to do this risky thing by an anxious nation. Then Webb would have the leverage, down the road, to claim the backing needed to see the agency through harder times. (Indeed, five years later, Webb wrote to then-President Johnson: "You will remember that in the sessions you had in 1961 with your advisers and Congressional leaders, I was quite reluctant to undertake the responsibility of building a transportation system to the moon and that you had to almost drive me to make the recommendation which you sent on to President Kennedy." For at that May 3, 1961, meeting, LBJ was "close to demanding that NASA recommend for Apollo.")[53]

The next day Webb wrote LBJ that he was ready to climb on board. "I think I can say also that my main effort yesterday was to be certain you and the Senators were under no illusions whatever as to the magnitude of the problems involved in carrying out this decision and the absolute necessity, in my opinion, for a decision to back Secretary McNamara and myself to the limit. . . ." Congress and the press would ride NASA "like two packs of hounds." He must be sure of the President and the Vice-President. Thrice he insisted, and thrice he referred to McNamara.[54]

Why McNamara? Why did the Secretary of Defense endorse a NASA moon program? He knew the importance of prestige and of a "peaceful, civilian" space effort. But the value of a moon mission was not self-evident. McNamara's internal studies in fact convinced him that some of the more "way out" space programs were ripe for the axe, and as late as March 23 he expressed preference for a "normal rate of investigation" and "stated rather emphatically that he would accord a higher priority to items included, about to be excluded, and already excluded from the Defense budget than he would to the programs in question."[55] But there were still larger considerations, including the very health of the aerospace industry. McNamara knew that the missile gap was dubious and believed that defense R & D was out of control and that the aerospace lobby was the main obstacle to his plans for modern cost-accounting. But the sort of cutbacks he envisioned might damage an industry that had expanded tremendously in the 1950s on the strength of jet aviation and the space boom. What was more, government financing of idle plant capacity, cost-plus contracting, and "progress payments" (or "get paid as you go") to contractors were all being phased out. These measures hit the industry just as new plant, equipment, and personnel were needed to participate in the space revolution. The result was a profit squeeze despite record sales and a *tenfold* increase in aerospace debt in the decade after 1957. Aviation was never an easy business in which to make money, but in the early Space Age the industry superimposed high financial risk upon high business risk, a classic violation of sound corporate finance.[56]

The health of the aerospace industry was as much a government worry in 1961 as in 1947. Fairchild, General Dynamics, mighty Lockheed, and Douglas were already in trouble, the latter having lost over $100 million in 1959–60. McNamara's pet reforms, including fixed-price contracting, cost-benefit analysis even on R & D programs, and insistence on high definition even of early design work all promised to trim further the narrow margin the industry trod.[57] In this context an expensive space program, rich in new technology but lying outside the reach of the USAF and his own budget, must have seemed to McNamara, on second thought, felicitous. Giving Apollo to NASA would please the aerospace lobby and Congress, while the USAF, bereft of its allies, would nurse its jealousy alone.[58]

Now everyone was on board. The Saturday after Alan Shepard's flight found NASA, DoD, and BoB delegations gathered to discuss the least mundane of political topics: going to the moon. In the morning session Webb and McNamara exchanged reports done at LBJ's behest. In the afternoon they tackled Apollo. McNamara approved it. He believed the USAF was "out of control": Apollo would help relieve military, industrial, and congressional pressure on him so that he could get on with the tough decisions on defense programs.[59] He also believed, and everyone concurred, that large space projects "reflect the capacity and will of the nation to harness its technological, economic, and managerial resources for a common goal."[60] Apollo was proposed; no one dissented.

The secret "Webb-McNamara Report" originally contained a preamble, drafted by John Rubel of the DoD, adumbrating McNamara's references to the USAF and aerospace industry.[61] It was cut in favor of the habitual rhetoric justifying spaceflight on the basis of science, commerce, defense, and especially prestige. Soviet attainments, the report suggested, were the result of a program planned and executed at the national level over a long period of time, while the United States had "over-encouraged the development of entrepreneurs and the development of new enterprises."[62] The United States must kick its tendency to embellish its designs. "We must insist from the top down that, as the Russians say, 'the better is the enemy of the good.' " Buried in the report was this conjecture: "It is possible, of course, that the Soviet program is not actually the result of careful planning toward long-range goals. . . . Perhaps luck played an important part. . . ." But the evidence pointed "dramatically" in the other direction. "Of all the programs planned, perhaps the greatest unsurpassed prestige will accrue to the nation which first sends man to the moon and returns him safely to earth."

Kennedy first saw the report on May 8 and met with Cabinet members on the tenth. The BoB still worried about setting dates for a moon landing and projects aimed at prestige rather than technological advance per se, not to mention the costs.[63] Economic advisers and Secretary of Labor Arthur Goldberg even denied that the space program would stimulate the economy. But, as McGeorge Bundy recalled, "the President had pretty much made up his mind to go," and, as Wiesner recalled, when McNamara showed the President that *without* Apollo a definite oversupply of manpower would exist in the aerospace industry, "this took away all argument against the space program."[64]

Johnson returned from a trip to Southeast Asia on May 24. A letter from Webb awaited him: "The President has approved the program you submitted, with very few changes, and the message will go up on Wednesday."[65]

That message asked Congress to spend upward from $20 billion on command technology for a political goal. Compared with 1946, when the Atomic Energy Act was assailed as totalitarian, or 1948, when funding for the NSF was restricted by law, or even 1958, when Ike shied from racing in space and struggled to restrict the terms of the education act, Apollo signaled a new age. The technology race that began with weaponry now extended to a civilian pursuit, held in turn to be a symbol of overall national prowess. Where the Eisenhower men doubled and tripled spending on science, education, and R & D, it was their intention to contain as far as possible the effects on traditional values and social institutions and the relationship of the public and private sectors. The men who launched Apollo came to office dissatisfied with existing state management of the national treasure and talent, and began to view the space program as a catalyst for technological revolution, social progress, and even the "restructuring of institutions" in ways that were dimly foreseen but assumed to be "progressive."

How this change occurred in so short a time is not a mystery, but rather that most vexing of historical problems, the "overdetermined event."[66] New men arrived and brought with them those ideas of the "seed time" of the 1950s. Among those ideas were the notions that the

"Fill 'Er Up—I'm in a Race"

Herblock, May 24, 1961. Copyright 1961 by Herblock in the *Washington Post.*

Third World was the main theater of the Cold War and that in that contest prestige was as important as power. Their new ideas validated a far greater role for government in planning and executing social change. The new men also cared more for imagery and felt increasing pressure to display their control over affairs in the wake of early setbacks in foreign policy. Finally, each major figure in space policy—Kennedy, Johnson, Webb, Dryden, McNamara, Welsh, Kerr, and others—saw ways in which an accelerated space program could help them solve problems in their own shop or serve their own interests. This is not to say that they were petty; it is to say that they were technocratic, applying command technology to political problems.

As for contrary arguments, they were disposed of, one by one. Nixon himself abandoned the original Republican skepticism toward space races even before the campaign. The USAF view that the main threat was military got nowhere with the image-conscious civilians. The scientists' argument against prestige-oriented manned spaceflight was bulldozed. When the Soviets weighed in by orbiting Gagarin, and the Shepard flight confirmed NASA's contention that the mission was feasible, all barriers came down. All, that is, except cost, and that, too, was less important in the new White House. We will probably never know precisely what was in Kennedy's mind when he decided that Americans should go to the moon. What may have tipped the balance for him and for many was the spinal chill attending the thought of leaving the moon to the Soviets. Perhaps Apollo could not be justified, but, by God, we could not *not* do it.

Of all those who contributed to the moon decision, the ones farthest in the background were the engineers of Langley and Goddard and Marshall, many of whom devoted their lives to spaceflight, designing dreams. Their reports and studies were necessary buttresses to the political arguments: they had to persuade that the thing could be done. Otherwise, they were absent. Some of their visionary talk about exploration and destiny found place in political speeches, but their efforts to stretch the minds and hearts of their fellows, to sow wonder for its own sake, got lost in their very adoption by the technocratic state. What Constantine's conversion did to the Christian church, Apollo did to spaceflight: it linked it to Caesar. The new faith might conquer the empire, but its immaculate ability to stir hearts was accordingly diminished. Of course, it could not have been otherwise.

CHAPTER 16

Hooded Falcons: Space Technology and Assured Destruction

That American administrations undergo a long learning process and ought not make irrevocable decisions in the first months in office are axiomatic. Yet Kennedy's men from the Ivy League, think tanks, and the big foundations, sure of their acumen and wanting to dissociate themselves from their predecessors, made decisions in their first four months that shaped the space program for a decade and the U.S. strategic posture for over two. The result was a rapid buildup in missiles that preceded even a settled strategy for deterrence or defense, that created a severe missile gap on the Soviet side, that placed military policy almost completely in civilian hands, and that partially militarized the civilian economy. Whether these decisions were nefarious or salutary, avoidable or inevitable, are questions blowing in the wind. For if the Soviet buildup beginning in the mid-1960s was indeed a response to the U.S. buildup, then perhaps the Kennedy team threw away the last chance to halt the missile race at a primitive level. If, on the other hand, the Soviets were committed to superiority, then the Kennedy-McNamara policies bought the United States another fifteen years of security. Either way, the process brought the tools of technocracy fully into the Pentagon, and the name it went by was "the McNamara revolution." Finally, the Apollo decision and the "civilians' strategy" of assured destruction sharply circumscribed the missions that USAF enthusiasts would be allowed to perform in space.

In fact, there were two McNamara revolutions. In less than a year the President and his Secretary of Defense (who had unusual confidence in each other and co-authored defense policy) recast U.S. strategy and weaponry as well as the machinery of the DoD. The former involved the biggest, fastest program of nuclear deployment to date, combined with upgrading of conventional and counterinsurgency forces. The latter completed the managerial shift from the uniformed services to the civilian bureaucracy fanning out from the Office of the Secretary of Defense

(OSD). Both revolutions grew out of the critiques of Eisenhower policies, but both were made possible by R & D and DoD reforms dating from the previous decade.

By the fall of 1960 the Kennedy transition team thought it knew what to do. Arbitrary budget ceilings and Ike's use of committees and "czars" to transcend interservice rivalry must be scrapped. The "military-industrial complex" did have inordinate influence, not by prying too much money from the government but by forcing its misallocation. So McNamara and his whiz kids insisted on a coherent national strategy at the highest civilian level, from which definable missions could be derived to serve in turn as yardsticks for the "cost-effectiveness" of competing weapons systems. The PSAC and DDR & E, set up after Sputnik, were more in line with the sort of civilian guidance McNamara envisioned.[1]

Another focus of criticism was massive retaliation. It seemed to give the United States no options in the face of Communist mischief below that of nuclear brinksmanship. The alternative, "flexible response," would enable the United States to respond with proportionate force to guerrilla war, conventional war, tactical nuclear war, and various options for strategic nuclear war. Flexible nuclear response emerged in a series of studies in 1959–60 at RAND and elsewhere, apparently with Eisenhower's approval. It went by the name of "counterforce" or "no-cities" strategy. If the Superpowers spared urban areas in a nuclear exchange, up to 150 million American lives could be saved.[2] McNamara learned of counterforce his first week in office and was "immediately impressed."

What sort of force structure best served such a strategy? A December 1960 report from the Weapons Systems Evaluation Group (WSEG) considered the ideal mix of delivery systems, which it deemed to be the currently scheduled number of B-52 wings, about forty Polaris submarines, and about 900 Minuteman ICBMs. This triad of delivery systems was made necessary by Soviet advances in antisubmarine warfare and air defense and by the various advantages offered by the three. Land-based missiles were the most accurate and packed the biggest punch (but were themselves stationary targets); sea-based missiles were elusive (but less accurate and harder to command and control); manned bombers could stay on airborne alert, choose targets in flight, and be recalled in need (but were slower and most vulnerable to defenses). The WSEG report recommended, on the basis of cost-benefit analysis, the cancellation of the B-70 and Skybolt air-launched missile and the phasing out of the B-47s and B-58s. Similar analysis dictated an ICBM force of 800 to 900 mobile, or 900 to 1,000 fixed, Minutemen, no more, no less. In every way, the coincidence of the WSEG projections with eventual U.S. deployments is extraordinary.[3]

These numbers, which resembled those emanating from RAND and the Kennedy transition team, dwarfed the Republican post-Sputnik plans for 255 Atlas and Titans, 400 stationary and 90 mobile Minutemen, and

19 Polaris boats. Such numbers would have assured the United States a comfortable lead until the middle of the next decade, but the USAF and some Democrats considered Eisenhower negligent and demanded missiles in the thousands.[4]

Acceleration of the defense effort was a high priority for Kennedy. He lambasted the "lack of a consistent, coherent military strategy," instructed McNamara to reappraise the entire U.S. posture, and pledged to speed up the entire missile program willy-nilly. McNamara groused about having to condense the study of fifteen years into six weeks, but Kennedy nonetheless felt justified by March in making "urgent and obvious recommendations." Before a joint session of Congress the President described American strategy as one of deterring attack by making clear that retaliatory forces existed to survive and deliver unacceptable losses to the enemy. But "our arms will never be used to strike the first blow in any attack." He stepped up the laying of Polaris keels from five to twelve per year, changed three mobile Minuteman squadrons to fixed mode, cancelled two Titan squadrons, and phased down the B-47 and B-70.[5] Thus after just two months Kennedy and McNamara made many of their "hard decisions." The United States would deploy Polaris and Minutemen more quickly and probably in greater numbers but deemphasize manned bombers and eliminate the big, liquid-fueled ICBMs with their larger throw-weight. In the course of his instant learning process, McNamara faulted SAC headquarters in Omaha on several counts and determined to reform military intelligence, strategic command and control, and the Single Integrated Operational Plan (SIOP) that targeted all U.S. nuclear forces.[6]

These measures were in the nature of a "quick fix." The FY 1963 budget was the chance to set the new plans in cement. As McNamara put it, "I equate planning and budgeting, and consider the terms almost synonymous, the budget being simply a quantitative expression of the operating plans."[7] The navy presented no problem. More interested in saving the surface fleet than building submarines, it thought forty to fifty Polaris boats sufficient for "minimum deterrence." By September McNamara had settled on forty-one. The USAF was another story. Those closest to the R & D cycle, like the Ballistic Missile Division, thought a production run of 1,000 to 1,500 Minutemen most economical. "Blue sky" generals like LeMay resisted the temptation to request a plethora of missiles for fear they would threaten the survival of manned bombers. But the SAC wanted 10,000 ICBMs! At the other end of the spectrum was the PSAC, the BoB, and the relatively dovish White House staff. Presidential military adviser Maxwell Taylor also favored minimum deterrence based on "a few hundred reliable and accurate missiles."[8] When McNamara met Kennedy to settle the budget, however, he confessed that 950 Minutemen was the lowest he could ask for in Congress and "not get murdered."[9]

By that time the services' total budget requests were in, and McNamara launched his second revolution. The requests were broken down in traditional fashion into personnel, operations, maintenance, procurement. But the OSD whiz kids took in hand the pillars of paper, amounting to some $60+ billion in line items, and reprogrammed all three service budgets in terms of overall *missions* and the programs meant to fulfill them. The staff could then make direct comparisons of the cost-effectiveness of programs competing for the same mission. Some 620 changes resulted in the first run-through alone, amounting to OSD improvements on the military plans made by the military professionals. The 2,500 Minutemen the USAF eventually requested shrank to 1,200 as an "eventual goal," with 800 funded through FY 1963. Still, the budget came in at $51.6 billion. When added to the emergency appropriations made earlier in the year, it meant that over $4 billion had been added to strategic programs alone in the first year of the new administration.[10]

The next two budget cycles completed the strategic buildup: 41 Polaris submarines and exactly 1,000 Minutemen. But all the major decisions had been made by the end of 1961: Titan deployment frozen, the B-70 cancelled, Polaris numbers fixed, the Atlas and B-47 condemned to phase-out, and the magic number chosen of 1,000 ICBMs in a fixed mode. All told, the Kennedy-McNamara team committed the United States to a force of 600 B-52s, 656 Polaris missiles (41 boats with 16 each), 1,000 Minutemen, 54 surviving Titans, and (until 1965) 126 Atlas. How is one to account for the size and suddenness of this program? One prevalent hypothesis held it all to be a hasty but understandable response to fears of a missile gap. A second explained it as a considered derivative of the new flexible counterforce strategy. But the first hypothesis is false, and the second, if true, casts an embarrassing light on the decision makers, for within eighteen months they had retreated from the strategy that hypothetically justified their actions.

The administration itself put to rest the missile gap before the major funding bills went to Congress. On February 6, 1961, McNamara staged an off-the-record news briefing and was asked if the USSR really had more combat-ready missiles than the United States. After sidestepping, the new Secretary admitted that each side appeared to have about the same small number.[11] Two days later Kennedy fought off the same question: studies were underway, but it was premature to say whether there was or was not a "gap."[12] Republicans ironically "congratulated" Kennedy on closing the missile gap in just eighteen days. For six more weeks reporters dogged the President, who took refuge in his "uncompleted," then "in progress," then "soon to be completed" study. In time the rush of events, including the Bay of Pigs and space race, buried the issue for good. Within the intelligence community, opinion still varied: the CIA and army believed the USSR had 125 to 150 missiles, the USAF 300 at least, the State Department 160, the navy only 10.[13]

DISCOVERER 25 AND SAMOS 2

By New Year's 1961 two U.S. spy satellite programs verged on brilliant success. Thanks to the energetic advocacy of Richard M. Bissell, Jr., of the CIA, Discoverer survived its long bout of frustration and brought number 14 back to earth, possibly with the first space-based photographs of the USSR. But the original Agena system was not equipped for a full reconnaissance of the Soviet land mass. In the winter of 1960–61 the USAF hurriedly readied a new Agena-B to be mated with an Atlas or upgraded Thor. It made its maiden flight on December 7, 1960, as *Discoverer 18,* orbited for three days, then fell earthward until a circling C-119 transport plane made an airborne snatch of the film capsule at 14,000 feet.

Samos, meanwhile, was a Polaroid wonder that developed its own film, scanned it electronically, and radioed the pictures to ground stations. Outgoing Secretary of Defense Gates authorized $200 million for Samos, then $84 million more, as the spy satellite best able to settle the missile gap with one successful mission. *Samos 2* orbited on January 31, 1961, and made some five hundred passes before its transmitters were switched off. Months would be needed to analyze all its photographs of likely Soviet missile locations . . . and indeed it was June when the CIA estimate of Soviet deployment suddenly and without explanation dropped by half, from 120 to 60.

Discoverers 20 to 24, testing out a new spacecraft, booster configuration, and Honeywell gyro stabilization system, all had problems. But *Discoverer 25* of June 16, 1961, orbited the earth thirty-three times and returned with what must have been a precious cargo. When it missed the recovery area and plunged into the sea, navy frogmen labored to retrieve it. Two more film-toting Discoverers were recovered by mid-September . . . and again the United States revised its count of Soviet missiles, to at most fourteen.

The first satellite sparked the missile gap; the first reconnaissance satellites snuffed it out. Joseph Alsop, previously a noisy patron of the "gap," broke the news on September 25, 1961: new intelligence revealed that something less than fifty Soviet missiles existed. In October Deputy Secretary Gilpatric admitted that the United States would have more missiles even after absorbing a surprise attack than the USSR had available for a first strike. In November the New York *Times* made it official. The United States had already some 233 missiles capable of reaching Soviet territory to some fifty or less Soviet ones able to reach the United States: "The 'missile gap,' like the 'bomber gap' before, is now being consigned to the limbo of synthetic issues, where it has always belonged."[14]

McNamara suspected that the missile gap was a myth from the start of his term. By the time his FY 1963 budget went to the Hill, everyone knew it was a myth. To be sure, Khrushchev was still blustering and bragging, and if the United States *had* stocked its armory in fear of a gap, the Soviet leader would have only himself to blame. But it did not. McNamara wrote in 1967 that the numbers chosen for the U.S. deterrent were based on "worst plausible case" analysis not of what the Soviets had then, but of what they might have five years hence. The result was that the U.S. force turned out "both greater than we had originally planned and in fact more than we require."[15] Of course, by 1967 the Soviets had begun to deploy at a furious pace, and it seemed fortunate that the United States had purchased a cushion. As McNamara told Congress in 1963, strategic programs must anticipate production decisions that the adversary may not even have made yet.[16] Soviet force levels were a factor in the Pentagon's calculations, but were "not the most important by any manner of means."[17] In other words, U.S. deployment followed its own logic, and that implied an a priori strategy.

That strategy was supposedly flexible response. Since the Soviets would soon have missiles of their own, the United States must build a sizable "second strike," or retaliatory, force. What was more, since the United States shunned the bombing of cities, it again needed more—and more accurate—missiles so as to threaten Soviet launch sites, air bases, submarine pens, and other military targets.[18] This was *not* a hawkish position. The larger force was necessary to reduce chances of either side ever using their weapons, to reduce civilian casualties in case of war, and to reduce the pressure on either side to launch an all-out "spasm attack." By contrast, the "dovish" position of minimum deterrence depended entirely on terror: the threat of atomizing populated cities. Liberal congressmen, of course, denounced the arsenal as "overkill," but a small deterrent was vulnerable to a preemptive strike unless the United States launched its missiles "on warning" and thus increased the chances of accidental war. There was little talk at the time of the morality of holding civilians hostage in a balance of terror. But the "dove" position— counting bombs and assuming that the fewer the better—actually embraced a techno-logic more extreme than that involved in counterforce. It assumed that technology could substitute for diplomacy in the prevention of war. It was this *ultima ratio* of democratized war—"We're only going to build a few missiles, but if you attack we'll annihilate your society"—that counterforce was designed to avoid.

The least terrible of nuclear worlds, therefore, required that the United States have a force second to none, or, in Secretary of State Dean Rusk's words, "a very large overall nuclear superiority with respect to the Soviet Union."[19] But not for massive retaliation; rather, the Kennedy administration reversed the roles of American forces. Under massive retaliation, U.S. conventional forces abroad were the nation's shield, the SAC was its sword. Under flexible response, the SAC became the shield and beefed-up conventional forces the sword, by which the United States could take initiative in Third World conflicts or along the Iron Curtain.

In March 1961 McNamara ordered the JCS to draft appropriate doctrine for a "no-cities" strategy, while RAND veterans Alain Enthoven, Daniel Ellsberg, and Frank Trinkl revised the Basic National Security

Policy and SIOP to set five nuclear options before the President *in extremis.*[20] In January 1962 McNamara introduced the "no-cities" strategy to Congress, warning that it called for a larger force than would otherwise be necessary. In June he explained it to the public in a commencement address at the University of Michigan (hence "the Ann Arbor strategy"). The purpose of war was still the destruction of enemy armed forces, not population; the purpose of counterforce was to give "a possible opponent the strongest imaginable incentive to refrain from striking our own cities."[21]

All this made sense—to Americans in 1962. Unfortunately, counterforce was something less than a counsel of perfection for the rest of the world. Europeans counted on the United States to deter or repel a Soviet invasion; second-strike strategies did not suit their needs. The Soviets, in turn, claimed the numbers and accuracy of U.S. missiles looked very much like a first-strike force. But worst of all, even the advantages of the McNamara strategy obtained only so long as the Soviets remained outgunned. Almost as soon as he had enunciated his strategy in September 1962, McNamara conceded it was "doubtful" that the United States could maintain its current superiority. After the Cuban missile crisis he went further: the Soviets would soon have full retaliatory power, and "when both sides have a secure second-strike capability, then you might have a more stable balance of terror."[22]

What then would become of the vaunted "options"? The answer to this question, though shrouded by increasing talk of "damage limitation," was another rubric: "assured destruction." Even in the face of a large Soviet missile force, the United States retained the capacity to rain assured destruction on the heads of the aggressor. By March 1964 McNamara had ordered the services and the WSEG to revise doctrine on this basis. The result confirmed that current force levels were sufficient to deliver "unacceptable damage" on the attacker and that more missiles would not add significantly to that sufficiency. But this meant a return to simple deterrence! Indeed, the WSEG concluded that a more cost-effective use of new resources would involve damage-limiting measures such as antisubmarine warfare, fallout shelters, and perhaps limited antiballistic missiles (ABM), but not more ICBMs.[23] In coming years even this damage-limitation aspect faded and assured destruction came to the fore. If by 1967 McNamara was still adhering to the Ann Arbor strategy, he would have to admit that the United States was underarmed. Clearly he had dropped that strategy. By the time of his retirement from the Pentagon he publicly advocated the environment of *mutual* assured destruction (MAD). Neither side had a first-strike capability, both had a secure second-strike capability, and this stability ought to be nurtured by both sides.[24]

When the tilt occurred is hard to pinpoint—strategy in these years was as "shifting sands."[25] The deployment decisions through 1962 fit counterforce, but other considerations also impinged in 1961: the belief that arms spending would help in the recession; the tactic of using one weapons system (Minuteman) to shoot down another favored by lobbies (B-70); the play of bureaucratic interests; the desire of the Kennedyites to make a clean break with their predecessors.[26] But whatever the mix of motives, by late 1963 the strategy for which the missile decisions were made was already becoming obsolete. The United States was left with a nuclear panoply outsized and ill-suited to minimum deterrence, but not large enough for first-strike or pure counterforce. The "doves," having lost the deployment battles, won the strategic battles. Assured destruction required no new missiles—and no new technologies, in space, for instance—that might "destabilize" the balance of terror.

Like Webb at NASA, McNamara believed that the drag on American efforts was not lack of money or brainpower, but poor management. Civilian experts, versed in computers, cost accounting, and systems analysis, must take over the biggest enterprise in the world, the Pentagon. In 1961 McNamara instituted five year plans for weapons development, R & D, and cost reduction, and imposed on all the first Planning-Programming-Budget System (PPBS) in the government (under LBJ all agencies were ordered to adopt PPBS). This brainchild of ex-RAND economist and Assistant Secretary of Defense Charles J. Hitch was a master plan in which "budgets, weapons programs, force requirements, military strategy, and foreign policy objectives are all brought into balance with one another." Combined with the five year plan, PPBS was the most important management tool for the Secretary of Defense.[27]

Five year plans cannot function, however, without suppression of particular interests (potential "wreckers" in Soviet parlance). In the Pentagon, these were the jealous services; the antidote was to pull all strings into the OSD. Consequently, McNamara expanded or created: the DDR & E to centralize all military R & D decisions in the hands of civilians; a Defense Intelligence Agency to override the intelligence arms of the services; a National Military Command at the level of the JCS to preempt interservice rivalries in operations; special civilian assistants to supervise strategic mobility and counterinsurgency; a unified Defense Supply Agency to take logistics out of the hands of the services; and a Defense Communications Agency to centralize C^3I (communications, command, control, and intelligence), meteorology, and administration for all the services.[28] In every functional pyramid, new layers of centralized, civilian bureaucracy splayed out from the organizational box of OSD in 1961.

Already in the 1950s military officers protested interference by civilian "amateurs." Now McNamara's whiz kids were everywhere, removing every vestige of independent authority and, with it, much of the pride of career officers. McNamara justified it by saying that in the modern age, economic, political, and technical considerations must supplement, not "downgrade," military advice.[29] But the uniformed military in the United States, as in the USSR, found their profession hostage to technocrats who, in one congressman's words, "believe we can settle all by a computer or a slide rule."[30]

Did not the technological revolution demand a managerial revolution? Until World War II, Enthoven explained:

> Both soldiers and statesmen could learn most of what they needed to know about military power and the relationship of weapons systems and forces to national security from their own direct experience and by reading history books. . . .
>
> But something new had been happening in the past twenty years. Science and technology have gone through a "take-off" and they are now in a period of rapid, accelerating, and apparently self-sustaining growth. Nuclear weapons, nuclear power, computers, large-scale rockets, and space flight are but the most spectacular examples of a revolution which has been led by both military men and civilian scientists. Before World War II, we did not plan on technological change; we merely adjusted to it. Now we are planning on it.[31]

Experience and history were no longer reliable guides! Enthoven went on to call for "a new analytical approach or discipline" in defense planning to adjust to rapid changes in technology.[32] The message was clear. If the uniformed services wished to preserve their role in their own profession, they, like the young technical officers in the USSR, must learn the new discipline. That new discipline was systems analysis.

Before the Congress, McNamara offered, by way of illustration, what he called an "oversimplified" example of systems analysis:

> Whether we should have a 45-boat Polaris program, as the Navy has suggested, or a 29-boat program, as the Air Force thinks, is in part affected by the decision we make on the Air Force Minuteman missile program. . . .
>
> A major mission of these forces is to deter war by their capability to destroy the enemy's warmaking capabilities. With the kinds of weapons available to us, this task presents a problem of reasonably finite [sic] dimensions which are measurable. . . .
>
> [First was to determine the number, type, and location of enemy targets; second to determine the number and yield of warheads needed to destroy them.]
>
> The third step involves a determination of the size and character of the forces best suited to deliver these weapons, taking into account such factors as (1) the number and weight of warheads that each type of vehicle can deliver; (2) the ability of each type of vehicle to penetrate enemy defenses; (3) the degree of accuracy that can be expected of each system . . . ; (4) the degree of reliability of each system; (5) the cost/effectiveness of each system, i.e., the combat effectiveness per dollar of outlay.[33]

But since the enemy could strike first, one must also calculate the size and effect of such attacks, based on estimates of the number and quantity of his missiles and the vulnerability of one's own competing systems.

McNamara admitted that every one of these factors was estimable only within a range of probability. Still, they could all be quantified, fed into a computer, and results compared. One analyzed each weapon in terms of its whole system. The cost of a B-52, for example, was not the unit cost of planes off the assembly line, but the total expense of a wing of B-52s from day one until their obsolescence, including R & D, economies of scale, manpower and airbases required, support systems, maintenance, and so on. When reduced to a computer read-out, such complicated issues became manageable and systems analysis a powerful prophylactic against boondoggles. But as the primary method of planning national defense, it was prone to the computer's bane of "garbage in, garbage out," while it hid from the layman many of the assumptions built into the analysis. How were alternatives—X bombers versus Y missiles—proposed in the first place? What reaction would each provoke from the enemy? How does one quantify the willingness of Congress and the voters to pay for given levels of risk? Rational comparisons might be made at low levels, but the parameters of the "system" were artificial: each system was itself a subset of a larger system, where important choices were made according to different criteria.[34]

Systems analysis was especially problematical in planning for R & D. How could one reliably estimate costs of developing new technologies or the benefits growing out of basic research? If a given weapons program had no mission under current strategy, it could be scrubbed without regret. But what if its utility was uncertain, if strategy should change, if the adversary might surprise, if spin-offs of incalculable value might flow from the technology? All told, the civilians' strategy of stable, assured destruction, and their management tool of systems analysis, boded ill for USAF space cadets. While NASA pushed manned over unmanned spaceflight, the DoD favored unmanned missiles over manned bombers and was likely to find little value in costly, "destabilizing" space weaponry.

The predictable military reaction to all this was that the McNamara revolutions emasculated the military profession and damaged U.S. defense even more than Ike's penury. In April 1964 SAC commander General Thomas Power spoke for many when he lambasted Pentagon leadership before three hundred officers and visitors. Those who advocated disarmament were fools; the test-ban treaty was a great mistake; the only way to avoid war was to remain strong; the United States would be foolhardy to ignore the military uses of space—the Soviets certainly would not. Who was to blame for selling the wrong plans to the White House and Secretary of Defense? The moralists who could not abide the use of force, who had given Hitler his start and now gotten us into the messes of Cuba and Vietnam, and the "computer types" who arrogantly ignored military judgment and "don't know their ass from a hole in the ground."[35]

Be that as it may, the computer types ultimately shaped the American response to the shock that began with Sputnik: a large, prestige-oriented space program, a strategy of stable assured destruction, and a methodology premised on the belief that Space Age management and technology could best determine the use of that technology itself. Each of these, in its own way, stymied the one use of space technology that its original patrons of the 1950s considered most important of all—active military operations in orbit.

The Kennedy team, with its gift for publicity, found the formula that had escaped Eisenhower's for presenting the American position on militarization of space: "The United States has space missions to help keep the peace and space missions to improve our ability to live well in peace." It was an appealing restatement of the old distinction between military and aggressive—*all* U.S. space programs, military or civilian, were of peaceful intent.[36] Nevertheless, when the whiz kids asked how much should be spent on military space, who should do the work, and what sorts of missions should be approved, they applied their tests of systems analysis and strategic stability to the various R & D programs under review and came up with read-outs that said: "Hood the falcons!" American strategy dictated that outer space be preserved as a sanctuary for passive military satellites and remain off-limits to "destabilizing" active weaponry. Although the Soviets contradicted this formula in doctrine and in practice, they were still years behind in the technology. Perhaps U.S. forbearance might restrain the Soviets, too, from extending the arms race into space and preserve the inviolability of spy satellites as well. The sanctuary notion therefore came to dominate U.S. military space policy for two decades.

The first problem of military space was how much to spend. In his March defense message Kennedy sought an additional $226 million, making a FY 1962 total of $850 million for DoD space. This reflected the natural growth of programs begun in the 1950s and now entering the testing or operational phase. But at the same time, McNamara pledged rigorous civilian management of R & D based on objective managerial tools, while the thought of generals with their "fingers on the button" so shocked some that Kennedy promised all decisions involving nuclear weapons would be made by civilian authorities.[37] So even as they won bigger budgets for missiles and space, the military professionals paid for them with a further loss of control over their own R & D and deployment.

The second question—who should do the work in military space— had been settled early on in favor of the USAF, where 87 percent of the effort already resided.[38] The ARPA survived but came to concentrate on ballistic missile defense, nuclear test detection, and materials and propellant research. But what of NASA and the military implications of manned spaceflight, orbital rendezvous, comsats, or big liquid boosters? While the USAF failed to demonstrate what military man might do in space, the argument could well be made that it had never had the chance to do so. The Soviets clearly thought manned spaceflight good for something, and USAF advocates argued the value of "having a man in the loop" to add flexibility and speed to reconnaissance or satellite inspection missions. But McNamara resisted a hasty move to manned military systems. Rather, he adopted the competing "building blocks" notion: the United States must overtake the USSR and prevent technological surprise, but the building blocks of space expertise could accumulate just as well in NASA as in the DoD. Gemini and Apollo were a benign means of learning about manned spaceflight, whatever its eventual uses.

The manned versus unmanned issue fed into the third question—what sorts of military space programs ought to be pursued? After all, preventing technological surprise could justify anything: in the 1950s the army and USAF even spoke of military bases on the moon as "decisive," while Gavin, British analyst Michael Golovine, and others predicted that the vicinity of earth would soon be replete with orbiting missiles, antimissile missiles, anti-antimissile missiles, decoys, satellite inspectors, orbiting chaff, and support spacecraft.[39] New ARPA Director Jack P. Ruina told the House Space Committee that the main consideration should not be one of cutting back on *outré* R & D or eliminating duplication, but just the opposite: "This sin of duplication is one of our least sins. . . . Much more is the question of what we aren't doing. . . ." Von Braun thought it shortsighted to say "we want to close our eyes to all military aspects of outer space just because the idea may be frightening. . . . The question of space superiority is just as important today as the concept of air superiority was ten years ago."[40] Generals Ferguson, LeMay, Power, and Schriever all insisted on the potential of space power to dominate earth. The very preservation of free and open space depended on the ability of the United States to prevent its military domination by others.[41] And yet the precise missions never seemed to materialize, so civilian leadership sounded much different. Harold Brown, the DDR & E, admitted that "we cannot now identify or much less describe what the military requirements for very large payloads might be." McNamara granted missions in communications and navigation, but "the requirements 5, 10, or 15 years from now are not at all clear to me."[42]

The civilian insistence on clearly defined missions cut the legs from under basic space R & D in the DoD. ARPA had been founded on the understanding that "if an end requirement, be it military or any other, must be established before we embark on research, then by definition it is no longer research."[43] Yet the solution of the 1960s was to let NASA do exploratory research and to oblige the USAF to show precise requirements. The phrase describing McNamara's approach, oft repeated, was that space is not a mission, or a program, or a cause; it is just a place. Some things could be done better there, others not. The job was to identify the former, and do them only. But this was easier said than done, and it obliged a few civilians to make educated guesses about where to place their bets. To cover all the numbers guaranteed waste; to cover only some, and choose wrongly, might expose the nation to peril. How much more comfortable to make R & D decisions according to some overall strategy. So it was a fourth question—would a given system stabilize or destabilize deterrence?—that came to dominate the counsel of the Kennedy years. A DoD Assistant Secretary explained:

> Soviet record on test ban negotiations makes it clear that, though we may earnestly hope that space will be used only for peaceful purposes, we cannot base our national security on hope alone.
> The resolution of this dilemma stands on three elements: first, continued full pursuit by the military of those missions in space which are intrinsically peaceful and stabilizing; second, development by the military of the basic building blocks of further space capability as insurance against contingencies; and third, continued pursuit of a broadly based national program in space technology.[44]

Space systems that served stable deterrence were judged to include satellites for early warning, test-ban verification, and C^3I. On the other hand, space weapons capable of destroying hostile satellites or missiles in flight might contribute to a first-strike capability and thus were judged destabilizing. Similarly, orbital bombardment systems were bad, since they might provoke the other side to develop space weapons to blunt the Damocles' sword of "bombs-in-orbit." Research on ABM or antisatellite systems should continue, but not be pushed to testing and deployment.

Programs dating from the 1950s entered testing at a growing pace in the new decade and gave evidence of the early imagination and foresight of American engineers. Reconnaissance, as always, stood highest in the ranks of military space programs. Following the *Samos 2* and *Discoverer 25* flights that put paid to the missile gap, an Atlas-Agena combination in December 1961 orbited a radio-transmission type satellite that dipped as low as 145 miles, improving photographic resolution to ten feet. In May 1963 a Thor-Agena D, with three "strap-on" rockets and a lighter, more compact camera from Eastman Kodak, permitted more film and still longer life at lower cost. New computers financed by DoD funds processed the images; CBS Laboratories provided the film scanner/converter for the spacecraft; Philco-Ford, the electronic signal processors. The new Agena D also boasted a restart capacity, so it could be moved in orbit to concentrate on suspicious areas. Later in the year a new generation of recoverable film-pack satellites arrived as well, courtesy of General Electric's Space and Missile Division. The new "bird" homed in on sites detected by its radio-transmission type cousin and returned high-resolution photos to earth in three to five days. By 1964 both types were orbiting routinely from Vandenberg AFB.[45]

Still, the revolution had just begun. By the mid-1960s the USAF, the CIA, and industrial contractors were designing systems that could penetrate clouds, ground cover, camouflage, and the darkness of night. One new design used multispectral photography, a cluster of filtered lenses whose images, examined together, could reveal the exact frequency-absorption properties of objects, exposing false foliage, tarpaulin, silo covers, or even factory productivity or drought and disease in Soviet farmlands. Another novelty was the infrared scanner, which could see in the dark or reveal the heated water expelled by the reactors of a nuclear submarine. By 1969 such technology followed the laws of evolution toward the complexity and giganticism of Program 467—Big Bird.[46]

The navy built on its success with the Transit navigation system, whose seventh sister went into polar orbit in June 1963. All the services worked on comsats. Project Westford of October 1961 aimed at exploding 400 million tiny copper dipoles into the upper atmosphere, an experiment to improve on Nature's ionosphere as a reflector of shortwave radio signals. The first mission failed, but in spite of severe foreign criticism, *Westford 2* went ahead in 1963. Similarly, following the Soviets' unilateral renunciation of a nuclear test moratorium in 1961, the AEC and the USAF staged Project Dominic, a series of eight high-altitude nuclear tests launched by Thor rockets from Johnston Island in the Pacific. Operation Beanstalk experimented with emergency communications by satellite; Project Anna was a joint Army-Navy-NASA-Air Force geodetic satellite; Surcal, a secret surveillance calibration "black box" connected to the spy satellite program; and Injun, a "starfish" radiation data collector, presumably connected with Dominic. Following the Partial Test Ban Treaty, which banned tests in aerospace, *Vela 1* and *2* were launched to detect nuclear explosions in near space.

Behind these exotic and top-secret programs were the real visible pillars of a USAF space capability. Solid-rocket propulsion, granted to the USAF under the Webb-McNamara agreement, led to the *Titan 3*, Martin's liquid-fueled ICBM muscled up with two huge strap-on solid boosters, developing over 3 million pounds of thrust and capable of launching twelve and a half tons into space. Another big mission was active defense—systems to inspect and, if necessary, destroy hostile spacecraft. Two such projects were Saint, a satellite able to rendezvous in orbit, and a simpler ground-based interceptor that worked like an ABM. Finally, there were alternate propulsion systems, especially the nuclear rocket. Though managed by the AEC, this program was of interest to the DoD as a future means of boosting very heavy payloads, like manned space stations, into orbit and beyond.

In the first eighteen months of the new administration, all these programs, plus the initial budgetary injections, encouraged USAF space managers enough to submerge the Air R & D Command into the new Air Force Systems Command under Schriever. It included a division devoted solely to space—the first U.S. "space command." DDR & E also gave signals to the USAF Science Advisory Board that a positive attitude on military space informed the OSD and that the USAF would have its role in manned spaceflight both through Dyna-Soar and NASA's projects.[47] Three months later General James Ferguson looked forward with confidence to nuclear power for space maneuver, a small satellite-inspector system, Dyna-Soar test flights, and R & D on nonnuclear space weapons.[48] Even Abe Hyatt of NASA predicted that space operations would have a major impact on the outcome of a war by 1965 and be decisive by 1975.[49] Meanwhile, NASA and the USAF respected each other's turf. Even *Aviation Week and Space Technology*, chronicler of discord, wrote of a "gleeful conspiracy" between the two agencies to advance military space capabilities "under NASA's strong financial shelter in the Apollo program."[50]

Then McNamara's civilians lowered the boom. Under the strategy of assured destruction, passive military satellites suddenly emerged as "good" systems and active space weapons as "bad" ones. Military applications satellites had already proved their worth and were stabilizing. To throw away their tacit safety in order to play with unproven, destabilizing, and expensive weapons seemed foolish. Better to use diplomacy to educate the Soviets to that effect. First came a September 1962 disavowal by Ross Gilpatric of nuclear weapons in orbit. This killed U.S. research on orbital bombardment. Next came the push for a Partial Test Ban Treaty that forbade nuclear explosions in space. This killed research into antisatellite systems with nuclear warheads, as well as the Orion Project for nuclear rocket propulsion.[51] Then the Soviets called off their campaign against spy satellites, and Project Saint was downgraded and later cancelled. Finally, and most bitterly for the proponents of active military systems in space, the USAF manned space vehicle also died a premature death.

The saga of the X-20 Dyna-Soar, a case study in R & D policy, deserves a telling. It began in 1943 when von Braun suggested attaching wings to his Peenemünde rockets and Eugen Sänger and Irene Bredt designed the "antipodal bomber" that so intrigued Stalin after the war. It got nowhere, but in the 1950s Walter Dornberger, von Braun's old boss, joined Bell Aircraft and began to pester the USAF with ideas of a manned space vehicle, Bomi, capable of bombing or reconnaissance up to 3,000 miles. In 1955 Bell and its USAF friends won approval from the brass—it was still two years before Sputnik. Now Bell had $1 million in Pentagon funds and another $2.3 million wheedled from six other firms eager for a piece of the action. When the USAF came to think of the spaceship mostly in terms of reconnaissance, Bomi became Robo, and then, a week after Sputnik, the Air R & D Command mated Robo with the X-15 concept, spawning the X-20. Still a rocket plane, but launched from the ground like a satellite, it was to reach a velocity of 25,000 feet per second and glide about the earth ("dynamic soaring") sixty miles high. They dubbed it the Dyna-Soar. In May 1958 NACA was brought in, contracts awarded to Martin and Boeing, and another $15 million anted up.[52]

Thus the origins of a typical big project: demonstration of technical feasibility, privately funded research and salesmanship leading to military acceptance, extrapolation of existing technology, contrivance of plausible military missions, the savor of "technological sweetness," and finally the Sputnik panic. By 1960 Dyna-Soar took shape as a low, delta-winged spaceplane to be launched on a Titan rocket but land like an airplane—a "shuttle craft" to orbit. The DoD gave it $100 million for FY 1962.

When the new administration accelerated military space programs, it budgeted $921 million for Dyna-Soar development through 1969. But the civilian authorities, even as they approved the program, refused to recognize the X-20 as a weapons system. Another year passed and the political weather changed again. NASA moved ahead with Gemini for rendezvous and docking, and McNamara's analysts showed that a modified Gemini might perform military functions in space better and more cheaply than the X-20.[53] Dyna-Soar was now in serious danger. A funding compromise left a low $130 million for Dyna-Soar in 1963 and 1964, bumping its maiden flight to 1966.[54] McNamara then turned around and recommended that the USAF be given an equal or dominant role in managing Gemini. Webb exploded at this open assault on NASA, which, he said, would jeopardize his ability to meet the lunar landing deadline and signal to the world a militarization of the U.S. space program, with serious consequences.[55] Webb and McNamara then agreed to let DoD experiments ride piggy-back on NASA's flights under the rubric Blue Gemini.[56] The frustrated USAF space men now labored frantically to find a military mission for the X-20. But in July 1963 NASA revived the notion of space stations. Would not an orbiting laboratory based on Gemini be able to determine if a military mission existed for manned spaceflight? This thought gave birth to the Manned Orbiting Laboratory (MOL). Quickly the PSAC reviewed the merits of Gemini, Dyna-Soar, and MOL—Dyna-Soar lost again. Congressional support began to erode. Even the chairman of the Senate Space Committee, Clinton Anderson, wrote that "many sincere Senators are worried over the fact that Dyna-Soar is still going strong." There was no longer, he added, enough money to go around.[57]

In October 1963 McNamara bypassed Dyna-Soar and requested instead a military follow-on to Gemini. NASA countered by suggesting a military MOL. The USAF then played its last card, touting Dyna-Soar as a supply vehicle to dock with the MOL in orbit. But the game was up. In December McNamara met with the new President and recommended termination of the X-20.* The next day MOL was announced instead.[58]

Conceived in 1943, again in 1950, designed in 1954, adopted in 1957, funded for seven years to a total of $400 million, Dyna-Soar died in the battle of "fiscal 1964." It had been "reviewed, revised, reoriented, restudied, and reorganized to a greater extent than any other Air Force program."[59] Yet for all that it was not atypical. It was a bastard child of the rocket revolution, an idea too good to pass up, if only because it promised spaceflight without dispensing with wings or a pilot. The USAF liked that. It was wet-nursed by industry and raised by the military on the vaguest of pretexts. Once subjected to McNamara's systems analysis, Dyna-Soar took sick. But its demise exposed the other side of the new technocracy. More money was being spent on R & D than ever before, but there were still "ceilings." The difference was that now civilian quantifiers cut the pie, not uniformed chiefs.[60] Civilians overcame USAF pressure for manned spaceflight by "shooting down" Dyna-Soar with MOL, only to shoot down MOL later on. The irony of Dyna-Soar, or bigger boondoggles like the nuclear airplane, was that in previous times the "Colonel Blimps" were accused of being backward-looking; now, at the very moment when the military managers became the most avid of innovators, they were stripped of their authority by stuffy civilians who demanded a political-technical goal be proven before they would "authorize" progress. But such is the pattern of technocracy; to us it no longer seems ironic at all.

Strategy, diplomacy, and bureaucratic politics all converged to make 1963 the first great turning point in the U.S. military space program, just as 1961 had been in the civilian program. USAF manned spaceflight was again postponed, and all space weapons programs but one were cut back or cancelled. The exception was a crude antisatellite (A-SAT) project based on Johnston Island. Beginning in March 1964 Thor boosters began to test spacecraft that simulated interception of satellites in low orbits. This program, and another with the army's Nike-Zeus on Kwajalein Island, amounted to little more than insurance against Soviet surprises.[61]

The "space as sanctuary" policy assumed that the Soviets would follow suit and come to appreciate the techno-logic of stable deterrence. Yet these were the very years when the military took charge of the Soviet space program, stepped up research on antisatellite and fractional orbital bombardment, in addition to passive military satellites, and argued the future decisiveness of space weaponry. Nor did any Soviet leader ever suggest that technology could, or should, stand still. Indeed, technical advances already on the horizon threatened to destabilize the balance of terror: multiple (independently targeted) reentry vehicles (MRV and

---

*Dyna-Soar was much loved and lamented. See note 60.

MIRV), permitting several warheads to ride on one missile; improved accuracy, increasing temptation to strike at the enemy's silos; and ABMs, which were said to encourage a first strike by protecting oneself against retaliation. Such breakthroughs would be a constant menace unless diplomacy somehow froze the arms race for good.

The idea of a special agency to promote arms control originated under Eisenhower but was consummated by Kennedy and veteran statesman John J. McCloy. Thus the ACDA emerged just as the United States entered its big missile buildup. This discrepancy could be variously explained. First, the United States must negotiate from strength if it was to expect a forthcoming Soviet posture. Second, the United States stressed arms control over disarmament: it did not seek to eliminate nuclear weapons but rather to preserve stable deterrence. Third, the ACDA was another important contribution to the image of the United States, whether or not it achieved anything.[62] To some conservatives the ACDA amounted to institutionalized treason. In the words of one JCS staffer, the ACDA assumed that nuclear weapons caused tensions, whereas in truth politics caused tensions, which a strong defense prevented from degenerating into war. According to General Power, "the U.S. is attempting the exercise of trying to dress and undress at the same time."[63]

Even assuming domestic support and Soviet likemindedness, how could the "strategic moment" be frozen? ACDA research suggested treaties to outlaw weapons in outer space and missile R & D altogether. A draft treaty prepared in 1962 went so far as to recommend monitoring of all missile production, stockpiling, and testing. A JPL study concurred: the world was best served by stable deterrence, space weapons were a threat, and any new technology had a "built-in multiplier effect as regards inherent instability." Space weapons in particular would have a "further destructive effect on an already depressed world opinion."[64] But such controls on missile technology might go so far as to shut down civilian space exploration as well! The American treaty tabled at Geneva in April 1962 proposed that signatories halt development and testing of new missiles and permit space rockets only under an international monitoring agency. ACDA researchers concluded moreover that space programs would

afford a potential violator more of an opportunity for improving specific subsystems and developing new systems for use in a ballistic missile system than does the missile production program, since the space programs can be expected to be developmental in nature. The space booster program should probably be as closely monitored as the weapons program.[65]

The sad fact was that space technology was almost identical to that of military missiles. Attempts to control the greatest terrors made by man might also shut down the greatest adventure undertaken by man. NASA agreed to permit inspection of its facilities as part of a treaty (how that must have smarted), but thought even inspection inadequate to prevent violations. The DoD agreed to a ban on new deployments but flatly rejected inspection or a shutdown of R & D.[66] In early 1964 Herbert Scoville, Jr., of the ACDA asked NASA to report on which of fourteen anticipated missile developments targeted for a freeze would inhibit NASA programs. They made a menu of the next generation of strategic systems:

1. Hardened launch sites; 2. large booster development; 3. small booster development; 4. solid propellant boosters; 5. improved guidance systems; 6. terminal guidance systems; 7. multiple warheads (MRV); 8. individually guided warheads (MIRV); 9. heavier reentry vehicles; 10. penetration systems improvement; 11. retargeting in flight; 12. simultaneity of MRV arrival; 13. improved fuzing; 14. mobility of launch sites.[67]

NASA replied in distress that ten of the fourteen innovations were definitely or plausibly important to the progress of space technology. Only items 1, 12, 13, and 14 were of no interest to NASA.[68] A JPL report elaborated, warning that "nearly all NASA programs could be inconvenienced in one way or another by such agreements."[69] Competition of all sorts—military and civilian—must live or die together.

Of course, the Geneva talks were sterile. U.S. and Soviet R & D teams pressed ahead quickly toward ICBM innovations that threatened stability far more than the space weapons programs still in their infancy. Whether the Soviets ever accepted assured destruction as anything other than a convenient encouragement to the United States to stand down while they pressed ahead is a live question. But for the time being, given their inferior resources and need to catch up with the United States in missile forces, the Soviets, too, might have been pleased to place space weaponry on the back burner. The American decisions of 1963 did seem to preserve outer space as a sanctuary well into the 1980s. But the wisdom of those decisions will hang on the ultimate efficacy of the civilians' strategy, second-strike deterrence and assured destruction: the balance of terror.

CHAPTER 17

Benign Hypocrisy: American Space Diplomacy

James Webb, in his first pitch to JFK, stressed the need for perceived U.S. supremacy in space if the emerging world order were to conform to American values and interests. He was not alone. Scholars and politicians, journalists and science fiction writers all popularized the notion that space technology would prove a powerful force for peace and prosperity, knitting the nations together as ideologies never could. The task of the United States in this age of "challenge and opportunity" was to prevent the surviving ideological empires, the USSR and China, from extending their sway by force or persuasion, so that neutral nations, and maybe even the Communists, could be folded into a new order by (in Webb's words) "the emerging world force" of new technology.

There was a fly in this mellifluent ointment. For whence came this power to "make the world safe for diversity," as JFK put it? From the same thing the Soviets summoned as evidence of their own superiority—technology again. Rostow's model predicted that societies inclined toward pluralism, but only after they struggled to a certain stage—of technology. Kennedy spoke of our power to cure or kill the world at our current level—of technology. The United States preached freedom and democracy, but the preachments usually got translated in terms of technology—to defend, to prosper, to educate. Ever since Truman, but especially since 1961, the United States turned to the world a profile, in Harvard Professor Stanley Hoffman's phrase, of "technological anticommunism." As Halberstam wrote of the Kennedy men: "Were we much richer than they, and more technological? No problem, no gap in outlook, we would use our technology for them. Common cause with transistors. Rostow in particular was fascinated by the possibility of television sets in the thatch huts of the world, believing that somehow this could be the breakthrough."[1]

The prospects of "doing good by doing well" seemed especially rich in space technology. Communications and weather satellites were tools that fit perfectly the hand of technological anticommunism. Though almost transparent in the glint of Apollo, these space applications, and others soon to come, were to be the global analog of the railroads, the nineteenth-century "wedding bands" of the disparate German states, far-flung Canadian provinces, and the United States themselves. New global technologies must spawn international organizations, in which the very tendency of bureaucracy to stretch its tentacles might prove a force for transcendence of jealous nationalism, for "functional integration" on a multilateral basis, for world peace.[2]

These hopes, by no means mysterious in a world beset by fears of nuclear war, poverty, and strife, nevertheless smacked of millenarianism. For a television in every hut could be a vehicle for education and understanding, or just as easily (and more likely) a tool of propaganda and hatred. But recognition of the damage done if new technology "fell into the wrong hands" made it all the more incumbent on the United States to take the lead and exert a telling influence over its use. Such hegemony, however, was likely to breed resentment abroad unless U.S. cooperation meant real transfers of technology and the raising of other countries to a competence approaching the American. Yet such cooperation would only undermine the leadership by which the United States hoped to ensure a benign technical regime in the first place!

Then there was the USSR. The whole American enterprise rested on the assumption that the Soviets were leading in space. If so, how could they be denied equal influence over space law and international management? If the United States did "leapfrog" the Soviets, or if real cooperation put an end to the race, then how could NASA continue to justify big budgets? If the United States then cut back on space budgets, it would only provide the Soviets and commercially minded Europeans and Japanese with a "window of opportunity" to catch up in selected fields and even smash U.S. hegemony over time. In short, fulfilling the dreams of the globalist space enthusiasts required (1) continued U.S. superiority in most branches of space technology and (2) acceptance of other nations of this superiority, or (3) wholesale transfer of U.S. launch and satellite technology to others. The first would be difficult to maintain, the second contrary to human nature, the third an international act of faith that involved giving away the store.

These facts of life informed an American space diplomacy with Soviets and Europeans alike of cooperation in science and competition in engineering; of agreements in areas where the United States was safely dominant and advocacy of laissez-faire in areas where there was still a race. It all suggests a "benign hypocrisy," a hard-headed approach not quite up to American rhetoric. Yet U.S. statesmen would have been derelict in their national duties had they acted differently.

Space diplomacy in the 1960s moved ahead on four axes: protection of U.S. military space programs, cooperation with the USSR in space arms control and space science, cooperation and competition with European allies, and organization of a regime for the operation of communications satellites.

The first problem, involving military satellites, was a delightful one to have. The Kennedy administration inherited a strategic posture and a reconnaissance capability better than it dreamed possible. But it also inherited the sticky problems of how to cope with the diplomatic niceties of space-based espionage. Should the United States publicize its capability or play it down? Develop antisatellite weapons or try to establish the legality of satellite snooping? In the fall of 1960 the State Department predicted that the Soviets would stiffen opposition to arms control, at least in the short run, because of the spy satellites, and Kennedy and McNamara learned during the transition of possible Soviet reprisals.[3] American inconsistency regarding publicity continued to embarrass, as newspapers reported that Samos would "take photos of every inch of communist territory," or heralded the extra $83 million appropriated for it.[4] Hence in January 1961, National Security Adviser McGeorge Bundy opted for a blackout on all reconnaissance launches. The USAF was henceforth forbidden to refer to satellite observation in public and, in effect, not to admit that such programs even existed.[5]

Space was also declared of "peculiar" concern to Foggy Bottom. The State Department urged recognition of the vast opportunities opened up by space technology and thus a program explicitly "for the benefit of all and with opportunity to cooperate with all." It sought to "implant the image of what the U.S. would like ultimately to see" and take the lead on freedom of space. Of course, peaceful uses of space might seem inconsistent with military uses:

our national security also requires that we undertake in a deliberate and calculated way the research on military space weapons that might become necessary . . . if effective arms control measures cannot be put into force. We must, however, be wary of impelling or appearing to impel an arms race in outer space. . . . Military space programs, even in research and preliminary planning stage, should be played down publicly.[6]

The blackout did not muzzle the trade press or service journals, but daily newspapers and magazines did cease to print news of military space missions. Soon enough they had space news of more moment anyway: Mercury launches, astronauts' wives, a race to the moon. In the UN, the United States supported the requirement for satellite registration, but in practice this meant only that the United States began to submit "blank" designations for classified satellites at the same time the Soviets launched their Kosmos series. The blackout seemed prudent—as State Department legal adviser Abram Chayes explained, "to the extent that reconnaissance satellite activity gets out in public, it forces the Russians

---

MERCURY-ATLAS 6

Ten months after Gagarin's orbital mission, the hulking, steaming, and decidedly dangerous-looking Atlas ICBM took an American on a similar trip. The first two manned Mercuries, Shepard's and Virgil Grissom's, went on up-and-down suborbital flights. But John Glenn was going into orbit. After three tours about the earth, during which he maneuvered his capsule when it drifted off course and fretted out other malfunctions, Glenn brought down his own craft, *Friendship 7*, and landed in the sea 800 miles southeast of Bermuda.

The United States had matched the Soviet feat! And all the suppressed emotion of the Space Age—stirred up by Soviet space shots, Khrushchev's shoe-pounding and rocket-rattling, talk of fallout shelters and strontium-90 in mother's milk, above all perhaps the nagging suspicion that can-do Yankee ingenuity had had its day—all this came tumbling out in a national catharsis unparalleled in the quarter century of the Space Age. Not the first U.S. satellite, or the first flight of the Space Shuttle in 1981, or even the landing on the moon—all occasions for proud and tearful celebration—matched the social release into which John Glenn, after five hours in space on February 20, 1962, incredulously stepped. It seemed that he had given Americans back their self-respect, and more than that—it seemed Americans dared again to hope.

---

to make a challenge of some kind. . . ."[7] In fact, the blackout only underscored the duality of the U.S. program. The Soviets knew what the United States was doing, foreign experts knew, and most world "opinion makers" were eager to assume the worst about U.S. intentions. Only the American public was being fooled.

What was to be done? Secretary of State Dean Rusk and Adlai Stevenson at the UN thought the blackout at best a stop-gap. John McCone of the CIA demanded continued and absolute secrecy.[8] In the midst of such uncertainty, with the United States embarking on the moon race, the administration ordered a major review of U.S. space policy. The incentives were suitably dramatic: the latest eruption of the Berlin crisis, Khrushchev's harangues and resumption of nuclear tests, the need to reassure nervous Europeans . . . and the greatest triumph yet in the civilian space program, the flight of *Friendship 7*.

As Glenn rode into orbit, the State Department, apparently goaded by the Pentagon, hit on a ploy that might mitigate adverse publicity about recon satellites. The UN regulation on spacecraft registration applied only to objects in "sustained orbits," which was held to mean two weeks. This stipulation was meant to "protect U.S. freedom of action . . . to minimize vulnerability to hostile counter-action." It also meant that since film-pack recon satellites stayed in orbit for a shorter period, they might escape official registration altogether. To establish this loophole, the

United States neglected to register Glenn's brief three-orbit flight at the UN! But all this reaped was ridicule, and State gave up the charade.[9] Meanwhile, Kennedy and Bundy had ordered a complete review of U.S. space policy—a full year after the major program decisions had already been made. State's U. Alexis Johnson chaired a task force whose stated goal was to keep "spy sats from being shot down by political action."[10] The NASC also held the Eisenhower policy documents to be "inappropriate and out of date."[11] The result was NSAM 156 of May 1962. The blackout on spy satellites would continue, but State's anxieties would also be addressed: the United States would support a ban on nuclear tests and weapons of mass destruction in space, but "oppose, and not accept any prohibition of military uses of outer space other than the foregoing, except in the case of agreement on a treaty on general and complete disarmament." The document called for further guidelines for the protection of military space activities, while the existence of such activities should in time be "acknowledged publicly" and "explained in terms of the overall objectives of, and necessity for that program. . . ."[12]

Hence the line to be adopted at the UN against Soviet verbal assaults: the United States would admit to having military programs in space, but stress their importance to peace and contrast American candor to Soviet secrecy. Even liberal Senator Albert Gore (D., Tenn.) concurred: "no workable dividing line existed between military and non-military uses of space," and "observation from space is consistent with international law, just as is observation from the high seas."[13]

By late 1963 the hand-wringing ended. As has been seen, the Soviets began launching their own spy satellites, called off their diplomatic offensive, and agreed to principles of behavior in space that, by neglecting to ban military activities, legitimized observation from space. The United States achieved a great deal in having military space systems treated not as a special category but as any other military systems, to be promoted or discouraged not because of the medium in which they operated but because of their contribution to security and stability. And yet, to the uninformed, the United States appeared to be the dangerous power that openly advocated "militarization" of the cosmos.

The second axis of development for U.S. space diplomacy involved cooperation with the Soviet Union on arms control and space science. The former, as noted in the last chapter, had serious implications for the future of space technology. Bombs in orbit, the scariest potential application of spaceflight, made little military sense. Land-based ICBMs were cheaper, more accurate, and more secure than satellite-bombs. It might prove possible to "harden" orbiting weapons against attack, or employ electronic countermeasures and decoys, but all that might spark an expensive arms race with little in the way of a prize. Orbiting bombs would also be subject to accidents during launch and hard to maintain once in orbit. In any case, they could not be simply "dropped" on the enemy, but must be decelerated with retrorockets and precisely guided back to earth. All in all, space was not a stable platform for nuclear artillery. Renunciation of such weapons allowed both the United States and the USSR to polish their images,[14] even as the State Department deflected any UN action to ban military space activities per se.[15] In September 1962, therefore, Deputy Secretary of Defense Gilpatric declared:

Today there is no doubt that either the U.S. or the Soviet Union could place thermonuclear weapons in orbit, but such an action is just not a rational military strategy for either side for the foreseeable future. We have no program to place any weapons of mass destruction in orbit. An arms race in space will not contribute to our security. I can think of no greater stimulus for a Soviet thermonuclear arms effort in space than a U.S. commitment to such a program. This we will not do.[16]

The peaceful resolution of the Cuban missile crisis then led quickly to those pacts that so frustrated the USAF: the Partial Test Ban Treaty, the UN Principles on Outer Space, the ban on orbital weapons of mass destruction. But the continued efforts by the United States to appear as the advocate of "space for peace" while owning up to extensive military operations in space exposed the country to charges of hypocrisy. That such charges were often born of pacifism, anti-Americanism, or just ignorance of the distinction between passive and active military systems was unimportant. The fact was, the Soviets simply lied about their military uses of space and reaped propaganda rewards for decades.

Nevertheless, détente seemed to open prospects for broader U.S.-Soviet cooperation, something to which the United States was triply committed: first, because of the Space Act of 1958 and congressional enthusiasm; second, because the nation's image depended on advocacy of cooperation; third, because Kennedy's advisers and enlightened opinion believed cooperation in space to be a force for global unity. Was it not only natural that man's escape from earth should foster a one-world consciousness? As the astronauts observed, from space the earth appeared as a breathtaking balloon of greens, browns, and blues, streaked with white clouds, with no political boundaries in sight.

The trouble with this lofty approach was that cooperation in space, if it involved giveaways of technology, ran counter to American interests, and was not a strong United States in the world's best interest? In any case, how could cooperation be folded into Kennedy's competitive race policy? The answer was that it could be, if held within strict limits and subordinated to the dictates of prestige. Whether or not significant cooperation were achieved, the United States must be perceived as desiring it. That could be accomplished through scientific cooperation, even as the United States shied from sharing the engineering know-how that was the real basis of space power. The Soviets knew this as well.

Academician Sedov admitted that "cooperation should be in science programs. If we really cooperated on man-in-space, neither country would have a program because the necessary large support in money and manpower was only because of the competitive element and for political reasons."[17] NASA's Director of International Affairs, Arnold Frutkin, agreed that emphasis should be placed on the "objectives of space science rather than the tools of space exploration" and advised the Kennedy administration, as he had Eisenhower's, to build an extensive record of overtures while scolding the Soviets for not responding.[18] Beyond space legal talks at the UN, American initiatives extended to space science, meteorology, and even a 1960 offer to place its Minitrack system at the disposal of the Soviet man-in-space program.

Superpower rivalry ensured that cooperation would emerge only as a function, not a cause, of détente. Yet "visionary" critics still challenged Washington's "conservative" policies for space cooperation. Frutkin in particular was said to exhibit a "hard-headed businesslike approach," while a truly "innovative" world view, according to one academic critic, would stress the inspirational quality of spaceflight and the chance it offered to ease Cold War tensions, spark functional integration, provide a moral equivalent of war, liberalize the USSR, pry open the Iron Curtain, and energize the UN.[19] But it was never clear how an "innovative" program might be realized. Given the divergent booster and spacecraft designs of the space powers, a forced merging of their programs would only burden the engineers and result in slower progress at higher cost. For the most part, Kennedy followed instead the pragmatic line first drawn by Glennan and Frutkin: "Wherever we can find an area where Soviet and American interests permit effective cooperation, that area should be isolated and developed."[20]

Such areas were scarce, but "innovative" rhetoric still found its way into public pronouncements, raising expectations that could not be fulfilled. Kennedy's inaugural address ("Let both sides seek to invoke the wonders of science instead of its terrors. Together let us explore the stars. . . .")[21] and State of the Union message ("This Administration intends to explore promptly all possible areas of cooperation . . . [and] remov[e] these endeavors from the bitter and wasteful competition of the Cold War")[22] were just exercises at image building, given the known barriers to collaboration and Soviet truculence. Indeed, the actual fostering of cooperation was less important, according to Webb, than "the effort to project an image of the U.S. as a nation leading in this field and willing to share this knowledge with other nations."[23] As usual, RAND analysis had anticipated U.S. policy, listing the following goals for cooperation in space in 1961: (1) acquisition of tracking stations; (2) collaboration in data collection; (3) sharing of costs on expensive missions; (4) political gains through sharing of prestige; (5) evidence of peaceful intentions; (6) an international forum to ascertain the plans of others and spread the truth about our own; (7) a model for cooperation in other fields; (8) forestalling restrictive controls on unilateral activity; (9) aid to developing countries; and (10) opening up of Communist bloc.[24]

So U.S. policy was not insincere but was overstated, and thus encouraged Americans and foreigners alike to view national space programs as manifestations of dangerous rivalry rather than hearty incentives to technological progress. Once Cold War fears subsided, citizens were naturally led to ask: if we cannot cooperate in space, why do it at all?

Nonetheless, the United States *was* more forthcoming than the USSR. At the Vienna summit Kennedy even spoke again of going to the moon together, an idea buried in a long list prepared by the PSAC but the one most likely to "win points" in world opinion.[25] Khrushchev replied that cooperation was impossible before disarmament. In his note of congratulations after the Glenn flight, Khrushchev mentioned "pooling our efforts," and NASA quickly drafted plans for a global weather satellite system, tracking stations on U.S. and Soviet soil, comsats, and exchange of scientific data.[26] But the Soviets refused to recognize the legality of Tiros weather satellites and shrank from a global comsat system that might break their monopoly of information in the East bloc. Khrushchev responded with his own list, but cautioned that "[c]onsiderably broader prospects for cooperation and uniting our scientific-technical achievement . . . will arise when agreement on disarmament has been achieved."[27]

Meetings followed between NASA's Dryden and the Academy of Sciences' Anatoly Blagonravov. The range of agreement eroded steadily through concessions to reality and Soviet backpedaling until a Memorandum of Understanding of June 1962 included only an exchange of meteorological data and a communications test using the *Echo 2* satellite. But the Cuban crisis overshadowed even this little initiative. Khrushchev never publicly acknowledged the accords, and Kennedy expressed disappointment in them.[28] Frutkin summed up:

> They provide for coordination rather than integration of effort, in other words for a kind of arm's length cooperation. . . . No classified or sensitive data to be exchanged. No equipment . . . no funds are to be provided by either side to the other. NASA is not required to embark upon any new programs or to modify existing programs.[29]

Had the Soviets been amenable to programs that did include classified data or exchanges of hardware, what were the chances the United States would have been willing to take part? Most NASA officials, not to mention the Pentagon, probably sighed with relief when the Soviets balked at "exploring the stars together." Considerations of economic and military security were already demonstrating that the United States was unlikely to give away technology even to its friends, much less its rivals.

The third and fourth axes of U.S. space diplomacy involved not the

Soviet Union but Western Europe. NASA had made clear that cooperation was impossible with states that had nothing to contribute. Thus the main candidates were limited to the United States' industrial allies. By the early 1960s they had begun to take steps to make themselves *bündnis-fähig*—worthy of being a partner—through national and European-wide efforts to assimilate the basics of space technology. But French, British, or European space programs only meant duplication of American expertise in the short run and competition in the long run. To be sure, if the Europeans restricted their demands on the United States to scientific payloads launched for hire by NASA rockets, then genial cooperation could proceed. But if—as would soon happen—European governments began to shift their efforts toward commercial applications such as comsats, then the United States would be undercutting its own position by transferring technology. European and French efforts to build their own space boosters raised even more sensitive questions. U.S. aid in that field might contribute to proliferation of nuclear missiles and complications within NATO.[30] In the end, U.S. policy toward Europe came to resemble that toward the USSR: cooperation in science, decided aloofness in engineering.

Space science pleased scientists; it did not please European aerospace firms eager to break the American monopoly in boosters and satellites, or European parliaments looking for economic and political returns from the money they voted for space. The limits of U.S. cooperation, and the apparent necessity for Europeans to make common cause in the face of U.S. superiority, were revealed most clearly in the dawn of the age of satellite communications.

Arthur C. Clarke joked that he gave away a multibillion-dollar idea for forty bucks in 1945. It was then that the eminent science fiction writer published his prediction of global communications systems based on satellite relays orbiting at the incredible altitude of 23,000 miles above the equator. Such geosynchronous comsats would rotate with the same angular velocity as the earth itself, hence would remain fixed above the same point on earth and have line-of-sight to a third of the earth's surface.[31] As Clarke admitted, however, the comsat's true parents were the engineers, including John R. Pierce of Bell Labs,[32] whose integration of rocketry, transistors, computers, and solar cells made the comsat a reality. By 1960 basic research on three types of comsats (low-orbit passive, a simple reflector; low-orbit active, an amplifier and repeater; and the geosynchronous active) had advanced to the point where policy decisions were necessary. RCA, AT&T, GE, and others angled for government support, while Hughes Aircraft designed a brilliant spin-stabilization system for geosynchronous comsats and began development at its own expense.[33]

NASA took an early interest in comsats, even though state-funded R & D in a commercial field violated the pattern of the American

ARIEL AND ALOUETTE

At a COSPAR meeting in March 1959, six months after the birth of the agency, NASA officials announced their willingness to cooperate with other nations in space. By September NASA delegates were touring Europe to explain the principles and restrictions of their policy. One of the rules of the game was that the United States would deal only with formal, governmental space agencies. The British accordingly worked through Sir Henry Massie of their National Committee on Space Research and concluded with NASA the first agreement for joint space exploration. In April 1962 the deal was consummated when a Delta booster, emblazoned with the Stars and Stripes and Union Jack, fired the *Ariel UK-1* satellite into orbit from Cape Canaveral. The satellite shell was American, the experiments inside designed by scientists from three British universities. They provided the first "topside" view of the effects of solar radiation on the ionospheric environment. *UK-2* followed in 1964, setting a pattern of friendly and efficient bilateral cooperation in space science.

The Canadian-built *Alouette 1* soared into polar orbit from Vandenberg AFB atop a Thor-Agena in September 1964. The Canadian satellite measured hourly electron densities in the ionosphere. U.S. and Canadian scientists quickly agreed on a series of four more spacecraft to study the ionosphere during the period of maximum solar activity. Italy got into the act with *San Marco 1* in 1964, and many other foreign experiments piggy-backed on American scientific satellites. The United States acquired its desired reputation as a fair and dependable provider of launch services for other nations, providing they restricted themselves to space science and released their data to all the world. In areas removed from strategic technology, the United States lived up to its principles of cooperation and openness in space.

communications industry.[34] A special assistant to the NASA administrator, Robert G. Nunn, reported his findings as follows: comsat systems must be international; many government agencies would have jurisdictional claims; the government must decide whether to fund comsat R & D through NASA or leave it to the private sector; a national policy of some kind was now mandatory. Why? Because the world was in the midst of a communications explosion, with overseas telephone calls alone growing at 20 percent per year. RAND estimated that a low-orbit comsat system would relieve the glut for about $8,500 per channel per year, or less than a third the costs of oceanic cables.[35] Comsats would be big business—and a shiny tool of U.S. foreign policy.

Eisenhower gave NASA the go-ahead on R & D, but he and Glennan saw no reason to change the pattern of privately financed carriers competing with each other.[36] Ike directed only that government and private enterprise work together for early establishment of a commercial comsat system.[37] NASA quickly agreed to a joint demonstration with

AT&T called Telstar, but invited open bids on another system called Relay. This policy might conceivably have ushered in two or more competing comsat systems regulated by the Federal Communications Commission (FCC). But it was not to be. For President Kennedy, in the same speech in which he aimed for the moon, requested $50 million more for comsat R & D, then turned again to the NASC and Vice President Johnson for means to satisfy his desire for "the earliest, optimal use of comsats" serving all the world, especially developing nations.[38] NASC staffers consulted NASA, the DoD, State, the AEC, Justice, the FCC, the BoB, and PSAC, papered over their differences of opinion, and presented a unanimous report to the President in July.[39] A week later Kennedy announced his policy. There was to be a single global comsat system; the U.S. share would be privately owned, but on the condition that the system be made operational as soon as possible, serve regions of the earth that were not profitable, encourage foreign participation, allow nondiscriminatory access, preserve competitive bidding for contracts, comply with antitrust regulations (!), and develop an economical system reflected in low rates. On its side, the government would conduct R & D, supervise international agreements, control launches, give the system its own business, supervise radio frequencies, and provide technical assistance to developing countries. Kennedy invited all nations to join "in the interest of world peace and closer brotherhood among peoples throughout the world." In fact, the need for speed was motivated in part by the desire to preempt the Soviets, and in part by the urgent needs of the military.[40]

How could all this be accomplished? The FCC, taking its jurisdiction for granted and perhaps promoting the interests of the giant common carriers (hence its own regulatory power), recommended a single comsat corporation operated by AT&T, ITT, RCA, and Western Union—that is, a "common carriers' common carrier." Aerospace firms howled that this plan meant vertical integration in the new industry, shutting out other potential suppliers of space hardware. The Justice Department agreed, but the carriers backed the FCC plan, promising to divide up the board of directors, facilities, expenses, and receipts according to their respective investments.[41]

These complex issues meant another headache brought on by state-funded R & D. Should comsats be set up as a government utility, befitting the public investment, or as a condominium of existing carriers, or as a privately owned chartered company? A state-owned system meant "creeping socialism," but an outright gift of comsat technology to private carriers meant politically sanctioned monopoly. Giving comsats to the common carriers was also hazardous, since they were the owners of the oceanic cables with which comsats were to compete. By the time the administration bill hit Congress in February 1962, sixteen pieces of legislation had been introduced, ranging from Senator Kerr's, siding with the carriers, to Senator Estes Kefauver's (D., Tenn.), stipulating state ownership. Again the arguments recalled the debates over atomic energy in 1946 and NASA's patent policy in 1958.

Kennedy's bill, justified by the need for speed, sired a new animal in American political economy: a chartered company of the sort founded by European princes in the age of the mercantilist power-state. But as the proposed Comsat Corporation was in fact a government-created monopoly, one could not in good conscience limit its ownership. Hence the new company would be capitalized by Class A stock sold to the public at not less than $1,000 per share and Class B stock available only to existing carriers. The latter conferred no voting rights or dividends, but would be included as an expense in the companies' rate base for other international services. No investor was permitted to own more than 25 percent of Class A stock or to control more than two members of the board.

No legislation during the Kennedy years sparked a firestorm of the sort that seared the Senate during debate on this Communications Satellite Act of 1962. Kerr assumed the role of administration advocate, his pro-carriers bill apparently a ploy to make Kennedy's seem moderate.[42] His most powerful witness was none other than the executive vice president of AT&T, who argued that the big carriers alone had the expertise and reputation abroad needed to instill confidence. Indeed, he said, comsats themselves were only an alternate means of transmitting signals from one continent to another and were worthless until integrated into the existing telephone networks owned by the carriers (here he produced an impressive map of the global Bell system). Why should there be "broad public ownership," since the carriers had agreed that no firm would dominate? If citizens wanted a piece of the space action, let them buy stock in AT&T or ITT. What was more, he hinted, the whole affair might not prove profitable. Large investments must be made, risks taken, and stock issues might be undersubscribed and the carriers left holding the bag. He wanted abolition of Class A stock.[43]

Senator Kefauver manned the opposite salient. But far from standing up for competition against the carriers, he urged complete government control! The battle over the comsat act was not one between free enterprise and a looming technocracy, but between styles of technocracy. Turning over all the technology developed at public expense, said Kefauver, amounted to an unprecedented giveaway. And to whom? To heretics who say that comsats are nothing more than another way to perform existing functions: "If there is, in fact, a new frontier today, it is the frontier of space." Kefauver urged senators to support a bill for a public Comsat Authority cosponsored by six fellow liberals. But to turn comsats over to AT&T was "no more appropriate than defining free enterprise like the elephant dancing among the chickens, who shouted, 'It's every man for himself.' "[44]

Neither AT&T nor Kefauver got his way. The Kerr Committee reduced the price of Class A stock, enlarged the board of directors to include three appointed by the President, struck out public auditing of the books and State Department control of foreign negotiations—and reported the bill out favorably. Kefauver took his brief to the Commerce Committee, where he denounced the gift of $685 million (through FY 1963) of NASA and DoD R & D.[45] Frustrated there, he then called for hearings in his own subcommittee even though no one had referred the bill to it. When the administration issue hit the floor of the Senate, liberals turned to other tactics, first loading the bill with amendments (each tabled by the chair) and finally resorting to filibuster. When all efforts at pacification by the majority leader proved fruitless, the Senate reluctantly moved for cloture—for the first time since 1927. The comsat act passed both houses and was signed by the President on August 31, 1962.[46]

One can only guess at the consequences had a free market in comsat technology, or outright state ownership, prevailed. But the "neither fish nor fowl" solution that did exemplifies the domestic adjustments forced by the growth of state-funded R & D. The comsat act met Kennedy's desiderata, themselves a function of the race for prestige, and the Comsat Corporation proved a remarkable success. AT&T's financial worries were chimerical: by the end of 1963 it was clear that only $200 million was needed, while the stock that went on sale in May 1964 at $20 per share was the most *over*subscribed issue in the history of Wall Street. AT&T netted 29 percent (it tried for 40) and ITT 10.5 percent. In time, AT&T even spent $70 million to lay another transoceanic cable.[47]

The Comsat Corporation's very *raison d'être* was to set up an international comsat system as soon as possible. Yet when it confronted foreign powers, it adopted the same policy as NASA regarding cooperation: give up no vital engineering expertise and conclude agreements that served U.S. national interest. In fact, Comsat Corporation directors initially wanted to work through many bilateral agreements, but the State Department, sensitive to the Europeans' mood and the complications of many separate pacts, pushed a multilateral plan.[48] The Europeans themselves decided the issue by forming the European Conference of Post and Telecommunications Administrations, the better to win leverage against the giant carriers that made up the American comsat entry.

The first priority, allocation of frequencies, inspired an extraordinary gathering of the International Telecommunications Union in Geneva. There the Soviets gave the first indication of their lack of interest in a truly global network by opposing allocation of any slice of the spectrum for the exclusive use of comsats. The Soviets also insisted that the ITU, with its large bloc of Communist members, regulate all space communication. Why would anyone, the Americans retorted, invest in comsats if their very freedom to transmit was subject to the whim of a politicized international bureaucracy? Non-Communist members then secured a

"Say, How Come You Don't Like This Guy?"

From *Straight Herblock* (Simon & Schuster, 1964). Originally appeared in the *Washington Post*, August 15, 1962.

frequency band sufficient for the next decade.[49] But Khrushchev showed no interest in a system in which the USSR would have a small minority interest and the United States a monopoly of technology; he denounced the planned consortium as a capitalist tool.[50] While Comsat Corporation was pleased not to have to deal with Soviet obstruction, negotiations with the Europeans revealed that even those nations prepared to participate would not do so in a spirit of deference and gratitude. In this, at least, AT&T executives were right: satellites did not inspire global fraternity just because they were in outer space.

In July 1963 a full-fledged European Conference on Satellite Com-

TELSTAR

One of the biggest hit records of 1962 was an instrumental by a British pop group called the Tornados, whose muted electric guitars and organ echoed the rolling, upbeat sound of the California surfer bands. This recording began with a roar—like rockets at lift-off—that blended into ascending chords, ever higher, ever more strained, as if fighting gravity in the quest for some great perch, the serenity of orbit. The last chord, highest on the scale, then revealed itself as the first tone of a relaxed and haunting psalm of achievement, but also of yearning—in a minor key. They named it "Telstar."

In July 1962 the United States had answered the Soviets with the orbital flights of Glenn and Scott Carpenter. The race was not over, but for the first time since Sputnik it all seemed less fearful. The wonder of spaceflight shouldered in beside Cold War emotions, and somehow a modest test comsat came along at this moment to pluck imaginations like no other unmanned satellite of the decade. *Telstar!* Built by AT&T, launched by NASA, it broadcast the first live television between continents and symbolized like nothing else the potential of space technology to unite the world. A mere thirty-five pounds, only three feet in diameter, it was the first active repeater, the first satellite of any kind built with private funds, and the first to transmit hundreds of communications between ground stations in Maine, New Jersey, Britain, France, Italy, and Brazil. It made the sort of impact that the administration hoped for from the entire comsat enterprise.

Two cousins of *Telstar* soon joined it in orbit. RCA and NASA launched *Relay 1* in December, an active repeater with built-in redundancy. The experiment was timely, not only because backup systems were a needed hedge to make comsats commercial, but also because *Telstar* itself was knocked out after six months by radiation from the American high-altitude nuclear tests.

The greatest of all were Hughes Aircraft's brash Syncom prototypes of the geosynchronous active comsat. Only three such birds could do the work of dozens of low-orbit satellites. True, they needed more powerful amplifiers, but their lofty orbits (one-tenth of the way to the moon!) meant that their charging solar cells spent little time in the earth's shadow. Tracking was also a breeze with a satellite that never seemed to "move." *Syncom 1* had electronic problems, but *Syncom 2,* boosted into geosynchronous orbit by a Thor Delta in July 1963, was a total success. The dreams of Clarke, Pierce, and the Hughes engineers came true.

munications huddled to plot strategy vis-à-vis the Americans. When the two sides met in early 1964, the French and British dragged their feet—they, too, after all, had to worry about their oceanic cables—until the Americans threatened to deploy a trans-Atlantic system with or without them.[51] The problem was that ownership of the system would reflect national usage as well as technological contributions. Either way the

United States claimed a massive majority interest. Europe insisted on a certain percentage of capitalization and control and that the initial convention be of limited duration so that in time it might break the U.S. monopoly. But for now it had no choice but to swallow the pill. In Britain's view, ". . . the only way of preventing an American monopoly . . . is to join a partnership . . . and so secure the right to influence the course of events."[52] At London in April it was agreed that the Comsat Corporation should have a 61 percent share of the international system, the Europeans 30.5 percent, the rest to Canada, Australia, and Japan. Other nations would be welcome, but the U.S. share must not fall below 50.6 percent. On important issues a two-thirds vote was required, but the United States still had veto power. The Europeans wanted an interim agreement lasting three years, the United States ten; they settled on five. Both sides agreed not to deploy competing systems. Finally, management of the system was vested in the only body able to do the job—Comsat Corporation.

The International Telecommunications Satellite Consortium, or INTEL-SAT, was initialed on these terms in July 1964.[53] U.S. policy, as defined by these objectives, prevailed: (1) develop a global system quickly; (2) realize the resulting economic and technical benefits; (3) enhance national prestige; (4) strengthen U.S. relations with developing nations.[54] Goals 1, 3, and 4 evinced the same spirit that informed Apollo: do something great in space, do it before the Soviets, and aim it in part at the Third World. As Senator John Pastore (D., R.I.) crowed over the first INTELSAT spacecraft in April 1965: ". . . *Early Bird* is orbiting and is controlled by the Comsat Corporation and can become the show window through which America can and will be seen throughout the world. And indeed it will be the mirror of our image."[55] Goal 2, however, struck a discordant note. The United States did reap the commercial benefits of its leadership: Comsat Corporation both managed and had majority interest in INTEL-SAT; the U.S. share of contracts for its first generation of satellites was virtually 100 percent; granting Europeans a politically determined share of the business would only mean slower and costlier development in the short run, outright subsidy of a competitor in the long run. But this very dominance in a growing market and exploding technology galvanized the Europeans into raising a challenge to the United States in the future. Indeed, before the ink was dry on INTELSAT, the Europeans determined to do precisely that. Competition, not cooperation, would define the relations of states even in this most "uniting" enterprise of all.

All told, the American crusade on behalf of "space for the benefit of all mankind" cloaked diplomacy that was self-interested, and perhaps necessarily so. If liberal senators fulminated against a "giveaway" of federal technology to private industry, how much greater would the outcry have been against a giveaway to foreign competitors, not to mention the Soviets? The inconsistency in U.S. space diplomacy lay in its treating the space program, and comsats in particular, as a glorious advertisement of American virtues. The Soviets were hardly going to be budged by such imagery, the Europeans knew better, some Third World leaders realized that space technology was irrelevant to their needs, and others indulged in exaggerated expectations of what the United States and space could do for them. The only customers for the unrealistic rhetoric, the only ones who may have believed in it, were the American people themselves.

In substantive ways, U.S. diplomacy was quite successful: the legitimation of spy satellites, INTELSAT, the UN Resolution on principles governing the use of space. But the image of American space programs as open and altruistic and able to spread brotherhood and prosperity to a world tempted by communism amounted at best to a benign hypocrisy. In later decades, after the U.S. monopoly in space was shattered by multiple competitors and Third World blocs were agitating for massive transfers of technology from the "exploiters," U.S. diplomats would silently rue the image making of the nation's first decade in space.

CHAPTER 18

Big Operator: James Webb's Space Age America

Lyndon Johnson, Speaker Sam Rayburn, Congressmen Albert Thomas and Olin Teague of key committees—Texans all; Bob Kerr, James Webb—both Oklahomans. Clichéd though it be, the immense tablelands on either side of the Red River do nurture big thinkers and doers. A Walt Disney comic book of the early 1960s hit the mark when it depicted Scrooge McDuck as a self-satisfied Big Operator, investing billions to develop the oil of some generic Middle Eastern desert and then, as an afterthought, to pipe in water and turn the desert into a garden for a tribe of midget Third World ducks. Thousands of miles of gargantuan pipeline, huge earth-movers, the latest technology arrived in convoys of merchantmen, while Donald Duck and his three nephews tagged along in dumb admiration, hoping for the chance to prove they, too, had the stuff to be Big Operators.

Jim Webb, the Carolina lawyer transplanted to the Okie oil fields, was a budget-cutter for Truman and suspicious enough of creeping socialism to slash funds for the TVA. As NASA chief, he was fond of recalling that misstep to balky congressmen, for he had been quickly brought to see that it was a false economy. An accomplished administrator, he then went West and learned, in his years with Kerr-McGee Oil and McDonnell Aircraft, the virtues of thinking big. By 1960 he had also learned the potential of progressive big government from his service on the Ford Foundation's commission on urban problems, the Educational Services Inc. Commission on post-Sputnik science curricula, the board of the Oak Ridge Institute of Nuclear Studies, and the National Cancer Advisory Council. Webb was not a scientist, but a lawyer and manager who administered the power inherent in applied science. Once given the astounding task of organizing an expedition to the moon, Webb must have felt that, improbable though the whole thing was, few men were as able as himself to tackle it. He was a Big Operator.

Apollo is often dubbed a peculiarly "American" enterprise.[1] But why? Great treks and explorations have been a part of many nations' history;

obsession with prestige has afflicted most Great Powers, not to mention big management of resources by government. The uniquely American flavor of Apollo, rather, smacks of mesquite: not just the big capitalism of Scrooge McDuck, or just the big government of technocratic social planners, but an amalgam of the two, infused with the spirit of the Southwest frontier. Here was limitless space, limitless opportunity, limitless challenge, a land where giants and giantism could live without trampling the liberty of the little. A Johnson or a Webb did not see a conflict between technocracy and freedom, between the expansionist state and the striving individual. In America's vastnesses they could coexist and feed on each other. The activist state fulfilled the individual through education, welfare, incentives, new technology; the individual, in turn, lent his ambition and skills to the collective effort. Apollo never lacked for analogies, but to the Big Operators who made it, the moon program was less like the voyages of Columbus or even the privately built railroads than it was like the Panama Canal, the TVA, or the magnificent interstate highway system. Like these earlier mobilizations of men, machines, and mortar under federal direction, Apollo would open up new realms for the individual in stimulation of the economy and elevation of the human spirit. What was more, the space program grew on its own movers and shakers until it outgrew the space race itself and seemed a model for a society without limits, an ebullient and liberal technocracy—not Space Age communism, but Space Age America.

Webb inherited a small federal agency composed for the most part of idiosyncratic scientists and engineers—brilliant or competent or merely eccentric. His total manpower in early 1961 was 6,000. Thanks to Apollo, the agency grew tenfold, but it still contained only a tenth of the personnel working on the space program. The other 90 percent were in private firms under contract. They doubled each year to reach a peak of 411,000 in 1965. NASA appropriations rode the same graph: less than a billion dollars in FY 1961, they doubled and redoubled to reach $5.1 billion in 1964.[2] But perhaps the immediate 1961 shock was the hardest to absorb, for then NASA received an overall increase of 61 percent in Webb's first fourteen weeks on the job.

The flux inherent in such growth reinforced other currents of instability: about how to get to the moon, how to make sound technical decisions in advance of hardware design, how to enumerate the tasks and steps (literally millions) to be trod from the first brainstorming sessions to the landing on the moon, how to divide up and coordinate the labor of NASA's several centers, contractors, subcontractors, and university labs. Glennan had opted for a small in-house capacity and reliance instead on the aerospace firms. Webb seconded this motion and thus gave the contractor system a permanent lease on life. Glennan considered it a lesser evil, but Webb made a virtue of it. NASA was to be a model of

1: Konstantin Tsiolkovsky (1857–1935), the "pragmatic dreamer" whose pioneering theoretical and mathematical formulations earned him the epithet of "father of cosmonautics." In the last decades of tsarist Russia, revolution and rocketry first converged in the personality and writings of Tsiolkovsky, an antetype of Bolshevik technocracy to come. *Courtesy of Sovfoto.*

2: Yuri A. Gagarin (1934–1968), the first man in space (*left*) and Sergei P. Korolev (1907–1965), the man who launched Sputnik and the Space Age in 1957. After decades in and out of Stalinist prisons, Korolev continued to endure enforced anonymity even after becoming the Chief Designer of Soviet rocket programs. *Courtesy of Novosti Press Agency.*

3: The rocket that launched the Sputniks—"Ol' Number Seven"—on the pad at Tyuratam. It was also the world's first test ICBM, but was not put into military production as Khrushchev claimed during the "bluff." Instead, it became the basis for the world's most prolific space booster. *Courtesy of TASS from Sovfoto.*

4: (*right to left*) Soviet Premier Nikita S. Khrushchev (1894–1971) and Leonid I. Brezhnev (1906–1982) talk to orbiting cosmonaut Valery F. Bykovsky during his *Vostok V* mission in June 1963. *Courtesy of TASS from Sovfoto.*

5: The first rocket test from Cape Canaveral, July 24, 1950. The missile is a captured German V-2 with a small WAC-Corporal second stage perched on top. The launch was the seventh in the Army's Project Bumper (based at White Sands, New Mexico), but the first conducted at the USAF Eastern Test Range, Florida. *Courtesy of U.S. Air Force.*

6: "Flopnik," "Kaputnik," or "Stayputnik": the humiliating failure of Vanguard, the Naval Research Laboratory's and Martin Aviation's satellite booster. On December 6, 1957, instead of launching the first American satellite, TV-2 settled back on the pad in a fiery cloud. Vanguard's later successes never undid the adverse publicity from this most disappointing space countdown. *Courtesy of NASA.*

7: (*left to right*) Hugh L. Dryden, President Dwight D. Eisenhower, and T. Keith Glennan after the White House swearing-in ceremony for Glennan as the first Administrator, and Dryden as the first Deputy Administrator of the new NASA, August 19, 1958. *Courtesy of NASA.*

8: An Atlas-Agena configuration blasts off from the USAF Eastern Test Range, Cape Canaveral, on May 24, 1960. The cargo was the *Midas 2*, the first test satellite to orbit in the USAF program designed to provide infrared early warning of foreign missile launches. *Courtesy of U.S. Air Force.*

9: (*left to right, facing camera*) President John F. Kennedy (1917–1963), Vice President Lyndon B. Johnson (1908–1973), and NASA Administrator James E. Webb (1906–) converse with Mercury astronauts (*backs to camera*) at the NASA launch complex, Cape Canaveral (later to be named the Kennedy Space Center), in 1962. *Courtesy of NASA.*

10: "Earthrise," the sublime vision of home that greeted *Apollo 8* astronauts Borman, Lovell, and Anders as they completed mankind's first journey around the moon on Christmas Day, 1968. This "holistic" view of our planet as a beautiful blue but fragile bubble in the black and barren sea of space became an icon of high-tech space enthusiasts and anti-tech environmentalists alike. *Courtesy of NASA* (in more ways than one).

11: "On the Moon." The first footprints on another celestial body, placed there by Neil A. Armstrong and Edwin F. "Buzz" Aldrin on July 20, 1969. The American flag, made of rigid material since no lunar breeze existed to make that "star-spangled banner yet wave," was a testimony to both the will and expertise of the American people, but also to the rout of "progressives" who predicted or hoped for internationalization of space programs. Well into the second quarter-century of spaceflight, competition remains the major fillip for national technocracies to spend large sums on space development. *Courtesy of NASA.*

12: "First Look," an ink-on-paper creation by artist Mitchell Jamieson on commission from NASA. It is said to depict an astronaut's awe upon gazing for the first time into space. The portrait seems rather to capture the stark terror of a civilization whose very technical glory has finally forced it to confront the Unending and Unknowable. Moses was permitted to peek only at the backside of God— is this what God would see if He let us look Him in the face? *Courtesy of NASA.*

13: "First Steps," an acrylic-on-canvas impression, also by Mitchell Jamieson, of astronaut Gordon Cooper leaving his Mercury capsule. He seemed "larger than life," like "one of the gods of Olympus whose real home was in the skies." And yet, Jamieson's almost cubist representation captures the fate of technocratic man, blending into machine, fading into his manufactured surroundings, while the machines in turn become more like men. *Courtesy of NASA.*

14: "The New Olympus," an oil-on-canvas allegory by artist Alden Wicks, commissioned by NASA. It depicts the half-completed Vehicle Assembly Building at Cape Canaveral, the largest enclosed space in the world, as a "suitable temple for the new race of gods," with a storm raging within and Renaissance statues without. Systems integrator Dean Wooldridge sees the human being as nothing but a machine—are the statues there to remind us, like museum pieces, of how we once, long ago, used to look to ourselves? *Courtesy of NASA.*

integration, a team effort drawing confidently on government, industry, and university. But the politico-philosophical justifications came later. At first, Webb had merely to get on with the job.

Two weeks after the May 25 call to arms, NASA organizational experts recommended reforms to accommodate the lunar mission. A manned spaceflight center must be established, ably staffed, and coordinated with the whole—it would, after all, absorb more funds than any other center except Marshall, the home of von Braun's rocket team. Liaison among the centers and headquarters must be tightened, for no more would NASA serve up a smorgasbord of projects each associated with one center. Everybody must have a piece of Apollo, hence center directors should report not to the program directors, whose interests were parochial, but directly to Associate Administrator Robert Seamans, whose new programming office could then oversee the entire budget and promote functional integration among all teams in the field. But internal organization paled in comparison to the national. The immediate and primary concern, according to the June study, was NASA's dependence on industry, the DoD, and the universities.[3]

Heretofore the manned program (Mercury) had resided in the Space Task Group at Langley. After the acceleration NASA was authorized to construct a massive Manned Spaceflight Center for the tasks of managing spacecraft R & D, training astronauts, and directing flight operations. In late 1961 the Army Corps of Engineers broke ground near Rice University in Houston. Also that summer Cape Canaveral was chosen as permanent launch site, and 80,000 more acres acquired on Merritt Island, Florida, astride the Banana River. Moon rockets would be so large that they could only be assembled near the point of launch. But the Florida coast was subject to heavy rains and salt air. So in addition to the gargantuan gantries and creeping transporters to move the Saturns to the pad, there arose on the cape the largest enclosed space in the world, the Vehicle Assembly Building. In September NASA elected to renovate an unused government plant near New Orleans for use by Saturn contractors, and so the Michoud, Louisiana, facility was born. But engines had to be test fired. The Marshall center at Huntsville had neither big enough test stands nor the remoteness from populated areas to accommodate the clusters of F-1 engines that would roar to life in a few years. Hence the Mississippi Test Facility, near Michoud at Bay St. Louis.

By the end of 1961, when Congress was still in a "runaway mood," when Johnson ran the NASC, Webb ran NASA, Kerr the Senate Space Committee, Overton Brooks and Olin Teague (D., Tex.) the House Space Committee, and Albert Thomas (D., Tex.) the Appropriations Committee, the Apollo complex took the shape of a crescent moon running around the Gulf of Mexico from Texas to Florida. There were sound reasons for it: year-round warm weather, deep-water transport for the big rockets (imagine the Soviet difficulties in moving big rockets on rail beds across soft country), and contiguity. But the political aspect is undeniable. Apollo was the largest single civilian project in history, and the rolling of the pork barrel began the instant Kennedy approved the Webb-McNamara report. On May 23, 1961, when Webb wrote LBJ that the moon message "will go up on Wednesday," he got down to business without breaking for a paragraph. "Considerable interest has been expressed in this program by members of the Congress," he began, "following your consultations with them, and as I have followed up." Congressman Thomas had "made it very clear that he and [Rep.] George [E.] Brown [D., Cal., on Space Committee] were extremely interested in having Rice University make a real contribution to the effort . . . some 3800 acres of land had been set aside by Rice for an important research installation." Apollo research had to be done somewhere, and, Webb continued, "we have looked carefully at Rice, and at the possible locations near the Houston Ship Channel. . . . I believe it is going to be of great importance to develop the intellectual and other resources of the Southwest in connection with the new programs which the Government is undertaking."[4]

Texas was especially attractive, given that Lloyd Berkner, who had pressed the Space Science Board to endorse Apollo, was establishing a $100-million research center in Dallas, while Senator Kerr and his "interested parties in the Arkansas, White, and Red River systems," were pushing to open the whole Arkansas/Oklahoma area and develop potential for Mississippi. The Dallas-Houston axis "would provide a great impetus to the intellectual and industrial base of this whole region." Webb then imagined a national complex including centers of gravity

in California, running from San Francisco down through the new University of California installation at San Diego [LaJolla], another center around Chicago with the University of Chicago as the pivot, a strong Northeastern arrangement with Harvard, M.I.T. . . . some work in the Southeast perhaps revolving round the research triangle in North Carolina (in which Charlie Jones as the ranking minority member on Thomas's Appropriations Subcommittee would have an interest), and with the Southwestern complex rounding out the situation.[5]

Ex-Administrator Glennan guessed what was afoot. When his governor asked him to help prepare a proposal to put the Manned Spacecraft Center in Ohio, Glennan laughed: "You know, I suppose that there are 25 states doing just this at the present time, and I'll lay you a year's salary that that Center is going to Houston."[6]

Similar considerations informed the choice of industrial contractors for Apollo-Saturn. In this, California congressmen took special interest, given their state's dominance in the aerospace industry. Kerr, the *eminence grise* of the Apollo acceleration, and Representative George P. Miller (D., Cal.), who assumed the chair of the House Space Committee after Brooks

Reprinted from the *Times-Picayune/The States-Item,* August 25, 1963.

died in September 1961, were especially interested in the North American Aviation plant at Downey, California. In July NASA briefed 300 companies on the tasks ahead and in September invited a few of the largest aerospace firms to bid on the Apollo command module. NASA then tested the five proposals according to complicated quantifications based on cost, strength of design, capabilities existing in the firm, and so on. Martin won; the contract, awarded in October, went to North American. Webb and Dryden explained that overall corporate resources and past performance tipped the balance, but muckrakers attributed it to the ostensible influence of Kerr, North American lobbyist Fred B. Black, and the shady crony of Kerr and LBJ, Bobby Baker. Kerr received promises from Black that benefits would be forthcoming for Oklahoma, Baker was granted a million-dollar concession for vending machines in North American plants, and Black, unrepentant, recalled: "North American was five times as large as Martin, so when they got the contract you had five times as many happy people as you had unhappy people. That's democracy in its finest form."[7]

That political plums ripened on NASA's bursting branches should not surprise.* Interesting congressmen in an expensive, discretionary project is not only smart but mandatory. These plums were special, however, in two ways. First, the NASA of Apollo was a creature of the image makers—not through the fault of its own excellent engineers but through political imprimatur. The agency was supposed to provide the nation with an example of efficiency and innovation that was above politics. It was to show that free people could pull together and outperform the totalitarians at their own game. NASA was to be clean, technically perfect, and meritocratic, the bearer of a myth. What is surprising about the agency's performance through the years is that gap between image and performance was not wider than it was.[8]

The second idiosyncrasy of the NASA plums was that Johnson, Webb, and perhaps others like John Stennis (D., Miss.) and Donald Fuqua (D., Fla.) viewed the political distributivism not as palm-greasing but as the economic and intellectual component of the Second Reconstruction. Of course, far more money and brokered placement of installations had flowed South through the Pentagon than through NASA. But the post-1961 NASA construction went beyond the pork barrel into the realm of social planning. Technological infusion was to call to life a New South, and the space program thus addressed several large items on the national agenda all at once.

A second management study, dating from late in 1961, sketched out new lines of authority in NASA. The central problem in Webb's eyes was how to coordinate Project Apollo's many constituent research teams and prevent them from flying off in their own directions, yet somehow provide the flexibility that centers and contractors needed to perform with imagination. The new recommendations urged that "program imbalance" be squarely faced by making the Manned Spaceflight Office a "state within a state." Second, NASA must somehow cut through the traditional layers of bureaucracy so that high-level attention could be brought to bear quickly on local problems. Third, NASA must endow individuals with the responsibility to oversee and direct all activities related to given missions *wherever* they might be taking place; that is, functional managers should be able to ignore bureaucratic "turf." If changes in policy or program were indicated, they could be taken directly to the associate administrator, to whom all center directors also reported. Fourth, Apollo-Saturn relied on systems integration like no other program save the ICBM: the administrative art of conceiving a whole, breaking it into subsystems, nursing along the R & D, testing, and evaluation of each like a cook with six dishes on the stove, and finally making sure

* The other major contracts went to Boeing (for the pentagonal array of F-1 engines for Saturn's first stage), North American again (for the second stage), and Grumman (for the lunar excursion module). Overall systems integration fell to von Braun's Marshall Spaceflight Center.

that each "interfaced" properly with all the others when time came to put the meal on the table. A design change in one subsystem often meant changes (and likely engineering headaches) in one or more other systems.[9]

All this required flexibility, coordination, oversight, cost control, and especially program monitoring. This last inspired PERT (program evaluation and review technique), first devised by the navy in its Polaris program. Though introduced too late to affect some NASA programs, PERT permitted computerized analysis of variables in the systematic completion of tasks. A project manager could survey all his checkpoints, hurdles, and schedules at once, and derive a realistic view of the progress of all his teams. Yet while NASA developed innovative, even experimental managerial techniques in the interest of efficiency, it also had to draft procedures for safety and quality control, clearance, and approval that met simply the highest standards imaginable. Webb was asked to do a surrealistic task, within a constricted time frame, and to do it in a glass bubble surrounded by Congressmen, reporters, and the public.

The fruits of the 1961 management studies, implemented in November, included abolition of the old program offices and creation of four new ones for Advanced Research and Technology, Space Science, Manned Spaceflight, and Space Applications. Above them reigned Seamans, the in-house management virtuoso, Deputy Administrator Dryden, handling scientific matters and technical judgments, and finally Webb, who was left free to integrate the integrators and deal with other agencies, Congress, and the White House. Over time, Webb aimed for a vast R & D complex with management flexible enough to innovate constantly yet structured enough to run on a daily basis while Webb and his lieutenants "worked to extend NASA's influence into economic and political spheres hitherto untouched by the space program."[10]

Once NASA had geared up, it faced the most treacherous hurdle of all (and one that absorbed a million man-hours before it was cleared): choosing the proper mode for going to the moon. Two 1961 studies concluded that a lunar landing was attainable before the end of the decade, but one opted for direct ascent and the other for earth-orbit rendezvous (EOR). The former mode was simplest—one rocket to blast off from earth, fly directly to the moon, land, and return home. But it required an immense booster, the Saturn follow-on called Nova consisting of perhaps eight F-1 engines in the first stage and developing 21 million pounds of thrust. Nicholas Golovin and DoD's Laurence Kavanau, authors of a second study, doubted that Nova could be ready in time. They favored EOR, whereby the lunar fuel tanks would be launched into earth orbit, with the astronauts to follow in a second rocket, rendezvous, and take off for the moon in a more compact spaceship. This multiplied the chances of launch failure and involved tricky orbital maneuvers, but the boosters would be smaller (perhaps four F-1s in the first stage), cheaper, and more likely to meet the deadline.

Throughout 1961 the Manned Spacecraft Center under Robert Gilruth, the Director of Launch Vehicles and Propulsion, Milton Rosen, and the Huntsville rocket team under von Braun pored and feuded over data favoring EOR and direct ascent. But the pacing element, clearly, was booster development. When in December 1961 the Huntsville rocketeers decided that the next step beyond Saturn 1 was to be a five-engine first stage arrayed like the five spots on a die, Nova was out, Saturn 5 was in, and the question then became how to get to the moon with such a rocket. Meanwhile, engineers at Langley and at Chance, Vought Inc. plumped for a third mode, lunar-orbit rendezvous (LOR), unaware that the Russian Kondratyuk had suggested it forty-five years before. The spokesman for the Langley group was John Houbolt, who took his presentation on the road seeking to persuade others that LOR was, in some respects, the best of both worlds. It needed only one launch from earth and still offered weight savings stemming from the separation of craft during the journey. Only a lunar module would descend to the moon, only a portion of it would return to dock with the command module, and only the Apollo capsule itself, with heat shield, would fall back to earth. It promised 10 to 20 percent in cost savings and the benefit of rendezvous techniques. First Houston, then Huntsville, came around to LOR and agreed to recommend Houbolt's dark horse to NASA headquarters on June 7, 1962.[11]

The ultimate variable, of course, was not feasibility or time or cost, but safety. Which mode promised the best shot of getting astronauts back alive? EOR supporter Golovin was the reliability expert at headquarters, and he made himself obnoxious and eventually unemployed with computer-derived safety estimates damning LOR. But now PSAC entered the fray. Having opposed the moon decision and lost, Wiesner formed an Apollo Committee, hired Golovin, and led the PSAC into battle against LOR. Scientific opinion again registered displeasure with Apollo's stress on prestige and big engineering and opportunism (get us to the moon, in whatever fashion) over pragmatism (develop useful earth-orbit capability). Wiesner and von Braun even crossed swords during a presidential visit to Huntsville, tarnishing somewhat the shining patina of scientific planning that was meant to dazzle a deferential world.[12] The PSAC proved incorrigible, but NASA stuck by LOR and awarded the lunar module contract to Grumman in October 1962. When the Cuban missile crisis erupted, preoccupying the President, Wiesner resigned himself to another defeat.

The lunar mode decision completed the restructuring of NASA. Following Mercury, an interim program, Gemini, would provide NASA with experience in multimanned spaceflight, rendezvous, spacewalks, and ground control of multiple spacecraft—all the skills needed for LOR.

Many "building blocks" of spaceflight would be hewn after all.[13] But the PSAC dissent stood—LOR was a "technological dead end" of limited future value to the space program; once current interest in lunar voyages flagged, the Saturns and Apollo craft would become extinct.[14] In retrospect, one can only applaud the success of LOR. It may indeed have been the only way to reach the moon by 1970. The Soviets, by contrast, never

---

MERCURY-ATLAS 8 AND 9

The Mercury series of "lucky 7" capsules seemed to be pulling the United States closer in the space race. In August 1962 Wally Schirra made final preparations for a six-orbit mission. Then suddenly *Vostok III* and *IV* orbited together on August 11–12. The Mercury program office weighed (literally) the idea of adding a near-rendezvous mission (with a passive target) to the flight plan for MA-8. It meant an extra 400 pounds of hardware and fuel, on a mission that already called for relaxation of the rules on oxygen reserves and a push beyond known limits of Mercury's endurance. NASA, unlike the Soviets, refused to improvise in response to the opponent.

Another obstacle popped up on the increasingly complicated vertical frontier. The Dominic nuclear tests had created another band of radiation below the Van Allen belts, necessitating a delay. It was not until October 3, 1962, that the Atlas fired Schirra's *Sigma 7* into the standard Mercury orbit, where it spent nine hours in weightlessness, while below NASA personnel completed paperwork for approval of LOR and Soviet ships unloaded missile components in Cuban harbors. Schirra executed yaw maneuvers by hand, was loath to return to automatic controls ("the chimp configuration"), and reported the amazing detail he could make out with the naked eye: "might as well be in an airplane at 40,000 to 50,000 feet altitude."[15] *Sigma 7* reentered perfectly to be picked up by the aircraft carrier Kearsarge and win the epithet of "textbook flight."

The final Mercury mission was a genuine capstone. The *MA-9* spacecraft ("capsule" had been dropped from the lexicon) would be an advanced design suitable for a whole day in space. Gordon Cooper dubbed it *Faith 7*, expressing his "trust in God, my country, and my teammates," in retaliation for the atheistic quips of the cosmonauts. There was so much photographic gear on board when Cooper went up on May 15, 1963, that he called it "practically a flying camera." His insistence that he could make out objects as small as trucks trailing dust on a Texas highway or chimney smoke from Tibetan huts must have piqued interest in USAF planners scratching to justify a manned military presence in space.[16] Cooper descended after twenty-two orbits. Again, it seemed not to bear comparison to *Vostok III*'s sixty-four orbits, or *Vostok V*'s eighty-one a month later. But it punctuated the Mercury series not with a period but an ellipsis leading to the more complicated Gemini, for which more men and resources were already mobilized—eighteen months in advance—than were employed in all of Mercury.

got to the moon, but their EOR mode blended into an evolving space station program that sustained overall progress through the 1970s.

By 1963 the U.S. civilian space program reached $3.7 billion per year, of which almost $2.5 billion went for R & D, 69 percent of that for manned spaceflight. That left some $750 million for nonmilitary, unmanned R & D.[17] In later years critics would demand cuts in prestige flights in favor of "practical" programs for science and applications. But in the first such reevaluation it was the other way around: Kennedy instructed Webb to report on the implications of reaccelerating Apollo to aim for a 1966 moon landing, while cutting back on other space programs. Webb replied in discouraging terms. A late 1966 landing, he wrote, "would require a crash, high-risk effort," with only a 50–50 chance of success and an additional $1.7 billion.[18] In November 1962 Kennedy raised the matter again. Perhaps the money to get Apollo over with quickly could be found in other NASA programs—the scientific satellites to study the earth, sun, planets, and stars, satellites to test new technology and applications, international cooperative programs. Webb responded in a tactful brief that preeminence in space required more than just going to the moon. To be sure, Gemini and Apollo were the natural focus of the program, but there were "many significant events by which the world will judge the competence of the U.S. in space." More important than accelerating Apollo was understanding that even the scientific programs contributed to technical progress, while technology in turn opened new vistas for space science. Nor could prestige be separated from the other things NASA did. In sum, the United States must build for activities beyond Apollo, requiring "that we pursue an adequate well-balanced space program in all areas, including those not directly related to the manned lunar landing."[19]

Webb won his case, but the next spring, in the midst of congressional criticism, Kennedy queried Johnson on the real value of Apollo. What were the "most salient differences" between his space program and Eisenhower's? What were the economic benefits he could point to in defense of the big space program? How might industry, government, and education suffer from continued high spending in space? Finally, to what extent could the program be reduced and not affect the Apollo schedule?[20]

These repeated inquiries mark a turning in the U.S. space program, for once the Cold War seemed to subside, Kennedy needed to show that Apollo was not just a race with the Russian bear, but a powerful engine of progress relevant to other social goals. Webb eagerly complied, for he came increasingly to see NASA's real importance not in its space achievements but in its managerial effort and as a model for planned social change. At first, he wrote, the goals might have been set by the Soviet situation, but Apollo was not an end in itself. As progress was made, it became possible to enlarge and crystallize national goals. The

Kennedy administration could be proud of having seen the danger of conservatism regarding new science and technology. "We must not repeat the failure to exploit the Wright Brothers . . . or Goddard." Of greater significance for the United States than proving its leadership in science and technology was "the motivation and drive of both the individuals and institutions of our society toward adaptation to modern requirements and self-improvement. Education will feel perhaps the greatest impact."[21] In lectures around the country Webb went further. The space program was indispensable in all its aspects.

[T]he nations of the world, seeking a basis for their own futures, continually pass judgment on our ability as a nation to make decisions, to concentrate effort, to manage vast and complex technological programs in our own interest. It is not too much to say that in many ways the viability of representative government and of the free enterprise system in a period of revolutionary changes based on science and technology is being tested in space.[22]

According to Webb, the space program required nothing less than the mobilization of the nation to a war footing in peacetime. Society (as he later wrote), had "reached a point where its progress and even its survival increasingly depend upon our ability to organize the complex and do the unusual. We cannot do these things except through large aggregations of resources and power." Whether we liked it or not, we were "in the midst of a crucial and *total* technological contest with the Soviet Union." What was being done in space could be done elsewhere in society, to engage "by all means available, in an endless search for new food for thought processes, new information and knowledge. . . ." Bigness was essential. Webb's "prototypes for tomorrow" included the Rockefeller oil empire, U.S. Steel, General Motors, the Panama Canal, TVA, and the Manhattan District. Large-scale endeavors were "adaptive, problem-solving, temporary systems of diverse specialists, linked together by coordinating executives in organic flux."[23]

Space Age systems outgrew the resources of any institution and required the forging of a "university-industry-government complex"[24] for the waging of "war" on the technological frontier. This complex was not a necessary evil to Webb, but a positive boon. The product of large-scale endeavors was change—in the attitudes, interests, and concepts of reality of the people. "It follows that the larger the effort in science and technology, the larger those changes will be, and the more rapidly they will occur."[25] At NASA, Webb said, "they sought to minimize the disruptive effects of such a vast undertaking. For instance [I] rejected a proposal from G.E. that the lunar effort be turned over to it as prime contractor for the entire job, or from UCLA to become the 'concessionaire' for all scientific research involved in the space program."[26] Instead, NASA purposefully spread the wealth—90 percent of its funds flowed into the private sector—and even pioneered the noncompetitive contract in order to save time and foster specific skills in a number of firms throughout industry. Likewise NASA "refused to go along with the old concept that scientific merit was the only determinant of who got a grant," and favored second-rate universities even if the scientific merit of their proposals was "less than that of Caltech or MIT."[27] Thus, NASA pioneered reverse discrimination in order to foster expertise in more regions (and please more congressmen). Webb granted that "traditions were broken," but found it good: "In the working partnership between universities, industry, and government . . . each of the three has retained its traditional values. . . . I believe that each has become stronger because of the partnership."[28]

How was the country "stronger"? Not only because of economic spin-offs from large-scale endeavors, but because the government/industry/university team force-fed such spin-offs into the mainstream of society. Free enterprise might do that on its own, he granted, "but can society afford to wait for or to rely solely upon the workings of such a slow and uncertain process?"[29] Then there was the way in which new technologies altered human values. Webb quoted Raymond Bauer, whose study of the space program suggested that its effects "may include changes in man's conceptions of himself and of God. . . ."[30] The space program promised a new era of great advances in the way large-scale efforts were managed, the encouragement of multidisciplinary efforts, new techniques and tools for the conduct of research in the social as well as physical sciences, and the manner in which they were applied to the solution of age-old problems.[31] The burden fell on the administrators, like himself, to help subordinates to see the "totality of the job," the relationship of their particular jobs to that totality, "to bring all their inner resources to bear in effective ways to help get the big job, the total job, done."[32]

What began as an extraordinary governmental initiative to reassure the world that individualism, free enterprise, and limited government were still superior came in Webb's mind to be a vehicle for "revolution from above." NASA's destiny was to serve as prototype for reallocation of national power for social and political goals. "The technological revolution that is now fully upon us," he wrote,

involves all areas and disciplines. It is the most decisive event of our times, and keeping ahead of it is essential. No nation that aspires to greatness, or to use its power for good, can continue to rely on the methods of the past. Unless a nation purposefully and systematically stimulates and regulates its technological advances and builds the fruits of those advances into the sinews of its system, it will surely drop behind. . . . The great issue of this age is whether the U.S. can, within the framework of existing economic, social, and political institutions, organize its development and use of advanced technology as effectively for its goals as can the Soviet Union. . . .[33]

The preceding quotations date from 1968–69, at the end of Webb's time at NASA. But as early as 1961 he had written Glennan that the basic institutions of the United States would have to be rethought in light of technological revolution. In 1963 he believed that

every thread in the fabric of our economic, social, and political institutions is being tested as we move into space. Our economic and political relations with other nations are being reevaluated. Old concepts of defense and military tactics are being challenged and revised. Jealously guarded traditions in our educational institutions are being tested, altered, or even discarded. Our economic institutions—the corporate structure itself—are undergoing reexamination as society seeks to adjust itself to the inevitability of change.[34]

Nor was Webb alone in his belief that space technology would stimulate changes far beyond its sphere. Dryden predicted "a great variety of new consumer foods and industrial processes that will raise our standard of living and return tremendous benefits to us in practically every profession and activity."[35] Congressman Miller could "not think of any other aspect of our space program that could better justify our space expenditures to the average taxpayer than industrial applications. . . . His return will be a wide variety of new or better products, at reasonable cost, which in turn will give rise to a greater consumer demand and economic stimulus."[36] But space did more than spin off products like sparks from a flywheel. According to Edward Welsh of the NASC: "The aerospace revolution is a rebellion only in the sense that the forces at work do rebel against the old, the worn out, and the wholly ineffective methods and techniques. This is not change for the sake of change. . . ." Space technology did not permit even the slightest chance of unanticipated error, thus paper and pencil could no longer be tolerated where computers could work better and faster. "Now the process of systems analysis and the great experiments which have led to amazing advances in managerial competence are solving problems we previously thought impossible of solution. We are indeed enjoying a technological revolution and must not let it slow down." Space, said Welsh, would pay a greater dividend than anything else in which the United States engaged.[37]

That the technocratic faith outgrew both the White House lust for prestige and the self-promotion of the Big Operators is indicated by the explosion of overall federal support for science, education, and R & D—and by the almost complete lack of congressional inquiry, oversight, or opposition to it. By FY 1964 the total NASA budget passed $5 billion, five times what it was three years before. Counting the DoD, the AEC, and other agencies, the government spent $6.8 billion on space, or a little over one percent of GNP. Total R & D performed in the United States rose over the three years from $14.3 to $18.9 billion, and now comprised 3 percent of the GNP. But government R & D rose from $9.3 to $14.7 billion, 58 percent above 1961 and now one-eighth of all federal spending. The trend begun under Eisenhower had quickened to the point that by FY 1964 over 78 percent of all American R & D was purchased by the federal government.[38]

The sharp rise in federal R & D was concentrated, of course, in defense and space. So unless the United States were to transmogrify into a high-tech military economy while civilian industries went begging, the bene-ficiaries of the boom must demonstrate that R & D for space and defense also energized the larger economy. One result was that no agency studied its social and economic impact more vigorously than Webb's NASA. Regional development, technology transfer, the impact of R & D on growth, information diffusion systems, total employment generated—all in all, $35 million were spent in NASA's first decade on research in the social sciences.[39] The results, accumulating in the later 1960s, were generally disappointing and mostly predictable. Economists demonstrated again that economic growth depended on too many things for the correlation between R & D and growth to be quantifiable. But the connection was intuitive, and Kennedy's sense that this was something the United States had to do was shared by many, even if the rewards might come in unfathomable ways. Perhaps they were right—unfath-omable effects can hardly be shown *not* to exist. But to maintain space budgets and sustain Webb's larger plans for social renewal through large-scale management, justifications of the space program drifted naturally into the material realm.

The role of the space program in stabilizing the delicate aerospace industry was obvious—McNamara saw this in 1961. But Webb expected the space program, as integrator of social forces, to contribute to overall growth and stability. The kernel of the NASA university program, for instance, was mobilization of professorial talent for a conscious assault on social problems. Eisenhower had feared that the NDEA would outlast its temporary status and lead to federal funding and regulation of education at every level. But Webb—and the growing educationist establishment—affirmed such involvement and believed that strewing "seed money" would cause "excellence" to sprout and spread like hardy ground cover across the length and breadth of the country.

One of the persistent criticisms of Apollo was that it sucked up the nation's technical talent to the detriment of research in other fields. So Webb determined to give back what NASA was taking up, "to do all I can to build up the university research, teaching, and graduate and postgraduate quality and quantity of education."[40] The PSAC Panel on Scientific and Technical Manpower had called for the United States to graduate 7,500 scientists and engineers by 1970. By 1963 NASA's Sustaining University Program alone projected support for 4,000 graduate students per year, an effort equal in size to that of the NDEA science program. Only one or two congressmen suggested that NASA had exceeded its authority by embarking on any such program.[41] But the real aim of the university program was social uplift. Webb wrote Lee DuBridge, president of Caltech, that as much as $35 billion over ten years would be channeled through NASA (much of it to the West Coast) to advance science and technology at the most rapid rate, and "to feed it back into our national economy and the fabric of our national life." He sought to interest DuBridge in NASA-derived technology for urban development, water resources, energy, communications, management, and life sciences. "I know that the lunar objective causes some problem to you," but it would prove its worth "even if we never make the lunar landing."[42] In January 1962 Webb promised Killian that NASA would not take professors away from their campuses, but find ways to use their talents while they remained with students and labs. He met with governors, university presidents, regional business delegations, and set up an ad hoc group to hold "think sessions" every month to plot the future of NASA/university/industry projects.[43]

Did all this not smack of elitist *dirigisme*, or at least a failure to recognize that technology alone could not provide the value judgments on which wise direction of national energies should be based? Bush made this point, as did a generation of skeptics informed by C. P. Snow's *The Two Cultures*. Webb acknowledged that value judgments were involved, "but the result is also dependent upon the skill of the social organization in resource allocation," that is, the means must be developed for social mobilization before the question of ends was more than academic. The important point was that

we are so operating as to spread all of the problems, scientific and technical, over a very large number of able minds in educational institutions, in the media of communications and throughout industry. We might develop a body of doctrine here which indicates that the role of our senior leadership group is to so do this as to keep a fluid enough situation so that new concepts have an opportunity to flourish ... and to be incorporated in the operating doctrines of our democracy through our present democratic processes or some incremental improvement or evolution of them.[44]

The tool's the thing, the "very large scientific and technical capability" now in place through NASA's "ongoing contacts with industry and universities."[45]

Those contacts went beyond hiring faculty as consultants and funding graduate students. NASA also sponsored grants for the construction of space sciences laboratories. University administrators made pilgrimages to Washington to solicit "brick and mortar" money. By the mid-1960s, the Office of Space Science and Applications devoted $30 million per year to university research, while the sustaining university program dispensed another $25 million. Twenty-seven facilities grants were made, while professors, administrators, and students learned to ride the wave. The likes of Governor Otto Kerner of Illinois, the president of McDonnell Aviation in St. Louis, and a delegation from Los Angeles petitioned for space institutes at the University of Illinois, Washington University, and UCLA. When Professor Samuel Silver of Berkeley arrived to request a space science lab, Webb stipulated that the new center take on two economists to study the feedback of science and technology into the community, to solve social problems. By mid-1962 Webb made it a condition of support that universities seek "ways and means to assist its service area or region in utilizing for its own progress the knowledge, processes, or specific applications arising from the space program."[46]

Such melding of theory and practice, of pure and applied research, of natural and social science, while hardly the same as the Soviet practice, nevertheless clashed with the traditional autonomy and values of American universities. In truth, the latter may never have taken the "social impact" clause of their contracts seriously, regarding it as congressional eyewash (this was NASA Space Science Director Homer Newell's suspicion). But that in itself suggests how far the country had come from 1945 when Congress *rebelled* against plans for central coordination of research to attack social problems. Now Congress *expected* such plans to justify funding. Even Newell granted that to place scholarly expertise "on ready call to be applied on command to problems of someone else's choosing . . . would destroy the very independence that generated the unique expertise in the first place."[47] But when academics dragged their feet on NASA-university assaults on local problems, Webb considered them "outrageously callous and irresponsible." For Webb, the organized pursuit of social objectives fixed by government was the whole point of the program.[48]

Still, universities fell into step behind the aerospace firms and politicians, eager to accept federal largesse in the name of Webb's Space Age America. Conceived and promoted by politicians and intellectuals, endorsed by corporations, universities, and bureaucracies that stood to gain, not only military but civilian technocracy spread across the land with no more philosophical debate than accompanied the triumph of nuclear deterrence. Perhaps the United States had to give up much that was dear to meet the Soviet challenge. But why did so many make such a virtue of that necessity?

In April 1961, in the midst of the cloistered debate on whether to go to the moon, Senator Kerr left the Space Committee for a trip back to the hustings. It was Annual Law Day at the University of Oklahoma, and the Sooner State's favorite son was the honored speaker. His fortune was in oil, but his roots and votes were in the soil of '89—which blew away in '36—which made every Oklahoma politician's first responsibility the garnering of federal water projects. Indeed, his favorite oratorical subject, said Kerr, was Land, Wood, and Water. But since chairing the Space Committee, he amended his longstanding slogan to Land, Wood, Water, and Space. "I believe this new version accurately describes a new horizon of great promise for the pioneering spirit of Oklahoma." Sooners had never lost the spirit of youth, optimism, and fearlessness toward what lay over the horizon: ". . . the qualities needed to explore the universe are the selfsame qualities that inspired our fathers and grandfathers to rush into Oklahoma's space and settle on what was, to them, a new and strange land."[49]

Kerr slid easily into the big thinking inspired by the biggest frontier of all. But whence came James Webb's bold, technocratic vision? He was no midwestern Progressive, or "socialist of the chair" at some Ivy League university, or even a member of his party's left wing. Rather, Webb was a rural populist, who grew to know the workings of government and industry and their potential for huge concentrations of power. His own motive was service, his ability surpassing, and like LBJ he did not fear power. He had come to believe, like Kerr, that the United States was shaped by its frontier. Since 1896, the historians said, the domestic frontier was closed. But space was a new frontier in which American society might recover the rugged virtue that had almost been forgotten in the twentieth-century's mundane exploitation of the tamed land. Of course, space as the new frontier was a cliché of Kennedy's Camelot. But it was more than that for the Big Operators of the Southwest, the land of the Hoover Dam, Alamogordo, the Red River Project, White Sands, and now the Manned Spacecraft Center. Even as New Frontier slogans grew stale in public, they grew rich in meaning and implication for Webb until the whole nation, all its regions and institutions, became as a single band of homesteaders riding on rockets like Conestoga wagons.

In the midst of the Johnson Senate hearings after Sputnik, columnist James Reston's intuition served him well: the U.S. and Soviet governments alike would now have to prepare to deal with Texas.[50] Later on even von Braun, who exhibited the passion many Germans share for the lore and land of the American West, had to remark that "space is bigger than Texas."[51] And then, in 1964 Webb discovered the theories of his namesake, Walter Prescott Webb, on the shaping of the American character on the frontier. In early 1965 he wrote to Horace Busby, a presidential aide, that problems such as Vietnam and racial integration and the other priorities on the national agenda were all of a piece. "I have several times around the country," he continued,

pointed out the frontier thesis. . . . based on the "wild and unperturbable" forces of the frontier, which show no mercy and no compassion, must be harnessed and utilized by the pioneer and in the process have the feedback effect of generating in the pioneer those qualities which have made for the American democratic system, the same kind of analogy may be considered in connection with our large-scale organized efforts such as those in space. Here an entire nation is developing technology which puts it, as an organized entity, very much in the same position as the pioneer was individually on the frontier.[52]

A world democracy could theoretically emerge as nations plied the frontier together—except that the Communists seemed disinclined to cooperate. "Whether we can force them to cooperate by developing so much power that there simply isn't any alternative is the real question before us *at this time.*"[53]

Here was an updated American vision—a Space Age America that was the pioneer writ large. "This may well put us into a position where we are more in control of the destiny of the world than we have been since the early years after World War II." If the Communists sought to fill vacuums like Korea, Vietnam, the Congo, then

we ourselves might then have a strong obligation to use our technology to begin to fill these vacuums. . . . Now what this may mean is that the government-industry-university team we have developed under the NASA system, which has been so effective in marrying science and technology and mobilizing large resources for focusing on limited but important objectives, might become the pattern needed by this nation. . . .[54]

Webb discovered (albeit erroneously) that the original "frontier thesis" of Frederick Jackson Turner was published in 1903, the year of the Wright brothers' flight. Through such theses Webb found a role for NASA as heroic as that of the pioneers of aviation. Turner proclaimed the closing of the frontier, the end of that confrontation with impersonal, impassive, wild, gigantic Nature. In 1951 Walter Prescott Webb expanded the notion: all western civilization faced a closed frontier, and as a result "we have a great pain in the heart, and we are always trying to get it back again." But after Sputnik, he wrote with delight to fellow Texan Lyndon Johnson that he had apparently settled on the closing of the frontier too soon.[55] James Webb took it to heart: the Space Age changed everything. But where Khrushchev and the Soviets expected the new age to fulfill their dream of a future utopia, Webb and the American technocrats expected it to restore a past apotheosis. Thus when Webb spoke of total technological competition, unlimited power, and mobilization of the entire nation, it was not with the foreboding of the Yankee Henry Adams, who had foreseen and dreaded the twentieth century, but with the boundless enthusiasm of a Boomer Sooner and the lofty certainty of a Texas Longhorn.

CHAPTER 19

Second Thoughts

The esteemed editor of the *Territorial Enterprise and Virginia City News* (Nevada) was ahead of his time in 1960 with his tongue-in-cheek lambasting of the Mercury program. No one but *Life* magazine, he told his little flock of readers on the nether side of Lake Tahoe, "gave a honk in a windstorm about the horse's ass-tronauts." Never had there been a sillier, more childish and futile grab for the taxpayers' billions: "Let's Scrap the Space Crap."[1]

He was not heeded, or at least not until a new President had aimed at the moon and enjoyed a long honeymoon on the space program. But beginning in late 1962, and cresting the following autumn, criticism of Apollo surfaced in various quarters, invoking a national debate and placing NASA's budget in jeopardy for the first time. Roughly three views competed for public support: the pro-Apollo technocratic argument, the anti-Apollo but still technocratic arguments, and the residual anti-technocratic argument. But the latter had little resonance, for most observers just lumped it together with those opposing the moon race for other reasons. Few challenged any more the notion of state responsibility for directing progress in science, technology, and education, for setting social priorities, and forging technical tools to achieve them. Left and Right, hawk and dove, by 1964 most Americans had opted for technocracy. Thus even as the space program was called into question, the new mode of governance for which it served as symbol was not. If there were still lingering doubts about the nature and future of Space Age America, the 1964 election put them forever to rest.

The scientific community was generally skeptical of manned spaceflight from the start, even as its decade-long campaign to pry open the Treasury succeeded beyond hope. Having urged the federal government to pay for training, facilities, and applied R & D, scientists were now chagrined to find so much of the effort dictated by political expedience. Whether or not they had foreseen the danger (their elders had in 1945), scientists gradually found their voices in 1962–63 and began to speak out in journals like *Science* and *Bulletin of Atomic Scientists*, and in Congress. The space program was fine, but it had grown too large and dominated by engineering. Apollo's scientific potential was nowhere near the cost, said the critics, or it drew funds and manpower from more worthy pursuits, or large-scale applied research ought to concentrate on problems "here on earth": medicine, Third World development, urban renewal. What was more, Apollo had little military value and was a dead-end even in terms of spaceflight. A straw poll by the director of the geophysical lab of the Carnegie Institution found non-NASA scientists opposed to Apollo by 110 to 3. While he granted that the poll was untrustworthy, he put more stock in the fact that his anti-Apollo editorial in *Science* drew only three mildly irritated letters. He was especially concerned with the way NASA tempted young technicians away from other fields. Consider, he said, the scientists at work on weapons. They scarcely chose this pursuit out of self-interest, for they were gravely aware of the consequences of their work, felt the public and collegial opprobrium, worked under secrecy, and could not share their findings in articles or scholarly meetings. Now the space program offered another patriotic career that was open, acclaimed, respectable, and paid up to 25 percent above government scale. Neither the weapons labs nor the universities could hope to compete: "the first thing to do would be to see to it that NASA is not permitted to go hogwild."[2]

The chairman of Columbia's physics department believed the space program would contribute little to the lives of Americans. We should be attacking air and water pollution, he thought, and mental illness, ocean-ography, solar energy, and education. Joshua Lederberg of Stanford lambasted the subordination and stereotyping of university research by government.[3] A poll of physicists, conducted under a NASA grant by the new Berkeley Space Science Lab, found that the vast majority believed propaganda and military aims to be the driving forces behind the space program, and the nation was not even getting its money's worth.[4] Scientists associated with NASA responded as best they could. Prestige racing might be foolish, said Lloyd Berkner, but was a fact of life. Capitalism must be proven superior if it was to survive. The conflict between science and engineering was artificial, since the two progressed together. Exploration was an expression of man's desire for knowledge, "one of the deepest driving forces among an exceptional group of human beings," according to Frederick Seitz, president of the NAS. Indeed, the space program went far to define mankind in the first place, and revealed "whatever meaning our own human existence has." The United States could not afford *not* to have a space program, and the moon was an excellent goal.[5] Phrased thus, in the negative, the argument cast different shadows. As Berkner remarked, the argument that space money should be spent instead on medicine or low-income housing or urban transit could be made as well against defense spending, or gambling, liquor, cigarettes, or chewing gum. Was society to devote itself solely to utilitarian concerns, to raise the standard of living of the masses some marginal amount? Or should the human race aspire to know the nature of the universe, the origins of the earth or of life itself?

How to measure the public worth of the space program? Was it good politics but bad science? Or, as Walter Lippmann wrote, a silly Cold War stunt but a useful engine to increase knowledge—good science, but bad politics?[6] As long as science and technology were the business of autonomous institutions and individuals, value judgments were matters of private conscience. Once the state, in all its sovereignty and might, took up the pursuit of knowledge, value judgments were politicized. Money and minds for space or for "down-to-earth" problems? Such questions whipsawed congressmen, but they could not ignore them once the Apollo honeymoon ended—and it was ended in part by ex-President Eisenhower.

In August 1962 Ike asked in the *Saturday Evening Post*, "Why the great hurry to get to the moon and planets?" He endorsed space research, but not a "fantastically expensive crash program." If we sought prestige, then "let us point to our industrial and agricultural productivity; why let the Communists dictate the terms of all the contests?"[7] The Republican Party followed up, forming a space advisory committee that condemned Apollo and called Kennedy's failure to build a strong military presence in space "perhaps the most disastrous blunder by any government since the last war." Democrat George Miller of the House Space Committee said the report "sounds like the mouthings of an Air Force jingoist," and Republicans who supported Apollo were embarrassed.[8]

Still, the GOP assault continued under Goldwaterites, who deflected the Eisenhower critique into a less profound struggle over goals and management of the space program. Senator Barry Goldwater (R., Ariz.) himself made a variation of the "priorities" complaint when he urged a shift of emphasis to the military. He had already spoken with eloquence against the fixation of prestige. Addressing the Air War College in 1960, he "count[ed] on finding here a greater urbanity toward world affairs than one would encounter at a meeting, say, of the Committee for Achieving World Peace by Making Democracy Work in the Congo. . . . Only the vain and incurably sentimental among us will lose sleep simply because foreign people are not as impressed by our strength as they ought to be." Goldwater attributed this American lust for world approval to a guilt complex. Americans were embarrassed by their world power after 1945: "In order to prove that we were unlike our predecessors in power—selfish, ambitious, warlike—we began to lean over backwards, and to gear our policies to the opinions of others. . . . Call into question any aspect of American policy and the argument you will hear after all others have been laid to rest is some variation of the world opinion theme."[9]

Why was deference to world opinion so harmful? First, thought Goldwater, it was self-defeating in that the very respect you covet is denied the moment you go out and beg for it; second, because trying to prove your worth paradoxically raises doubts about it; third, because in choosing world opinion as your rudder, you lean on the one most open to manipulation by your enemies. "World opinion" really meant intellectuals, journalists, the organizers of street mobs—all prime targets of Communist influence. Americans seemed to have a vague feeling that power was immoral. But power, explained Goldwater, was an inevitable product of the human condition. Someone had to have it.[10]

When it came to space, Goldwater voted for Apollo but preferred a vigorous USAF program instead. Only U.S. military superiority could ensure the peaceful uses of space.[11] Instead, we were "moonstruck"; six full fiscal years into the Space Age and "we have not authorized a single military space weapon." Only space weaponry gave hope of abolishing war on *earth*. But the "building blocks" approach prevented the drafting of blueprints, strategic concepts, or even a definition of the military role in space. Did we think that the other side would proceed in like fashion? Goldwater quoted Sokolovsky on space warfare, and added, "I am told that Nikita Khrushchev keeps on his desk a piece of metal scarred by a laser beam. . . . Perhaps the Russians are doing far more than speculate. We are not."[12]

The military critique of Apollo became a main plank in the platform under construction by conservative Republicans in 1963. In May their Senate policy committee called the moon race adolescent, and a task force chaired by Representative Louis C. Wyman (R., N.H.) called for a shift of emphasis to the Titan 3, Dyna-Soar, Midas follow-ons, a military space station, all coordinated by a Strategic Space Command.[13] *Reader's Digest* cried, "We're Running the Wrong Race with Russia,"[14] and conservative pundit William F. Buckley, Jr., asked "The moon *and* bust?" He derided this "most public relations conscious administration in the history of this country." One Negro riot, mourned Buckley, "and we fall back abashed before the judgment of the men who sent tanks into Budapest. . . ." Why not tell the Soviets at the UN: "Very well, you have reached the moon, but meanwhile here in America we have been trying, however clumsily, to spread freedom and justice?"[15]

Such critiques, and McNamara's cutbacks, prodded the USAF and aerospace interests into another offensive. They warned that "dangerous things can happen on the way to the moon." By spring 1963, with the Dyna-Soar in trouble, the USAF and even civilian strategists like Alton Frye and David Robison broke with the trend toward "stable deterrence" by advocating aggressive pursuit of military advantage in space. Otherwise, self-restrictions in space only threw away the U.S. bargaining position. Goldwater warned that "All Russian orbits, all Russian space endeavors are conducted under the military. They have no peaceful purposes in space." He even assailed the UN Principles on Peaceful Uses of Space as a ploy to kill U.S. military space programs, "the last great bulwark against communist domination of the world." Kennedy felt the heat and in July 1963 asked LBJ to mobilize the NASC in support of the position that Apollo did have major, if indirect, military value.[16]

Likewise, the campaign for a shift of R & D to domestic priorities gained momentum in 1963. *Newsweek* writer Edwin Diamond ridiculed the space race as a "potlatch ceremony" in which neighboring chiefs vied for prestige by throwing more valuables into a fire.[17] Sociologist Amitai Etzioni called Apollo a "moon-doggle."[18] Both challenged the idea that military and space R & D stimulated economic growth and held the space program responsible for unemployment and inflation (running at the dizzy rates of 5.6 percent and 1.3 percent respectively!). But Etzioni, speaking for many academic and congressional liberals, did *not* suggest that the government cut back on spending. Rather, he called for moon money to be reallocated, for more deficit spending, wage and price controls, aid to the underprivileged, and federal retraining of workers. The government "must learn to shift people from one part of the country to another, from one sector of the economy to another" and to help dislocated workers make "psychological and social adjustments." As for science, Etzioni did not suggest the defederalization of R & D to prevent "moon-doggles," but rather its total control through a Science-for-Development Agency to bring "the power of mass science to the assistance of development of 'have not' regions here and abroad." A major division of each wing of this SDA would be devoted to rehabilitation of criminals, psychological/educational research, a "less unhappy on-the-job world," and more. All this would follow from the "breaking of the lunar spell"—but in the meantime Americans still suffered the "gigantic investments made in charting the moon [which] serve those who seek to preserve the America of yesterday as it is confronted with the problems of tomorrow." Americans were apparently "emotionally not ready to accept the consequences of peaceful coexistence."[19]

Liberal congressmen could not, or would not, go so far. But just as conservatives urged a shift of priorities to defense, many liberals wanted a shift to social spending. The real question before Congress, said J. William Fulbright (D., Ark.), was that of priorities. There was a "dangerous imbalance between our efforts in armaments and space on the one hand and employment and education on the other." Senator Joseph Clark (D., Pa.) called domestic well-being a considerably higher priority, and Senator William Proxmire (D., Wisc.) denounced the space program as "corporate socialism." The United States must maintain a high rate of research, but in civilian industries lest they soon be shut out of international markets.[20]

Kennedy struggled to ward off such critiques. He refused to list priorities, reminded the Congress of its nearly unanimous support for Apollo, and insisted that the country had enough money to do everything that needed to be done in education, for instance, as well as space. "I know there is a feeling that the scientists should be working on some other matter, but I think that this program—I am for it and I think it would be a mistake to arrest it."[21] But House Minority Leader Charles Halleck (R., Ind.) quoted Eisenhower to the effect that the NASA budget was "downright spongy," and Fulbright, chairman of the Senate Foreign Relations Committee, was the first to counsel abandonment of Apollo on May 3, 1963. Ike spoke up again in June when he called spending $40 billion to reach the moon "just nuts."[22]

The NASA FY 1964 budget was in committee when another blow fell on the shoulders of Apollo. Sir Bernard Lovell, director of Britain's Jodrell Bank Observatory, returned from a visit to the USSR with the impression that there was no Soviet manned lunar program; Keldysh told him that the USSR was concentrating on space stations and unmanned exploration. Widely publicized, Lovell's report reinforced doubt about the wisdom of Apollo.[23] Kennedy again insisted that the United States go ahead: its "building blocks" program should not be diverted by a newspaper story.[24] After all, there was no assurance that Keldysh spoke with authority or Lovell with understanding, and if the Soviets were trying to turn U.S. opinion against the lunar program, Keldysh's remarks were ideal for the purpose.

The House Space Committee reported out the NASA bill with a cut of a half-billion dollars and a minority report for still greater cuts tied to a stretch-out of Apollo beyond the end of the decade.[25] Space had become a source of embarrassment to the President—who would have thought it?—but the reason was embedded in U.S. space policy and the split between the need for competition (as a spur) and the desire for cooperation. Why did the Soviets send signals such as those conveyed by Lovell? Frutkin believed their aim might be to elicit new cooperative proposals from the United States, maintaining their image as space leader to whom even the United States petitioned. If so, it worked, for two days before a presidential address on space to the UN General Assembly, Senator Hubert Humphrey (D., Minn.) urged the White House to make another declaration of its willingness to cooperate, and Arthur Schlesinger inserted the following paragraph in Kennedy's prepared text:

Why, therefore, should man's first flight to the moon be a matter of national competition? Why should the U.S. and Soviet Union, in preparing for such expeditions, become involved in immense duplications of research, construction, and expenditure? Surely we should explore whether the scientists and astronauts of the two countries—indeed of all the world—cannot work together in the conquest of space, sending some day in this decade to the moon not the representatives of a single nation, but the representatives of all our countries.[26]

Go to the moon together! Would Kennedy make this appeal before the UN? On September 17, the day before Humphrey's speech, Dryden reported his own thoughts on a meeting with Blagonravov. The Soviets were discussing the value of a manned lunar landing, but it would be dangerous to conclude that they currently had no such program. After all, the Academy of Sciences did not run their program, the military did. The next day McGeorge Bundy briefed the President. He was concerned about a New York *Times* story that implied the United States was the foot-dragger on cooperation. Soviet disinformation, such as it was, was working. "The obvious choice," wrote Bundy, "is whether to press for cooperation or to continue to use the Soviet space effort as a spur to our own." He advised a bold stroke to smoke out the Soviets.[27]

Kennedy then confronted Webb. He did not ask NASA to approve a joint lunar voyage but did ask that NASA not undercut such an initiative. Webb assured the President that he "had sufficient control to see that he was not undercut. He said 'Thank you very much.' I went on to Missouri where I got a call from Bundy saying Kennedy was going ahead. I phoned around the [NASA] centers with direct instructions to make no comment of any kind on this matter." A prudent move—only days before the manned spaceflight chief Robert Gilruth had spoken with trepidation of the problems of systems integration, classification, language, and politics involved in a joint flight to the moon.[28]

Kennedy's astounding proposal ("Let us do the big things together. . . ."), made at the UN on September 20, 1963, elicited no response.[29] Initial Soviet editorials either ignored the moon offer or dismissed it as "premature." Then *Pravda* reprinted a Walter Lippmann column implying that the speech was an effort by *Kennedy* to squirm out of the lunar race, considered by Lippmann to be a "morbid and vulgar stunt."[30] A month after the speech, on October 25, Khrushchev in turn squirmed out of the Kennedy offer in that conversational news conference: "It would be very interesting to take a trip to the moon, but I cannot at present say when this will be done. We are not at present planning flight by cosmonauts to the moon. . . . We have a frequently quoted joke: He who cannot bear earth any longer may fly to the moon. But we are all right on earth." The U.S. press, sensitive to the Apollo debate, translated these vague remarks into sensational headlines. Khrushchev soon turned around and teased reporters for reading too much into his words—he had not said that the USSR had dropped out of the moon race—but neither did he grasp Kennedy's outstretched hand.[31]

The heavy assault on Apollo had a curious combination of effects. Critics of Apollo leaped on the joint lunar proposal as an argument against "crash program" funding. The Senate stuck to its half-billion-dollar cut in NASA's FY 1964 budget, causing the projected first moon landing to slip from 1967–68 to 1968–69. But Fulbright's proposal for a further 10 percent across-the-board cut lost 36 to 49, while the Senate passed a rider expressly forbidding a joint lunar landing with any country without consent of Congress.[32] Kennedy's dramatic UN speech actually helped to steel the congressional will to see an American flag planted, alone, on the moon. NASA would never again be free from budgetary pressure, but Apollo was never again in danger of extinction. It must have amused the President and gratified him as well. Kennedy once remarked, without explanation, that he himself might not live to see completion of the moon project. But not to fret, he said, for he would be up there somewhere, in his rocking chair, and he would have the best view of all. Nine days after the Senate rider passed, Kennedy was dead.

RELAY 1

The first television pictures transmitted across the Pacific Ocean were carried by the new comsat *Relay 1* on November 22, 1963. The program was to open with greetings from President Kennedy, but his remarks were deleted after the events in Dallas a few hours before. Four days later the state funeral of the fallen President beamed up to space and back again to Japan, as well as Europe and the USSR. Never before had so many people around the world been able to view history unfolding.

In the midst of the great debate on Apollo one must strain to hear echoes of the arguments that boomed forth in opposition to state direction of science, R & D, education, and ultimately social change in 1944, '46, '48, '52, and '58. The only issues for most orators in 1964 were whether or not James Webb's America (itself an echo of Senator Kilgore's America) should dedicate itself to space and nuclear hegemony, global prestige and cooperation, or social welfare. Few asked whether technocracy in *any* form was what the United States ought to embrace. One of the few who did was John Medaris, now in retirement, who still stood up against the compromise of values, invitations to waste, corruption, and uncapitalist manipulation of the contract system. Another was Vannevar Bush, who had gone on record against manned spaceflight in 1960, warning that even its propaganda value was transitory. "I can see the time coming when results are few and far between, and simply bore the public, which is after all fickle on such matters."[33] By 1963 Bush believed more than ever that the public mood would someday reverse sharply on space, perhaps after the death of some astronauts, perhaps out of boredom. But more to the point was that the nation could not afford such projects. Budgets had been unbalanced for years, the gold flow was serious, high taxes impeded commercial vigor, the danger of inflation was genuine. Finally, there was the moral dilemma raised by "big science." The scale of Apollo meant that "nearly every man who could speak with authority on the subject has a conflict of interest." He did not mean that all scientists suppressed their better judgment for gain. Rather, they were caught in a gigantic national trend pushed on them by government, their own institutions, and colleagues. Who could stand against it? They just consoled themselves, wrote Bush, with Cromwell's admonition—"I beseech you, bethink that you may be mistaken"— and then joined the crowd.[34]

Then there was Eisenhower, speaking like a ghost from his Gettysburg farm. In the 1964 campaign he heard the Republican Right claim that the military-industrial complex was *under*nourished, while Democrats urged the extension of centralized state management to far corners of American society. In the *Saturday Evening Post,* Ike asked voters to consider basic philosophies and explained the principles of his own Republicanism. For years, he wrote, the obvious drift had been toward centralization of power in Washington, the sacrifice of individual liberty and responsibility to the lure of the "easy way," dependence on government to provide a risk-free, Poppa-knows-best future. Federal subsidies of all kinds—to business, labor, universities, or individuals—had a cloying effect and sapped the nation's moral fiber like a narcotic. Local responsibility for education, housing, health, transportation, and agriculture gave way to federal direction. Widely distributed R & D gave way to gigantic "crash" federal stunts such as the race to the moon. Sound fiscal policy gave way to experimental and dangerous overspending, which inevitably brought inflation and eroded the basic right of every citizen to have a dollar worth as much today as it was worth yesterday. Flashy public relations, observed Ike, persuaded Americans that labels were somehow solutions: New Frontiers and Wars on Poverty. Such panaceas usually turned out to be new channels to siphon off power to the federal government. The space program had started with a step-by-step approach and was now blown out of all proportion by hysterical fanfare. These were the trends, and soon it would be too late. Many people would grow up under paternalistic federal control and never know the fundamentals of Republican philosophy. To these people such a philosophy would only seem an attempt to "turn back the clock." But one day historians would record that "here was where the U.S., like Rome, went wrong—here at the peak of its power and prosperity when it forgot those ideals which made it great."[35]

Few listened anymore, or if they did, they themselves were too reliant on federal programs to support dismantling the system.[36] Sputnik, the missile gap, and the education gap had seemed to justify forced change under the auspices of the government. The "total Cold War" known as "peaceful coexistence" extended the foreign challenge from defense to image making, and thence to economic growth, housing, medical care and other standards of comparison. The domestic rebellion against segregation and inequities of opportunity pushed in the same direction. If subsidies were to flow to certain groups, science students or aerospace firms, then all groups should receive federal assistance.

After Kennedy's death the Cold Warrior/social liberal alliance found its natural leader: the man who had led the surge into space, who tied his party's fortune to it, who viewed it as a paradigm. He was the biggest of Big Operators—Lyndon Johnson. As Harold Brown recalled, LBJ did not have the same kind of personal interest in science or intellectual curiosity as Kennedy. But he always looked at science and technology in terms of what they could do to solve problems—military, political, any kind. No vice president acceding to the presidency had as little time as Johnson to prepare his own election campaign, but none was better able to stitch his own banner. He had run the Senate from 1954 to 1960 (Clark Clifford thought, thankfully, that he had run the *country* in those years). As vice president he was frustrated but still dominated space policy and, like Webb, saw it as vindication of his older faith in centralized management of the social "machine." According to Eugene Rostow (Undersecretary for Political Affairs), Johnson's eye was always on the long term. According to Dean Rusk, "You could never get President Kennedy to think beyond what he had to do at nine o'clock tomorrow morning, whereas with Johnson it's always, 'Well, where are we going to be ten years from now?' "[37]

Only six weeks after inheriting the White House, Johnson presented his blueprint for the campaign and the nation in his State of the Union Address. "We have in 1964 a unique opportunity and obligation—to prove the success of our system; to disprove those cynics and critics at home and abroad who question our purpose and our competence. . . . This administration today, here and now declares unconditional war on poverty in America." The government programs to wage the war added up to a morally inspired but technocratic vision of a "nation that is free from want and a world that is free from hate—a world of peace and justice, and freedom and abundance, for our time and for all time to come."[38] By summer, the War on Poverty was caught up in an even larger theme. The United States, under the leadership of the federal government, led in turn by the Big Operators, was to create a Great Society.

The 1964 campaign was one of the most significant in recent U.S. history, not because its result was in doubt, or because it provoked national debate on fundamental principles—but precisely because it was not and did not. To be sure, Goldwater stood four-square for self-reliance, balanced budgets, and a rollback of government intervention. That was what "In Your Heart You Know He's Right" was all about. But he married those principles to an all-out faith in unbridled military technology not surprising for an Arizona general in the Air National Guard. He saw that the Soviets would never rest content with inferiority or with an interim stage of technology that made for "stability." But that very perception did injury to his august principles. Goldwater's was not, therefore, a libertarian platform.

In any case, he was isolated. The Apollo debate of 1963 had been almost exclusively between competing versions of technocracy, and the 1964 campaign consecrated the new political terrain. Like the authors of

THE UNISPHERE

The basic purpose of the New York World's Fair is to help achieve "Peace Through Understanding," its major theme being "Man's Achievements in an Expanding Universe."

. . . The potential of this fair for promoting international goodwill and understanding is apparent. I believe participation in the fair by the Federal Government is essential.

The theme, "Challenge to Greatness," proposed by a citizens' advisory committee established by Secretary [of Commerce Luther H.] Hodges, will enable us to present to the world not a boastful picture of our unparalleled progress but a picture of democracy—its opportunities, its problems, its inspirations, its freedoms.[39]

With these words President Kennedy signed the bill authorizing $17 million for a federal pavilion at the New York World's Fair. All told, the fair was expected to cost some $60 million for construction and set-up, $77 million for operations, and $58 million more for demolition and restoration. It involved not only the largest global exhibition in history but also a massive renovation of Flushing Meadows. It included funding for Lincoln Center for the Performing Arts, the Van Wyck-Long Island Expressway, the permanent Science Center facility, Shea Stadium, and a new marina beneath the Grand Central Parkway. Thanks to contributions from thirty-three states, dozens of the largest corporations in the United States, more than forty foreign countries, the ideas of 183 architectural and design firms, the breezy leadership of "Big Operators" Robert Moses, banker David and governor Nelson Rockefeller, and Mayor Robert Wagner, and a no-strike pledge from construction unions, the fair opened in April 1964, to present the story of freedom and democracy as expressed in, and proven by, high technology.

Now, every such extravaganza since Britain's Crystal Palace Exhibition has been a paean to industrial progress, but never before did the United States seem at once so persuaded of the superiority of American material civilization and of the imperative to display it. Whence came that superiority? From liberty, to be sure, but this time there was scant mention of the past, with its lone inventors like Bell and Edison, and instead an embrace of the future, with its government projects and corporate inventiveness: General Motors' "Futurama"; New York State's "Tent of Tomorrow"; Westinghouse's "Time Capsule" with its forecast of life in the future; the Transportation and Travel Pavilion with its entire second floor given over to Project Apollo and a simulated moonscape; West Virginia's "Radio Astronomy Sky"; a "Sermons from Science Pavilion" sponsored by the Christian Businessmen's Committee and Moody Institute of Science; the Hall of Science, a "central, cathedral-like hall" with a clear eighty-foot-high span "suggestive of the challenges of space and the horizons of science" and featuring Martin-Marietta's simulation of rendezvous in space; and on and on. The Federal Pavilion, with its theme "Challenge to Greatness," celebrated the "pioneer spirit of America, its present and its future" (not its past).

Finally, there were the symbols and pride of the fair: the "Lunar Fountain," the "Fountain of the Planets," and the "Unisphere," a 140-foot-high hollow steel globe, strutted by longitude and latitude lines and ringed by three "orbits" symbolic of the atom or the space satellite, take your pick, but in either case a homogeneous metallic world, lassoed by technology.[40]

The Ugly American, who had punctured American efforts to win "hearts and minds" in Southeast Asia in 1958 only to urge the United States to find *better* ways of boosting their prestige, or C. S. Lewis's teacher friend who became so fed up with bureaucratic interference with his school that he repented of his socialism, only to ask if Lewis might be able to get *him* a bureaucratic job in the ministry of Education!—so the contending factions in 1963–64 debated technocratic priorities without challenging the principle of state-directed change in the first place.[41] The question was simply who could best manage the machine for which national goals?

In space policy Johnson benefited in the campaign from the aftereffects of Kennedy's joint moon proposal. He, too, pledged to seek cooperation with the Soviets, but otherwise he would see Apollo through to the end. Goldwater, on the other hand, called cooperative moon shots "too ludicrous for comment" and Apollo itself a waste. In his "realistic space policy," all manned spaceflight would be run by the military and the United States would launch a crash program to develop ABM and laser weapons. The latter, prophesied Goldwater, might prove "one of the most *practical* defensive devices ever conceived by the brain of man." The laser would have many applications, he predicted, "but would be well worth the money and effort if it did no more than guarantee us military control of access to space."[42] Late in the campaign, Johnson defused the issue by making public the "fact" that the United States already had means of intercepting hostile spacecraft. This ran counter to official policy and was not really true (the Johnston Island tests hardly yielding an operational "space war" system), but it was good politics.

Meanwhile, Johnson exhorted federal agencies to "think ahead, plan ahead," and link their ideas for new programs to the campaign theme. He regarded the NASC a "key in-house resource, not limited to space alone, but to which we can look for initiatives on education, manpower, foreseeable changes in American life, implications and opportunities of international scientific cooperation, etc." The President, as his aide wrote, hoped

to discuss in speeches the building of what he refers to as "The Great Society." An obvious component of this theme is the vast array of implications of our present R & D activity. . . . Another area on which we are attempting to develop information is the American West. The horizons of the science frontier carry many implications not yet appreciated for the old and disappearing land frontier.[43]

In 1960 the Democrats hit the space and missile "gap"; in 1964 it was the education and opportunity "gap" to be closed by efficient, revolutionary, federal programming along the lines of the space program. The pitch seemed to work. By mid-October 77 percent of Americans thought Apollo should continue at the present pace or be speeded up, and 62 percent wanted the same or more money spent on space.[44] Johnson himself had a huge lead in the polls, while his rival hit hard on the President's alleged failure to understand the requirements of combating the USSR. In fact, it was Goldwater who failed to grasp that the Cold War now pervaded everything: the United States could not hold all else constant in its society while gearing up militarily. Johnson said as much in his final campaign swing.

It was in St. Louis, after a week that had brought the fall of Khrushchev, a Conservative defeat in the British elections, and the first Chinese atomic test, that LBJ elaborated "total Cold War." The events of the week, he began, brought to the forefront the underlying crisis of our period in history: the struggle between those who wish to be free and those who want to enslave mankind through a Communist world revolution. There were no easy answers, but two main approaches. One was to smash the Communists through military means—nuclear war, hundreds of millions dead, and the rest organized on a totalitarian basis. The other approach was to draw on "the forces of freedom":

First, we fight Communism through a defense establishment. . . . Second, we fight Communism by strengthening the defense capacity of other nations. . . . Third, we fight Communism by maintaining superiority in every field of science and technology. . . . This applies to the exploration of outer space. . . . Fourth, we fight Communism by giving our support to the concepts of liberty . . . to be fair, to be just, to ensure that all of our citizens, regardless of creed, religion, national origins, or color, enjoy the liberties which are inherent in our notion of freedom. Fifth, we fight Communism by building a Great Society here at home. . . . The Great Society is a powerful weapon against man's enslavement. Sixth, we fight Communism by uniting nations and uniting people in their determination to be free. . . . Seventh, we fight Communism by supporting the Peace Corps. . . . Eighth, we fight Communism by expanding world trade. . . .

So ladies and gentlemen, we have the capitalist, we have the manager, we have the worker, and we have the Government—all working together shoulder to shoulder, not fearing, not doubting, not hating, but hoping and believing and producing and leading the rest of the world. We are going to fight Communism by building a family of free men. . . .[45]

In Los Angeles a week later, Johnson was in an even more expansive mood. He told the crowd:

I remember an old man in my town in Texas. We lived way out in the country. Finally we got a little railroad 30 miles from where I lived. When it got there, the old man said it would never work, and so forth. Finally, the day came . . . and the train started off for San Antonio. The old man said, "Well, I have been saying now for 3 months that they would never get her started and I don't think now that they will ever get her stopped."

Now that kind of faith in the future is not what built California . . . the greatest educational State in the Union . . . the greatest space and the greatest aeronautical and the greatest missile, the greatest technological State in the Union. . . .

But we are not satisfied. We are going ahead full speed. We are going to a future of horizons that are unlimited. . . .

If we are going to compete with the Soviet Union, we are not only going to have to have the best heels and the best hearts that we can, but we are going to have to have the best heads. . . .[46]

Presently a heckler interrupted, and the President retorted: "It is that same old crowd." He recalled how they always cried "Socialism!" at social security, the minimum wage. "If they get a great thrill out of hollering 'Socialism,' let them be happy." Whereupon he returned to his "horizons." You could not be first on earth and second in space. Space was now the decisive medium, and would "determine how we live." But there was so much more to do, "to act now to control and to apply what we learn to improve our daily lives": to beat heart attacks, strokes, and cancer; rebuild the cities, purify the air, improve mass transit:

If John Glenn can go around the world, we have to find some way to get from a suburb to our plant. . . .

. . . our program has to be to desalt the seas and rebloom the deserts. There are a hundred other ways that we can make our world a better place to live. So why do we want to go around being grouchy?[47]

Conclusion

The Kennedy years saw the quiet withdrawal of American political resistance to technocracy. Its triumph was all the more fundamental for its being so subtle.

In the late 1950s the principle of limited government still clung to life. Federal "brick and mortar" money for schools, aid to students in science and foreign languages, R & D for national defense and even for prestige: these were still defined as special violations of the hoary boundary between federal power and local freedoms. At the same time, the notion that these were not violations of that boundary, or indeed that the boundary itself was a superstition, made steady progress among northern Democrats, liberal Republicans, academics, journalists, and even conservative Cold Warriors—all of whom favored a technological "race posture" against the USSR, domestic social problems, or both. In 1961, as JFK proclaimed, the torch passed to the younger generation. Far from holding back the varied constituencies of an enlarged federal role, Kennedy relied on their coalescence and then picked from them the advisers that led the United States into the 1960s. An alliance of military hawks and social activists, they endorsed federal action on the principle that the state could best foster new knowledge, power, and economic growth through planned management. Once in power, the constituencies began to quarrel, each favoring vigorous federal action in some areas and not others. But by 1964 the fundamental question—whether a society shaped by state planning and spending was consonant with American freedoms—got lost. Liberals wanted less hardware and more social engineering; mainstream Democrats insisted that the United States could afford both; even Goldwaterites granted government's redoubled responsibility for Space Age technology, if only for national defense. The debate revolved around the purposes to which the technocratic tool ought to be put, not the political or moral costs of brandishing the tool in the first place.

The roots of American technocracy were as varied and deeply buried as its visible branches were varied and widespread. The doctrine of efficiency dated from the Progressive Era, the protest against the insecurities of the free market from the depression, the temptation to forge government/industry/university teams for federally directed progress from World War II, the demand for equality for the underprivileged from the postwar years. Perhaps these tendencies would have nudged the United States into technocracy in any case, according to the American tradition of softening social or ideological discord through expansion and growth—the frontier policy. But the challenge of Sputnik, because it symbolized Soviet parity in strategic weapons, and Khrushchevian peaceful coexistence (read "nonviolent competition"), because it pushed Cold War rivalry into every arena of national achievement, surely helped to trigger the rapid political change of 1957 to 1964. Propaganda to the effect that racism did not exist in the USSR, that Soviet women were liberated, that its economy would soon surpass the American, that the social product was equally distributed among Soviet citizens, might all be cynical lies. But Americans had difficulty damning the USSR while headlines proclaimed riots in Birmingham and the rhetoric of every disaffected social group. Above all, the Soviet claim on the future through scientific and technical supremacy could hardly be gainsaid while cosmonauts beat the Yanks to every spectacular.

Throughout the century, Soviet leaders declared that the Western countries might be ahead for the moment but were run by and for monopolists who lived off imperialism foisted on others by their high technology. The USSR, on the other hand, was a people's state . . . and in any case would soon catch up. Now the United States was saying that the Soviet Union might be ahead for the moment but was run as a slave state in which resources could be mobilized to serve militaristic ambitions. The United States, on the other hand, was a free country . . . and in any case would soon catch up. In so saying and doing, American leaders bought the Communist line that technology was both a symbol of social superiority and the main agency for making one's own preferred image real. The Midas touch of government-directed R & D, education, and social engineering could turn base society into a Great Society, eliminate want, inequality, injustice, and even, in LBJ's vision, dispel hatred.

Through all this, U.S. space policy changed little under Kennedy. Government assumed a larger role in comsat deployment, but positions on space law and cooperation—and the contradictions in those positions—remained. The military program fell under tighter wraps and civilian control. The American agenda at the UN triumphed in the Principles of 1963: national sovereignty and nuclear bombs banned in space, but no inspection of space launches to ensure that they were for "peaceful purposes," no demilitarization, no UN space agency, and no definition of where outer space began. American (and Soviet) policy makers judged all this prudent and were probably wise to preserve a loose regime in which spaceflight could advance most rapidly and serve the national interest. Similarly, U.S. agencies continued a deliberate approach to cooperation that stemmed from the inherent difficulties of international R & D and the fact that competition was the motive force for space spending in the first place. The preferred posture was cooperation in science, competition in engineering. This did little to please Europeans, who were coming to believe Superpower rhetoric about space R & D as a forcing house of technology and growth, and resented U.S. dominance. The Kremlin never showed serious interest in cooperation. Kennedy's call for a joint moon program reverberated, but its sounding boards were the UN and Congress, not the Kremlin.

Thus the space technological revolution did not change world politics or usher in an age of global integration or even an abatement of the Cold War. The international system absorbed space just as it absorbed the atom. Nor did the Space Age oblige traditional elites to surrender policy to a technical elite. Ike's fears of a military-industrial, scientific-technological complex were not confirmed in the way he expected. The danger was never a *technocracie* of technicians, but a technocracy of politicians, arrogating to government the right to fix a national agenda and order fabrication of techniques, both hardware and management, for its fulfillment. Even NASA ceased to be an agency run primarily by scientists and became a juggernaut of the politicians and engineers. The military chiefs lost what remained of their autonomy in military strategy, R & D, and procurement to the civilian managers. University scientists, businessmen and technicians in the corporations, these, too, were integrated in varying degrees into a national complex for the promotion of new techniques, paid for, directed by, and dependent on the whim of the federal government.

Kennedy knew little about space, but he bowed to politics, as if intuitively, and pointed to the moon. Within two years he became a believer. "But why, some say, the moon? Why choose this as our goal?" he asked an audience at Rice University in September 1962. "Why climb the highest mountain? Why, 35 years ago, fly the Atlantic? Why does Rice play Texas?"[1] To aspire, to strive, to slay giants. The challenge of the frontier now lay in technology and, physically, in outer space. If the effort required the nation to become as one man, like Lindbergh fixed to his machine like a knight errant, or as a collective, like a well-oiled, inspired football team pitted against a bigger rival—then so be it. At least we, in the United States, would do it voluntarily, and not by coercion.

James Webb, too, became a believer. Once he moved enough rocks from his imposing pile to see that a moonflight might just be achieved on schedule, he marveled at the power of organization it revealed. LBJ, in Webb's judgment, was not a good administrator. Nor were Kennedy or Eisenhower.[2] But Webb himself was, and like McNamara he believed that the real frontier was a management frontier. The space program promised many ineffable but one certain benefit—perfection of the management of large systems. Apollo was not an end, but a means that brought order out of the chaos of institutions and that could be applied to a host of social projects. Webb himself may never have chanted, "If we can send a man to the moon, why can't we . . . ?" But if others did, calling for Apollo-type assaults on everything from medical care to synthetic fuels, it was because the James Webbs had, by their talent and energy, made command innovation look easy—and "American."

Soon after Johnson's landslide, Webb meditated on technocracy American-style. "Progress in science, in technology, and in efficient organization of large-scale effort is of itself not sufficient; any great society must as well believe in the values of intellectual vigor and the creativity of the individual." Good Americanism, but how was this to be preserved? "The work of individuals," Webb wrote the President,

must be organized within a framework. . . . The leisure of the past is gone; where it once took hundreds of years to fully exploit a new technology, we are now capable of taking enormous strides in a very short time. . . . The earth orbiting satellite has become the first tool of the human race that is not limited by the fuel it can carry, the boundaries of other nations, or by the earth's atmosphere or its oceans. The fact that it can work for any nation over which it passes is a symbol of a great society of the world, and, perhaps more than any tool before, a portent of a more universal society.[3]

Space spending would not detract from other things in the building of a great society.

In fact, greater support of the space program will undoubtedly yield greater support for these other needs by demonstrating what can be done when we find new and more effective ways to involve large numbers of imaginative, gifted, and innovative people in an area that has almost explosive potential. . . . In reality, the space program lies in your first area of building the great society.[4]

Johnson himself was far less moved by the adventure and wonder that moved Kennedy—the wonder that still, in the National Air and Space Museum, brings tears to the eyes of more visitors than the Declaration of Independence or Lincoln Memorial. To LBJ the space program was a model of the role government should play in society, and the role technology should play in government, an expression as well of new and apparently limitless power, as if handed down by some Promethean party boss in the form of command technology and federal management. For the War on Poverty and Great Society, as much as Apollo or Vietnam, were Cold War phenomena, but they were not only that, far from it. They were born of a moral vision in which men of power and charity sought to use their gifts for the less gifted. It *was* possible to eradicate poverty, crime, ignorance, whip the Communists, and develop the Third World, or so thought LBJ. The power existed and needed only to be grasped. Indeed, it must be grasped—by good men— lest the other side do so first. To Eisenhower, the essence of courage was to resist the temptation to use dangerous tools; to Johnson, the essence of courage was to dare to take them up in a good cause. Whether in decaying cities, outer space, or Third World jungles, American technology would overwhelm the enemies of dignity.

For several years after the 1964 election, technocratic enthusiasm spread with the appeal of a perpetual motion machine. Adlai Stevenson called it "our almost miraculous capacity to use existing technology to create new technology" and declared that "Science and technology are making the problems of today irrelevant in the long run, because our economy can grow to meet each new charge placed upon it. . . . This is the basic miracle of modern technology. . . . It is in a real sense a magic wand that gives us what we desire. Don't let us miss the miracle by underestimating this fabulous new tool."[5]

Miraculous, magical, fabulous—it was not the scientists and engineers who were euphoric at the prospect of thaumaturgy, but the politicians and their allies in academics, business, and journalism. The euphoria fed a sixfold increase in federal R & D over a single decade, then an expansion of social engineering that proved as irreversible as a techno-logical revolution. Civilian space spending peaked in FY 1965 ("Some people's feet got tired," recalled LBJ). But space spending was no longer the measure of the technocratic trend, for the mentality and methods of the space program had taken hold elsewhere: "We realized, after *Sputnik I* and *II*, that we had been challenged and challenged successfully. . . . Our people are slow to start but hard to stop." Those with tired feet "raised the question, 'Well, if we can go to the moon, why don't we take that money and do some of the things that need to be done here?' " Until Sputnik, said Johnson,

the Federal Government hadn't passed any education bill. We didn't have any Federal aid for education. . . . So we started passing education bills, we made a national effort in elementary education, a national effort in higher education, where two million students were brought into our colleges. And they said, "Well, if you do that for space and send a man to the moon, why can't we do something for grandma with medicare?" And so we passed the Medicare Act, and we passed forty other measures. . . .

And I think that's the great significance that the space program has had. I think it was the beginning of the revolution of the '60s.[6]

PART VI

The Heavens and

the Earth:

The First

Twenty-five Years

Once again two brave Americans have carried the quest for knowledge to the threshold of space. . . . In this struggle all men are allies and the only enemy is a hostile environment. —LYNDON B. JOHNSON, 1965

Preserving its aggressive nature . . . imperialism is putting great emphasis on the aspects of the struggle of the two world systems connected with the development of the economy, science and technology, and education.
—ALEXEI KOSYGIN, 1969

In view of the way in which science seems to condemn us to live in a world of rapid social change, we may have to get used to [flexible] constitutional systems. Perhaps indeed a nation can be free only if it is not in too great a hurry to become perfect. —DON K. PRICE, 1965

I plead guilty to having placed the idea of man above the idea of mankind.
—RUBASHOV to Stalinist inquisitor, 1938

That's one small step for man, one giant leap for mankind.
—NEIL ARMSTRONG, Tranquillity Base, 1969

"THE HOPES and fears of all the years are met in thee tonight"—it was Christmas Eve, 1968, a year that began with the Tet Offensive, took form in the Robert Kennedy and Martin Luther King, Jr., assassinations, dragged on through race riots and political quakes in Chicago, China, and Czechoslovakia, then ended with the fantastic flight of *Apollo 8*. It had been twenty-three months since a fire in the command module of Apollo unit 204 killed Virgil Grissom, Edward White, and Roger Chafee, forcing a stand-down that lasted until October 1968, when a Saturn 4B finally put a manned moonship through its paces in earth orbit. The Soviets, too, had known disaster when their first manned Soyuz, fouled in its parachute, crashed into the ground, killing Vladimir M. Komarov. But now, it seemed, the Soviets were up to something. In March 1968 they launched a new Zond rocket toward the moon; in the fall two more Zonds circled the moon and returned to soft landings on earth. The Soyuz program was only just resuming flight-testing, and the super-booster had not appeared, but perhaps the Soviets had one more card to play: launching a cosmonaut around the moon might steal some of Apollo's thunder and permit them to claim a victory of sorts.

James Webb was gone. After numerous budget cycles and the Apollo fire his credibility with Congress waned. He also claimed to know as early as August 1967 that LBJ would not run for reelection. So Webb stepped down, and Thomas O. Paine took the baton for the final lap of the moon race in October 1968. The next flight would be only the second manned Apollo and the first to use the full Saturn 5 system. But buoyed by safety assurances and troubled by the mystery of Zond, Paine went for the moon. On December 21, astronauts Frank Borman, James Lovell, and William Anders achieved earth orbit, then, while reporters and VIPs milled around waiting for buses at Cape Canaveral, loudspeakers piped in the spacetalk: "Apollo? Houston. You are go for TLI." Comprehension crept slowly through the crowd: TLI, Trans-Lunar Insertion. The first men in history were about to depart for another celestial body.[1]

Apollo 8 reached lunar orbit on December 24; the next day, while millions prayed for their safe return, they kicked their spacecraft on course for home, and on the twenty-seventh they splashed down. In their cameras was a present, the sublime photo "Earthrise" of our cloud-streaked droplet ascending into the black above the rim of an arid moon. President Johnson sent copies to every head of state in the world, even Ho Chi Minh, while the luxuriant ecology movement gained an icon by grace of the very technology it denounced.

What hopes and fears of all the years hung on the spaceships racing for the moon? In the USSR, to be sure, the people boasted of their deeds and mourned their dead spacemen as any nation. But when *Luna XV*, an unmanned probe meant to scoop up soil on the moon, crash landed even as Neil Armstrong and Edwin (Buzz) Aldrin walked the surface and a lonely Michael Collins orbited overhead, there was no national lamentation, only the customary retreat into secrecy. Soviet officialdom pinned no "hopes" on space technology, only expectations. Mastery of space would come in time as part of the inevitable Communist conquest of nature; value judgments were not officially involved. "We have no contradictions in the Soviet Union," Kosygin had said, "between appropriations for space research and for the needs of the population. . . ."[2]

In the United States, hopes and fears were the heartbeats of the quest, and they surfaced in the shouts and tears of millions when *Apollo 11* landed in July 1969. Even Norman Mailer, sometimes radical man of letters who styled the space program proto-fascist, found the feat a credit to the purposefulness of WASP culture and a rebuke to his debauched counterculture. "You've been drunk all summer . . . and *they* have taken the moon."[3] Black activist Ralph Abernathy led a protest march to Cape Canaveral only to succumb to the awe-inspiring launch and bless the enterprise. Gurus of the New Age claimed the Apollonian metaphor in pleas that we devote as much attention to "inner space" and the maintenance of "Spaceship Earth." LBJ, in Texan retirement, thought the space program the catalyst for the Great Society, and President Nixon thought those days in July the greatest week since Creation—ridiculous perhaps, but also the most honest image of the hopes and fears of technocratic man: men as gods, creators in their own right with all the glory and tragedy of divinity.

Senator Anderson recalled the polls of 1960 that showed the world believing that the future belonged to the Soviets. Now polls showed confidence in the United States as "without peer in power and influence." It was not Vietnam that caused the change, he said, nor diminishing foreign aid, nor internal strife and violence: it was the space program, "above all the Apollo lunar landing—the victory of man over his terrestrial domain—these are the events that moved the minds of men around the globe."[4] By 1969 Anderson's colleagues mostly had other hopes and fears than those of 1961: the fear of social disintegration and the hope that the Apollo method might help alleviate poverty, pollution, decaying cities. The hope that rode on Apollo was the hope for human adequacy in the face of awful challenges. NASA had whipped the Soviets, and now technocracy—state-managed R & D, state regulation, state mobilization, and systems analysis—could be applied to "down-to-earth" problems. Now that the technocratic method was proven out, space travel was becoming dispensable. But the United States had done something unique—it spent billions on a peaceful, exploratory effort, done it in the public eye, and done it without overarching ulterior motives. As Mailer wrote, a bureaucracy for the first time had embarked on a surrealistic adventure! But the first irony of Apollo was that over time, the means had become more important than the end, even though that means—technocracy—was to prove inapplicable to most of the items on the new national agenda. Going to the moon was an engineering problem; eliminating discrimination or poverty or even urban blight was not. By the 1980s Americans could ponder *two* decades of federally managed change and wonder if the tangible benefits outweighed the intangible costs, and how to fashion a balance scale to make such judgments at all.

"We came in peace for all mankind" read the plaque Armstrong and Aldrin left on the moon—the second irony of Apollo was that this grandest of all space missions was not even central in shaping the role of space technology in international politics. The moon was not what space was all about. It was about science, sometimes spectacular science, but mostly about spy satellites, and comsats, and other orbital systems for military and commercial advantage. "Space for peace" could no more be engineered than social harmony, and the UN Outer Space Treaty, entering into force almost exactly ten years after *Sputnik I*, fixed the environment of future spaceflight as one of competition among national technocracies, while the apparent force of "targeted R & D" drew many nations into the hunt for advantage, not integration, through spaceflight. The fruits and foibles of technocracy matured and spread, ensuring that the next decades in space, those of routine exploitation, of space stations and shuttle craft, new space weapons and new space powers, would finally alert the world to the fact that the Space Age would neither abolish nor magnify human conflict, but only extend politics-as-usual to a new realm. The international imperative remained, as did human imperfection. Space technology might, through its institutional offspring, alter societies and economies, but it could never change its parents: the international system of states and the curious, aspiring human spirit.

Perhaps maturity in the Space Age, as in any technological revolution, can be measured by the growing realization that these latest creations of man do not and will not change man himself.

CHAPTER 20

Voyages to Tsiolkovskia

President Johnson was in the habit of meeting every Thursday for lunch with his top aides in defense and foreign policy. The conversations ranged from Vietnam to nuclear proliferation, to ABM, and—in early April 1966—to outer space. Walt Rostow proposed a presidential initiative for a treaty to solemnize the 1963 UN principles for space law, establishing the United States as the leader of "space for peace" and demonstrating détente with the USSR despite the war in Southeast Asia.[1] When rumor had the Soviets about to announce their own plan for a space treaty, Rusk drafted a presidential statement, cleared it with McNamara, Webb, and UN Ambassador Arthur Goldberg, and urged its immediate release. LBJ, at home for the weekend, summoned reporters to the ranch on the Pedernales and affirmed his nation's desire that exploration of the moon and planets be for peaceful purposes only: "I believe that the time is ripe for action."[2]

In fact, preparations for such a treaty had been carried on for eighteen months. Adlai Stevenson declared that the UN space principles of 1963 were not the last word on space law, but "one of the first," and throughout 1964 and 1965 the State Department sounded out other agencies on the shape of a space treaty.[3] After all, it was the logical next step, suggested by the impending moon journeys, the threat of space weaponry, and momentum derived from the Partial Test Ban treaty. But in another sense one could ask, "Why bother?" For the interagency review process made clear that the treaty would be, at best, redundant. The DoD insisted that treaty language make explicit the retention of national prerogatives in space and national ownership of space hardware on the moon, and enjoin lunar explorers to release "data" (but not "all data") to the world.[4] NASA suggested that the treaty apply only to the moon, the JCS that it not outlaw military activities, provide for verification and a right of withdrawal if national security were deemed threatened. What was more, said the Joint Chiefs, a nation developing resources on a celestial body must have some right to their use free of encroachment.[5] In the end, the State Department held to a formula (based on the 1959 Antarctic Treaty) that banned military fortifications, maneuvers, and any type of weapons on the moon, in the belief that this was "less likely to be interpreted" as prohibiting military surveillance and other passive systems. A preliminary draft treaty won DoD acceptance on March 11, 1966, paving the way for Rostow's suggestion and Johnson's speech.[6]

The press was enthusiastic. Editors draped the familiar slogans on "space for peace" in new cloaks of urgency, as if in the absence of a treaty squads of astro- and cosmonauts, armed with flags, ray guns, and theotolites, would ascend on the moon in colonial warfare.[7] But one columnist took a jaundiced view: since national claims and nuclear bombs in space were already rejected as impractical, a space treaty could only be a facade to make the Cold War rivals look good without constraining them from doing anything they might really want to do.[8] Indeed, Eilene Galloway, the leading congressional staffer on space law, thought a treaty might involve new restrictions only if a ban on *all* weapons tests and maneuvers (e.g., the planned MOL) were applied to *all* of outer space. Such a result was surely not U.S. policy.[9] This concern grew in June when Andrei Gromyko presented the Soviet draft treaty— it explicitly applied to all of outer space. Since the USSR claimed not to have a military space program, and since verification of space activities was a chancy business, the U.S. military clauses combined with the enhanced Soviet scope for the treaty might only highlight U.S. "militarization" of space. And even if Moscow was willing to compromise, the United States still faced the self-righteous ire of neutrals, like Egypt and India, that held that any meaningful treaty must outlaw all military activity in space.[10]

Negotiations opened in Vienna in July 1966. The State Department instructed Goldberg to agree to extend the treaty to all of outer space, but to follow closely the 1963 language, which did not ban all military activity in space. He should also avoid raising the issue of ownership of lunar resources and trust that the Soviets had the sense to do the same.[11] Within three weeks Goldberg and Soviet diplomat Platon D. Morozov agreed on a text encompassing all of outer space, on the freedom of scientific investigation for all states, the barring of claims to sovereignty, and the conduct of exploration in accord with the UN Charter.[12] Differences arose over the American language on use of military equipment for peaceful purposes and over European claims to equal rights in space for international or private organizations. Brazil, speaking for Third World countries, pushed for space activity carried on only "for the benefit of all mankind." The United States acceded to this so long as specific references to property or economic rights were excluded. Finally, Morozov tossed in a demand that frightened not only pro-Western countries in the COPUOS (dubbed "The Friendly Fifteen" by the United States) but even the neutrals. The USSR, stuck high up in the Northern Hemisphere and short of alliance partners around the world, interpreted "equal access to space" and "freedom of scientific investigation" to mean that it should have equal access to foreign soil for the basing of tracking stations. If Kenya or Australia let NASA on to its territory, it must do the same for the Soviet Union![13] Rusk absolutely rejected such an interpretation and wired Goldberg to find some anodyne formula that might appease the Soviets.[14] Nothing worked. Morozov stuck to his demands, accused the Swedes and Australians of "distortions and fantasy" and Goldberg of trying to shift blame for the stalemate. The committee adjourned on August 4.[15]

The State Department reviewed the chessboard and prepared, over three feverish weeks, various tactics for dealing with each contested issue . . . except tracking stations.[16] When the COPUOS reconvened in New York, Morozov still had instructions to stand pat.[17] But on October 4 (the anniversary of Sputnik again), he suddenly acknowledged that tracking facilities must depend on bilateral agreement (hence host countries retained a right of refusal), admitted international organizations in space (so long as their member governments bore their legal obligations), adopted the U.S. clause enabling military personnel and equipment to take part in peaceful exploration, and indicated a willingness to ignore "subjects lying outside the space field," presumably the bombing of North Vietnam.[18] The logjam broke, a final draft won unanimous General Assembly approval in December, and LBJ thanked Goldberg by telephone for "the nicest Christmas present I could have."[19]

The Outer Space Treaty, signed by sixty-two nations, went to the U.S. Senate on February 7, 1967. The President's letter of transmittal called it "a first step, but a long step" toward peace, and Secretary Rusk's letter of submittal cited it as "an outstanding example of how the law and political arrangements can keep pace with science and technology."[20] Ratification was fairly certain, but Johnson liked unanimity. So the State Department managed a brief, self-confident rush to judgment. Only Goldberg, McNamara, Webb, and General Earle Wheeler testified. But each agency prepared bluebooks in anticipation of difficult questions, including why the Soviets were so eager to conclude a treaty. First, thought the State Department, they probably figured to gain from the amiable provisions for assistance to astronauts, liability, and so on. Second, they clearly hoped to win greater access to other countries for their tracking network. Third, they might hope to forestall an arms race in space, given their staggering arms budget. Fourth, they saw a chance to isolate China, which denounced the treaty as collusion with the imperialists. Fifth, the treaty made the USSR appear equal to the United States, and sixth, it enhanced the Soviet image as a benevolent space power.[21]

Rusk opened debate by heralding the treaty as "an impressive model for further cooperation among the nations—a cooperation that is essential if the world is going to escape destruction by conflict and if it is going to make headway in conquering disease and poverty, in relating population rationally to means of decent livelihood, and in offering all men proper scope for their talents and energies." It is hard to see how anything in the treaty made it such a "model," or indeed what Rusk's words even meant. But they conveyed all the right images and technocratic hopes that teased the first decade of the Space Age. Goldberg, wearied by an eleven-day tour of Asia, then adumbrated the content of the treaty. He assured the committee that the major clauses were not new, but drawn from past UN resolutions. Still, senators wondered if the United States wanted to make these principles binding. What was meant, for instance, by "for the benefit and in the interests of all countries"? Did this mean, asked Senator Gore, that the United States was obliged to make outer space available to all? Not quite, said Goldberg, it merely stated a "goal subject to further refinement. . . . It surely had a meaning in broad perspective, not intend to not mean [sic] that as a general principle outer space shall be carried out [sic], exploration should be carried out, for the benefit and in the interests of all countries."[22] Forgiving the ambassador his jet lag, one concludes that the phrase did *not* oblige parties to the treaty to share their technology or its fruits with others: no "international socialism" in space.

Moving to disarmament, Chairman Fulbright clarified that while *all* weapons were prohibited on the celestial bodies, only *weapons of mass destruction* were banned from earth orbit. Furthermore, the treaty right to inspect foreign facilities applied only to the celestial bodies, not to all space vehicles. Therefore, the treaty effectively demilitarized the moon but specifically *sanctioned* the militarization of orbital space. Without inspection, Bourke Hickenlooper (R., Ia.) asked, how could we know whether the Soviets were even refraining from "bombs in orbit"? Rusk assured him that such systems could be detected over time, and that in any case the treaty did not prohibit an antisatellite capability should it become necessary.[23] Senator Gore then returned to the ticklish matter of "space for all mankind." Did it mean, for instance, that the United States must give comsat facilities to all the world? If not, then what did it apply to? "The terms of the treaty are indeed indefinite. I can almost use the word fuzzy. And I wonder why we would negotiate [such] a treaty. . . ?"[24] Six days later Goldberg tersely explained that space as the "province of all mankind" was meant as a sort of "freedom of the seas" clause, no more.[25] This satisfied senators—it would not satisfy Third World UN members in years to come.

General Wheeler's appearance on April 12 climaxed the hearings. He assured the committee that the JCS supported the treaty, but rather than enabling the United States to cut back on military space R & D, it required "intensified efforts to develop capabilities to detect the orbiting of nuclear weapons. . . ." You mean more budgetary support for military space R & D? asked John S. Cooper (R., Ky.). "That is correct, Senator," Wheeler replied.[26]

Thus the UN Outer Space Treaty of 1967, ratified by a vote of 88 to 0 on April 25. It denuclearized outer space and demilitarized the moon. But it did not demilitarize outer space. As for space being "for the benefit of all peoples irrespective of the degree of their economic and scientific development," the negotiators described it as a vague principle with no foreseeable application. In terms of the "space for peace" globalists, therefore, the space treaty was all show and little substance.

One must ask, first, how much substance there *could* have been? This was no League of Nations, conceived in irenic idealism and thrust upon senatorial mossbacks. Rather, the space treaty was drafted with congressional and military opinion in mind at every juncture. Even if LBJ (and Brezhnev) had embraced demilitarization and global sharing in space, what chance would such a dreamy program have had of ratification? The DoD would not sit still for a ban on all military uses of space, given American reliance on a technological edge and the secrecy of the Soviet program. Nor would the American people sit still for an obligation to give away their expensive technology to all comers. Second, one can ask how much substance there *should* have been? If the treaty defined an essentially laissez-faire regime, was this not a triumph for American diplomacy, achieved because the principles of space law were laid down while the United States still had overwhelming influence over what was done in space? If so, then the administration's sense of "urgency" becomes more explicable. To be sure, LBJ may have wanted a treaty prior to moon landings, or proof that détente could survive Vietnam. But James Gehrig, chief of staff on the Senate Space Committee, suggested that the United States was moving hurriedly to codify space law while it still had the clout of a duopolist.[27] Of course, the Soviets were indispensable partners in the task, but they had no more desire than the Americans to wait until new space powers diluted their influence over space law. Better for both to draft a "freedom of space" charter and do it quickly, even if, as one critic wrote, it was merely an empty box "wrapped in many silken flags and tied with much gold braid."[28]

Did this mean a nefarious U.S. plot to dominate the heavens? Or a cynical Superpower condominium ganging up on the rest of the world? Or was it a prescient effort to preserve space from the grip of international bureaucrats? For "internationalism" in its UN guise was already coming to mean regulation, restriction, and redistribution: such international technocracy did not complement so much as contradict national technocracy. If international laissez-faire was a *conditio sine qua non* for space development, then the vacuous language of the space treaty was a boon. It established the only sort of regime in which spaceflight might rapidly flourish. It was also appropriate to the technology. How could governments pretend to regulate the growth of space technology while it was still in its infancy? Space technology was moving very quickly; international legal committees move slowly. As specific uses and interests emerged, then nations could hammer out temporary conventions. But of the general charter for space, the spirit, not the letter, was the essence.[29]

Finally, the space treaty was an offspring of two abiding American mentalities. The first might be termed the Wilsonian, stressing liberalism and the rule of law. Moral metaphors dominated the Wilsonian vision. The international order was a postlapsarian jungle teeming with suspicion and fear, which in turn bred militarism, imperialism, and tyranny. The United States should promote the rule of law, whereupon cooperation, trust, and disarmament might remind man of the harmonious aspirations for which he was made. It was such a hope that Rusk expressed before the Foreign Relations Committee and that Johnson implied when he counted the space treaty among his greatest efforts to shrink the arena of conflict and mistrust between the Superpowers. The second strain might be termed the Hooverian, stressing engineering and material prosperity. Here managerial and medical metaphors dominated. Poverty and ignorance weakened bodies politic and made them susceptible to the diseases of tyranny, communism, and war. Unbridled growth, fashioned by financial and technical engineers through trade, investment, and technology, could eliminate the conditions that bred political disease. LBJ embodied the Hooverian strain as well—whether in the Great Society, the Third World, or global boosterism in space.

Yet the two strains meshed roughly. Wilsonian law was meant to melt fearful competition, but in a technocratic world it was such competition that stimulated the technology necessary to the Hooverian. The legacy of the space treaty was a world of competing national technocracies in which even cooperation was an avenue for lesser powers to become competitive. Yet its globalist patina only encouraged idealists—and Third World governments—to expect a ceiling on arms racing and shared space technology to foster rapid development. Indeed, the real gainer in the space treaty was space technology itself. It grew and spread around the world, to Europe, Japan, China, India, and elsewhere by the 1980s, force-fed by national technocracies, in targeted competition, pushing into the heavens in new ways for new purposes with new organizations. If unbridled progress is salutary, then the wisdom of the space treaty is manifest.

American post-Apollo planning began in 1964, when LBJ instructed NASA to form a Future Programs Task Force. At the time Webb spoke vaguely of unmanned missions to Mars and use of the Apollo-Saturn system in the earth-moon neighborhood. Over the next five years vagueness became the pattern. Webb apparently thought it poor strategy to promulgate long-range plans, when the costs of new programs were unpredictable and sure to seem excessive in the Vietnam/Great Society era. Inevitable tinkering with long-range plans might also make NASA appear to be confused or fishing for big money. In any case, the NASA centers themselves disagreed sharply on what to do.[30] As years passed and NASA budgets started down the curve, Webb's appeals in Congress became more frantic. He mixed his frontier thesis and technocratic ideology with warnings of new Soviet surprises, while NASA reported ways that its technology and management skills might serve the war effort in Vietnam, or community planning, police, fire-fighting, education, resource management, medical care, and transportation, and help to end airline hijacking, drug traffic, crime, and pollution.[31]

All this, but no plan for space—and citing the hypothetical benefits of state-managed R & D was probably a mistake. If NASA had forged new technocratic tools for organized social change, as Webb claimed, and if the lasting benefits of the space program were in earth applications, then why not turn R & D money and management directly toward those programs? The tools were made, the crucible could rust. Inside NASA the only consensus was that it would be a shame not to exploit the operational capability achieved through Apollo. George Mueller of the Office of Manned Spaceflight wanted a manned mission to Mars, but his own planning staff, the Office of Space Science and Applications, and von Braun himself all preferred an orbital workshop, or space station, and reusable space shuttles to supply it. Meanwhile, the PSAC and Space Science Board made skeptical reports in 1967 calling for deemphasis of manned spaceflight,[32] while budgetary pressures, inflation, and the environmental movement discouraged coherent planning for space. When Webb stepped down he predicted (with some sour grapes) that the United States would lose its leadership in space in the 1970s. The first moon landing still lay in the future, but it was already clear that the American hare had stopped again to take a nap.

President-elect Nixon named a Space Task Group chaired by Spiro Agnew. It reported in September 1969 on three possible long-range space programs, each with its own budgetary level. The most ambitious called for a manned mission to Mars by the mid-1980s, an orbiting lunar station and a fifty-man earth-orbiting station served by a reusable shuttle; funding would reach $8 to $10 billion per year. The second plan postponed Mars until 1986 and kept funding under $8 billion per year. The third involved only the space station and shuttle, with annual spending between $4 and $5.7 billion.[33] Nixon chose option 3 in March 1970, then deferred the space station pending development of the shuttle. The last two Apollo flights were scrubbed and the Apollo Applications Program shrank to a single Skylab. Nixon even cancelled the MOL, again frustrating the USAF. In all this Nixon carried out a liberal agenda, just as he did by expanding social entitlement programs and imposing wage and price controls. Nixon also abolished the PSAC and Federal Council on Science and Technology, while Congress reorganized the standing space committees out of existence. Henceforth space became the purview of Senate Commerce and House Science and Technology subcommittees.

It is easy to attribute the collapse of interest in the space program to the growth of more pressing problems and to relaxation of Cold War tension. But it would be wrong to consider Vietnam, the Great Society, and other developments as isolated or in opposition to the space effort. They were all of a piece—a package that Americans purchased after Sputnik in the belief that the United States must adopt the technocratic model to get back on top. First out of the package was the space program, but Kennedy and Johnson encouraged the nation to believe not only that it could send men to the moon but that it could eradicate poverty, resist Communist expansion, and promote development abroad, to the point where the country's reach exceeded its technical and financial grasp. In time the original model for civilian technocracy, the space program, became dispensable.

Another source of the space bust was the very thing that sparked its boom—the Kennedy Effect. The bold lunar goal seemed just the fillip the United States needed at the time. But it may also have done the space program grave harm in the long run. It encouraged Congress and the nation to believe that Apollo *was* the space program. Once the race was over and won, Americans could turn back to their selfish pursuits.[34] Another Apollo was out of the question—the press and Congress laughed when Agnew, in the flush of *Apollo 11*, called for men on Mars—but so was a gradual, substantial campaign to push back the frontier in space, in the Soviet image. By the time of the last Apollo flights even men on the moon were boring. The American style of "panic and response" contributed to disillusionment in other ways. In the early 1960s Americans were told that their educational system was second rate, both in size and rigor. By the end of the decade universities were bloated and marred by declining standards and indiscipline. Science and engineering became less popular, not more, while students and the public embraced astrology, superstition, and hedonism. In the early 1960s Americans were told there was a dangerous shortage of science and engineers. By the end of the decade thousands were unemployed. In the middle 1960s Americans were told their technological superiority was assured. By the early 1970s the U.S. balance of payments was in such crisis that Nixon took the dollar off gold, while foreign cars, steel, and electronics invaded the American market. In the early 1960s Americans were told that the Third World would go communist unless the United States made emergency efforts. By the early 1970s those efforts only bought frustration in Vietnam, Third World vitriol, and the Nixon Doctrine, a formula for retreat. How relevant, then, were lunar flights of fancy?

In the end, the periodic problem of an ailing aerospace industry gave the space program back a future. By 1971 the NASA administrator feared the industry could not survive another year of diminished space work.[35] The only major new program still in the running was a space shuttle, meant to provide routine access to earth orbit at a much lower cost-per-pound than expendable rockets. NASA pushed hard for this Space Transportation System (STS) without success until 1972, when the White House became sensitive to the electoral logic of aerospace depression in states like California, Texas, and Florida. After North American Rockwell provided exaggerated estimates of the employment it would stimulate, Nixon and his adviser John Ehrlichman approved the STS.[36] And so the Space Shuttle emerged, but no decision on the goals of future spaceflight. Apollo was a matter of going to the moon and building whatever technology could get us there; the Space Shuttle was a matter of building a technology and going wherever it could take us.

Even then, NASA had to cut its initial cost estimates in half to win approval of the STS. First the payload was sharply reduced, then the orbiter redesigned according to Pentagon specifications, the operating envelope was lowered, even the "fully reusable" feature was jettisoned, and the STS evolved as an unlikely configuration consisting of the delta-winged orbiter, its nonrecoverable external tank filled with LH_2 and LOX, and two strap-on recoverable solid rocket boosters. It would lift payloads of up to 65,000 pounds into low (100+ mile) orbits suited especially for military missions. Spacecraft requiring higher orbits, up to and including geosynchronous comsats 22,000 miles out in space, required an entirely separate "space tug," or inertial upper stage, to lift satellites to their apogee, whereupon still another rocket attached to the satellite would nudge it into a circular orbit. The STS appeared to be a handy tool for lifting great weights into low earth orbit, but clumsy for anything else. It was this fact that European, especially French, competitors perceived when the American post-Apollo program took final form in 1972.

Other national technocracies were already lining up in the marketplace of space. Fifteen years before (and only eight months after the launch of *Sputnik I*) Charles de Gaulle returned to power and pledged to restore French *grandeur* through military and economic independence. But in the Space Age, even *la gloire*, that most abstract of goals, was a function above all of permanent technological revolution.[37] De Gaulle and the missile age arrived together. Now that the Soviets could directly threaten the U.S. homeland, was the American nuclear umbrella still credible? Would Americans risk Chicago or New York to save Berlin or Paris? This question reinforced French determination to pursue their own nuclear research in the teeth of American nonproliferation policy. When de Gaulle began pronouncing on NATO and plans for a nuclear *force de frappe*, his rhetoric was aimed in every case at the United States, not the USSR.[38]

But technical independence also meant a revolution at home. Under de Gaulle the state assumed the role of a managerial czardom, folding nuclear, aeronautical, and rocket agencies and private firms and universities into national teams for the force-feeding of high technology. In

1960 France exploded its first A-bomb; by 1967 solid-fueled IRBMs entered testing; by 1972 land-based missiles, nuclear-armed submarines, and Mirage jets completed a little triad of nuclear forces. As early as 1960, de Gaulle also announced plans for a French space program. The Algerian proving ground, opened in 1947 for France's share of captured V-2s, became the busiest rocket range outside those of the Superpowers. Two French space agencies gradually crept up on an orbital capacity with a series of ever more precious rockets: the Agate, Topaze, and Rubis solid-fueled stages, then the liquid-fueled Émeraude first stage, the Saphir two-stage rocket, and finally the Diamant. Its first stage was propelled by the exotic mixture of nitric acid and turpentine, the second and third were solid-fueled, and together they developed 107,000 pounds of thrust, roughly the force of the Jupiter-C that launched the United States into space in 1958.[39]

Why a French space program? Surely if prestige was a primary object of Gaullist policy, then space beckoned irresistibly. If military rocketry was to proceed anyway, then the additional effort of adapting missiles for spaceflight was surely worthwhile. But the fundamental reason for a French space program was the centrality of space technology in the drive for permanent technological revolution. France would live again in command technology, especially nuclear, aviation, space, and computer, "because their labs and their inventions provide a spur to progress throughout the whole of industry. . . ."[40] Of course, France could not compete with the Big Two in space, but such was not its purpose. Europe would inevitably enter the high-tech age as well, and France would position itself in targeted markets for leadership in the competition certain to develop in Europe.[41]

De Gaulle's hybrid, statist economy was uniquely adapted to the age of technological revolution. He explicitly rejected pure capitalism, which bore within it the seeds of a materialistic and individualistic "moral sickness." On the other hand, communism was a tyranny that plunged life "into the lugubrious atmosphere of totalitarianism without achieving anything like the results, in terms of living standards, working conditions, distribution of goods, and technological progress which are obtainable in freedom."[42] So de Gaulle anathematized both systems and sought a *juste milieu*. Competition was indeed the engine of progress, as the capitalists claimed, but it was also the solvent of community, as the Marxists claimed. Hence the competitive stimulus must be international, while domestic institutions united for technical dynamism. So Gaullist France, too, founded a unique style of technocracy and based its legitimacy on a vision of France "in the year 2000." The R & D budget quintupled in five years, and of the 5.43 billion new francs spent on R & D in 1962, 76.6 percent was governmental. As Gaullist minister Michel Debré explained in the first five-year plan for science, these funds were a *masse de manoeuvre* with which the state could target selected areas for growth,

ASTERIX-1

It is a far cry from Cape Kennedy. There are no neon signs, no drive-ins—and no night clubs. There are only some scattered huts and towers, lost in a desolate flatland as big as New Jersey, its pebbly red floor covered with a pale green haze after a spell of rain. In the huts, which are filled with electronic equipment, one can hear, almost any morning, a calm young voice on a loudspeaker saying, "dix, neuf, huit, sept. . . ." In the distance a needle with a tail of fire slowly rises above the desert and roars into the sky.[43]

The site was called Hammaguir, an adobe village where sheep, goats, and a small herd of dromedaries nosed about in the brittle weeds. The nearest town, Colomb-Béchar, lay 80 miles to the north, itself 700 miles into the Sahara from Algiers. The Treaty of Évian ended French rule in Algeria in 1962 but reserved to the metropole, for a time, its proving ground at Hammaguir, where French technicians, some in burnooses like a cosmic Foreign Legion, labored to get a French satellite into orbit before *FR-1*, another Gallic spacecraft, went aloft aboard an American Scout rocket. The French national space program was in competition, significantly, with its own cooperative program—and with other Europeans to become the third nation in space.

By mid-November of 1965, the NASA launch of *FR-1*—and a French presidential election—were only three weeks away. Now the identity of Gaullist France, wedded to technological dynamism more consciously even than Kennedy's America, rode on the outcome; ". . . *trois, deux, un* . . ." the countdown ended on November 26, 1965. Pre-set charges exploded the bolts holding down the sleek cylinder, its one large exhaust nozzle fired up to full thrust and pushed the rocket skyward. Soon France's own network of tracking stations reported in. *Asterix-1*, the modest 42-kilogram satellite named for the red-whiskered Celtic barbarian of French comics, was transmitting from orbit. Its primitive chemical batteries quit after just two days, but the Diamant had glistened, and *Le Monde* proudly proclaimed "La France Troisième 'Puissance Spatiale'!"[44]

while the spin-off effects of R & D would stimulate the economy across the board.[45]

The British, too, awoke to the cockcrow of Sputnik. "We are now in the Space Age, whether we like it or not," intoned Tory back-bencher David Price. "Our public policies, in everything from defence to education, have got to be shaped so as to be able to accommodate the sudden changes in our human environment. . . . Viewed historically, Europe dare not stand apart from the space race." But where France was in resurgence, Britain was in retreat, and Price saw European collaboration as the only road to the new frontier. At least Europe need not start from scratch, for the British Blue Streak IRBM, marked for cancellation by the government, could serve as the first stage of an all-European satellite launcher! Price also foresaw coordinated European space research, joint testing and launch facilities, new comsat systems, and R & D in other space systems neglected by the Superpowers. Was this fantastic? Not at all, for Western Europe had a combined GNP greater than the Soviet and over half the American. Without the burden of military or expensive manned space programs, the Europeans could surely compete in selected technologies of scientific and economic potential.[46]

These prognoses had a willing audience in a Europe searching for its place in the postwar, postimperial world. When Minister of Aviation Peter Thorneycroft officially offered the Blue Streak to Europe in 1961, the French, Germans, and lesser powers hastily formed a European Launch Development Organization (ELDO) to fashion their own satellite booster. Meanwhile, Pierre Auger, Sir Henry Massie, Eduardo Amaldi, and other scientists beat the drum for cooperative space research, in part to plug the "brain drain" of scientific talent to the United States. So ten nations joined in a European Space Research Organization (ESRO) in 1962, pledged to support joint R & D in space science and technology, and budgeted $306 million for the first eight years.

The numbers were acceptable to budget-minded parliaments. But was it enough? European aerospace industrialists understood better the cost and frustrations of large-scale R & D and cried out for a bolder program through an industrial lobby of ninety-nine companies called EUROSPACE. In the words of its president Jean Delorme, "Unless the European countries wish to join the ranks of the backward and underdeveloped countries within the next fifty years, they must take immediate steps to enter these new fields." A low-orbit launcher, scientific satellites, and half a billion dollars were not enough for what he called "a matter of survival."[47] EUROSPACE boldly preached a kind of Euro-Gaullism. It frowned on importing U.S. space systems even if permitted, so that Europeans might acquire experience in R & D. The payoff was in the technocratic means, not uncertain commercial ends. "European industry," according to EUROSPACE, "never considered space as a money-making activity. Of course, it cannot operate at a loss forever, but its main initial motive was to improve its technology so as to remain competitive in world markets. Space was thus visualized by the firms as a means of forming or retaining qualified teams capable of developing advanced items of equipment and also—and perhaps above all—to manage the joint development of complex subsystems or systems." No little brother role—rather "the target for European industry is clearly to acquire prime contractor ability for all space applications systems."[48]

Prime contractor status for all space applications! What did this imply for European-American cooperation? The United States was eager to share in scientific endeavor, but refused to provide launch vehicle data that might find its way into French military projects or launch services and technical aid for comsats liable to undercut INTELSAT. Europe needed its own booster, or, to paraphrase JFK, whatever the U.S. undertakes, Europeans must fully share.[49] But the EUROSPACE recommendation of an additional £218 million did not please European treasuries. So ELDO and ESRO, underfunded and poorly conceived, went down as textbook examples of how *not* to generate high technology. In addition to start-up difficulties, ESRO members quarreled over the distribution of contracts (the issue of *juste retour*) as the French, true to their intent, garnered a percentage of ESRO contracts up to twice the level of their contribution. Efficiency demanded that business go to the most qualified firms, but politics demanded "affirmative action" for countries playing "technological catchup." Either the poor subsidized the rich, or the rich subsidized mediocrity in the short run and new competition in the long run. Not until 1967 did the first *ESRO-1* satellite go into orbit, courtesy of NASA.

Meanwhile, ELDO's Europa booster underwent several redesigns that upgraded it from a low-orbit launcher to one capable of boosting comsats into geosynchronous orbit. The confusion attending these changes meant more delay and waste. Systems integration is always a challenge; a multinational rocket was a boondoggle. Every technical hurdle had to be surmounted by an international committee, and the babble of tongues only complicated the habitual lack of communication among scientists, engineers, and bureaucrats.[50] By 1969 the rocket still had not flown, though its budget had more than tripled, and Britain and Italy threatened to pull out.

By the late 1960s the European space program was a shambles—and this at the very peak of panic over the "technology gap," "brain drain," and "industrial helotry," all presumptive products of explosive American technocracy. European industry tried to capitalize on this mood, best expressed in Jean-Jacques Servan-Schreiber's *The American Challenge*, by urging larger space budgets. EUROSPACE warned that if Europe did not regain its place in the first rank of technological civilization it would soon be too late. The Germans expressed this as *Torschlusspanik*: Europe must jump through the door to the Space Age before it slammed shut. The Italian government called for a "technological Marshall Plan" and British Prime Minister Harold Wilson for a European Technological Community. Americans, of course, believed their own advertising. The Atlantic Institute, Organization for Economic Cooperation and Development, and other Euro-American institutions made earnest inquiries into how to bridge the technology gap, while Robert McNamara insisted that the real gap was not in technology, but in systems management as practiced by NASA, the USAF, and U.S. corporations. Then-professor Zbigniew Brzezinski thought that "all inventions for a long time will be made in the U.S. because we are moving so fast in technology and large-scale efforts produce inventions."[51]

In this stressful mood the Europeans passed through a whorl of confusion while the United States stumbled toward the shuttle. The Nixon administration eased its restrictions on launching foreign commercial satellites and made concessions in negotiation of the permanent INTELSAT convention. Europeans and Third World members could now outvote the Americans, on placement of contracts for instance, and planned to terminate the management of INTELSAT by the U.S. Comsat Corporation. But Europe could take little advantage until it acquired its own booster and state-of-the-art comsat technology. In 1969 the United States added insult to injury by selling the Thor-Delta booster outright to the Japanese, something it had always refused the Europeans. (Japan, unlike France, promised not to use the technology for military purposes or to compete with INTELSAT.) Meanwhile, the *Europa-2* failed four times to get something into orbit and ELDO finally collapsed in dissension.[52] To be sure, Nixon offered the Europeans a share of the STS, at first the inertial upper stage until the USAF appropriated it, then a scientific workshop called Spacelab designed to fit inside the Shuttle. The Germans were enthusiastic, but hitching a ride on the Shuttle implied a second-fiddle position that the French especially despised.

All was not bleak. France and Germany collaborated on a test comsat called *Symphonie* and pushed their national programs. The French learned through the IRBM effort and ELDO apprenticeship, and constructed an advantageous equatorial launch site at Kourou, French Guiana, to replace the Algerian facility. Above all, the European Space Conference meditated on the lessons of ESRO and ELDO and on means to ensure a unified, long-range program with guaranteed budgeting, centralization of management and systems integration, and smorgasbord participation by which member states could elect to share in some major programs but opt out of others. In December 1972, after five years of uncertainty, the conference announced a new European Space Agency (ESA) to absorb ESRO and ELDO, coordinate all European space R & D, and aim at targets of opportunity in boosters and comsats. For the Space Shuttle, impressive as it was, did not markedly improve U.S. ability to launch commercial comsats into their high orbits.

The ESA compromise fulfilled the French wish for development of an independent launch capability—the L3S rocket, dubbed Ariane—while Germany got its major cooperative program with the United States—the Spacelab—and Britain its pet project—a marine comsat. The research centers inherited from ESRO and the administrative budget remained common responsibilities. But "big R & D" was now a mixture, not a solution, of the national inputs. ESA won the loyalty of the various governments only through a partial nationalization of its international program.

On Christmas Eve, 1979, seventeen years after the birth of ELDO, the Ariane placed a European satellite into orbit from Kourou. Since then the Ariane has had mixed success, but it is the first non-Superpower launcher declared operational for competitive commerce. The French (with a 59-percent interest) promptly incorporated a "private" company, ARIANESPACE, to market launch services to the world, and won bids for comsat launches from INTELSAT, ESA, South American and Arab states, and even a few American firms. Gaullism and Euro-Gaullism coexisted, ushering in the current era of neo-mercantilist competition in space.[53]

During its decade-long snooze, when NASA spending fell to 36 percent of its Apollo peak in constant dollars, other technologies began to slip away from the United States. British, Canadians, Japanese, and Germans all sought niches for themselves in future comsat markets, while the French developed the SPOT remote-sensing satellite designed to compete with the American remote-sensing spacecraft LANDSAT in providing data for minerals prospecting, fishing, land use, mapping, and soil management. Policy straitjackets and funding cutbacks stymied the United States in exploitation of its own technology. The Space Act of 1958 gave NASA no mandate to be an operational agency, hence delay and confusion resulted in Washington over how to market services such as LANDSAT or even the Shuttle. Similarly, the separation of military and civilian space inhibited commercial use of technologies developed by the Pentagon. One by one, problems unforeseen or unresolved at the creation of U.S. space policy came to vex the would-be leader as space technology diffused around the world.[54]

The greatest handicap borne by the United States is its almost sole responsibility for the strategic defense of the non-Communist world. Even the stress on civilian over military spaceflight, a legacy of Apollo, ended in 1981, when the DoD space budget surpassed that of NASA. Less money was available for American civilian R & D in space even as the Europeans and Japanese reached technological parity in this or that space market, and Chinese, Indians, and others entered the field.[55] These perplexing problems for the competitive stance of the United States arose in more and more sectors as state-driven technological change pushed foreign governments further away from the free market that flourished in the merely industrial age. Like Britain in the late nineteenth century, the United States in the late twentieth still has an overall technological lead, but finds itself outmaneuvered by determined rivals in this sector or that market around the world. Intellectuals tended to assume that the postcapitalist age would usher in socialism. Instead it ushered in national technocracy, whether of an American, Soviet, or Gaullist variety, characterized by mobilization at home and Darwinism abroad. The lost monopoly of space expertise is symbolic of the relative decline of the United States in a neo-mercantilist world, even as its growing concentration on military space applications symbolizes its continuing struggle to stay abreast of the first and abiding technocracy—the Soviet Union.

Beginning in 1958, the first full year of the Space Age, the United States launched many times more spacecraft than the USSR. By 1966 the American lead reached 437 to 197. Then, in 1967, Soviet launches rose 54 percent in one year and continued to climb until, in 1973, the USSR placed 124 spacecraft in orbit or beyond (one every three days) to just 23 for the Americans, and surpassed the United States in the total score. To be sure, most American satellites had greater capabilities and longer lives than their Soviet equivalents, but the trend continued throughout the 1970s, demonstrating a continued high level of Soviet space spending. What was more, almost two-thirds of all Soviet spacecraft were presumptively military, compared to about half of U.S. satellites. By the early 1970s the USSR made up for its late start in applications by deploying new systems in surveillance, communications, meteorology, navigation, electronic ferreting, and varieties of space science. By any measure, the USSR never stopped racing in space.

Luna XVI (launched September 1970) and its follow-ons saved some face for the USSR on the moon. Unmanned landers scooped soil from the surface and returned it to earth. Later missions included a miniature dune buggy called *lunakhod* that crawled away from its landing site. Soviet spokesmen played up mechanical exploration, while insisting that the rational approach to manned spaceflight was to routinize earth orbit operations. The ill-starred Soyuz began again to carry men into orbit in October 1968, achieved the first Soviet space docking in 1969, then engaged in a bizarre triple flight during which three spacecraft and seven men were in orbit at once, though none docked. In April 1971 the heralded Salyut space station (a "salute" to Gagarin's flight on its tenth anniversary) went into orbit atop the Proton booster. But the Soviets were still in a slump. The first cosmonauts to dock successfully with Salyut turned up dead upon landing, suffocated as the air leaked out of their Soyuz. In 1973 two more stations failed to achieve orbit. *Salyut III* and *IV*—one scientific, one military—orbited successfully in 1974. But intermittent failures to dock still plagued the program until the long lifetime of *Salyut VI*, launched in 1977, established Soviet primacy in manned orbital operations. By the end of a decade that had seen no Americans in space after 1975, the Soviets had launched over forty manned Soyuz, shuttled cosmonauts and supplies routinely to the Salyut, and supported crews for as much as six months in space. There they performed patient experiments in space biology, gardening, materials processing, and military applications. In the early 1980s new evidence appeared of Soviet R & D on a Saturn 5-class launcher and a miniature version of a shuttle. This decade of intense activity passed almost unnoticed in the American press. The Soviets had been given a reprieve during which they might, once again, venture to catch up with the capitalists.[56]

For Americans wanted to judge the Cold War a thing of the past. First the Partial Test Ban Treaty, then the Outer Space and Nuclear Non-proliferation treaties, then the SALT negotiations, and the "end" of the space race all seemed to promise a relaxation of the relentless Soviet technological pressure. The SALT-1 Treaty of 1972 was a milestone, not only because it enshrined MAD (the balance of terror) by banning ABM deployment but because it prohibited signatories from interfering with the other's "national means of verification" (i.e., spy satellites). Thus the use (but not R & D or deployment) of antisatellite (A-SAT) systems was implicitly outlawed, propping up the U.S. policy of keeping outer space as a sanctuary for passive military systems.[57] The Superpowers fleshed out the law of space in other ways. Clauses in the space treaty inspired the Agreement on the Rescue and Return of Astronauts and Objects Launched Into Outer Space, concluded in 1968, the Convention on International Liability for Damage Caused by Space Objects of 1973, and the Convention on the Registration of Objects Launched Into Outer Space of 1976. The International Telecommunications Convention of 1973, which possesses treaty status, regulated the use of comsats and radio frequencies. The Nixon-Brezhnev Accords of 1972 provided for sharing of scientific data, exchange programs, and, of course, the Apollo-Soyuz Test Project (ASTP), in which astronauts and cosmonauts docked in orbit and exchanged "handshakes in space."[58]

None of this did much to hobble Soviet technocracy. The ASTP was obviously a creature, not a cause, of détente, and it gave Soviet technicians the chance to traipse through U.S. space facilities and study the hardware and flight operations firsthand. The integrating power of communications did not function, since the USSR fashioned its own INTERSPUTNIK system rather than participate in INTELSAT. Soviet-American exchange programs only allowed Soviet postdoctoral technicians to study critical fields of technology at eminent U.S. universities, while the KGB apparently increased its industrial espionage in the U.S. computer and microprocessing industries. Otherwise, Soviet cooperative policies have included little besides the launching of satellites for friendly Third World countries, like India, and "guest cosmonauts" from East Bloc countries (and a Frenchman in 1982) on visits to the Salyut, where their activities were restricted.

During the decade of détente American military spending fell sharply as a percentage of GNP and the federal budget. The United States froze its nuclear missile systems at 1,054 ICBMs and 656 SLBMs, while its number of operational bombers fell from 650 to 316 (by 1981). The United States did "MIRV" its missiles, so that three warheads could be carried by one vehicle, and in the 1980s began to replace its superannuated submarine fleet. But the arsenal remained the same size. By contrast, the Soviets increased their ICBM array from 500, when SALT talks began in 1967, to 1,500 in 1974. New generations of missiles followed hard upon each other, including the gigantic SS-18s with as many as ten warheads each. From 100 SLBMs in 1967 the Soviets increased their sea-based deterrent ten times by the early 1980s. Intercontinental bombers held steady at about 100, but the shorter range Backfire bomber added versatility. Even the American lead in total warheads, the nuclear stockpile, fell from being three times the Soviet to near parity. The Soviets claim greater needs since they must contend with the Chinese, British, and French in addition to the Americans, but they nevertheless used the era of détente to build a nuclear force far exceeding the demands of mere deterrence.[59]

Do the Soviets envision nuclear war as a possibility? as a legitimate extension of policy? as "winnable"? Soviet military leaders do seem to advocate cold-blooded preparation for all-out war, while civilian leaders generally hold that nuclear war is unthinkable. Yet the push for ever higher levels of technology, in the oldest tradition of the Bolshevik state, still operates. Indeed, the "Apollo Revolution" in the United States only strengthened the Soviet commitment to the S-T Revolution. President of the Academy of Sciences Keldysh warned that a worldwide acceleration in the application of research was occurring, that the imperialists were mobilizing for "stormy technological progress" through state organization of science. He and Kosygin concluded in time-honored fashion that the USSR must borrow from the West, hence détente, but strive at the same time to achieve "the very highest level in the world in the decisive fields of science and technology."[60] The imperialist West, said Kosygin, is "putting great emphasis on the aspects of the struggle of the two world systems connected with the development of the economy, science, technology, and education.... The contemporary bourgeois state ... uses the most diverse means of ensuring adequately high and stable rates of economic growth and the further development of scientific-technological progress." The Soviets' USA Institute concluded that "the internal economic mechanism of capitalist society has succeeded on the whole in adapting to the requirements of the contemporary S-T Revolution."[61]

After 1965, Soviet R & D spending continued its upward spiral, with increases averaging 9.3 percent through 1968. Science spending rose 17 percent in 1970 alone, and 5 to 10 percent annually throughout the decade. By 1980 the USSR invested 21.3 billion rubles on science, or about 3.5 percent of GNP, almost twice the U.S. effort.[62] After 1968, Brezhnev, like de Gaulle, named the space/defense sector a "technological dynamo" for the entire economy. He told the Twenty-fourth Party Congress in 1971 that "taking into account the high scientific-technical level of the defense industry, the transmission of its experience, inventions, and discoveries to all spheres of our economy acquires the highest importance."[63]

During the 1970s, therefore, the tension continued in the USSR between Traditionalists, emphasizing imperialist hostility, autarky, and home-grown technology, and Nontraditionalists, emphasizing the dynamic nature of capitalism and the tactical wisdom of détente and trade. The latter group won some victories in policy—deals with Nixon's United States and Willy Brandt's Germany provided access to Western technology—but under Brezhnev and Party theorist M. A. Suslov the Traditionalists carried the ideological field. Cooperation in space, such as the ASTP, was a double boon, since it appeared to restore the USSR to a level of equality in space after the defeat on the moon and also provided access to American technology.

And yet the space program remained under wraps. Launch failures continued to be covered up and new R & D projects kept secret pending their success. But the large military component also required secrecy. In the 1970s the Soviets tested satellite interceptors, and evidence grew of expensive Soviet research into "directed energy" technology—lasers and high-energy particle beams—that promised a "ray gun" capable of destroying satellites and even ICBMs at the speed of light. Both the United States and the USSR studied lasers, particle beams, and ABM systems generally, if only to avoid technological surprise. In 1978 the Superpowers discussed a ban on A-SATs, but the Soviets insisted that the Shuttle was itself a potential A-SAT system in need of control, and after the invasion of Afghanistan, the talks broke down.

By 1980 both powers had come to rely heavily on passive military systems in space for C^3I. Once vital military assets were located in space—or anywhere else—they became inviting targets, and methods of satellite defense had to be devised. Space as a sanctuary and even MAD itself is now challenged by the prospect of laser ABM systems operating from the ground or from orbiting battle stations. It is at least conceivable that in the twenty-first century strategic defense may again predominate and render the balance of terror a relic of the early nuclear age.[64]

The loudest opponents of an "arms race in space" have been the Third World members of the UN. They seized on the rhetoric of the 1960s as indicative of the Superpowers' intention to pursue spaceflight in the interest of global material welfare. The First UN Conference on the Peaceful Uses of Outer Space, in Vienna in 1968, encouraged them in this belief. American delegates especially conveyed the impression that space communications and remote sensing could work wonders for developing countries. In the 1970s these states mobilized around the banner of "anti-imperialism," variously defined, and when instant wealth through space technology did not materialize, they angrily denounced the First World for reneging on its promises. New sources of discord arose as well. Third World delegates in COPUOS and other bodies claimed that data from remote-sensing satellites might be used against them by profiteering multinational corporations or hostile neighbors; that direct-broadcast television satellites might flood their countries with American westerns and Coca-Cola commercials, raising material hopes that could not be fulfilled; that a first-come, first-served allocation of radio frequencies or equatorial orbits discriminated against nonspace powers. Even the moon itself became a disputed hunk of cheese as less developed countries insisted that the Draft Treaty on the Moon (1980) prohibit lunar exploitation pending an agreement for the sharing of proceeds among all nations—analogous to the draft treaty on the deep sea bed.

The United States found itself in the minority, if not isolated, on such issues, while the USSR, which shared American concern about the dangers of UN regulation, stood aside and let the Americans play the role of "heavy." The freedom of space as promoted by the United States from 1958 has survived, but at a heavy cost in adverse propaganda. Above all, the UN membership assailed the militarization of space, which, in its view, threatened peaceful uses and absorbed funds that should go to Third World applications. Again, since the Soviets claim they have no military space program, the United States absorbs the ire of the nations. This was especially evident at the second UN conference on peaceful uses, UNISPACE, in 1982.[65] Of course, the goal of many Third World governments, some Marxist, most authoritarian, is to frustrate the potential of space technology to break down their monopoly of information within their own countries. Furthermore, the sort of controls they advocate might dry up investment in spaceflight for anything *but* military and scientific missions. Still, the United States reaps the bitter harvest of its "benign hypocrisy" in space diplomacy.

The 1980s are a decade in which the political patterns of space technology are in greater flux than at any time since 1961. The U.S. Space Shuttle made its maiden voyage in 1981, rekindling national interest and seeming to restore American leadership in space. President Ronald Reagan then emulated Kennedy by calling for a civilian space station within a decade, against the counsel of his scientific, budgetary, and military advisers. Yet military space budgets surpassed that of NASA in 1981 for the first time since the 1950s, and Reagan's Strategic Defense Initiative sharply increased funding for research on space-based ABM systems able to destroy attacking missiles in their boost phase. The SDI, dubbed "Star Wars" by its critics, touched off a debate that may last a generation. It reverses twenty years of U.S. adherence to assured destruction and its corollary doctrine that space is to be a sanctuary for passive military systems and off limits to active weaponry. The SDI drew strength from new technological possibilities, from continued Soviet ABM and directed energy research suggesting that they had never embraced assured destruction, and—ironically—from the moral condemnation of nuclear weapons made by American Catholic bishops and the "peace movement" at large. But opponents of SDI insisted that space weaponization would be unworkable and enormously expensive, and would only provoke an arms race in space.

The new U.S. assault on the "High Frontier," however, belied the failure of the Space Shuttle itself to provide economical, routine access to space. After five years of operation, shuttle managers never approached their ambitious flight schedules (totaling twenty-four flights rather than twelve to twenty-four per year) and cost estimates ($650 to $2,300 per pound of payload rather than the $258 [1985 dollars] predicted). Embarrassment then gave way to tragedy in January 1986 when the Challenger with its seven crew members blew up on launch. Reliable Titan and Delta launchers then exploded in the spring, the former wasting a critical spy satellite. Instead of rocketing ahead, the U.S. space program went on hold for over a year.

Nevertheless, American public and political support held, while the accidents triggered a bureaucratic as well as technical debugging of NASA and inspired rational rethinking of the place of the shuttle and expendable launch vehicles in the space program. Far from sounding a retreat, the National Commission on Space plotted a sanguine course in 1986 for the next twenty years, including an aerospace plane, space station, moon base, and manned mission to Mars. Other Americans, such as Gerard K. O'Neill and Brian O'Leary of Princeton, Carl Sagan of Cornell, ex-governor Jerry Brown, and numerous and growing clubs such as the L-5 Society, continued to preach the gospel of a Promised Land in space. They want to mine the moon for minerals, capture asteroids and nudge them toward earth with "mass drivers," build colonies at the gravitational libration points of the earth-moon system, plant crops for a hungry earth in gigantic hydroponic farms in orbit where the sun always shines, build square miles of solar arrays in orbit to beam down in microwaves the energy demanded by twenty-first-century civilization. Meanwhile, the Soviets' new Mir ("Peace") space station, with its six docking ports for "tinker-toy" expansion, testifies to their dogged drive toward permanent habitation and a veritable Kosmograd, or space city, realizing Tsiolkovsky's dream of the conquest of gravity. It all amounts to the technocratic vision in cinemascope, the dream of voyages to that utopian realm we might dub Tsiolkovskia, where earth, then solar system, then galaxy are given over to the pious purveyors of power, where mankind's social imperfections are attacked and vanquished, one by one, until none remain.

Will any of this come about except under the impulse of competition? Must great nations always ape their rivals? Whatever the USSR undertakes, must the United States fully share? Must France follow the United States? And Japan France? and China the USSR? and India China? Surely a nation may choose to ignore the barbarians and their clever ships and guns, but it must then become a helpless giant. Rather, it is in the very structure of international politics in our age that states must, in their own ways, fashion national technocracies, the better to compete, adjusting inherited institutions and values as required, and so embark on their unique voyages to Tsiolkovskia.

CHAPTER 21

The Quest for a G.O.D.

Technocratic methods first appeared in American government in the nineteenth century and became widespread in the military emergency during and after World War II. But technocratic *ideology* captured the country only after Sputnik, when a new willingness to view state management as a social good and not a necessary evil turned a quantitative change into a qualitative one. How can we judge the effects of such change on American life? Its proponents point to all manner of marvels to win our fealty: from spaceflight to breakthroughs in medicine, energy and ecology to scientific management and systems analysis. But the irony of it is that such analysis, while it may help us choose among competing projects, cannot help us assess technocracy itself. The reason is that public programs rest on value judgments, and their costs involve such unmeasurables as loss of individual or institutional autonomy and the amorphous "quality of life." The cost of new missile systems, automobile emission standards, promotion or harassment of nuclear plants, is obviously more than the dollars or man-hours invested in programs. Society itself has to adjust to bring them about. Are such adjustments "good" or "bad" or "worth the results"?

At the peak of enthusiasm for technocracy, these questions were seldom asked. Americans looked to the future with impatience, or rued, like Benjamin Franklin, that they were born too soon to see the wonders that lay beyond the horizon. The American Academy of Arts and Sciences asked Daniel Bell to lead a Commission on the Year 2000 to help assess the impact of government programs. Think tanks followed suit, and all forecast tremendous change in government services, social structures, international organization, psychological and sexual "fulfillment"—all by extrapolating the expansion of man's knowledge and power. It seemed humanity's destiny to chase and embrace the new.[1]

The USAF, too, learned to dance to McNamara's beat. Its young technical officers pushed to completion the so-called Big L-systems begun after Sputnik, including satellite surveillance, ballistic missile warning, intelligence data processing, and overall continental air defense. In the 1960s the proliferating new technologies were linked, or "capped," through the new North American Air Defense Command, carved out of Cheyenne Mountain in Colorado, and the National Military Command System, vast networks of interlocking computers. Nuclear defense and war-fighting seemed to demand the most rigorous integration and centralization of systems, not least to preserve civilian authority and reduce chances of accidental war. Whether the sophisticated systems would survive a single nuclear blast, however, was an unsettling question.[2] The USAF also institutionalized ten-year plans for technological revolution under Project Forecast, in order "to go beyond the traditional expression of general objectives of our democratic form of government."[3]

NASA in turn recommended systems approaches for poverty programs, pollution, crime, housing, and transportation.[4] But such optimism reflected one or more planted axioms: that the benefits of technocracy would outweigh its costs, that political direction to see that they did was no great task, or that change was simply inevitable and that grousing about it was as futile as a dog's barking at the moon.

By the late 1960s reports on the second-order consequences of the R & D boom began to appear—and were almost universally nasty. For what had the United States invested $146 billion in government R & D from Sputnik to the moon landing? A mammoth pork barrel. Seven out of every eight scientists who ever existed were alive and working in the 1960s, yet no national plan for R & D existed, no articulated goals by which progress might be measured. In the eye of these critics, a veritable "R & D cult" had grown up on the promise that more power was the solution to all ills. Politicians applauded, for it absolved them of the responsibility to examine or challenge national values and institutions.[5]

The first line of defense against such criticism was the "obvious" importance of R & D to the economy. One study concluded that two-thirds of all economic growth from the Crash of 1929 to Sputnik was traceable to new technology; the average return on R & D spending was 100 percent. Critics retorted that, barring the war, that had been private R & D aimed at civilian markets. By contrast, the government funneled 35 percent of the national effort into aerospace by 1960, even though it accounted for only 4 percent of the total "value-added" by U.S. manufacturers. Core industries like mining, food processing, and textiles, which accounted for 60 percent of value-added, attracted only 8 percent of the R & D.[6] True, the economy grew rapidly in the 1960s, but this could hardly be traced to an R & D boom that had not yet had time to produce. The soaring sixties were, if anything, feeding off the breakthroughs and carefully husbanded capital of the Eisenhower years. In any case, GNP was an unreliable guide, since it lumped together growth factors with nongrowth factors like service industries, defense spending, bureaucrats' salaries, and inflation. Government spending did not necessarily yield growth, it only padded GNP numbers.[7]

Did not R & D produce valuable spin-offs? One study from the mid-

1970s claimed an overall return on NASA spending of 43 percent. But even sophisticated econometric models rested on dubious assumptions (how might funds have otherwise been spent by the government? by private citizens? how might given tasks have been otherwise performed?). The General Accounting Office thought the 43-percent figure unconvincing. Other studies spoke of phenomenal benefit-to-cost ratios in such things as increased rice production or malaria control in Southeast Asia, in order to justify a space station or LANDSAT system. But they ducked the problems and costs of implementing vast social programs, no matter how good the satellite-derived data. NASA tried to push technology transfer on to U.S. industry, yet even in point sectors like computers NASA's needs were too specialized to contribute much to civilian innovation.[8]

In any case, spin-offs were a weak justification for federal R & D. Secretary of Commerce Luther Hodges thought space and military programs too concentrated and specialized to foster overall growth. Or as Wolfgang Panofsky of Stanford quipped: "If you want the by-product, you should develop the by-product."[9] Spending billions for space or weapons might be necessary for other reasons, but to do so in hopes of serendipity was an irresponsible roll of the dice.

The R & D boom surely transformed the aerospace industry, but not, according to the critics, in any way that made it a model for the whole economy. Aerospace was an industry given to "gold-plating," making products more complicated and expensive, because it had no incentive to lower unit cost. It had an unusually high percentage of skilled employees and an R & D wage scale for researchers almost double that of agriculture. It thrived on international discord, yet in times of peace must maintain excess capacity in case of emergency, saddling firms with inordinate fixed costs. It suffered a militant Machinists Union that learned to exploit its indispensability to national defense in the same manner as its employers. Its market was an oligopsony, in which only a couple of buyers (e.g., NASA and the DoD) provided both the sales and the R & D funds needed by firms to stay alive. So aerospace had to be an unabashed suitor of the technocratic state. The Europeans, with their smaller scale, simply gave up on competition and collapsed almost all their firms into semipublic behemoths (British Aerospace, French Aerospatiale, Italian Aerospaziale, the German merger of Messerschmidt [MBB] and United Aviation [VFW]). But in the United States, as economist Horace Gray put it: "We socialize the financing of research, but permit private monopolization of its output. . . . The end product is an institutional monstrosity—a bastard form of socialism crossbred with a bastard form of capitalism."[10]

Yet consider the pressures on an aerospace executive. Technology for tomorrow elbows out systems for today. New systems are either not deployed at all or have shorter production runs ("If it works, it's obsolete

. . .''). Company health depends increasingly on R & D contracts, which in turn means premature decisions as to the feasibility and value of new ideas and an incentive to oversell new projects.[11] In the past firms invested their own capital in design competitions in hopes of steady future profits from a production run. By the 1960s R & D was an end in itself, with virtuosity, not price, the main variable. Several large R & D tasks might be undertaken simultaneously, straining design departments even as factory space lay idle. Requirements might be changed in midcourse, adding to costs and demanding flexibility. Contracts became fewer and larger, so that the balance sheet for years to come might hang on a single program—and on persuading Congress to continue funding. Executives must make overlapping decisions on whether to bid or not to bid, go for a prime or subcontract, plan for short- or long-term allocation of resources, meet current needs or create new demand, evaluate the assets of competitors and the likely calculations of government agencies choosing among bidders, specialize in a technology, shift specialties, or diversify according to trends in technology and government programs?

Government in turn must feed the aerospace industry in slack times, spread the wealth so as to foster specific expertise in two or more firms, and subsidize research. In 1964 NASA and the DoD revised their regulations to permit companies to write off a share of their own long-range management planning, especially for dislocations caused by altered government programming. McNamara favored low-bid or fixed-fee contracts to prevent cost padding, but that only exposed well-intentioned firms to the pitfalls of "unk-unks" (unknown unknowns), unanticipated hurdles on the technological frontier. An innovative, risk-taking firm, therefore, might get punished, while more favored brethren, with cost-plus contracts, were free to underestimate costs in the design stage and pad them in the performance stage. But to the extent that government reduced such risks, then technocracy turned the state into the role of venture capitalist and the private firm into the performer of the public service![12]

All in all, aerospace was a hard business in which to make an honest dollar. Aviation firms had to "bet the farm" every time they designed a new airplane for the civilian market and play politics on every government contract. To escape this, firms seized the rare moments when they had available cash to diversify (North American with Rockwell Standard and Martin with Marietta cement), or merge (McDonnell and Douglas) or both (General Dynamics out of Convair and Material Services). The restructuring of the industry proceeded with federal approval, for it meant a cushion against fluctuations in government spending.[13]

How could the government police the billions of dollars that flowed to private firms? Expanding the arsenal system seemed to border on socialism. But the alternative "contract state," according to critics, prevented socialism only by corrupting capitalism. Congressional inquiry into the USAF's "sweetheart deal" with Ramo-Wooldridge on the Atlas ICBM did create enough heat in the late 1950s so that NASA chose to retain some in-house capacity to set standards of cost, speed, and quality in command R & D. But over time even NASA obliged Huntsville and the JPL to contract out more and more of their work. Political scientist H. L. Nieburg called this "throwing away the yardstick." Bereft of in-house talent, government agencies contracted with another parasite of technocracy, the consulting firm, to oversee the contractors, as in NASA's Apollo deal with Bellcomm.[14] Even James Webb confessed that "where 90 percent of the funds appropriated to us are spent with non-governmental entities, you have a fairly tenuous control of these activities through normal contractual relationships."[15]

The picture emerging from such critiques was of a bloated R & D complex propagating a technocratic ideology to milk agencies incapable of judging even their performance, much less the social value of their products. How much blame should rest with greed, how much with a system that multiplied temptations, and how much with a technocratic age that broke down categories of public and private, socialism and capitalism, is a matter of temperament. To inquire whether an arsenal system, or any system, might have improved matters would call for a systems analysis of systems analysis, and still rest at bottom on a value judgment.

The explosion of state-funded R & D was not confined to aerospace. A great society needed great technology, which needed great science. It was common currency in the 1960s that basic research contributed to the "storehouse of knowledge" and somehow fueled progress automatically. Lloyd Berkner claimed that for each new Ph.D. we could employ five to ten engineers and for each engineer ten to fifteen skilled workers.[16] Yet by the end of the decade thousands of Ph.D. technicians went begging, while government R & D had little to offer unemployed workers in industries like steel or autos. Said the critics, the big R & D of the 1960s had been gilding the lily while the vegetable patch dried up.

Basic science, however, was above all a university matter. In 1960 Seaborg had called for more and more scientific spending with no end in sight. That was good news for administrators who built enrollments, laboratories, and research "empires" at public expense and for professors encouraged to spend their time on grantsmanship and consulting rather than teaching. Like defense contractors before them, academics learned to hustle, with a nose for the smoldering issues on Capitol Hill and an ear for the current buzzwords. Federal support rose from $170 million in 1955 to $1.1 billion in 1965, with an additional $640 million allotted to university-managed government labs, or 75 percent of all university research.[17] Could such a torrent be neatly funneled into the most deserving projects? Could such a plethora of worthy projects exist? The critics said no, that waste, shoddy work, duplication, and diminishing returns were bound to creep in. Universities "kept up with the Joneses" in expensive physics and astronomy facilities, and expanded their charge to the government for overhead—everything from bookkeeping to janitors—until it ate up an agreed-upon one-third of all monies. By 1963 the government supplied 88 percent of the entire Caltech budget, 66 percent of MIT's, 59 and 56 percent of the University of Chicago's and Princeton's, and a 25 percent chunk of Harvard's and Stanford's.[18] Such campuses inevitably compromised their role as detached centers of inquiry and were politicized by military and social involvement alike. Once universities took federal funds for any purpose, they exposed themselves to regulation, investigation, and coercion on matters of academic freedom, classification, affirmative action, labor relations, and even curriculum, obviating meritocracy, peer review, and self-governance.

Social scientists resented the prizes heaped on their colleagues and questioned a society that valued technical more than "human-oriented" research . . . at least until the Great Society cut them in as well. Federal grants for social science rose from $35 million in 1960 to $222 million in 1967, making a cottage industry of jargon-laden, often redundant, usually quantitative or prescriptive studies of social problems. The Office of Science and Technology insisted that the academic community learn to blend technical judgments on poverty, birth control, and education with those from economics, sociology, and the behavioral sciences. Jerome Wiesner urged a bridging of the communications gap, since social scientists would increasingly be using tools and concepts of the natural scientists, while the latter must become increasingly concerned with the social effects of their labors.[19] Sociologist Kingsley Davis testified that "the first nation which breaks through the barrier and manages to put social science on a footing at least as sound as that of the natural sciences, will be way ahead of every other nation in the world. I would like to see the U.S. be that nation. . . ."[20]

Even assuming that scientists, engineers, and social engineers could integrate their disciplines for the coherent planning of social change, was such practice compatible with American values? French futurist Bertrand de Jouvenel glimpsed three ages of history: those ruled by priests, by lawyers, and by scientists. The politics of the first age derived from sacred scripture and a presumption of popular ignorance; the politics of the second from human scripture and the presumption that We The People could judge matters of common concern; the politics of the third age were anomalous. *Demos* still had the responsibility, but had lost the competence to judge matters of arcane technology. "This great age of science is, by way of corollary, an age of personal ignorance."[21] Must policy, therefore, fall into the hands of a technical elite? It was this that troubled Eisenhower, not to mention Stalin, and was currently inspiring Mao's Cultural Revolution. American critics spoke of scientists as "the new priesthood."[22] Yet even Eisenhower had known the quandary of conflicting "expert" advice, as well as the need to override scientific advice for political reasons. The scientific community itself proved to be a "complex and variegated social order," falling more into division and impotence than concentrated power.[23] The logic of technocracy suggested that if knowledge is power, then the pursuit of knowledge is *ipso facto* a political activity. Technicians in the United States, though not to the same extent as in the USSR, became creatures of the politicians, another skill group at the beck and call of government, like the military officer corps. (Strategist Werner Schilling even saw an affinity between scientists and generals, since they were both predisposed to action and took the "whole problem" approach.)[24] But if the experts cannot, or do not, run society, then Jouvenel's anomaly stands: ours is a society run *for* the acceleration of change, *on* analytical principles, but *by* politicians and pressure groups incompetent to judge the extended effects of change.

Thus the indictment of technocracy—justifiable, and yet still unfair to the thousands of honest, patriotic, and sometimes brilliant scientists, businessmen, and public servants who labored to conquer space, cure disease, or defend the Western world. Still, technocracy arrogated to the state immense power, first through the military-industrial-scientific complex, then through the welfare-warfare state—and did not even fulfill its promises. But what did the critics propose as solutions? The answer usually came back: more technocracy.

To wit: The government sought to cut back on cost overruns. But its solution was excessively complicated contracts, running to six or eight fat volumes, that only contributed to the overhead costs of R & D. It added compliance standards involving employment of minorities, distribution of subcontracts, relations with unions, and so on until the Armed Force Procurement Regulations, for instance, expanded from 100 to 1,200 pages by 1967.[25] The copious (David) Bell Report of 1962 addressed the oversight problem, but its recommendations included not only higher government wage scales to hold in-house personnel but controls on wages and benefits for all private contractors, more federal entities to monitor performance, and the expansion of the NSF and PSAC into super-agencies for the collation and diffusion of the entire national fund of knowledge.[26] For systems managers like McNamara or Webb and political scientists like Nieburg or Brzezinski the answer was more rigorous cost accounting. Seaborg insisted that the United States accelerate still more its cycle of knowledge creation and use, like the motorist who hopes to make his ragged engine purr by revving it all the more. Critic Michael Reagan proposed changing priorities: first priority should be "those social objectives which are defined as most urgent politically"; second was scientific education; third, undirected small-scale research; and last priority to Big Science like Apollo. But that only begged the question. What if some Big Science project was deemed most urgent politically? By the 1970s "technology assessment" was advanced to probe the unanticipated problems of growth. But a phrase is not a panacea, and benign efforts at such assessment only revealed the absence of shared values among scientists, engineers, businessmen, and bureaucrats.[27] All the consequences of new technology might be managed only by total control from the center, something Americans presumably would not want in any event. But feeding technical revolution without such control demoted the state to the role of sorcerer's apprentice.

Priorities, regulation, stricter application of management techniques, technology assessment: the solutions seemed a product of the same mentality as the problems. Occasional renegades said as much. Princeton researchers wondered if "the recent emphasis on cost-effectiveness may not be too much of a good thing"; USAF brass grumbled about "no new weapons systems since 1960" and began to look upon the frugal Eisenhower years as the good old days; a RAND analyst identified the divergent interests within business and government, but had no solution, since no measure of R & D "efficiency" existed.[28] Webb never really solved the dilemma at NASA. Planners, he said, interfered with daily operations "as if they were smarter than everyone else," but if they did not familiarize themselves with realities in the field, their planning existed in a vacuum. Webb tried a joint committee of planners and researchers, but it flopped: "People could not come in frankly to discuss their problems. They were just determined not to let a central planning destroy their own initiative."[29] Kennedy aide Carl Kaysen thought the whole systems approach wasteful; Assistant Secretary of Defense for R & D, John Rubel, railed against military monopolization of talent. Brodie of RAND even doubted the value of systems analysis in military strategy, since one had to make dispositions against a cunning and unknowable enemy rather than a predictable Nature. "Scientific" management only seduced its practitioners into thinking themselves objective.[30]

Such suspicion about technocratic methods, pressure to shift attention to social priorities, the rise of New Left academics, attacks on the military exacerbated by Vietnam, increasing demands for material results, technology assessment, and the environmental movement all combined to decelerate the R & D engine in the 1970s.[31] Military spending, some 50 percent of the federal budget in 1960, crashed to 23 percent in 1980, while health, education, and welfare spending rose from 22 to 49 percent in the same period. Federal R & D spending sagged, then rose more slowly than inflation until it amounted to 2.3 percent of GNP in 1977, compared to 3.5 percent for the USSR and 1.8 to 2.3 percent for the Europeans and Japanese.[32] The R & D lull clobbered aerospace for a few years, but it hardly signaled a return to smaller budgets, deficits, or government. On the contrary, the alleged cures for technocratic excess only applied new layers of government to the planning and management of R & D, subordinated the Treasury to even greater claims in the social sector, and re-created the abuses of aerospace in a "welfare industry" that had no more of a stake in actually "winning" its wars on poverty and discrimination than the Pentagon did in downgrading the Soviet threat.[33] Rather, big government got bigger, social groups dependent on the state proliferated, and techniques of managed change, problematical even in hard R & D, were applied to society as a whole.

The antitechnocratic backlash of the 1970s produced yet another phenomenon: a profound anti-intellectualism beginning, ironically, on the campuses themselves but spreading throughout the society. It manifested itself in rebellions against traditional curricula, against rationalism, against authority, and flirtations with gurus, cults, and other heterodox movements in the religious and secular realms alike. Why the backlash? A *prima facie* case certainly suggests itself that the rebellions and movements of the 1970s were a product in part of disillusionment, of the overselling of technocracy in the 1960s. Unfortunately, it appeared as if many American citizens, from "alternate life-stylists" on the Left to fundamentalists on the Right, had rejected science along with scientism, technology along with technocracy. And so respect for the principles and fruits of science and technology, for scientists and engineers, declined in the United States in the 1970s—a perverse product of the enthusiasm of the 1960s.[34]

Eisenhower was surely right—the American system was not set up for central planning, nor did its values condone it. But JFK was also right—the old invisible hand was no longer equal to the foreign challenge in an age of technological revolution. In the end, the United States got the worst of both worlds: a "free market" twisted at every turn by state intervention and a technocratic state incapable of managing the change it provoked. The Soviet state, by contrast, could make up in tyranny and disregard for deleterious side effects what its centralized technocracy lost in creativity, while France and Japan could plan with some success for specific commercial goals because their societies were relatively small, relatively free from military complications, and wedded neither to an individualist nor a collectivist ethic. But the United States was obliged to fight a total Cold War—like the USSR—yet retain democratic institutions—like France or Japan. It didn't work, and the technocratic machine soon broke down before imponderables ranging from the stubbornness of the North Vietnamese to the anti-intellectual rebellion on the campus, the public's boredom with space, the perturbing influence of special interests, and exploitation of the system by the mediocre, lazy, or corrupt.

What went wrong? Would better technocracy, more systems analysis make possible the methodical achievement of social goals? Or did the trouble lie in the system itself?

A satellite really is a little world, an enclosed space of some few cubic feet, nestled in the top stage of a rocket. It may have antennae and solar panels tucked away like the wings of a dozing bird. Otherwise, it is a fixed volume that its makers pack full of experiments, telemetry, and electronics, making rational trade-offs to squeeze the greatest possible performance from the system. To be sure, the satellite is part of a larger system, the rocket, but that is a thoroughly known quantity. Higher-energy stages might be developed, or a different booster altogether substituted, but the rocket-spacecraft system is still a billiard ball universe, in which every effect has a known (or in the case of malfunction, discoverable) cause and every cause a predictable effect. To be sure, the genius of the robot-masters at JPL, for instance, who spirited the Viking spacecraft to Mars and walked it through experiments a hundred million miles away, seems more like art than engineering. If so, it is still, as Hyman Rickover described systems analysis, "art based on the scientific method."

As systems expand, however, reality fades. Planning future R & D on launch vehicles calls for judgments of cost, feasibility, lead times, and what sorts of missions might be flown in the future. Those depend on planning for the whole space program, judgments as to how space technology may or may not serve economic, foreign, or military policy. At this point the space program becomes a subsystem of the state. How best can government allocate its scarce resources, and what does "best" mean in each case? The state in turn exists in a society, churning with factions, enthusiasms, and panics, and in the largest human unit of all, the international system, in which sovereign competitors behave in ways both predictable and perverse. Every decision at a higher level changes the decisional environment at lower levels until cross-currents race up and down like snow on a television tube, and systems managers find their rational analyses subject to so many contingencies that reason itself becomes as sounding brass.

Imagine playing a leisurely game of chess by mail and exulting over the board, "Aha! I've got him now. If he uncovers his pawn, I take it with my knight; and if he doesn't, I can attack his bishop with my queen." Now imagine a game in which you are obliged to write your orders twelve moves in advance, while other players, on bigger boards, are deciding that knights will not be funded next year or are upgrading bishops to move like queens; and players on still bigger boards are deciding to get everyone out of chess altogether and into parcheesi.

The holy grail in systems analysis is the closed system. But even assuming away extraneous inputs and presuming perfect internal control, one is still left with the question of values. What is it all for? Why strive to maximize that instead of this, or anything at all? Nieburg perceived that all modern nations were moving toward integrated state control on the assumption that "science and technology equals increased wealth and happiness." This myth was now so mighty that the values bred by four hundred years of history face a pervasive challenge in this century. The unfinished task of the present generation is somehow to preserve the irreducible human values of this heritage despite the inevitable necessity of enlarged public authority, the overriding demands of security and national interests, the encompassment of individual lives and fortunes by impersonal and relentless mass organizations.[35]

An elegant summary, but what are those irreducible human values threatened in our century—human dignity? charity? life, liberty, and the pursuit of happiness? truth, justice, and the American way? Nieburg does not say. Nor does he ask whether those values themselves validated the "inevitable necessities" of twentieth-century government. Is technocracy at war with inherited values, or an expression of those values?

Now, the material quantitative, utilitarian bias of systems analysis makes sense in designing a spacecraft, but it runs into snags as soon as someone asks, "Why build this spacecraft rather than another? or none at all?" and gets thoroughly tangled once second-order consequences are taken into account. Pesticides or natural predators are introduced into an ecology to eliminate swarming pests only to become pestiferous in turn. A ban on chemicals and a return to "natural foods" strokes environmentalists but condemns less privileged people to a poorer diet. Automation increases production, but also unemployment. Controls on energy production—nuclear plants or strip mining or oil refineries—cleanse the environment but withhold cheap energy from common people.[36] In every case, a laudable decision at one level of analysis becomes a clumsy intrusion at another.

Critics who look at the "big picture," like Alvin Toffler in *Future Shock*, or the Club of Rome in *Limits to Growth*, or Jeremy Rifkin in *Entropy*, trace the technocratic predicament to the ultimate finitude of resources and vulnerability of the biosphere.[37] But even a "holistic" observer, perceiving the entire system from "outside," still views the world and its inhabitants as a machine that would run smoothly if offending human subsystems (that overpopulate and burn fossil fuels) could be controlled. But that is not only another (larger) systems approach, it is also fallacious. Would reducing the human population to a tenth of its current number, or perfecting a limitless, clean energy like nuclear fusion, cover a multitude of sins? Certainly not, unless one can know what the privileged survivors chose to do with all that cheap energy. Chances are they would use it to build weapons, appliances, billboards, cities—and to support more offspring.[38]

The earth may be a holistic unit, but it is not a "system" in the sense that systems analysts use the word. For a system is a "set of components that work together for the overall objective of the whole."[39] What is the overall objective of the earth or the human race? To ecologist David Ehrenfeld, "Spaceship Earth" is the product of an engineering, not an ecological perspective. One can construct artificial life-support systems in a spaceship, independent of the astronauts who will live off them. But the systems that support life on earth are complex, natural, and self-regulating, and the lives they support are part of the whole.[40] Technocratic man is inside the system he means to improve from without.

A wise analyst would admit that government is not really a system either, since politics prevent a coherent "overall objective." Still, the analyst hopes that progress can be made against poverty, pollution, and so on, if only logic can triumph over politics, superstition, and other enemies of rationality.[41] Communist parties claim to embody that logic. But no one ever aspired to abolish politics in the United States. It seems rather that American leaders endorsed centralized mobilization of national energies on the assumptions that change meant progress; progress was good; and optimal distribution of its fruits would result from the natural play of pluralistic politics. The happy conclusion was that government need not worry much about questions of value: more R & D, education, social equality, and so forth would spawn more wealth and power, which in turn would provide the wherewithal for more spending on welfare, defense, . . . and R & D. Undesirable side effects could be redressed through the democratic representation of social groups, preferably under the next administration. C. West Churchman, a leading theorist on systems analysis, observed that the fundamental test of a system's objective is whether its managers will knowingly sacrifice other goals on behalf of the objective.[42] American technocracy, however, has always promised economic growth *and* clean air, support for excellence *and* mass education, a booming stock market *and* universal welfare, guns *and* butter. The objective of American technocracy was to avoid defining its objective.

Hence the United States got bad technocracy. But how many Americans would want to live in a "good" technocracy, a well-oiled totalism really capable of driving society relentlessly toward a preprogrammed future? Modern mythologists have always imagined such a system with horror: Franz Kafka's *The Castle*, Mary Shelley's *Frankenstein*, Karel Čapek's robot world of *R.U.R.*, George Orwell's *1984*, Aldous Huxley's *Brave New World*, C. S. Lewis's *That Hideous Strength*, J. R. R. Tolkien's *Lord of the Rings*.[43] The lesson is always that the technocratic promise is either a cheat or a nightmare, that to wield total power, even with the best of intentions, would be our undoing. In short, we dream of a system that gives us all we want but could not bear to live in a system powerful enough to do so.

So what is the point of it all? British science writer C. H. Waddington thought state research should aim at "ensuring that life is biologically enjoyable."[44] Is that what government is about, to make pets of its people? Even if we could agree on the goals of our social system, how can we be sure that policy A or program B really moves us closer to their fulfillment? What every system needs is something outside itself to ensure that its managers are gaining on their goals (i.e., that progress exists) and that their goals are worth pursuing (i.e., that good, or value, exists).[45] This "something outside" (in the words of Wroe Alderson and Churchman) is a Guarantor of Decisions, or Guarantor of Destiny: that is, a G.O.D. Every philosophy that makes room for human purpose needs its G.O.D.: the Chinese *I Ching* (Book of Changes); the Fates of Greek mythology; the Atman, or universal being, of Hinduism; the "Reason that governs the universe" in Spinoza's pantheism; the "propositions about reality that lie beyond doubt" to a Cartesian rationalist; or, for St. Augustine or Maimonides or Mohammed, the Divine Will itself.[46] To dispense with a guarantor is to sink into relativism and endless doubt. To ask whether one should trust in a guarantor is tautological—that is what the word "should" means. What is good is what is G.O.D. is what is good.

The Soviets, of course, have Marx's dialectical materialism, which interprets the flow of history, and Lenin's corollary, which holds the Communist Party to be incapable of acting other than as the agent of history. They would presumably condemn the critique of technocracy as (in Bukharin's words) "a new medievalism."[47] But what is the G.O.D. of the West? Ehrenfeld identifies it as our humanist faith in human reason, which he deems a full-fledged religion (of which Marxism is itself a denomination). "We have been fooled," he writes, "by our humanist cant into thinking that we are actually learning to steer the planet in its orbit." The assumptions permitting such folly are: (1) all problems are soluble; (2) all problems are soluble by people; (3) many problems are soluble are technology; (4) those that are not have solutions in politics, economics, and social science; (5) we will always apply ourselves in a crisis and achieve a solution before it is too late; (6) some resources are infinite, while those that are not have substitutes; (7) human civilization will survive.[48] These assumptions cut across all party lines, Left and Right, socialist and capitalist. But they add up to a solipsism: whatever advances human power is good for humanity. Man is the measure of all things, his own guarantor: "I look upon my work, and behold, it is good."[49]

Various groups of Americans threw themselves across the march of technocracy by the end of the first decade of the Space Age. Some accepted the humanist assumptions but criticized their shoddy implementation. Some joined the "counterculture," damning technocracy as alienating, inhumane, materialistic, macho, and polluting but meanwhile living off the system they "trashed" and practicing a deviant private morality that hardly recommended their paradigm for society. Mailer spoke of them as his "abominable army" that debauched and dropped out while "they," with cool discipline, have taken the moon.[50] A very few, in the "Thoreau to Woody Guthrie tradition," moved to sylvan cabins or remote farms. Most Americans, even when conscious of a profound undernourishment, did none of these things. After all, our social stability, our standard of living, possibly our national survival, depend on our keeping the system functioning, however awkwardly. How could Americans dare to dismantle technocracy so long as the international imperative impels it forward? Only a world government might remove the impulse, but to be effective a world government must itself be a technological, bureaucratic tyranny—the Soviet, not the Anglo-Saxon, model on a global scale.

The United States, and the world, is caught in a *Flucht nach vorn*, a flight into the future. What do we fear most, that technocracy will be perfected, or that it won't be? Americans delight in such futurist epics as *Star Trek* and *Star Wars* precisely because the human qualities of a Captain Kirk or Han Solo are always victorious over the very technological mega-systems that make their adventures possible. We want to believe that we can subsume our individualism into the rationality of systems yet retain our humanity still.

Perhaps we can. Apollo itself, as Churchman demonstrated under contract to NASA, could not be justified by cost-benefit analysis.[51] The symbol of the system failed the tests of the system, a hint perhaps that even technocratic man aspires to the frivolous. Lewis Mumford thought the space program "technological exhibitionism" and the analog of pyramid or cathedral building, requiring society itself to become a vast machine for the service of inert idols.[52] But how can this be? For analysis would never endorse a cathedral, or the moonship, in the first place. If the promises made for systems analysis are fraudulent, as this chapter has argued, then we must look elsewhere for our G.O.D.

CHAPTER 22

A Fire in the Sun

Out of the blue and into the black
You pay for this but they give you that
And once you're gone, you can never come back
When you're out of the blue and into the black.
—NEIL YOUNG, "Hey Hey, My My," 1979

"But what manner of use would it be ploughing through that blackness?" asked Drinian. "Use?" replied Reepicheep. "Use, captain? If by use you mean filling our bellies or our purses, I confess it will be no use at all." . . .

"Fly, fly! About with your ship and fly!" [cried the crazed castaway in that lightless realm.] "This is the island where dreams come true . . . dreams, do you understand—come to life, come real. Not daydreams: dreams."
—C. S. LEWIS, The Voyage of the "Dawn Treader," 1952

Yonder stands the orphan with his gun
Crying like a fire in the sun
Look out—the saints are coming through
And it's all over now, Baby Blue.
—BOB DYLAN, "It's All Over Now, Baby Blue," 1965

Put them in fear, O Lord,
That the nations may know themselves
To be but men.   —Psalm 9:20

In June 1983, the *Pioneer 10* spacecraft, 2.7 billion miles distant, left the solar system forever—mankind's first celestial homerun. It reminded me how a skinny infielder once discovered in himself, in his twelfth or thirteenth spring, the new-found power to reach the wall of his schoolyard outfield—an intoxicating onset to a sober adolescence. Is our own such an age of rapid muscular growth, restless rebellion, psychological depression, and undirected self-assertion? Or is man a perennial adolescent, striving for maturity, longing for childhood?

In this book I have tried to trace the political origins of space technology. Why end up padding about the Looking Glass gardens of systems analysis or philosophy? Because readers who have indulged me this far are not content to take refuge in mellonolatry or misoneism, the first being the worship of the future, the second the hatred of change. We must ponder the ambiguities of the Space Age, because we have to live in it, willy-nilly.

Our technological civilization has evolved for centuries. But the international rivalry of our own time, enhanced by the Communist victory of 1917 and culminating in Sputnik, induced a saltation in the politics of technology through the transformation of the state into the dominant promoter and manager of technological progress. Alexander Gerschenkron theorized that the more economically backward a country, the more the state must play a role in forcing change.[1] In the current age of rapid and perpetual technical advance, *all* countries have become "backward" on a permanent basis. Hence the institutionalization of wartime methods, the suspension of peacetime values, the blurring of distinctions between the state and society, and the apparent erosion of cultural differences around the world. History, as any alert undergraduate would attest, has been speeding up, and the leading nations justify their accelerating pace of innovation by the need to maintain military and economic security. Yet that very progress may undermine the values that make a society worth defending in the first place.

Prescient observers like Bertrand Russell, Aldous Huxley, and Leo Tolstoy sensed the dark side of state-supported "invention of the future" as early as the turn of the century.[2] Even Jules Verne saw it coming in his last *voyages extraordinaires*. What is it about the Space Age, then, that is novel? Surely not the urge to abolish the unknown. Long before Sputnik, African empire-builder Cecil Rhodes thrust his clutches skyward and cried, "I would annex the planets if I could!" and polar explorer Fridtjof Nansen shrugged: "It is therefore to no purpose to discuss the uses of knowledge—man wants to know and when he ceases to do so he is no longer man."[3] Is it that for the first time men believed it really possible to eradicate disease, poverty, inequality, or violence? Perhaps— but that very belief may be as dangerous as it is uplifting, for it leads to the land where nightmares as well as daydreams come true. Why is it, as Malcolm Muggeridge asked, that the quest for heaven on earth always ends in a *gulag*? Is it that human beings will indeed colonize the solar system and turn the earth into a grand public park, like some rehabilitated inner city? Perhaps—but there is no reason to believe, as Tsiolkovsky did, that space colonies would be free of greed, envy, politics, and war. Can the scientific knowledge or new perspectives gleaned from space exploration spawn a higher consciousness or wisdom and prepare a new, sublime culture? Perhaps—but way back in the introduction, we read that Daniel Bell suggested otherwise. Culture, he said, responded to the timeless questions of love and death and the meaning of life, which are impervious to technological change. But even if culture *is* independent of technics, might not the rationalist, materialist values of our technocratic age compete with or transform inherited values?

We tend instinctively to assume that technological progress is somehow "Western" and the Apollo program very "American." Daniel Boorstin dubbed us The Republic of Technology, and John William Ward (meditating on the meaning of Lindbergh's flight), a cult of both hero and machine.[4] Some scholars conclude that this technical bent in Western civilization stems in part *from* the Judaeo-Christian world view, which affirmed the reality and goodness of the material world and bade mankind to subdue the earth, or the rationalist theology of the medieval Schoolmen and the clockwork routine of the monasteries. But if this be so, why the conflict between science and religion in our own day? Historian Herbert Butterfield attempted to sort out this muddle by arguing that secularization of European thought and the demoralizing wars of religion combined to elbow theology out of its central position as the source of knowledge. In time, natural philosophy ceased to be a subset of moral philosophy and laid claim to a universal competence. Science, studying the nature of Nature, and religion, studying the natures of God and man, parted ways, and each was the poorer for it.[5] As Freud might have it, science, a child of religion, murdered its father and set itself up as master.

There is a quick way out of this predicament. It is to deny historic culture altogether, define material values as the only true values, and surrender to the machines. This sounds fantastic, but it has its adherents, and not only in the Soviet Union. Hear Dean Wooldridge, the cofounder of the firm that did systems integration on the Atlas:

> In the late 1960s we seem justified in the broadest possible application of what may be called the central thesis of physical biology: that a single body of natural laws operating on a single set of material particles completely accounts for the origin and properties of living organisms. . . . Accordingly, man is essentially no more than a complex machine.[6]

This was surely a blow to our self-esteem, but we would accept it, continued Wooldridge, and attain true rationality, including personality control and an end to superstitions about free will (which "simply doesn't exist") and a personal God (for which "there is obviously no room"). Governments would compensate by "investing their own institutions with a semireligious aura that the unanalytic can worship. . . . Whether this will make the citizen's life more or less pleasant is not our concern here. At issue is only the question of the continued moral behavior of individuals after they have learned that men are machines." Far from suffering, according to Wooldridge, "men who know they are machines should be able to bring a higher degree of objectivity to bear on their problems than machines that think they are Men."[7]

Buckminster Fuller shared this reverse conceit, defining man as "a self-balancing, 28-jointed adapter base biped, an electro-chemical reduction plant, integral with the segregated stowages of special energy extracts in storage batteries, for subsequent actuation of thousands of hydraulic and pneumatic pumps, [and so on, and so on]."[8] Bruce Mazlish wondered at our reaction once a computer finally spit out, "Cogito, ergo sum." It would, he wrote, be the Fourth Discontinuity in history. The first three were the Copernican, teaching that our earth was not the center of the universe; the Darwinian, arguing that we were not created by, or in the image of, God; and the Freudian, asserting that we were not even masters in our own house. The Fourth Discontinuity obliges us to ask what distinguishes us from our machines.[9] Science fiction writers have suggested that humanity is only a middle rung on the evolutionary ladder; once we create perfect self-replicating machines our day will be over, and we can sleep in the peace of extinction. Churchman himself, no fantasist, concluded that the real issue facing humanity is not "does God exist?" but rather "how to design a god?"[10]

Scholar-philosophers like Mumford and Jacques Ellul rejected such man-machine objectivism as neurotic mythologizing or as a disguise for the old lust for power. Two French critics called ours a new Dark Age in which the astronaut was the new knight, the space race the new chivalry, and space exploration the new imperialism that would bring cosmic tyranny. We were realizing the myth of Icarus, for whom it was not natural to fly and who was propelled only by incorrigible pride.[11]

Is the G.O.D. we wondered about at the end of the last chapter simply pride? If so, is it the pride that says we are *not* just machines, but special? Or the pride that says that we can conquer the universe, and ourselves, *with* our machines? Or is this whole line of reasoning a product of fallacious "nothing but" thinking? According to Aldous Huxley:

> Human beings, it is more or less tacitly assumed, are *nothing but* bodies, animals, even machines; the only really real elements of reality are matter and energy in their measurable aspects; values are *nothing but* illusions that have somehow got themselves mixed up with our experience of the world; mental happenings are *nothing but* epiphenomena, produced by and entirely dependent on physiology; spirituality is *nothing but* wish fulfillment and misdirected sex; and so on. The political consequences of this "nothing but" philosophy are clearly apparent in that widespread indifference to the values of human personality and human life, which are so characteristic of the present age.[12]

Huxley was writing in 1946 in the backwash of Nazi genocide and medical "experiments," Stalinist "liquidation" of class enemies, and Allied carpet-bombing of cities. He might just as well have been writing against a background of terrorism, street crime, infanticide, drug cultures, mental "therapy" for Soviet dissenters, and soon of gene-splicing, thought control, and space colonies. "Nothing but" reductionism characterizes apologies for technocracy, but more to the point, it even characterized the early responses to the Space Age made by representatives of organized religion. The orthodox Christian position, espoused by Pius XII at the dawn of the atomic age and iterated before and after Sputnik, held that technology was *nothing but* a tool for human use. It was morally neutral, and man, exercising free will guided by grace, used it either for good or ill. The pope blessed efforts to launch a satellite in 1956 and saw no conflict between science and religion as long as scientists and theologians pursued truth honestly within the limits of their own disciplines.[13] Later, six years after Sputnik, Pope John XXIII addressed the "problem" posed by modern science directly in his encyclical "Pacem in Terris." Scientific and technical competence and experience were not enough to elevate social relationships to a genuinely human level, that is, to an order whose foundation was truth. For that it was necessary to pursue science within a moral order, for each individual to live and act within his own conscience so as to synthesize technical elements and spiritual values. Nevertheless, granted the pope, economic organizations, trade unions, legal systems, political regimes, and more would all have to be adjusted in the era of the atom and the conquest of space.[14]

Lutheran Wernher von Braun, an orthodox layman, conveyed a similar message with the clarity of an engineer, the erudition of a classically educated aristocrat, and the authority of one who knew totalitarianism:

Technology and ethics are sisters. While technology controls the forces of Nature around us, ethics controls the forces of Nature within us.... I think it is a fair assumption that the Ten Commandments are entirely adequate, without amendments, to cope with all the problems the Technological Revolution not only has brought up, but will bring up in the future....

It has frequently been stated that scientific enlightenment and religious belief are incompatible. I consider it one of the greatest tragedies of our times that this equally stupid and dangerous error is so widely believed....[15]

Modernist Christians and Jews, however, could not accept the notion that ancient scripture was as relevant as ever. Since nuclear weapons and space travel had ushered in an age of history wholly different from those that had gone before, millennia-old teachings on the human condition were *nothing but* surviving lore from a less-enlightened age. The most iconoclastic "theologians," like Gabriel Vahanian, announced the Death of God, since Christianity no longer "informed" modern life. Ours was a "post-Christian era" in which "technological mastery is bound to affect man in a different relation, or the absence of a relation, to God."[16] Episcopal Bishop James Pike of San Francisco thought space exploration an expression of God's gift of reason, but one that enjoined clergy to

speak in the voice of our time, employing the techniques now at hand to probe the problems of individuals; for many of the things we think of as bad in man and once called "sinful," we now understand are due to psychological compulsions and conditioning. In the complexity of today's society, we cannot overcome this "sin" by making holy noises suitable for a past age. Counselling rather than preaching, treatment rather than penance are the more useful remedial procedures. Science has penetrated every aspect of our life and so must religion. Religion must enter the marketplace.[17]

The National Council of Churches of Christ believed Sputnik emphasized the necessity for faith and spoke vaguely of government working "more sensitively and comprehensively through appropriate bilateral and regional means." The Boston University School of Theology asked after "the place of religion in the satellite era" and found it in "the area of psychic-health-preserving cultural survival." Its professor of "Applied Christianity" thought the Space Age impelled us to "shake off the superstitions of the past" and "end divisive rivalries among the ethical religions of the world."[18] It is hard to find in these responses anything that religion offered beyond what psychotherapy, diplomacy, and foreign aid already promised. For the most part, the liberal theologians responded to the Space Age like the Wizard of Oz. Behind the opaque curtain of their seminarian phraseology lurked the same old social engineer. When cosmonaut Pavel Popovich sneered that "God is my co-pilot—and his name is Adrian Grigorevich Nikolayev!" it was left to a Muslim editor to retort: "How did [the cosmonauts] expect to see God? Sitting down? Standing up? ... We should warn them against worshipping their machines."[19]

The Soviets, of course, reflected the third position on the spiritual implications of space technology, the atheistic, insisting that space probes were the last nails in God's coffin. "When man conquers the universe," wrote Moscow historian Zheya Sveltilova, "he will learn to believe in himself. It will simply be ridiculous to rely on any force other than himself. People who now believe in God will reject him. Such belief won't be logical or natural. Man will be stronger than God."[20]

It seems to me that these reactions, ranging from the orthodox to the atheist, missed the point of Space Age technology. Let us take them in reverse order, since the atheistic is most readily disposable. If God exists, mankind can hardly be "stronger than Him" or He would not be God at all. Nor could He be contingent on the works or beliefs of man, or hover superciliously between the earth and moon, waiting for cosmonauts to come up and tell Him that He's a Communist. On the other hand, if God does not exist, no amount of scientific progress can ever demonstrate the fact, since by definition it cannot measure the supernatural. The materialist may ridicule the theist for his "blind" faith but is himself silent before the retort, "How do you know God *doesn't* exist?"

The liberal position retains belief in God but is so tinged with historicism that it tends toward surrender to the technocratic world view. If religions must adjust their fundamental creeds to technological change in order to be relevant, then they spoke to nothing sacred or eternal or "really real" in the first place. All the world's great religions admonish against bending to the "winds of doctrine." To deny the reality of sin, as Bishop Pike did in deference to psychoanalysis and sociology, is to deny both the soul and the need for salvation. To reinterpret karma or the Tao in terms of psychological adjustment, or the Law of Moses or the Koran as simply social management in a pastoral culture, is not to update religion but to strip it of what makes it religion.

There is another problem with the liberal theological position. It is the assumption that new technologies are in fact so mighty in the first place. Obviously nuclear energy, missiles, or computers are many orders of magnitude more powerful than their functional equivalents of earlier times. But religion ultimately hangs on the relationship between man and God. (We are assuming, with the liberal theologians, that God exists.) Can our inventions alter that a whit? The Mount St. Helens volcano erupted with the force of hundreds of megatons—an entire nuclear stockpile beneath one mountain. Migrating plovers navigate from arctic to antarctic and back with greater precision than a Minuteman missile. The nearest planet that might sustain life is at least five light-years distant—what is a moon voyage, then, but the poking of a toe on to our own front porch? The coming disastrous earthquake—and it is indeed coming, sooner or later, in California or Japan or elsewhere—should explode the myth of man's immense power as abruptly as the Lisbon earthquake demolished the eighteenth-century ideal of a benevolent, ordered Nature. At the first atomic bomb test, Robert Oppenheimer had "a searing flash of self-recognition" that recalled to him the *Bhagavad-Gita:* "I am become Death, the destroyer of worlds." These days we applaud his shame but ignore his bravado. He was no such thing, and no matter how intimidating (to ourselves) man's power may grow, it will always be picayune beside that of Nature.

To be sure, the scientistic world view tried to assimilate these facts by denigrating mankind and calling it humility. Modern astronomy, it was said, has debunked the fond hope that the earth, the sun, even our Milky Way galaxy is the center of the universe. Rather we are "living out our lives on a tiny hunk of rock and metal circling one of 250 billion stars that make up our galaxy in a universe of billions of galaxies."[21] Carl Sagan went even further, chiding mankind for its human chauvinism, earth chauvinism, even carbon chauvinism, since life might conceivably evolve elsewhere on a different organic chemistry. He urged a cosmic consciousness derived from the insight that we are all just "star stuff" fabricated from elements born in stellar reactions; "from dust to dust" become "from star-dust to star-dust." Arthur C. Clarke agreed that "Astronomy, as nothing else does, teaches men humility."[22]

Others preached the humility of the "cosmic clock." If the earth's age of 4.5 billion years is represented as twenty-four hours, then it was not until 10:51 P.M. that life emerged, twenty-two seconds before midnight that man stood upright, and all of recorded history occurred in the last three-tenths of a second. Clearly, it seemed, the unimaginably large, old, and dynamic universe plays out the real drama of existence with no more reference to its human ants than human historians make to microscopic plankton in their own accounts.

The "enlightened" response to astronomy, space travel, and their contemplation was thus one of *pride and wonder* at man's *works* (the universe is our oyster) and *contempt* for man *himself* (we are inconsequential dust). Here was the burden of the "nothing but" mentality: as man's power waxed, man himself was diminished. But it is likely that this enlightened response was doubly wrong. First, as we have seen, man's works are as nothing compared to the power trapped, for instance, just five or ten miles away from us—in the earth's crust. But however paltry human power compared to Nature's, by what logic can we say that mankind is insignificant? Are Leonardo da Vinci and St. Francis of Assisi unworthy of notice, because they were just two specks of organic matter on the edge of a minor peninsula off a nondescript continent in a cosmos beyond their imagining? Is a diamond less significant than a hunk of coal? And if man *is* so unimportant in the real scheme of things, why *not* blow ourselves up? Who cares?

The eventual impact of spaceflight, I imagine, will be precisely the reverse of this enlightened viewpoint. It will teach us to have *contempt* for our *works* and *value* for *ourselves*. That, it seems to me, is true humility.

Finally, there was the orthodox response to the Space Age, summarized by the Jesuit W. Norris Clarke:

> We can bear the responsibility of trying to become sons of God, and stewards under Him, of the cosmos that He planned and made for us. We cannot bear the burden of trying to be lonely gods of a purposeless universe we did not make, with no other place to go and no strength or wisdom but our own to rely on.[23]

The Space Age provided us with a "God-given challenge and opportunity" based on new power "as potent for evil as for good," but through repentance we might plunge fearless into the future, into our ever unfinished task of doing God's will.[24]

The orthodox response to the Space Age, though consistent, still appeared naive. Could technology really be morally neutral? To begin with, orthodox theology taught that mankind is a fallen race, consumed with pride. More power, therefore, only multiplied temptation and magnified capacity for destruction. The Archbishop of Canterbury went so far, after Sputnik, as to surmise that "it might be within the providence of God that the human race should destroy itself by nuclear weapons." The Bishop of Rochester thought total destruction a better fate than totalitarian serfdom: "in an evil world, war can be the lesser of two evils."[25]

Orthodoxy failed to consider the hypothesis that the pursuit of modern technology, whatever its ultimate use, "makes demands upon those who would use it and transmits values to those who embrace it. . . ." Technology had "a logic of its own."[26] The result was a growing "tension between the need to control technology and our wish to preserve our values. . . ." The very planning required by technology "demands explicit recognition of value hierarchies. . . ."[27] Technologically derived experience might have nothing to say to us about the timeless concerns of culture—love, death, justice—but it did inevitably attract people's attention away from those concerns! When the technocratic promise began to ring true, in the words of James Killian, the United States began "yielding on its commitment to freedom. In universities, communications, and other sectors of society, the essential qualities of adventure and freedom have been subject to growing constraints, especially through rampant government regulation. . . ." Killian saw "grave danger in overzealous efforts to seek a risk-free society. . . ."[28] If new technologies, however they were used, ipso facto seduced people into placing security above the love for truth, or material comforts above spiritual longings, then how could the orthodox position on technology suffice?

If the preceding survey is plausible, it appears that the historical conflict between science and religion has been fought out on false issues. The fact that the medieval church embraced Aristotelian positions in natural philosophy is irrelevant; the God of the Bible would not fall off His throne because Galileo looked through a telescope. Nor has modern physics confirmed the existence of a perfectly predictable, mechanical universe.[29] The "real sin" of Galileo, as Mumford perceived, was not just in overthrowing accumulated dogmas and doctrines of the church but in trading the totality of human experience "for that minute portion which can be observed within a limited time-span and interpreted in terms of mass and motion, while denying importance to the unmediated realities of human experience, from which science itself is only a refined ideological derivative."[30] In other words, to live one's life on the assumptions that nothing exists except what can be measured by our senses, and that no joy or meaning exist except carnal pleasure and psychological adjustment is to make as extreme an act of faith as do the religious. Even pragmatic philosopher William James shocked his turn of the century colleagues with this: "I can, of course, put myself into the sectarian scientist's attitude, and imagine vividly that the world of sensations and scientific laws and objects may be all. But whenever I do this, I hear that inward monitor . . . whispering the word 'bosh!' Humbug is humbug, even though it bear a scientific name. . . ."[31]

Yet even if one granted separate but equal validity to material and spiritual experience, the fact remained that "high technology" not only forced societies to adjust their politics, law, and institutions—a "social invention," according to Mazlish[32]—but also encouraged them to alter expectations. That is, the mere existence of technologies, regardless of how these are used, could create patterns of human behavior that render inadequate the simple sentiment: "technology doesn't hurt people, people hurt people." This is what weakens orthodoxy. But liberal theology, in turn, breaks down because of the impossibility of knowing whether technocracy will really "improve" human life, or whether a fallen race will instead blow itself up. Liberal theologian Harvey Cox, therefore, set aside the Judaeo-Christian traditions of teleology and apocalypse and invoked that of prophecy as the one that "accords well with the possibilities created by technology."[33] But where, and who, are the prophets?

We stand with our technology like Dylan's orphan with his gun, fatherless and dangerous, as inadequate before the cosmic presence as a fire in the sun. But I do suspect that long before our descendants are all psychically programmed test-tube clones, science will have reached absolute limits in its quest for the origins, extent, composition, and fate of the material universe. When that day arrives, the technocratic pump may cavitate, the human heart have a meltdown, and science become again a branch of moral philosophy.

Let me explain. Throughout the twentieth century the systematic cross-fertilization of science and technology has extended our senses almost to the ultimate. When we think of space science, "big ticket" items come to mind, like Viking and Pioneer. But the greatest payoffs have come from satellites costing a few millions of dollars: X-ray astronomy, for instance, culminating in *Uhuru*, launched from Italy's San Marco platform off the coast of Kenya in 1970. *Uhuru* sensed 161 X-ray sources in the galaxy, confirming that pulsating sources—pulsars—were in fact binary stars, one of which must be extremely small and dense, a white dwarf or neutron star (equal to 1.4 solar masses compressed into a ball six miles in radius) or even a black hole. "The existence of a black hole in the X-ray binary Cygnus X-1 has profound implications for all astronomy," wrote Homer Newell. "In black holes matter has returned to a condition similar to the primordial state from which the Universe was created. The potential scientific and intellectual returns from this research are clearly staggering."[34]

Cosmological debate, too, entered a new round with the recent hypothesis that background radiation equally distributed in space is nothing other than an echo of the Big Bang of Creation. The Space Telescope, scheduled for launch by the Space Shuttle in this decade, will "see" seven times farther than before, perhaps to the "end" of the universe—and illustrate what "see" and "end" mean in Einsteinian space. But soon will come a time when we cannot "see" any deeper into the past or into the distance. At the same time physicists continue to splice the atom into protons, neutrons, and electrons, muons, neutrinos, lambda and sigma particles, cascades, photons, pions, and K-mesons, quarks and charmed quarks, until there will be nothing left to splice. The director of the Oak Ridge Institute of Nuclear Studies (and Episcopal priest) William Pollard, thought it

inevitable that men should question why any atom possesses the given property under discussion and not some slightly different one. As soon, however, as this question is raised and an explanation for it found, it is clear that whatever will be the more basic entity which explains the property in question, it will not itself possess that property. Pursuing this line of reasoning step by step, we see that the only possible ultimate solution of the atomic quest will be to arrive at a really elementary component of all matter which, while accounting completely for all the manifold diversity of matter and its associated phenomena as we observe and experience it, will not itself possess any observable property whatever. . . .

The explanation of why the universe is the way it is and not some other way, the reasons for all this special particularity of creation must obviously be sought for outside creation. It is only our momentary enthusiasm over each new partial step towards, and achievement in, understanding nature which has given us a kind of myopia. . . .[35]

Whether we pursue the infinite or the infinitesimal, sooner or later we will arrive at a point where the next step will take us right through the boundary of space–time. Pollard cautioned that these undeniable limits cannot lead us to a knowledge of God.[36] It is mathematically possible to construct universes with more than three dimensions of space or more than one of time, to assume that natural laws are still sovereign even if we can never know them. But if this is what we must rely on to banish the supernatural, then we open a can of worms. For if other, unknowable laws of space and time can exist, then the known laws by which science debunks, for instance, the biblical account of Creation, may not be reliable. Once again, Bob Dylan:

> Then I awoke in anger, so alone and terrified.
> I put my fingers against the glass
> and bowed my head and cried.

My instinct tells me that our science and technology, feeble as they are in controlling Nature, are so acute in studying it that they will soon reveal their limits. It is then that man must confess the mortality of his works, without turning on them or himself with contumely. It is then that the orthodox message is a sure guide: God made us, is disappointed in us, but loves us anyway, by which we are redeemed. Technology is our subcreation. We made it, we will be disappointed in it, but we must love it anyway, or it cannot be redeemed. The temptation will grow to reject science along with scientism, reason along with the worship of reason. But Frankenstein's "monster" Karl, defective brain and all, ran amok only after people turned on him in hatred and fear.

How can we avoid either worshipping or hating technology—mellon-olatry or misoneism? We cannot banish the will to power by identifying it, naming it, digging up its roots in human nature and the international imperative. Somehow, rather, we must try to compete for power without making power an end in itself, pursue knowledge without mistaking knowledge for truth, cultivate our subcreation yet know its fruits as only part of reality, and remember that for everything we gain there is something lost.

> From these holy waters, born anew
> I came, like trees by change of calendars
> Renewed with new-sprung foliage through and through
> Pure and prepared to leap up to the stars.[37]

Dante's stars were in heaven. Ours are only in space, and it is probably only after our adventures in the stars that we shall find spring again. For we are a stiff-necked people. Once we have gone as far as we can on our own, we might consent to follow our friend *Eusthenopteron* and, as Plato imagined, rise up, like fishes peering out of the sea, descry the things there, and, if our strength can endure the light, know that there is "the true heaven, the true light, and the true Earth."[38]

Appendix

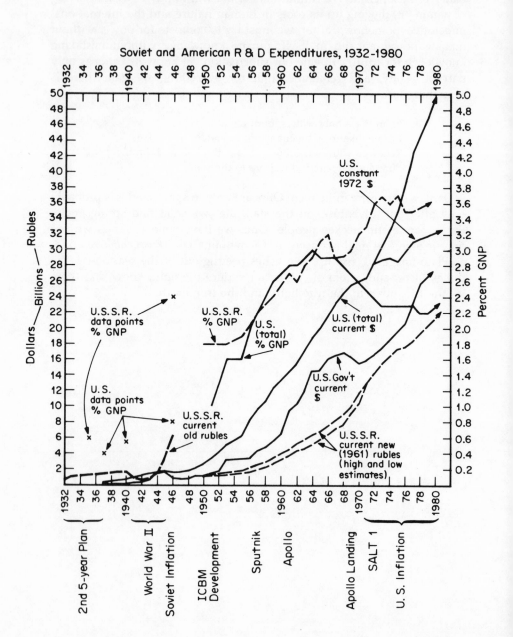

Soviet and American R & D Expenditures, 1932-1980

R & D Expenditures, 1927–1980: Data Points

Year	USSR (million rubles)	% GNP	U.S. Govt ($ million)	U.S. Total	% GNP
1927	≃120[a]				
1932	≃700[b]				
1935	≃1,100	0.6[b]			
1937			70[c]	est. 264[c]	0.4
1938			78		
1940			74[d]	est. 570	0.6
1941	1,651[e]		198		
1942			280		
1943			602		
1944	1,300		1,377		
1945	2,000		1,591		
				est.	
1946	6,300	2.4	918	1,600	0.8
1947			900		
1948			855		
1949			1,082		
1950	1,000[f]	1.8	1,083		
1951			1,301	3,200	1.0
1952			1,816		
1953	[h]1,130–1,360[h]	1.8[f]	3,101	5,100[g]	1.6[j]
1954	1,300–1,570		3,148	5,700	1.6
1955	1,500–1,810	1.9	3,308	6,200	1.6
1956	1,700–2,100		3,446	8,400	2.0
1957	2,000–2,410	2.2	4,462	9,800	2.2
1958	2,420–2,820		4,991	10,700	2.4
1959	[i]2,820–3,300[f]	2.4	5,806	12,400	2.6
1960	3,100–3,900		7,744	13,500	2.7
1961	3,800–4,500	2.7	9,287	14,300	2.8
1962	4,300–5,200	2.6	10,387	15,400	2.8
1963	4,700–5,800	2.8	12,012	17,100	2.9
1964	5,100–6,400	2.9	14,707	18,900	3.0
1965	5,900–6,900	3.1	14,889	20,000	2.9
1966	6,500–7,500	3.2	16,018	21,800	2.9
1967	7,200–8,200	2.9	16,859	23,100	2.9
1968	7,800–9,000		17,049	24,600	2.8
1969	9,300–10,000	3.0	16,348	25,600	2.7
1970	10,200–11,700	3.2	15,736	25,900	2.6
1971	13,000–13,000	3.3	15,992	26,600	2.5
1972	14,400	3.6	16,743	28,400	2.4
1973	15,700	3.7	17,510	30,600	2.3
1974	16,500	3.6	18,326	32,700	2.3
1975	17,400	3.7	19,590	35,200	2.3

Year	USSR (million rubles)	% GNP	U.S. Govt ($ million)	U.S. Total	% GNP
1976	17,700	3.5	20,668	38,800	2.3
1977	18,300	3.5	23,244	42,900	2.3
1978	19,300		25,851	47,300	2.2
1979	20,200			51,600	2.2
1980	21,300				

Notes: "Total U.S. R & D expenditures" include all private (corporations, foundations, and universities) as well as public outlays. U.S. spending in constant dollars after 1972 is from the National Science Board, *Science Indicators 1978* (Washington, D.C., 1978), p. 44. The seemingly meteoric increase of U.S. expenditures in the 1970s was due largely to inflation. The USSR in fact was spending not only a far higher percentage of GNP on R & D than the United States around 1980, but a larger absolute sum as well. Direct dollar-ruble comparisons are extremely tentative. The "R & D ruble" has been calculated to be worth as much as $4, which would suggest that the USSR was investing 68 percent more on R & D than the United States in 1978. Expert Robert Campbell "compromises" on 50 percent (or a Soviet investment 1.5 times that of the American), but is dubious even of this, given the lack of a common production function for the two countries (i.e., what results each country gets from a dollar or ruble spent). See Robert Campbell, NSF 80-SP-0727, *Soviet R & D Statistics 1977–1980* (Washington, D.C., 1980) pp. 26–34. The relentless Soviet R & D push in all phases of its history is undeniable, however, whatever the corresponding U.S. figures.

[a] Estimate based on an "almost sixfold" increase in R & D spending during the first Five Year Plan: Bruce Parrott (citing Robert A. Lewis), *Politics and Technology in the Soviet Union* (Cambridge, Mass.: 1983), p. 27.
[b] Estimate based on a "60 percent" increase in R & D under the second Five Year Plan and a level in 1935 twice that of the United States: Robert A. Lewis, "Some Aspects of the R & D Effort in the Soviet Union, 1924–35," *Science Studies* 2 (1972): 162.
[c] National Resources Committee, *Research—A National Resource, vol 1, Relation of the Federal Government to Research* (Washington, D.C., 1938), p. 69.
[d] National Science Foundation, NSF 78-300, *Federal Funds for Research, Development, and Other Scientific Activities*, vol. 26 (Washington, D.C., 1978), p. 3.
[e] Official Soviet science budgets: Parrott, *Politics and Technology*, pp. 100–01.
[f] National Science Foundation 78-SP-1023, by Robert W. Campbell, *Reference Source on Soviet R & D Statistics 1950–1978* (1978), p. 2.
[g] National Science Foundation, NSF 75-307, *National Patterns of R & D Resources. Funds and Personnel in the United States, 1953–1975* (Washington, D.C., 1975), pp. 18–19; and NSF 78-313, *National Patterns . . . 1953–1978–79* (Washington, D.C., 1978), p. 36.
[h] Estimates (in 1961 new rubles) are based on the known 1960 figures for R & D, and Parrott's calculation to the effect that R & D spending increases averaged between 15 and 18 percent from 1953 to 1960 (*Politics and Technology*, p. 157).
[i] Estimates are based on the official science budget and calculations of Nancy Nimitz, "Soviet Expenditures on Scientific Research" (RAND, 1963), contrasting with those of Campbell to produce the "range" shown on the graph. Soviet figures beginning with 1950 are calculated in 1961 "new rubles."
[j] NSF, *National Patterns of R & D Resources, 1953–1975*, p. 3.

Abbreviations Used in Notes

AFBMD	Air Force Ballistic Missile Division
AFCHO	Air Force Command History Office
AFSC	Air Force Systems Command
ARDC	Air Research and Development Command
Califano Treaty History	Office files of Joseph Califano, Celestial Bodies Treaty Negotiating History
CINCSAC	Commander-in-Chief, Strategic Air Command
DDE Library	Dwight D. Eisenhower, Library, Abilene, Kansas
FRUS	*Foreign Relations of the United States*
HO	History Office
HQ	headquarters
HST Library	Harry S Truman Library, Independence, Missouri
IMEMO	Institut Mirovoi Ekonomiki i Mezhdunarodnykh Otnoshenil
JFK Library	John F. Kennedy Library, Waltham, Massachusetts
LBJ Library	Lyndon B. Johnson Library, Austin, Texas
LC	Library of Congress
NA	National Archives
NASA HO	NASA History Office
NASM	National Air and Space Museum
Papers of the Senate Space Committee	Papers of the Senate Committee on Aeronautical and Space Sciences
PP of P	*Public Papers of the Presidents*
USIA	United States Information Agency

Notes

Introduction

1. Wernher von Braun proposed this analogy at the time of *Apollo 11*, while others spun variations on the theme of Columbian voyages and President Nixon spoke of the moon mission as the "greatest week since creation." See Bruce Mazlish, "Following the Sun," *Wilson Quarterly* 4 (Autumn 1980): 90–93.

2. Nicholas Daniloff, *The Kremlin and the Cosmos* (New York, 1972), p. 20*n*.

3. A pioneering overview of the political and social history of military technology is William H. McNeill's *The Pursuit of Power: Technology, Armed Force and Society since 1000 A.D.* (Chicago, 1982). Chaps. 6 and 7 trace the beginnings of the return to "command economies" in the Atlantic world under the pressure of international competition. The origins of command technology are briefly sketched in Maurice Pearton's new survey, *The Knowledgeable State: Diplomacy, War and Technology Since 1830* (London, 1982).

4. On the technocratic impulse of the Bolshevik Revolution, see especially Kendall E. Bailes, *Technology and Society Under Lenin and Stalin: Origins of the Soviet Technical Intelligentsia, 1917–1941* (Princeton, N.J., 1978), and Bruce Parrott, *Politics and Technology in the Soviet Union* (Cambridge, Mass., 1983).

5. For two extreme views on the implications of centralized state-supported R & D from the era of World War II, compare the optimistic, technocratic notions of J. D. Bernal, *The Social Function of Science* (London, 1939), who extolled the "science of science" and planned technology as practiced in the Soviet Union, expected such methods to solve the material problems of humanity, and predicted the coming of "the socialized, integrated, scientific world organization," with the liberal John R. Baker, *Science and the Planned State* (New York, 1945), who demonstrated in deliberate fashion that the central planning of science is essentially part of the totalitarian theory of the state (pp. 15–23).

6. The phrase "the American Century" originated as the title of a nineteenth-century magazine. Walter Lippmann adopted it and eventually became associated with it. By 1945 it seemed finally to have arrived. Americans always believed in their exceptionalism, but only after World War II was it clear that the United States had inherited from Great Britain the mantle of liberal, insular world power, acquiring the benefits and responsibilities of the world's technological and commercial leader and "balancer" of the world political system.

7. Liberal intellectual disparagement of Eisenhower throughout the 1950s and 1960s combined with lingering ultraconservative resentment to tarnish Ike's initial historical image. Recently that image has been polished up in light of the inflation, Great Society boondoggles, and continuing arms race that followed his term and of which he had warned. See, for instance, Robert A. Divine, *Eisenhower and the Cold War* (Oxford, 1981), and the citations in chapter 5, note 4.

8. Johnson as quoted by Richard Hirsch and Joseph Trento, *The National Aeronautics and Space Administration* (New York, 1973), p. 107.

9. Daniel J. Boorstin, *The Republic of Technology: Reflections on Our Future Community* (New York, 1978), p. 30.

10. In Bacon's seventeenth-century utopia, *The New Atlantis*, the research scientist took the place of the philosopher-king, the research institute the place of the church, and earthly utopia the place of a transcendental heaven. Bacon coined the phrase "knowledge is power" and foresaw the expansion of the human empire until all things were possible. See Arthur Johnston, ed., *The Advancement of Learning and the New Atlantis* [1627] (London, 1974), pp. 213–47. Nineteenth-century liberals, utopian socialists, and positivists, despite

their differing philosophies, embraced the promise of unending progress manifest in industrial revolution. Karl Marx, in *The Poverty of Philosophy* [1847] (New York, 1963), pp. 109–10, offered the pithiest distillation of a technological determinism: "The hand-mill gives you society with the feudal lord; the steam-mill society with the industrial capitalist." But recent scholars argue persuasively that Marx does not deserve the label of "technological determinist," and in fact developed a sophisticated and subtle approach to the role of technology in history. See Nathan Rosenberg, *Inside the Black Box: Technology and Economics* (New York, 1982), pp. 34–51; and Donald MacKenzie, "Marx and the Machine," *Technology and Culture* 25 (July 1984): 473–502.

11. See Lewis Mumford, *The Myth of the Machine*, vol. 2: *The Pentagon of Power* (New York, 1964). Mumford's 1930 classic, *Technics and Civilization*, betrayed a more optimistic view of human sovereignty:

> Our mechanical civilization . . . is not absolute. All its mechanisms are dependent on human aims and desires: many of them flourish in direct proportion to our failure to achieve rational social cooperation and integrated personalities. In the last century or two of social disruption, we were tempted by an excess of faith in the machine to do everything by means of it. We were like a child left alone with a paint brush who applies it impartially to unpainted wood, to varnished furniture, to the table cloth, to his toys, and to his own face. (pp. 426–27)

A similar view is expressed by Jacques Ellul in "The Technological Order," *Technology and Culture* 3 (Fall 1962): 394–421. See also Jacques Ellul, *Le système technicien* (Paris, 1977).

12. The "chicken and egg" problem is succinctly posed by Robert Heilbroner, "Do Machines Make History?" in *Technology and Culture*, ed. Melvin Kranzberg and William H. Davenport (New York, 1972), pp. 28–40.

13. Daniel Bell, "Technology, Nature, and Society," in *The Winding Passage: Essays and Sociological Journeys 1960–1980* (Cambridge, Mass., 1980), pp. 3–33. See also Daniel Bell, *The Cultural Contradictions of Capitalism* (New York, 1976), pp. 3–32.

14. Bell's trisection of human experience into society, polity, and culture begs comparison with Jacob Burkhardt's triad of the state, religion, and culture in *Reflections on History* [1905] (London, 1943). According to Burkhardt, one of the three "powers" *is* dominant in any age and conditions (or suppresses) the other two. Writing in the nineteenth century, he predicted the coming dominance of the state, the death of culture, the triumph of materialism and collectivism. So, too, did T. S. Eliot, "Notes Toward a Definition of Culture" [1949], reprinted in Eliot, *Christianity and Culture* (New York, 1968), pp. 79–202. But Bell's assumption of *discontinuity* among the three realms allows for optimism—the possibility, at least, of liberty and culture surviving in a mass technological age.

15. In this I differ sharply with sociologist William Sims Bainbridge, *The Spaceflight/Revolution: A Sociological Study* (New York, 1976). See the discussion in part I herein.

16. Bruce Mazlish, *The Railroad and the Space Program: An Exploration in Historical Analogy* (Cambridge, Mass., 1965), pp. 36–82. This formulation may not be what Daniel Bell had in mind by his separation of realms. It is more an expression of Arnold Toynbee's "challenge and response" or Leopold von Ranke's "primacy of foreign policy." But Bell himself acknowledged the role of international competition as a forcing house of domestic change in *The End of Ideology* (New York, 1960):

> Politics today is not a reflex of any internal class divisions but is shaped by international events. . . . The need for containment [of the Soviet Union] has set in its wake a whole consequence of political and social changes: the military build-up, regional military alliances, the creation of a "dual economy," a new role for science and scientists—all of which have reworked the map of American society. (p. 14)

17. Bell, "Technology, Nature, and Society," p. 7. Theda Skocpol, *States and Social Revolutions: A Comparative Analysis of France, Russia, and China* (New York, 1979), argues, despite an avowedly Marxist perspective, that the inability of the state to adjust to crises stemming from foreign competition is an important cause of great revolutions like the French, Russian, and Chinese. When popular revolution does *not* occur, it is because the challenged state has generally adjusted through revolution from above, or "White Revolution," in Henry Kissinger's evocative phrase.

Part I

Headquotes: Herzen in *From the Other Shore,* ed. Isaiah Berlin (Cleveland, 1963), p. 42; Konstantin Tsiolkovsky, "Investigating Space with Reactive Devices" [1911] in *Works on Rocket Technology,* ed. M. K. Tikhonravov [Moscow, 1947], NASA TT F-243, pp. 81–82; Lenin in *Polnoe sobranie sochinenii,* 5th ed. (Moscow, 1958–65), vol. 26, p. 116; Stalin quoted in Kendall Bailes, *Technology and Society under Lenin and Stalin, Origins of the Soviet Technical Intelligentsia, 1917–1941* (Princeton, N.J.: 1978), p. 160.

1. D. I. Mendeleev, *Arkhiv,* vol. 1 (Leningrad, 1951), p. 36, cited by Bailes, *Technology and Society,* p. 19. Other quotes from James H. Billington, *Fire in the Minds of Men: The Origins of the Revolutionary Faith* (New York, 1980), p. 387. On the tsaricide, see Adam Ulam, *In the Name of the People* (New York, 1977), pp. 326–56.

2. I. A. Slukai, *Russian Rocketry: A Historical Survey [Rakety i traditsii],* [Moscow, 1965], NASA TT F-426 (Jerusalem, 1968), pp. 8–9; N. A. Rynin, *Interplanetary Flight and Communications* [Leningrad, 1928], trans. Israel Society for Scientific Translations (Jerusalem, 1970), vol. 4, p. 37; Willy Ley, *Rockets, Missiles, and Men in Space* (New York, 1969), pp. 113–15.

3. Rynin, *Interplanetary Flight,* vol. 7, p. 29. The dream of flight in the atmosphere led mankind to envy the birds: "to fly like a bird"; "free as a bird." But Bob Dylan, meditating on freedom, asked if birds were really free from the "chains" of the sky. The dream of space travel transcends even the air and gravity in which birds and airplanes are "trapped."

4. Konstantin I. Tsiolkovsky, *The Call of the Cosmos* (Moscow, 1960), p. 80.

5. A. Kosmodemyanski, *Konstantin Tsiolkovsky: His Life and Work* (Moscow, 1956), pp. 14ff.

6. Tsiolkovsky, "Investigating Space with Reaction Devices," pp. 81–82.

7. In *The Spaceflight/Revolution: A Sociological Study* (New York, 1976), William Sims Bainbridge argues that spaceflight in the mid-twentieth century was almost accidental. It was not the product, he says, of natural military exigencies, but of a "social movement" of plucky visionaries. For instance, the Nazi regime did not take up the rocket and "thus produce the V-2, rather the Spaceflight Movement *caused the German military to be taken up by the rocket*" (p. 4). Later, he says, it was fortuitous timing that permitted Russian and American rocketeers to beguile their governments into the first space programs (pp. 1–11). This thesis, while puckish and illuminating, ignores the true relationship between the Soviet and German rocketeers and their totalitarian masters, the logic of Soviet military needs after World War II, and especially the elemental role of technological dynamism in Communist theory and organization.

Chapter 1

1. Jules Verne first published *De la terre à la lune* (From the Earth to the Moon) in 1865, and it was first translated into English in 1873. H. G. Wells published *The War of the Worlds* about a Martian invasion in 1898 and *The First Men in the Moon* in 1901 (it was serialized the previous year). Edgar Rice Burroughs began his "Barzoom" series on Mars on the eve of World War I.

2. Konstantin Tsiolkovsky, "Exploration of Cosmic Space with Reactive Devices," *Pioneers of Rocket Technology* (Moscow, 1964), p. 54, also cited by Nicholas Daniloff, *The Kremlin and the Cosmos* (New York, 1972), p. 13, and Wernher von Braun and Frederick I. Ordway III, *History of Rocketry and Space Travel* (New York, 1968), p. 41.

3. N. A. Rynin, *Interplanetary Flight and Communications* [Leningrad, 1928], trans. Israel Society for Scientific Translations (Jerusalem, 1970), vol. 3, no. 7, pp. 1–8. On Tsiolkovsky's biography, see also Tsiolkovsky, *Beyond the Planet Earth,* trans. Kenneth Syers (Oxford, 1960), pp. 1–16.

4. Rynin, *Interplanetary Flight,* vol. 3, no. 7, p. 2; A. A. Kosmodemyansky, *Konstantin Tsiolkovsky—His Life and Work* (Moscow, 1956), p. 8.

5. G. A. Tokaty, "Soviet Rocket Technology," *Technology and Culture* (Fall 1963): 516–17.

6. Konstantin Tsiolkovsky, *Works on Rocket Technology*, ed. M. K. Tikhonravov [Moscow, 1947], NASA TT F-243, pp. 81–82.

7. A. Kosmodemyansky, *Tsiolkovsky* [1960 ed], pp. 77–80, cited by Daniloff, *Kremlin and the Cosmos*, p. 20.

8. See Loren R. Graham, *The Soviet Academy of Sciences and the Communist Party, 1927–1932* (Princeton, N.J., 1967), pp. 1–10.

9. On the academy's policies, dating from Nicholas I's Minister of Education and President of the Academy Count Uvarov, see Nicholas V. Riasanovsky, *Nicholas I and Official Nationality in Russia, 1825–1855* (Berkeley, Calif., 1959).

10. Kendall Bailes, *Technology and Society under Lenin and Stalin. Origins of the Soviet Technical Intelligentsia, 1917–1941* (Princeton, N.J., 1978), pp. 25–26; William Blackwell, *The Beginnings of Russian Industrialization* (Princeton, N.J., 1968), pp. 328ff.

11. Nicholas Haus, *History of Russian Educational Policy, 1901–1917* (New York, 1964), pp. 238–240. By 1913 new enrollments in technical schools had climbed from 600 to 24,800. Over half were penurious children of peasants and tradesmen. The St. Petersburg Technological Institute harbored Lenin's first radical organization, and the Moscow Institute was a recruiting center for the 1905 revolution. Both campuses shivered with terrorists, informants, and police.

12. Zhores A. Medvedev, *Soviet Science* (New York, 1976), p. 4.

13. Bailes, *Technology and Society*, pp. 40–41.

14. Medvedev, *Soviet Science*, pp. 7–8.

15. Graham, *Soviet Academy of Sciences*, p. 39.

16. David Joravsky, *Soviet Marxism and Natural Science, 1917–1932* (New York, 1961), 150–97.

17. V. I. Lenin, *Polnoe sobranie sochinenii*, 5th ed. (Moscow, 1958–1965), vol. 26, p. 116.

18. Medvedev, *Soviet Science*, p. 9.

19. A. V. Lunachevskii (Minister of Culture), in *Revoliutsiia i Kultura*, no. 1 (1927), p. 29, cited by Bailes, *Technology and Society*, p. 48.

20. *Trudy GOELRO. Dokumenty i materialy* (Moscow, 1960), vol. 1, p. 595, cited by Bailes, *Technology and Society*, p. 416.

21. On *Krasnaia Zvezda [Red Star]* (St. Petersburg, 1908), see Neil Barron, ed., *Anatomy of Wonder. A Critical Guide to Science Fiction* (New York, 1981), p. 440. On Bogdanov and atomic energy, see Arnold Kramish, *Atomic Energy in the Soviet Union* (Stanford, Calif., 1959), p. 4. On Soviet science fiction in this period, see Rynin, *Interplanetary Flight*, vols. 1–4.

22. On amateur rocketry in the interwar years, see especially the excellent new summary by the National Air and Space Museum's Frank H. Winter, *Prelude to the Space Age. The Rocket Societies: 1924–1940* (Washington, D.C., 1983).

23. Hermann Oberth, *Die Rakete zu den Planetenräumen* (Munich, 1923), and Hermann Oberth, "Autobiography," in A. C. Clarke, ed., *The Coming of the Space Age* (New York, 1967).

24. Daniloff, *Kremlin and the Cosmos*, pp. 21–22; Albert Parry, *Russia's Rockets and Missiles* (Garden City, N.Y., 1960), p. 106.

25. *Tekhniki i Zhizn'* 12 (1924), in V. N. Sokol'skii, *A Short Outline of the Development of Rocket Research in the USSR* (Jerusalem, 1967), p. 12.

26. Tokaty, "Soviet Rocket Technology," p. 518; Slukhai, *Russian Rocketry*, pp. 10–20.

27. Michael Stoiko, *Soviet Rocketry, Past, Present, and Future* (New York, 1970), pp. 39–40.

28. V. N. Sokol'skii, "The Work of Russian Scientists on the Founding of a Theory of Interplanetary Flight," in A. A. Blagonravov et al., *Soviet Rocketry: Some Contributions to Its History* (Jerusalem, 1966), pp. 24–40.

29. Bruce Parrott, *Politics and Technology in the Soviet Union* (Cambridge, Mass., 1983), pp. 5–6ff.

30. On the relative freedom of technicians under the NEP, see Bailes, *Technology and Society*, pp. 44–68; on the Academy of Sciences in the 1920s, see Graham, *Soviet Academy of Sciences*, pp. 24–31.

31. Parrott, *Politics and Technology*, pp. 20–21.

32. J. Stalin, *Sochineniia*, 13 vols. (Moscow, 1946), vol. 13, pp. 38–39, cited by Jerry

Hough and Merle Fainsod, *How the Soviet Union Is Governed* (Cambridge, Mass., 1979), pp. 163–64.

33. Parrott, *Politics and Technology,* pp. 22–24.

34. The chief of the Gosplan remonstrated: "We must stubbornly struggle against 'hat-throwing,' presumptuousness, and relaxation on the basis of the successes achieved in fulfilling the slogan of overtaking and surpassing foreign advanced technology" (ibid.).

35. Quote from *KPSS v rezoliutsiiakh i resheniiakh s"ezdov, konferentsii i plenumov Tsk,* 8th ed. (Moscow, 1970–72), vol. 3, p. 247, cited in Parrott, *Politics and Technology,* p. 27. On trade figures, see Parrott, p. 29, based in turn on Alexander Baykov, *Soviet Foreign Trade* (London, 1946), p. 104; Alec Nove, *An Economic History of the U.S.S.R.* (London, 1968), p. 229; and Herbert S. Levine, "An American View of Economic Relations with the U.S.S.R.," *Annals of the American Academy of Political and Social Science* 414 (July 1974): 11.

36. Jane Degras, trans., *Soviet Documents on Foreign Policy* (London, 1952), vol. 2, pp. 300–01, cited by Bailes, *Technology and Society,* pp. 87–88. The discussion of the Shakhty affair is from Bailes, chap. 3, and Medvedev, *Soviet Science,* pp. 23–31.

37. Stalin, *Sochineniia,* vol. 12, p. 14, cited by Medvedev, *Soviet Science,* p. 23.

38. The accused "ringleaders" included the director of the Thermal Technical Institute in Moscow and the chairman of the Metallurgical Advisory Council of the Supreme Council of the National Economy (both professors at the Moscow Higher Technical School), the vice-chairman of the Production Sector of Gosplan (and professor of military aviation), chairman of the Fuels Section of Gosplan, the head of the Textile Research Institute (and professor of engineering), a technical director in textiles, an engineer in the All-Union Textile Syndicate, and the scientific secretary of the Thermal Technical Institute.

39. Bailes, *Technology and Society,* pp. 96–97; on mass arrests in the industrial Donbass region, see p. 150.

40. On American "technocracy," see Edwin T. Layton, Jr., *The Revolt of the Engineers: Social Responsibility and the American Engineering Profession* (Cleveland, 1971); Henry Elsner, Jr., *The Technocrats* (Syracuse, 1967); Samuel C. Florman, *Blaming Technology: The Irrational Search for Scapegoats* (New York, 1981), chap. 3; Daniel Bell, *The Winding Passage: Essays and Social Journeys, 1960–1980* (Cambridge, Mass., 1980), chaps. 3–4.

41. As early as 1927 Stalin forcibly disbanded a seminar on general questions of technology conducted by the former rector of the Moscow Higher Technical School, I. A. Kalinnikov. It was devoted to innocent discussion of the need for a philosophy of technology in the modern world and never attracted more than sixteen participants. Accusations of "treason" stemming from merely objective disagreements on technical policy are poignantly described in Arthur Koestler's novel *Darkness at Noon* [1941] (New York, 1981).

42. *Wells-Stalin Talk* (London, 1934), p. 11 (conversation of 23 July 1934).

43. Graham, *Soviet Academy of Sciences,* pp. 56–62. On Bukharin's theories on science and technology, see Stephen T. Cohen, *Bukharin and the Bolshevik Revolution* (New York, 1973), esp. pp. 352–53, and essays by N. I. Bukharin, et al., *Science at the Crossroads: Papers Presented to the Second International Congress of the History of Science and Technology* [1931] (London, 1971), pp. 11–33: "Theory and Practice from the Standpoint of Dialectical Materialism," by N. I. Bukharin.

44. See Graham, *Soviet Academy of Sciences,* p. 125. Of this charge, Graham writes: "Even the impassive historian of these events must ask himself if such charges should be dignified by repetition." Yet one may ask, in light of historical criticism of passive German professors in the Third Reich, why we should be so eager to insist on the academics' innocence? If the academy was *not* a center of resistance to the regime, it certainly should have been! Most likely, the gentle dons, too, were taken in by the balmy climate of the "Golden Age." After all a totalitarian state was a new animal of hardly predictable behavior. In any case, Stalin took for granted the danger of an autonomous intellectual center.

45. Graham, *Soviet Academy of Sciences,* pp. 80–113.

46. Parrott, *Politics and Technology,* pp. 49–53.

47. Bailes, *Technology and Society,* p. 220.

48. Parrott, *Politics and Technology,* pp. 27–28, 32–34. In the early 1930s Soviet citizens abroad on state business who chose not to come home rose from twenty-eight in 1928 to double or triple that figure in 1930, when one in twelve defected.

49. Ibid. See also Robert A. Lewis, *Science and Industrialization in the U.S.S.R. Industrial Research and Development, 1917–1940* (London, 1979).

50. Robert A. Lewis, "Some Aspects of the R & D Effort of the Soviet Union, 1924–35," *Science Studies* 2 (1972): 153–179. Figures are from pp. 162–64.

51. Parrott, *Politics and Technology,* p. 27.

52. Ibid., pp. 14–16 (on "organic" and "mechanistic" administration); pp. 57–70 (on the bureaucratic barriers to innovation).

53. On the Allied intervention in the Russian Civil War, see George F. Kennan, *Soviet-American Relations, 1917–1920* (Princeton, N.J., 1956); Richard H. Ullman, *Anglo-Soviet Relations 1917–1921,* 3 vols. (Princeton, N.J., 1961–72); Arno J. Mayer, *Politics and Diplomacy of Peacemaking: Containment and Counterrevolution at Versailles 1918–19* (New York, 1967).

54. On German-Soviet secret military cooperation in the 1920s, see Michael Geyer, *Aufrüstung oder Sicherheit. Die Reichswehr in der Krise der Machtpolitik 1924–1936* (Wiesbaden, 1980), pp. 149ff. See also Hans Gatzke, "Russo-German Military Cooperation During the Weimar Republic," *American Historical Review* 63 (1958): 565–97.

55. The Comintern, on orders from Moscow, denounced Social Democrats as "social fascists" and expected fascism to self-destruct, paving the way for communism. Only in 1935–36 did Stalin order European communists to join "popular fronts" against fascism.

56. Medvedev, *Soviet Science,* p. 34; Asher Lee, *The Soviet Air Force* (London, 1950), pp. 74–77.

57. Parrott, *Politics and Technology,* chap. 2. See also Robert A. Kilmarx, *A History of Soviet Air Power* (New York, 1962), pp. 75–117; Alexander Boyd, *The Soviet Air Force Since 1918* (New York, 1977), pp. 35–54; and Asher Lee, *The Soviet Air Force* (London, 1950), pp. 72–89. Bailes argues that the "organic" nature of military R & D is exaggerated, however (*Technology and Society,* p. 353), and that Stalin's "personal interventions" were often ignorant impositions (p. 356).

58. Seweryn Bialer, ed., *Stalin and His Generals: Soviet Military Memoirs of World War II* (New York, 1969), p. 169. On the Soviet air force in Spain, see Kenneth R. Whiting, "Soviet Aviation and Air Power under Stalin, 1928–1941," in Robin Higham and Jacob W. Kipp, *Soviet Aviation and Air Power. A Historical View* (Boulder, Colo., 1977), pp. 47–67; and Kilmarx, *A History of Soviet Air Power,* pp. 143–47. By the spring of 1937, over 90 percent of the Loyalist (Republican) air force consisted of Soviet planes.

59. Bailes, *Technology and Society,* pp. 381–84.

60. In June 1931 Harold Gatty and Wiley Post circumnavigated the globe by air, touching down at Moscow and proceeding by stages across Siberia to the Kamchatka Peninsula and Alaska.

61. Chkalov, in *Letchik nashego vremini* (Moscow, 1938), p. 315, cited by Bailes, *Technology and Society,* p. 393.

62. *Industria,* 28 Oct. 1938. On the aviation campaign, see Bailes, *Technology and Society,* p. 393, and Alexander Boyd, *The Soviet Air Force Since 1918* (New York, 1977), pp. 74–87.

63. Bailes, *Technology and Society,* pp. 388–91; Boyd, *The Soviet Air Force,* pp. 35–54.

64. Bialer, *Stalin and His Generals,* p. 168.

65. Alfred von Kesselring, *A Soldier's Report* (New York, 1954), p. 90.

66. Boyd, *The Soviet Air Force,* pp. 88–124; Lee, *Soviet Air Force,* pp. 72–95. Kilmarx, *History of Soviet Air Power,* curiously neglects the disaster of 1941 entirely. On the debate over Stalin's responsibility for the disaster of 1941 (made possible during the brief de-Stalinization period under Khrushchev), see Vladimir Petrov, ed., *"June 22, 1941." Soviet Historians and the German Invasion* (Columbia, S.C., 1968), a translation and discussion of the 1965 book by A. M. Nekrich.

67. Evgeny Riabchikov, *Russians in Space* [Moscow, 1970], trans. Guy Daniels (Garden City, N.Y., 1971), pp. 104–7; Tokaty, "Soviet Rocket Technology," p. 519; Stoiko, *Soviet Rocketry,* pp. 42–56.

68. Stoiko, *Soviet Rocketry,* p. 61.

69. Riabchikov, *Russians in Space,* p. 126.

70. The jet fighter was designed by V. F. Bolkhovitinov, A. Y. Bereznyak, and A. M. Isayev.

71. Riabchikov, *Russians in Space,* pp. 127–30.

72. Leonid Vladimirov, *The Russian Space Bluff,* trans. David Floyd (London, 1971), pp. 28–40.

73. Allan Monkhouse, *Moscow 1911–1933* (Boston, 1934), p. 265.

74. Alexander Solzhenitsyn, *The First Circle* (New York, 1968); see esp. pp. 170–73 for the exquisite portrait of scientific prisoner "Chelnov."

75. A. Sharagin [Georgy S. Oserov], *Tupolevskaya Sharaga*, trans. Zhores Medvedev, (Frankfurt/M, 1971), p. 36.

76. Medvedev, *Soviet Science*, p. 42.

77. Roy Medvedev, *On Stalin and Stalinism*, trans. Ellen de Kadt (Oxford, 1979), p. 133.

78. The 1941 plan called for 1,651 million rubles for R & D. The war wiped out this generous allotment to the point that in 1944—the next figure available—1,300 million were directed to R & D. We can safely conclude that this was the maximum the regime could manage and as much as the disrupted R & D sector could effectively use.

79. Tokaty, "Soviet Rocket Technology," p. 522.

Chapter 2

1. See especially the excellent narrative by Frederick I. Ordway III and Mitchell R. Sharpe, *The Rocket Team* (New York, 1979), and William Sims Bainbridge, *The Spaceflight/ Revolution: A Sociological Study* (New York, 1976), pp. 86–115.

2. Wernher von Braun and Frederick I. Ordway III, *History of Rocketry and Space Travel* (New York, 1968), p. 107; Ordway and Sharpe, *Rocket Team*, pp. 35–36.

3. On von Braun's arrest, see Ordway and Sharpe, *Rocket Team*, pp. 47–48. On efforts by various Nazi officials—Himmler, Goering, and Speer among them—to "annex" the rocket program, see Albert Speer, *Infiltration*, trans. Joachim Neugroschel (New York, 1981), esp. chap. 15.

4. Von Braun and Ordway, *History of Rocketry*, pp. 118–19.

5. G. A. Tokady, "Foundations of Soviet Cosmonautics," *Spaceflight* 10 (1968): 335–46.

6. United Kingdom, War Office, *Report on Operation "BACKFIRE,"* Jan. 1946:

The lesson of Operation "BACKFIRE" is that what Britain and the United States can do, other nations can do. No nation can afford to allow the development of long-range rockets to jog along as a matter of routine. There is need of all the imagination, drive, and brains that can be mustered. For the sake of their very existence, Britain and the United States must be masters of this weapon of the future. (p. 27)

7. Ordway and Sharpe, *Rocket Team*, p. 20.

8. Vojtech Mastny, *Russia's Road to the Cold War, Diplomacy, Warfare, and the Politics of Communism, 1941–1945* (New York, 1979), pp. 243–44.

9. G. A. Tokady, "Soviet Rocket Technology," *Technology and Culture* 4 (Fall 1963): 523.

10. Ordway and Sharpe, *Rocket Team*, p. 274.

11. Ibid., pp. 254–70. Quote from pp. 1–2.

12. James McGovern, *Crossbow and Overcast* (New York, 1964); Charles B. McDonald, *The Last Offensive*, U.S. Army in World War II Series, ed. Maurice Matloff (Washington, D.C., 1973), p. 392. On conditions at the V-2 plant, see Ordway and Sharpe, *Rocket Team*, chap. 5, and Speer, *Infiltration*, pp. 210–16.

13. Ordway and Sharpe, *Rocket Team*, pp. 278–87.

14. See Clarence G. Lasby, *Project Paperclip: German Scientists and the Cold War* (New York, 1971).

15. The British were also upset by the American monopolization of German rocket talent. They had painstakingly studied the V-2 for the better part of a year, overrun V-2 launch sites in the Low Countries, and now, like the Soviets, found that the Americans had beaten them to the "source." When British intelligence agents spotted V-2 material on the docks of Antwerp they contacted their superiors, who complained bitterly to Eisenhower about such unilateral requisition. By that time, the ships were at sea. See Ordway and Sharpe, *Rocket Team*, p. 282.

16. Ibid., p. 290.

17. G. A. Tokady, "Foundations of Soviet Cosmonautics," pp. 335–46; quote from p. 343.

18. Tokady, "Soviet Rocket Technology," p. 523.

19. Tokady, "Foundations of Soviet Cosmonautics," p. 343. As Tokady told one German: "We are going to win the jet-rocket race quite independently of you, the Germans. . . . We

will try to marry our own theories and projects with your production experience, and that, I think, will be enough."

20. See Irmgard Gröttrup, *Rocket Wife* (London, 1959). On the Germans in the USSR, see Ordway and Sharpe, *Rocket Team*, chap. 17.

21. On the Yalta Conference from the Soviet point of view, see William O. McCagg, Jr., *Stalin Embattled* (Detroit, 1978), pp. 151–56, and Diane Shaver Clemens, *Yalta* (New York, 1970).

22. During Sir Anthony Eden's trip to Moscow in early 1942, the Soviets demanded all the territories gained under the Nazi-Soviet Pact and extensive spheres of influence in North- and Southeastern Europe. But Vojtech Mastny, in *Russia's Road to the Cold War*, argues that their ambitions rose and fell with their fortunes on the battlefield.

23. See Charles S. Maier, ed., *The Origins of the Cold War and Contemporary Europe* (New York, 1978).

24. See, for instance, the unsympathetic view of Stalin in Adam Ulam, *Expansion and Coexistence: The History of Soviet Foreign Policy, 1917–67* (New York, 1968), and William H. McNeill, *America, Britain, and Russia: Their Cooperation and Conflict, 1941–1946* (New York, 1953); an interpretation based on the nature of the Soviet system in McCagg, *Stalin Embattled*; the critique of Truman and his advisers in Daniel Yergin, *Shattered Peace: The Origins of the Cold War and the National Security State* (Boston, 1977); the New Left critiques of American policies in William A. Williams, *The Tragedy of American Diplomacy*, rev. ed. (New York, 1962), Walter LaFeber, *America, Russia, and the Cold War, 1945–1967* (New York, 1967), Gabriel Kolko, *The Politics of War: The World and U.S. Foreign Policy, 1943–1945* (New York, 1968), Thomas G. Paterson, *Soviet-American Confrontation: Postwar Reconstruction and the Origins of the Cold War* (Baltimore, 1973), and Clemens, *Yalta*; the idiosyncratic interpretation of American imperialism in Franz Schurmann, *The Logic of World Power* (New York, 1974), esp. chap. 1; the "atomic bomb" interpretation in Gar Alperovitz, *Atomic Diplomacy: Hiroshima and Potsdam* (New York, 1965), and *Cold War Essays* (New York, 1970). Empirical and theoretical critiques of all these interpretations can be found in Robert J. Maddox, *The New Left and the Origins of the Cold War* (Princeton, N.J., 1973). Examples of the "broad-minded" or "longer-range" perspective are Louis Halle, *The Cold War as History* (New York, 1967), John Lewis Gaddis, *The U.S. and the Origins of the Cold War* (New York, 1972), and John Lukacs, *A New History of the Cold War* (Garden City, N.Y., 1961), which is clearly anti-Soviet but places the Cold War within a 200-year context of the American and Russian roles in the world.

25. Arnold Kramish, *Atomic Energy in the Soviet Union* (Stanford, Calif., 1959), pp. 6–23.

26. David Holloway, "Entering the Nuclear Arms Race: The Soviet Decision to Build the Atomic Bomb, 1939–45," Working Paper #9, International Security Studies Program, Woodrow Wilson International Center for Scholars (1979), p. 19. Holloway also notes that early Soviet estimates of the cost and time required to tap nuclear energy were pessimistic in any case. After June 1941, the USSR had other priorities.

27. Kramish, *Atomic Energy in the Soviet Union*, pp. 56–57.

28. I. M. Golovin, *I. V. Kurchatov*, 3rd ed. (Moscow, 1978), p. 58. See also Gordon Wright, *The Ordeal of Total War, 1939–1945* (New York, 1968), pp. 104–05.

29. Holloway, "Entering the Nuclear Arms Race," p. 34.

30. General Leslie Groves, head of the Manhattan Project, reported the following to Roosevelt and the Chief of Staff at the end of 1944: one 10-kiloton bomb possibly ready by August 1, 1945, and another five months later, but that problems with a larger implosion bomb were still intractable: *Foreign Relations of the United States* (hereafter *FRUS*): *Diplomatic Papers—the Conferences at Malta and Yalta, 1945*, pp. 357–58.

31. Harry S Truman, *Memoirs*, vol. 1: *Year of Decision* (New York, 1965), p. 458.

32. See Alan Moorehead, *The Traitors*, rev. ed. (New York, 1963), pp. 97–102; Kramish, *Atomic Energy in the Soviet Union*, p. 78.

33. G. K. Zhukov, *Vospominaniya: Razmyshleniya*, 2nd ed. (Moscow, 1974), vol. 2, p. 418. The English translation, *The Memoirs of Marshal Zhukov* (New York, 1971), p. 675, gives an altered account, omitting the sentence "They're raising the price . . ." and attributing the sentence about Kurchatov to Molotov.

34. Kramish, *Atomic Energy in the Soviet Union*, p. 89.

35. Quoted by A. Lavrent'yeva in Stroiteli novogo mira, *V mire knig* (1970), no. 9, p. 4, cited by Holloway, "Entering the Nuclear Arms Race," p. 41.

36. Telegram, Harriman to Secretary of State, 27 Nov. 1945: *FRUS*, 1945, vol. 5, 922–24.

37. Telegram, Kerr to Bevin (British Foreign Secretary), 3 Dec. 1945: *FRUS*, 1945, vol. 2, 82–84.

38. Holloway, "Entering the Nuclear Arms Race," pp. 45–46. Material from this article now appears in his book, *The Soviet Union and the Arms Race* (New Haven, 1983), pp. 15–23.

39. This is the judgment of Kramish, *Atomic Energy in the Soviet Union*, pp. 95–96, and Richard G. Hewlett and Oscar E. Anderson, Jr., *A History of the United States Atomic Energy Commission*, vol. 1: *The New World, 1939/1946* (University Park, Pa., 1962), chap. 16. Martin J. Sherwin, *A World Destroyed: The Atomic Bomb and the Grand Alliance* (New York, 1977), argues that American policymakers came to assume that the Soviets would surrender important political and territorial objectives in return for neutralization of the new weapon. They were wrong (pp. 237–38). But whether American presumption or Soviet ambition is held to be the more responsible for the failure of international control of atomic weapons after 1945, the technological fact of the bomb itself remained.

40. Bruce Parrott, *Politics and Technology in the Soviet Union* (Cambridge, Mass., 1983), p. 79.

41. Evgeny Varga, *Changes in the Economy of Capitalism Resulting from the Second World War* (Moscow, 1947). Chapters of the book were published in 1945 and its ideas appeared in *World Economy and World Politics*, the institute's journal, which Varga edited.

42. Molotov in *Pravda*, 7 Feb. 1946, and in *Bol'shevik* 21 (1945), 5–13, cited by Parrott, *Politics and Technology*, pp. 84–85.

43. *Pravda*, 10 Feb. 1946.

44. These are "old rubles"; for comparison with Soviet figures post-1961, divide by 10. Figures from *Istoriia Velikoi Otechestvennoi voiny Sovetskogo Soiuza* (Moscow, 1961–65), vol. 5, p. 409; M. Prots'ko, "Intelligentsiia strany sotsializma," *Bol'shevik* 6 (1949): 20, cited by Parrott, *Politics and Technology*, pp. 100–01. See also Louvan E. Nolting, *Sources of Financing the Stages of the Research, Development, and Innovation Cycle in the USSR*, Foreign Economic Reports no. 3, U.S. Department of Commerce (Washington, D.C., 1973), pp. 9–13, and Mose L. Harvey et al., *Science and Technology as an Instrument of Soviet Policy* (Miami, Fla., 1972), pp. 61–74.

45. There are serious difficulties in making international comparisons in R & D figures, especially for these early years. See the Organization for Economic Cooperation and Development, *Science Policy in the U.S.S.R.* (Paris, 1969), pp. 95–96. American R & D figures must combine governmental expenditures with those of universities and private industry. However, the breakdown of government R & D, military and civilian, and by agency, is reliable. The Soviet numbers provide a global R & D sum for the entire economy, but breaking them down involves guesswork. Significant military-related research is probably hidden in the regular defense and civilian R & D budgets. It is safe to assume that critical programs like atomic energy and missiles had all the funds they could absorb at a given point in time. During the period of greatest Soviet advance in rocketry, 1947 to 1954, American missile programs were starved for funds or actually shut down (see chapter 6). Zhores Medvedev attests that science leaders in the most important research fields in the Soviet Union, by comparison, usually had an "open account" in the state bank—in foreign currencies as well as rubles (*Soviet Science* [New York, 1976], p. 45).

46. Leo Gruliow, trans., *Soviet Views on the Postwar Economy: An Official Critique of Evgenii Varga's "Changes in the Economy of Capitalism Resulting from the Second World War"* (Washington, D.C., 1948). See especially M. N. Smit's invocation of Lenin's views on the war economies of 1914–18 (pp. 8–14) and Varga's general defense (pp. 114–20).

47. *Pravda*, 8 Feb. 1946, cited by McCagg, *Stalin Embattled*, p. 159. McCagg stresses the role of Varga's work in *revising* the Leninist theory of imperialism as a basis for Soviet foreign policy, while Parrott describes Varga more accurately as a "liberal" who attempted to understand the West on the basis of empirical research. The "encirclement" debate flared again after the Communist victory in China in 1949, with military R & D policy again in the balance.

48. Parrott, *Politics and Technology*, pp. 90–92.

49. *Current Digest of the Soviet Press* 19 (1949): 4, 9, cited by Parrott, *Politics and Technology*, pp. 93–94.

50. The size of the Red Army in Eastern Europe has, in any event, been exaggerated. By the late 1940s there were about 175 divisions under arms, many of which were "paper units" consisting of cadre. In 1960 Khrushchev reported that Soviet armed forces declined

from a high of 11.4 million men to 2.9 million in 1948. See Thomas W. Wolfe, *Soviet Power and Europe* (Baltimore, Md., 1970), pp. 38–40, Malcolm Mackintosh, *Juggernaut: A History of the Soviet Armed Forces* (New York, 1967), pp. 271–76, and Samuel F. Wells, Jr., "Sounding the Tocsin: NSC-68 and the Soviet Threat," Working Paper #7, International Security Studies Program, Woodrow Wilson International Center for Scholars (1969), p. 42. The thirty elite divisions that did patrol Eastern Europe were fully sufficient to threaten the thin Allied line in West Germany and deter inordinate conventional threats to the Soviet Union.

51. Tokaty, "Soviet Rocket Technology," p. 524.

52. G. A. [Tokaty-] Tokaev, *Stalin Means War* (London, 1951), pp. 91–108. Quotes from pp. 97, 104–5.

53. Ibid., pp. 109–21. Quotes from p. 115. Stalin's son Vasili had a reputation as an "imbecile" mostly interested in women, vodka, and foul language. He saw little future in the Sänger commission since Kishkin "has been abroad before [hence untrustworthy] and Keldysh is a Yid." But he was enthusiastic about rockets, "his main notion being that we should pack V-2s with atomic bombs and launch them at America and Britain" (ibid., p. 129).

54. Tokaty, "Soviet Rocket Technology," p. 281.

55. Zhores A. Medvedev, *Soviet Science* (New York, 1978), p. 44; Parrott, *Politics and Technology*, pp. 114–17.

56. Roy A. and Zhores A. Medvedev, *Khrushchev: The Years in Power* (New York, 1976), p. 38.

57. Ordway and Sharpe, *Rocket Team*, pp. 318–43. See also Irmgard Gröttrup, *Rocket Wife*, and Helmut Gröttrup, *Über Raketen* (Berlin, 1959).

58. Ordway and Sharpe, *Rocket Team*, p. 342.

59. Leonid Vladimirov, *The Russian Space Bluff*, trans. David Floyd (London, 1971), pp. 46–49.

60. Nicholas Daniloff, *The Kremlin and the Cosmos* (New York, 1972), p. 44; Robert A. Kilmarx, *A History of Soviet Air Power* (New York, 1962), pp. 233–36; Asher Lee and Richard E. Stockwell, "Soviet Missiles," *The Soviet Air and Rocket Forces* (New York, 1959), pp. 146–59.

61. Michael Stoiko, *Soviet Rocketry, Past, Present, and Future* (New York, 1970), p. 59.

62. Kramish, *Atomic Energy in the Soviet Union*, pp. 124–27.

63. Parrott, *Politics and Technology*, pp. 127–28.

64. Herbert S. Dinerstein, *War and the Soviet Union: Nuclear Weapons and the Revolution in Soviet Military and Political Thinking* (New York, 1959), p. 104.

65. Khrushchev's famous "secret speech" as released by the U.S. Dept. of State on June 4, 1956, is reprinted as Appendix 4 to Nikita Khrushchev, *Khrushchev Remembers* (New York, 1970), pp. 608–77.

66. See Roman Kolkowicz, *The Soviet Military and the Communist Party/Institutions in Conflict* (Princeton, N.J., 1967), pp. 181–92. Note that Kolkowicz's work was originally released as RAND R-446-PR (Aug. 1966). Pagination may vary.

67. N. S. Khrushchev, "The International Position of the Soviet Union," speech to the Twentieth Party Congress, 14 Feb. 1956.

68. N. S. Khrushchev, *Khrushchev Remembers: The Last Testament*, trans. Strobe Talbott (Boston, 1974), pp. 46–47.

69. *Pravda*, 28 Nov. 1953.

70. Martin Caidin, *Red Star in Space* (n.p.: Crowell-Collier Press, 1963), pp. 70–71.

71. Ibid., pp. 72–75.

72. James E. Oberg, *Red Star in Orbit* (New York, 1981), pp. 25–27.

73. Caidin, *Red Star in Space*, p. 79.

74. Oberg, *Red Star in Orbit*, pp. 29–30.

75. Walter Sullivan, *Assault on the Unknown: The International Geophysical Year* (New York, 1961), p. 1.

76. Oberg, *Red Star in Orbit*, p. 14.

77. Evgeny Riabchikov, *Russians in Space* (Moscow, 1970), trans. Guy Daniels (Garden City, N.Y., 1971), p. 146.

78. Sullivan, *Assault on the Unknown*, pp. 1–2.

79. James Reston Interview with Khrushchev, *New York Times*, 9 Oct. 1957.

80. Oberg, *Red Star in Orbit*, p. 33.

Part I Conclusion

1. "Had the [long-range rocket] group been allowed to continue its work without interference from outside, the USSR might well have succeeded in putting a Sputnik around the earth sometime in 1950–52" (G. A. Tokady, "Soviet Rocket Technology," *Technology and Culture* [Fall 1963]: 525). In the United States, von Braun evangelized for a satellite program as early as 1949. Even after nine years of postwar neglect he could have modified the Redstone rocket (a descendant of the V-2) for a satellite by 1956.

2. Evgeny Riabchikov, *Russians in Space* (Moscow, 1970), trans. Guy Daniels (Garden City, N.Y., 1971), p. 132.

3. Kendall E. Bailes, *Technology and Society Under Lenin and Stalin: Origins of the Soviet Technical Intelligentsia, 1917–1941* (Princeton, N.J., 1978), pp. 341–42.

Part II

Headquotes: R. P. Basler, ed., *The Collected Works of Abraham Lincoln* (New Brunswick, N.J., 1953), vol. 1, p. 109; James R. Killian, *Sputniks, Scientists, and Eisenhower: A Memoir of the First Special Assistant to the President for Science and Technology* (Cambridge, Mass., 1976), p. 75; John von Neumann, "Can We Survive Technology?" *Fortune* (June 1955): 106ff.

1. *Time*, 9 July 1945, pp. 58–59.

2. George Fielding Eliot, "Science and Foreign Policy," *Foreign Affairs* 23 (Apr. 1945): 378–87.

3. Richard G. Hewlett and Oscar E. Anderson, Jr., *A History of the United States Atomic Energy Commission*, vol. 1, *The New World, 1939–1946* (University Park, Pa., 1962), p. 341.

4. Compare comments by Truman, James Forrestal, and others in Hewlett and Anderson, *The New World*, pp. 415–21. This is also the main theme of Gregg Herken, *The Winning Weapon: The Atomic Bomb in the Cold War 1945–1950* (New York, 1980).

5. Daniel J. Boorstin, *The Republic of Technology: Reflections on Our Future Community* (New York, 1978), chap. 4 and p. 59.

6. If this seems like hairsplitting, it is only because the promise of American life has become vulgarized—"liberty" being gradually identified with, or supplanted by, the putative fruits of liberty: peace and prosperity. This process has been ingeniously traced by Michael Kammen, "From Liberty to Prosperity: Reflections upon the Role of Revolutionary Iconography in National Tradition," *Proceedings of the American Antiquarian Society* 86 (Oct. 1976): 237–72. But even so, most Americans in the mid-twentieth century still assumed that freedom and technology nurtured each other, just as Marxists assumed that communism and technology nurtured each other.

7. Charles A. Beard and William Beard, *The American Leviathan: The Republic in the Machine Age* (New York, 1931), pp. 5, 8, 9. Charles Beard's expectations were anticipated in many respects by other American intellectuals of European origin or inspiration in the decades preceding and following World War I.

Chapter 3

1. The classic treatment of American science policy is A. Hunter Dupree, *Science in the Federal Government* (Cambridge, Mass., 1957), which takes the history up to 1940. Dupree praises the support given by the federal government to research, especially applied research, throughout its history. But compared to European examples, especially since 1914, the United States appears unusually reticent to fund or direct research. Compare Daniel S. Greenberg, *The Politics of Pure Science* (New York, 1967), chap. 3, who describes American basic research as an orphan before World War II. See also Howard S. Miller, *Dollars for Research: Science and Its Patrons in Nineteenth-Century America* (Seattle, 1970), and Robert

Gilpin and Christopher Wright, eds., *Scientists and National Policy Making* (New York, 1964).

2. James T. Schliefer, *The Making of Tocqueville's Democracy in America* (Chapel Hill, 1980), p. 77. Tocqueville also observed Americans' restlessness with the technological status quo. When he asked a sailor why American ships were built to last only a short time, "he answers without hesitation that the art of navigation is every day making such rapid progress that the finest vessel would become almost useless if it lasted beyond a few years. In these words, which fell accidentally . . . from an uninstructed man, I recognize the general and systematic idea upon which a great people direct all their concerns" (Leonard S. Silk, *The Research Revolution* [New York, 1960], pp. 204–05).

3. Especially the Springfield Armory. See Felicia Johnson Dreyup, *Arms Makers in the Connecticut Valley* (Northampton, Mass., 1948); Robert J. Woodbury, "The Legend of Eli Whitney and the Interchangeability of Parts," *Technology and Culture* 1 (1960): 235–51; Merritt Roe Smith, *Harpers Ferry Armory and the New Technology* (Ithaca, N.Y., 1977). On the role of private foundations in American science, see Paul A. Hanle, *Bringing Aerodynamics to America* (Cambridge, Mass., 1982), pp. 2–10.

4. On World War I and federal science policy, see Dupree, *Science in the Federal Government*, pp. 302ff. When the American Chemical Society, for instance, offered its services to the war effort, Secretary Newton Baker "looked into the matter and found the War Department already had a chemist" (James B. Conant, *Modern Science and Modern Man* [New York, 1952], p. 9).

5. The War Department Chief of Staff concluded after the Armistice that "Nothing in this war has changed the fact that it is now, as heretofore, the Infantry with rifle and bayonet that, in the final analysis, must bear the brunt of the assault and carry it on to victory" (Greenberg, *Politics of Pure Science*, p. 58). See also James Everett Katz, *Presidential Politics and Science Policy* (New York, 1978), p. 6.

6. On the NACA, see Alex Roland, *Model Research: A History of the National Advisory Committee for Aeronautics, 1915–1958*, NASA SP-4103 (Washington, D.C., 1985), and George C. Gray, *Frontiers of Flight: The Story of NACA Research* (New York, 1948); also Frank W. Anderson, Jr., *Orders of Magnitude: A History of NACA and NASA, 1915–80*, 2nd ed. (Washington, D.C., 1981), chap. 1; Memo, Robert F. Freitag to George E. Mueller, "NACA Contributions to Aviation, 1918 to 1958," 27 May 1964, NASA History Office (hereafter NASA HO). I am grateful to Alex Roland for allowing me to see his completed manuscript in 1982.

7. Carroll W. Pursell, Jr., "The Anatomy of a Failure: The Science Advisory Board, 1933–1935," *Proceedings of the American Philosophical Society* 109 (Dec. 1965); Daniel Kevles, *The Physicists* (New York, 1978), chap. 17; Dupree, *Science in the Federal Government*, pp. 350–61. Karl Compton of MIT was Arthur Compton's brother.

8. On the National Resources Coordinating Board, see Greenberg, *Politics of Pure Science*, p. 65; on R & D estimates, see United States, National Resources Committee, *Research—A National Resource* [Washington, D.C., Nov. 1938], (reprint, New York, 1980).

9. Wernher von Braun and Frederick I. Ordway III, *History of Rocketry and Space Travel* (New York, 1968), pp. 44–56. John D. Clark, *Ignition!* (New Brunswick, N.J., 1972), never mentions Goddard in his thorough account of the history of rocket propellants except to say "practically nobody had heard of him" (pp. 5–6). The standard biography, Milton Lehman, *This High Man: The Life of Robert Hutchings Goddard* (New York, 1963), describes Goddard's obsession with priority. His suspicion of other rocketeers was hardened by the discovery that Hermann Oberth's *Die Rakete zu den Planetenräumen* (Munich, 1923) repeated many of the ideas in Goddard's own dissertation, which Oberth had requested of him several years before. See also Frank H. Winter, *Prelude to the Space Age. The Rocket Societies: 1924–1940* (Washington, D.C., 1983), pp. 74–78.

10. On the AIS in the 1930s, see Winter, *Prelude to the Space Age*, pp. 73–85; William Sims Bainbridge, *The Spaceflight/Revolution: A Sociological Study* (New York, 1976), pp. 125–32; and von Braun and Ordway, *History of Rocketry and Space Travel*, pp. 78–85.

11. Bainbridge, *Spaceflight/Revolution*, p. 129; Frank J. Malina, "The Origins and First Decade of the Jet Propulsion Laboratory," in Eugene M. Emme, ed., *History of Rocket Technology* (Detroit, 1964). The American Interplanetary Society became the American Rocket Society in 1934.

12. J. D. Bernal, *The Social Function of Science* [1939] (Cambridge, Mass., 1967), p. xxi.

13. On the origins of the JPL, see Clayton R. Koppes, *JPL and the American Space Program:*

A History of the Jet Propulsion Laboratory (New Haven, 1982), chaps. 1–2; on science in World War II, see Greenberg, *Politics of Pure Science*, pp. 79–86; A. Hunter Dupree, "The Great Instauration of 1940: The Organization of Scientific Research for War," *Twentieth Century Sciences*, ed. Gerald Holton (New York, 1972), pp. 445–54.

14. On military applications of scientific research during World War II, see Greenberg, *Politics of Pure Science*; Dupree, "The Great Instauration of 1940"; James Phinney Baxter III, *Scientists Against Time* (Boston, 1946); Irvin Stewart, *Organizing Scientific Research for War* (Boston, 1948); Vannevar Bush, *Science—The Endless Frontier* (Washington, D.C., 1946), and *Modern Arms and Free Men* (New York, 1949). R & D figures are from National Science Foundation, *Federal Funds for R & D*, Fiscal Years 1977, 1978, and 1979, vol. 27 (Washington, D.C., 1974), publication #79-310.

15. Greenberg, *Politics of Pure Science*, pp. 99–102; Daniel Kevles, "Scientists, the Military, and the Control of Postwar Defense Research: The Case of the Research Board for National Security, 1944–46," *Technology and Culture* 16 (Jan. 1975): 20–47.

16. U.S. Senate, Subcommittee of the Senate Committee on Military Affairs, *Hearings on Science Legislation*, 79th Cong., 1st Sess., 1945, p. 11.

17. Kevles, "Scientists, the Military, and Postwar Defense Research," p. 38.

18. Bush, *Science—The Endless Frontier: A Report to the President on a Program for Postwar Scientific Research*, July 1945.

19. Kevles, "Scientists, the Military, and Postwar Defense Research," pp. 39–43.

20. Milton Lomask, *A Minor Miracle: An Informal History of the National Science Foundation* (Washington, D.C., 1975), pp. 33–57 (on funding, pp. 66–67); Greenberg, *Politics of Pure Science*, pp. 121–22.

21. Kevles, "Scientists, the Military, and Postwar Research," p. 20.

22. Richard G. Hewlett and Oscar E. Anderson, Jr., *A History of the United States Atomic Energy Commission*, vol. 1, *The New World, 1939-1946* (University Park, Pa., 1962), pp. 411–15.

23. "Statement by the President Announcing the Use of the Atomic Bomb at Hiroshima," 6 Aug. 1945, *Public Papers of the Presidents* (hereafter *PP of P*), *Harry S Truman 1945* (Washington, D.C., 1961), pp. 197–200.

24. Henry D. Smyth, *A General Account of the Development of Methods of Using Atomic Energy for Military Purposes under the Auspices of the United States Government, 1940-45* (Washington, D.C., 1945), p. 165.

25. U.S. Congress, House Committee on Military Affairs, *Hearings, Research and Development*, 79th Cong., 1st Sess., May 1945.

26. Hewlett and Anderson, *The New World*, pp. 425–26.

27. Ibid., p. 432.

28. Ibid., pp. 445–48.

29. *New York Times*, 31 Oct. 1945.

30. Hewlett and Anderson, *The New World*, pp. 445–47. Quote from p. 447.

31. *Congressional Record*, 79th Cong., 2nd Sess., Appendix, pp. 2410–15, cited in Ibid., p. 508.

32. Ibid., p. 519.

33. Ibid., p. 527. The congressman was R. Ewing Thomason (D., Tex.).

34. This brief account of the Baruch plan follows Hewlett and Anderson, *The New World*, pp. 580–619, but compare Gregg Herken, *The Winning Weapon, The Atomic Bomb in the Cold War, 1945-1950* (New York, 1980), pp. 158–90; Daniel Yergin, *Shattered Peace: The Origins of the Cold War and the National Security State* (Boston, 1977), pp. 237–40; Larry G. Gerber, "The Baruch Plan and the Cold War," *Diplomatic History* 6 (Winter 1982): 69–95.

35. Memo for the Assistant Chief of Staff, G-3, 21 June 1945, cited by Edmund Beard, *Developing the ICBM: A Study in Bureaucratic Politics* (New York, 1976), pp. 20–21.

36. Memo, Joseph T. McNarney, Deputy Chief of Staff, to Commanding General AAF, 2 Oct. 1944, cited by Beard, *Developing the ICBM*, pp. 21–22.

37. On the "roles and missions" controversy during the period of military unification, see Demetrius Caraley, *The Politics of Military Unification. A Study of Conflict and the Policy Process* (New York, 1966); Perry McCoy Smith, *The Air Force Plans for Peace 1943-45* (Baltimore, Md., 1970); Vincent O. Davis, *Postwar Defense Policy and the United States Navy, 1943-46* (Chapel Hill, N.C., 1976).

38. Yergin, *Shattered Peace*, pp. 201–04.

39. David MacIsaac, "The Air Force and Strategic Thought 1945-1951," Working Paper

#8, Woodrow Wilson International Center for Scholars, International Security Studies Program (1979), p. 10.

40. U.S. Senate, Military Affairs Committee, *Department of Armed Services, Department of Military Security: Hearings,* 79th Cong., 1st Sess., 1945, pp. 291–92.

41. Charles D. Bright, *The Jet Makers* (Lawrence, Kans., 1978), pp. 2–13.

42. Letter, Truman to Secretary of War, 8 Aug. 1945, cited by Beard, *Developing the ICBM,* p. 45.

43. Bruce L. R. Smith, *The RAND Corporation* (Cambridge, Mass., 1966), pp. 30–65.

44. Memo, LeMay to Spaatz, 20 Sept. 1946, cited by Beard, *Developing the ICBM,* p. 39.

45. Memo, Brigadier General R. C. Coupland, Guided Missiles Division, 6 Mar. 1949, cited by Beard, *Developing the ICBM,* p. 29.

46. Yergin, *Shattered Peace,* pp. 213–14.

47. Telegram, Ambassador Smith to the Secretary of State, 5 Apr. 1946, *FRUS* 1946, vol. 6, p. 733.

48. Ernest May, *"Lessons" of the Past: The Use and Misuse of History in American Foreign Policy* (New York, 1973), chap. 2. "The Cold War: Preventing World War II," demonstrates how U.S. policymakers in the late 1940s and 1950s were influenced by their experience of Hitler and appeasement in the 1930s.

49. MacIsaac, "Air Force and Strategic Thought," pp. 14–15, based in turn on Robert Frank Futrell, *Ideas, Concepts, Doctrine: A History of Basic Thinking in the United States Air Force* (Maxwell AFB, Ala., 1971), vol. 1.

50. Ibid., p. 19.

51. From the Spaatz Board Report of Oct. 23, 1945, from which AAF approval of funds for the new RAND Corporation stemmed. See also Curtis LeMay, *Mission with LeMay* (Garden City, N.Y., 1965), pp. 394–400, cited by MacIsaac, "Air Force and Strategic Thought," p. 22.

52. Yergin, *Shattered Peace,* p. 265.

53. David Alan Rosenberg, "American Atomic Strategy and the Hydrogen Bomb Decision," *Journal of American History* 66 (June 1979): 62–87. On "Pincher," see p. 64. I am grateful to the author for providing me with a manuscript of this article in 1978. "Pincher" and other postwar JCS plans have now been described and analyzed in Herkin, *The Winning Weapon,* pp. 219–24, et seq.

54. Rosenberg, "American Atomic Strategy," pp. 64–66.

55. MacIsaac, "Air Force and Strategic Thought," pp. 17, 27.

56. Richard G. Hewlett and Francis Duncan, *A History of the United States Atomic Energy Commission,* vol. 2: *Atomic Shield 1947/51* (University Park, Pa., 1969), pp. 53–55.

57. JCS 1805, 23 Sept. 1947, cited by Rosenberg, "American Atomic Strategy," p. 67.

58. JCS 1745/15, 27 July 1948, cited by Rosenberg, "American Atomic Strategy," pp. 67–68.

59. Historical Documentation K105.5-24, 2 Feb. 1967, pp. 36–37: USAF HO; cited by MacIsaac, "Air Force and Strategic Thought," p. 32.

60. Yergin, *Shattered Peace,* pp. 339–40.

61. X (Kennan), "The Sources of Soviet Conduct," *Foreign Affairs* 25 (July 1947): 566–82 ("nook and cranny" and "containment," p. 575).

62. Yergin, *Shattered Peace,* p. 341.

63. JCS 1844/4, 6 May 1948, cited in Rosenberg, "American Atomic Strategy," p. 68.

64. Rosenberg, "American Atomic Strategy," pp. 71–72.

65. *FRUS,* 1949, vol. 1, pp. 481–82.

66. MacIsaac, "Air Force and Strategic Thought," p. 41. On Truman's drift into a strategy of atomic deterrence, see especially Lawrence Freedman, *The Evolution of Nuclear Strategy* (London, 1981), pp. 47–62.

Chapter 4

1. Edmund Beard, *Developing the ICBM: A Study in Bureaucratic Politics* (New York, 1976), pp. 52–55.

2. "DoD Obligational Program for Missile Systems, Fiscal Years 1946–1960," *Congressional Record,* 1 Feb. 1960, p. 1639.

3. Beard, *Developing the ICBM*, pp. 56–57.

4. U.S. Senate, *Hearings Before the Special Subcommittee on Atomic Energy*, Dec. 1945, cited in U.S. Senate, Committee on Armed Services, *Inquiry into Satellite and Missile Programs. Hearings Before the Preparedness Investigating Subcommittee*, 85th Cong., 1st and 2nd Sess., 1957–58 (Washington, D.C., 1958), pp. 822–23.

5. Vannevar Bush, *Modern Arms and Free Men* (New York, 1949), pp. 84–85.

6. Ibid., p. 2.

7. Beard, *Developing the ICBM*, pp. 62–66.

8. Daniel Lang, *From Hiroshima to the Moon* (New York, 1959), chap. 21.

9. Data on science fiction from William Sims Bainbridge, *The Spaceflight/Revolution: A Sociological Study* (New York, 1976), pp. 198–208.

10. Ibid., pp. 133–37. In 1963 the American Rocket Society merged with the Institute of Aeronautical Sciences to form the current American Institute of Aeronautics and Astronautics.

11. Gerry de la Ree, "Space Travel—When and How" (1953), NASA HO; portions appeared in the *Bergen Evening Record* (Hackensack, N.J.), 22 Aug. 1953.

12. Report cited in R. Cargill Hall, "Earth Satellites: A First Look by the U.S. Navy," *Fourth History Symposium of the International Academy of Astronautics* (Oct. 1970): 253. Other reports on German research circulated in various branches of the U.S. military, stimulating thought on the future of rocketry. See the bibliography of Frederick I. Ordway III and Mitchell R. Sharpe, *The Rocket Team* (New York, 1979), pp. 414ff.

13. Hall, "Earth Satellites," pp. 253–55.

14. Ibid., pp. 256–57. This material is also covered in Hall's article, "Early U.S. Satellite Proposals," in Eugene M. Emme, ed., *The History of Rocket Technology* (Detroit, 1964), pp. 67–93 (pp. 69–71).

15. LeMay and first Bush quotes from R. Cargill Hall, "A Chronology of Some Events in Early U.S. Satellite Studies During the 1940s," JPL/HN-7, Apr. 1970; second Bush quote from Bush, *Modern Arms and Free Men*, pp. 84–85.

16. Douglas Aircraft Company, Inc., Report No. SM-11827, "Preliminary Design of an Experimental World-Circling Spaceship," Abstract, 2 May 1946, "summary" page and pp. 1–7 ("conservative and realistic" quote, p. 4), NASA HO.

17. Ibid., "summary" page and p. 7.

18. Ibid., pp. 8, 9, 17, 23.

19. Robert L. Perry, "Origins of the USAF Space Program 1945–1956," PA 8-23, Air Force Systems Command Historical Publications Series no. 62-24-10, AFSC HO, Edwards AFB, Md.; Hall, "Earth Satellites," pp. 259–67.

20. On the debate over whether to develop the hydrogen bomb, see Robert Jungk, *Brighter Than a Thousand Suns: A Personal History of the Atomic Scientists*, trans. James Cleugh (New York, 1958); J. R. Shepley and C. Blair, Jr., *The Hydrogen Bomb* (New York, 1954); Robert Gilpin, *American Scientists and Nuclear Weapons Policy* (Princeton, N.J., 1962); Herbert York, *The Advisors: Oppenheimer, Teller and the Superbomb* (San Francisco, 1976); United States, Atomic Energy Commission, *In the Matter of J. Robert Oppenheimer* (Cambridge, Mass., 1971); Warner Schilling, "The H-Bomb Decision: How To Decide without Actually Choosing," *Political Science Quarterly* 76 (1961): 24–46.

21. Samuel F. Wells, "Sounding the Tocsin: NSC-68 and the Soviet Threat," Working Paper #7, Woodrow Wilson International Center for Scholars, International Security Studies Program (1979), p. 4.

22. David Alan Rosenberg, "American Atomic Strategy and the Hydrogen Bomb Decision," *Journal of American History* 66 (June 1979): 82–85. Quote from p. 84.

23. Ibid. Quote from p. 82.

24. "NSC-68," 14 Apr. 1950, *FRUS*, 1950, vol. 1, pp. 234–92. Quotes from pp. 234, 237, 285, 291.

25. *Wall Street Journal*, 20 Oct. 1950.

26. K. T. Keller, "Final Report of the Director of Guided Missiles, Office of the Secretary of Defense," 17 Sept. 1953, Harry S Truman Library, Independence, Missouri (hereafter HST Library), Keller Papers.

27. *New York Times*, 1 Nov. 1957.

28. I found nothing at the Truman Library to call into question Beard's judgment that the Keller appointment "may have been as much a public relations effort as a sincere and aggressive attempt at reorganizing and firmly coordinating the various guided missile programs of the Armed Forces" (Beard, *Developing the ICBM*, pp. 124–25).

29. Convair engineers quoted by Brig. General J. S. Sessums (ARDC) to Director of R & D, USAF Deputy Chief of Staff, 25 Sep. 1951, cited by Beard, *Developing the ICBM*, p. 134.

30. York, *The Advisors*, pp. 82–92. On the discovery of lithium-deuteride as a nuclear fuel, see also Kosta Tsipis, *Arsenal: Understanding Weapons in the Nuclear Age* (New York, 1983), pp. 29–38, 262–63.

31. Beard, *Developing the ICBM*, p. 143. CEP denotes the radius of the circle centered on the target within which 50 percent of the missiles can be expected to land. On the size and weight of hydrogen bombs, see Tsipis, *Arsenal*, pp. 37–38, 44.

32. Robert L. Perry, "The Ballistic Missile Decisions," American Institute of Aeronautics and Astronautics Paper 67-838, Oct. 1967, pp. 4–13. On the technical specifications of ICBMs, see Tsipis, *Arsenal*, chap. 5.

33. John T. Greenwood, "A Short History of the Air Force Ballistic Missile and Space Program 1954–1974," unpublished ms., AFSC HO; Beard, *Developing the ICBM*, pp. 180–81. On Ramo-Wooldridge's Space Technology Laboratories and the USAF "contract system," see also Michael Armacost, *The Politics of Weapons Innovation: The Thor-Jupiter Controversy* (New York, 1969), pp. 155–63.

34. Memo, H. A. Craig, for the Vice Chief of Staff, "Earth Satellite Vehicles," 8 Jan. 1948, NASA HO; see also Perry, "Origins of the USAF Program," AFSC HO, pp. 29–33.

35. Lockheed Space and Missiles Company, *History of the Agena Spacecraft*, cited by *Aerospace Daily*, 13 Nov. 1969.

36. Memo, G. M. Clement (RAND) to J. E. Lipp, 14 Apr. 1948; Letter, Brigadier General D. L. Putt, Director of R & D, Deputy Chief of Staff, Materiel, USAF, to F. R. Collbohm, Douglas Aircraft Company, 6 Oct. 1948, NASA HO.

37. Paul Kecskemeti, "The Satellite Rocket Vehicle: Political and Psychological Problems," RAND RM-567, 4 Oct. 1950.

38. The Soviet attacks appeared in *New Times*, 10 Dec. 1947 and 7 Nov. 1949: quotes from Kecskemeti, "The Satellite Rocket Vehicle," pp. 5, 9.

39. Kecskemeti, "The Satellite Rocket Vehicle," pp. 9–10.

40. Ibid. Quotes from pp. 13, 14, 15.

41. Ibid., pp. 15–17. Quotes from pp. 16–17.

42. Ibid., pp. 21–23. Quotes from pp. 21–22.

43. Memo, J. E. Lipp to R. M. Salter, Douglas Aircraft Company, 20 Oct. 1950, NASA HO; Perry, "Origins of the USAF Space Program," p. 34. See also lecture by L. R. Hafstad, "Introduction to Guided Missile Programs," Dec. 1946, HST Library, Hafstad Papers. In his lecture Hafstad was among the first to exclaim on the breadth of the rocketry revolution. So many branches of science were involved—aerodynamics, radar, electronics, telemetry, gyroscopy, computers, thermodynamics, combustion, metallurgy, propulsion, chemistry— as well as problems of management—division of labor, internecine feuds, new R & D concepts, classification—and technical hurdles—test facilities, supersonic wind tunnels, guidance and reentry, warheads, countermeasures—that a vast technological frontier would be pushed back in the process. In fact, the difficulties were *so* great that Hafstad thought "push-button warfare" was *far* from "just around the corner."

44. *Aerospace Daily*, 13 Nov. 1969; Perry, "Origins of the USAF Space Program," pp. 35–36, 42–44.

Chapter 5

1. Charles C. Alexander, *Holding the Line: The Eisenhower Era 1952–1961* (Bloomington, Ind., 1975), p. 27.

2. Ironically, George Kennan made this point himself in *Russia, the Atom, and the West* (New York, 1957), p. 93: "Armaments are important not just for what could be done with them in time of war, but for the psychological shadows they cast in time of peace."

3. Kennan has insisted that Stalin never considered an invasion of Western Europe and virtually all historians have assumed the same, though no evidence exists either way. Recently, however, Nikolai Tolstoy has suggested that Stalin *did* entertain the idea; see Tolstoy, *Stalin's Secret War* (London, 1981), pp. 359–61.

4. Almost universal criticism of Eisenhower by professional historians has recently given way to an appreciative revisionism. See especially the review article by Mary S. McAuliffe, "Eisenhower the President," *Journal of American History* 68 (Dec. 1981): 625–32, and Fred I. Greenstein, "Eisenhower as an Activist President: A Look at New Evidence," *Political Science Quarterly* 94 (Winter 1979–80): 577–86. On the Eisenhower administration generally, see the memoirs: Dwight D. Eisenhower, *The White House Years*, 2 vols. (New York, 1963–65); Arthur Larson, *The President Nobody Knew* (New York, 1968); Emmet John Hughes, *The Ordeal of Power: A Political Memoir of the Eisenhower Years* (New York, 1963); Sherman Adams, *Firsthand Report: The Story of the Eisenhower Administration* (New York, 1961). Recent scholarly works include Alexander, *Holding the Line;* Elmo Richardson, *The Presidency of Dwight David Eisenhower* (Lawrence, Kans., 1979); Robert A. Divine, *Eisenhower and the Cold War* (Oxford, 1981); Blanche Wiesen Cook, *The Declassified Eisenhower: A Divided Legacy* (Garden City, N.Y., 1981); Stephen E. Ambrose, *Eisenhower: Soldier, General of the Army, President-Elect, 1890–1952* (New York, 1983). The burden of the new work confirms in many essentials the message of the original memoir literature to the effect that Eisenhower was a strong, activist, and moderate president, and was by no means subservient to the "manichean" John Foster Dulles in the setting of foreign policy.

5. Quotes from Alexander, *Holding the Line*, p. 29. On Eisenhower's philosophy of the role of government in domestic policy, see Robert Griffith, "Dwight D. Eisenhower and the Corporate Commonwealth," *American Historical Review* 87 (Feb. 1982): 87–122; E. Bruce Geelhoed, *Charles E. Wilson and Controversy at the Pentagon, 1953 to 1957* (Detroit, 1979), esp. chap. 6; Douglas Kinnard, *President Eisenhower and Strategy Management: A Study in Defense Politics* (Lexington, Ky., 1977), pp. 7–8.

6. On NSC 162/2 and "massive retaliation," see Lawrence Freedman, *Evolution of Nuclear Strategy* (London, 1981), pp. 76–90.

7. "The Chance for Peace," Address delivered before the American Society of Newspaper Editors, 16 Apr. 1953; *PP of P, Dwight D. Eisenhower 1953,* pp. 179–88. Quote from p. 182.

8. Ibid.

9. On disarmament diplomacy since World War II, see especially U.S., Arms Control and Disarmament Agency, *Documents on Disarmament, 1945–* (Washington, D.C., 1960–). Scholarly works include Bernard G. Bechhoefer, *Postwar Negotiations for Arms Control* (Washington, D.C., 1961), a Brookings Institution study; Chalmers M. Roberts, *The Nuclear Years: The Arms Race and Arms Control, 1945–1970* (New York, 1970); Lincoln Bloomfield, Walter C. Clemens, and Franklyn Griffiths, *Khrushchev and the Arms Race: Soviet Interests in Arms Control and Disarmament, 1954–1964* (Cambridge, Mass., 1966); and the revisionist works by Richard J. Barnet, *Who Wants Disarmament?* (Boston, 1960), and Edgar M. Bottome, *The Balance of Terror: A Guide to the Arms Race* (Boston, 1971). A comprehensive account hostile to the policies of both Superpowers is Alva Myrdal, *The Game of Disarmament: How the United States and Russia Run the Arms Race* (New York, 1976).

10. James R. Killian, *Sputniks, Scientists, and Eisenhower: A Memoir of the First Special Assistant to the President for Science and Technology* (Cambridge, Mass., 1976), pp. 70–75. Quote from p. 75. Italics in original. The report was commissioned in April 1954 and presented to the NSC on 14 Feb. 1955.

11. Ibid, pp. 79–80.

12. Anthony Kenden, "U.S. Reconnaissance Satellite Programmes," *Spaceflight* (July 1978): 243. On the U-2, see John Taylor and David Monday, *Spies in the Sky* (New York, 1973). On the decision to build the U-2, see Killian, *Sputnik, Scientists, and Eisenhower,* pp. 81–82; Stephen E. Ambrose, *Ike's Spies: Eisenhower and the Espionage Establishment* (New York, 1981), pp. 265–78.

13. NSC 5440, 14 Dec. 1954, and NSC 5501, 7 Jan. 1955, National Archives (hereafter NA), Modern Military Branch. The first draft, NSC 5440, described Soviet nuclear plenty as "a peril greater than any the United States has ever before faced." The Bureau of the Budget succeeded in deflating this "categorical judgment" to merely "a grave peril."

14. Constance M. Green and Milton Lomask, *Vanguard: A History,* NASA SP-4202 (Washington, D.C., 1970), pp. 19–23. On the IGY, see Sydney Chapman, *IGY, Year of Discovery* (Ann Arbor, Mich., 1959); Walter Sullivan, *Assault on the Unknown: The IGY* (New York, 1961); J. Tuzo Wilson, *IGY: The Year of the New Moons* (New York, 1961).

15. A. V. Grosse, "Report on the Present Status of the Satellite Problem," 25 Aug. 1953, quotes from pp. 5–6, HST Library, Miscellaneous Historical Document File.

16. Erik Bergaust and William Beller, *Satellite!* (Garden City, N.Y., 1956), pp. 36–37.

17. Green and Lomask, *Vanguard*, pp. 17–18; Wernher von Braun and Frederick I. Ordway III, *History of Rocketry and Space Travel* (New York, 1968), p. 179.

18. Frederick I. Ordway III and Mitchell R. Sharpe, *The Rocket Team* (New York, 1979), p. 376. Italics in original.

19. NSC-5520, "Satellite Program," 20 May 1955, quotes from pp. 1, 3, 4: Dwight D. Eisenhower Library, Abilene, Kansas (hereafter DDE Library), Office of the Special Assistant for National Security Affairs, Box 3.

20. Ibid. Quotes from pp. 11, 13.

21. James Hagerty Press Release, 29 July 1955, a day following the oral briefing.

22. Besides Stewart, the committee included Charles C. Lauritsen, Caltech physicist; Joseph Kaplan, chairman of the National Committee for the IGY; Richard Porter, consultant to General Electric Company's Guided Missiles Division; George H. Clement of the RAND Corporation; Clifford C. Furnas, chancellor of the University of Buffalo; J. Barkley Rosser, Cornell University mathematician; and Robert McMath, University of Michigan astronomer.

23. Robert L. Perry, "The Origins of the USAF Space Program 1945–1956," AFSC Historical Publications Series #62-24-10, pp. 45–50, AFSC HO, Edwards AFB, Md.

24. Green and Lomask, *Vanguard*, pp. 41–48.

25. Ibid., pp. 48, 55.

26. Ibid., p. 36.

27. Ibid., pp. 52–55, quote from p. 55; von Braun and Ordway, *History of Rocketry and Space Travel*, p. 179.

28. Allen Dulles, *The Craft of Intelligence* (New York, 1963), p. 168.

29. Ordway and Sharpe, *The Rocket Team*, pp. 377–78. The President's son John recalled that Andrew Goodpaster, White House Chief of Staff, learned in March 1956 that the Redstone Arsenal could launch a satellite by the end of that year. Goodpaster then called Deputy Secretary of Defense Reuben Robertson, who said they would look into it. After two weeks Goodpaster called and asked, "How are you coming?" The Pentagon replied: "Well, there are all sorts of considerations here." And that was the last heard of it (John S. D. Eisenhower, Oral History, pp. 45–46, DDE Library).

30. For instance, Arthur Schlesinger, Jr., Daniel Bell, Max Lerner, Seymour Martin Lipset, Irving Kristol, Daniel Boorstin, Louis Hartz, and Richard Hofstadter could all be considered "consensus" historians and sociologists. They rejected mass movements, utopianism, populism (the "root of fascism"), and endorsed the American traditions of realism, pragmatism, and nonideological politics. See Alonzo L. Hamby, *Beyond the New Deal: Harry S Truman and American Liberalism* (New York, 1973). A critical view of the "New Liberalism" and consensus scholarship is Marian Morton, *The Terrors of Ideological Politics: Liberal Historians in a Conservative Mood* (Cleveland, 1972).

31. On the test ban debate in the 1956 presidential campaign, see Robert A. Divine, *Blowing on the Wind: The Nuclear Test Ban Debate 1954–1960* (New York, 1978), chap. 4.

32. Herbert Marcuse, *Eros and Civilization, A Philosophical Inquiry into Freud* (Boston, 1956); C. Wright Mills, *The Power Elite* (New York, 1956); John Kenneth Galbraith, *The Affluent Society* (Boston, 1958); and Paul Goodman, *Growing Up Absurd. Problems of Youth in the Organized System* (New York, 1960).

33. Quote from Alexander, *Holding the Line*, p. 190. On the New New Look, see Samuel P. Huntington, *The Common Defense: Strategic Programs in National Politics* (New York, 1961), pp. 88–113.

34. Dwight D. Eisenhower, *The White House Years*, 2 vols., vol. 1: *Mandate for Change, 1953–1956* (New York, 1963), p. 520.

35. Eisenhower's White House chief of staff, General Andrew J. Goodpaster, was asked about Ike's later statement that "we knew the Russians wouldn't accept [Open Skies]," to which Goodpaster replied: "No. The pessimism came after the proposal was rejected at Geneva, not before, when it was thought of as laying groundwork, a show of sincerity. . . ." (Goodpaster Oral History, p. 861, DDE Lib.).

36. "Annual Message to the Congress on the State of the Union," 10 Jan. 1957, *PP of P, Dwight D. Eisenhower 1957*, p. 26.

37. 12 Jan. 1957, U.S. ACDA, *Documents on Disarmament, 1945–1959*, vol. 2, p. 733.

38. 27 July 1957, ibid., p. 825.

39. Edmund Beard, *Developing the ICBM: A Study in Bureaucratic Politics* (New York, 1976), pp. 205–08.

40. "Blue ocean" traditionalists in the navy, like the "blue sky" bomber faction in the

USAF, resisted the Fleet Ballistic Missile program for fear that submarines would come to replace the surface navy, while even submariners preferred to sink ships with torpedoes in a "battle of wits," not blast inland cities with missiles. See Harvey M. Sapolsky, *The Polaris System Development: Bureaucratic and Programmatic Success in Government* (Cambridge, Mass., 1972), pp. 16–18. On Polaris development, see Sapolsky; James Baer and William Howard, *Polaris!* (New York, 1960); and Wyndham D. Miles, "The Polaris," in Eugene M. Emme, ed., *The History of Rocket Technology* (Detroit, 1964), pp. 162–75.

41. On the development of Thor, see Julian Hartt, *Mighty Thor* (New York, 1961); on its subsequent competition with the Jupiter, see Michael Armacost, *The Politics of Weapons Innovation: The Thor-Jupiter Controversy* (New York, 1969). On the ICBMs, see J. L. Chapman, *Atlas: The Story of a Missile* (New York, 1960), and Roy Neal, *Ace in the Hole: The Story of the Minuteman Missile* (New York, 1962). For a summary of all military missile programs, see von Braun and Ordway, *History of Rocketry and Space Travel*, chap. 6, and Robert L. Perry, "The Atlas, Thor, Titan, and Minuteman," in Emme, ed., *History of Rocket Technology*, pp. 142–61.

42. Green and Lomask, *Vanguard*, pp. 61–67.

43. Ibid., pp. 72–78, 87, 104–06.

44. Ibid., p. 106.

45. Ibid., pp. 129–30.

46. "Chronology of Significant Events in the U.S. Intermediate and Intercontinental Ballistic Missiles Programs," 8 Nov. 1957, DDE Library, Office of the Staff Secretary, Department of Defense (hereafter DoD) subseries.

47. John B. Medaris, *Countdown for Decision* (New York, 1960), p. 154.

48. Ibid., p. 155.

49. Ibid. On this episode, see also Ordway and Sharpe, *Rocket Team*, p. 382.

Part II Conclusion

1. "Memorandum of a Conference with the President" (Goodpaster Notes), 8 Oct. 1957, DDE Library, Ann Whitman File, Box 16.

Part III

Headquotes: Ernst Stuhlinger Memo, "Russian Comments to the American Satellite Project," 29 Oct. 1957, 8th Congress of the International Astronomical Federation, Barcelona, Spain, NASA HO; Bernard Baruch, "The Lessons of Defeat," *New York Herald Tribune*, 16 Oct. 1957; Victor Gilinsky (Caltech), Letter [of 8 Nov. 1957] to the *New York Times*, 13 Nov. 1957; Dwight D. Eisenhower, "Radio and Television Address to the American People on Science in the National Security," 7 Nov. 1957, *PP of P, Dwight D. Eisenhower 1957*, pp. 789–99.

1. Dwight D. Eisenhower, *The White House Years*, 2 vols., vol. 2, *Waging Peace, 1956–1961* (New York, 1965), p. 464.

2. Goodpaster Notes, "Paraphrase of Remarks by the President," 27 Nov. 1959, DDE Library, Ann Whitman File, Box 29.

3. Minutes of Cabinet Meeting, 3 June 1960, DDE Library, Ann Whitman File, Box 16.

Chapter 6

1. Quotes from Lyndon Baines Johnson, *The Vantage Point: Perspectives of the Presidency 1963–1969* (New York, 1971), p. 272; see also interview with Gerald W. Siegel, 25 June 1968, NASA HO.

2. Interview with Eilene Galloway, 9 Apr. 1974, NASA HO; Galloway interview with the author, 15 Feb. 1979. Galloway was in the Congressional Research Service in 1957 and

became LBJ's leading research assistant on missiles and space policy. She attests: "If [Johnson] looked at you, and told you something, or asked you a question, you felt that he was taking an X-ray and developing and assessing it all in that instant glance. And you felt you had to get on with it and get it done in a hurry" (NASA HO Interview, p. 9). For Johnson's reaction to Sputnik, see also Memo, Solis Horwitz to LBJ, 11 Oct. 1957, Lyndon B. Johnson Library, Austin, Texas (hereafter LBJ Library), Senate Papers, Box 355; and Doris Kearns, *Lyndon Johnson and the American Dream* (New York, 1976), pp. 151–52.

3. Interview with Glen Wilson, 15 Mar. 1974, NASA HO.

4. *New York Times,* 6 Oct. 1957. For compilations of American reactions to the Sputniks, see Martha Wheeler George, "The Impact of Sputnik I," NASA Historical Note no. 22 (July 1963); Lynne L. Daniels, "Statements of Prominent Americans on the Opening of the Space Age," NASA Historical Note no. 21 (July 1963), NASA HO.

5. See, for example, "A Propaganda Triumph," *New York Times,* 6 Oct. 1957, and "Beep, Beep, Beep . . . Its Global Effects," *Washington Post,* 7 Oct. 1957.

6. George, "Impact of *Sputnik I.*"

7. "Senators Lash Defense Policy," *Washington Post,* 7 Oct. 1957.

8. " 'Sputnik' and Freedom," *Washington Evening Star,* 7 Oct. 1957.

9. *Washington Evening Star,* 7, 8, 13 Oct. 1957.

10. Walter Lippmann column, *New York Herald Tribune,* 10 Oct. 1957.

11. U.S. Department of State, *American Opinion Reports,* 20 Oct. 1957.

12. *New York Times,* 20 Oct. 1957.

13. *Life* magazine, 21 Oct. 1957, pp. 19–35.

14. NSC-5520, "Satellite Program," 20 May 1955, p. 11, DDE Library, Office of the Special Assistant for National Security Affairs, Box 3.

15. Eisenhower, *Waging Peace,* pp. 211–12; James R. Killian, *Sputniks, Scientists, and Eisenhower: A Memoir of the First Special Assistant to the President for Science and Technology* (Cambridge, Mass., 1976), pp. 15–17.

16. *New York Times,* 6 Oct. 1957.

17. Memo, J. F. Dulles to James Hagerty, 8 Oct. 1957, NASA HO (from DDE Library).

18. Remarks at National Defense Executive Reserve Conference, 13 Nov. 1957, Firestone Library, Princeton, Dulles Papers, Box 359.

19. Wilson in *New York Times,* 9 Oct. 1957; Eisenhower in *Facts on File* 17 (9 Oct. 1957), p. 330.

20. Letter, Ernst A. Steinhoff to Quarles, 5 Oct. 1957; Letter, Symington to Eisenhower, 8 Oct. 1957; Letter, Doolittle to Eisenhower, 10 Oct. 1957; Memo, Javits to the President, 11 Oct. 1957; all in NASA HO.

21. Letter, Christian Sonne, Frank Altschul, et al. (Trustees of the National Planning Association) to the President, 15 Oct. 1957, NASA HO.

22. Memo, Lodge to the President, 16 Oct. 1957; Letter, Lodge to the President, 21 Oct. 1957, both in DDE Library, Ann Whitman File, Box 27; also Memo, McElroy for the President, 21 Oct. 1957, NASA HO.

23. "Text of Address by Nixon in San Francisco Assessing Challenge of Soviet Satellite," *New York Times,* 16 Oct. 1957.

24. Dulles News Conference, 16 Oct. 1957, Firestone Library, Princeton, Dulles Papers, Box 122.

25. The "basketball game" quip was authored by Bryce Harlow, something he laughingly regrets to this day. Interview with the author, 3 Feb. 1982. See *New York Times,* 20 Oct. 1957.

26. Review article, *New York Times,* 24 Oct. 1957.

27. Killian, *Sputniks, Scientists, and Eisenhower,* p. 10.

28. Wallace-Wilson exchange cited in Richard Witkin, *The Challenge of the Sputniks* (New York, 1958), pp. 47–48.

29. Oliver M. Gale, "Post-Sputnik Washington from an Inside Office," *Cincinnati Historical Society Bulletin* 31 (Winter 1973): 226.

30. Note, Reedy for LBJ, 17 Oct. 1957, LBJ Library, Senate Papers, Box 421.

31. Memo, Charles Brewton, LBJ Library, Senate Papers, Box 421. Italics in original.

32. Gale, "Post-Sputnik Washington," p. 227; Interview with Bryce Harlow, 11 June 1974, NASA HO.

33. "Arguing the Case for Being Panicky," by George R. Price, *Life* magazine, 16 Nov. 1957, pp. 125–28.

34. John B. Medaris, *Countdown for Decision* (New York, 1960), pp. 159–65.

35. Memo, Quarles for the President, "The Vanguard-Jupiter C Program," summary of documentation, 7 Jan. 1958, DDE Library, Ann Whitman File, Box 28; Medaris, *Countdown,* pp. 165–69.

36. Radio and Television Address, "Science in National Security," 7 Nov. 1957, *PP of P, Dwight D. Eisenhower 1957,* pp. 789–99. On Killian's appointment, see Killian, *Sputniks, Scientists, and Eisenhower,* pp. 15–30. Killian was inevitably dubbed the new "missile czar," presumably authorized to stand above the services and "knock heads together," which is what czars do in Washington. George Dixon's syndicated column tried to sort out the "missile mess," but found "More Czars in US than Romanoffs had":

President Eisenhower announced the appointment a few weeks ago of Dr. James R. Killian. He was promptly identified as our new Missile Czar. But seemingly he is only Ike's Czar. The Pentagon says its missile czar is William Holaday.

Dr. John P. Hagen is described as the Navy's Satellite Czar and Major General Donald J. Keirn as the Air Force's new Nuclear Projects Czar. And now Secretary of Defense Neil McElroy says he is looking for an overall head of space projects who will indubitably be hailed our Space Czar.

Chairman Lyndon Johnson of the Senate Preparedness Subcommittee says he is trying to find out who is our Missile Czar, but he himself is being described as our Missile Investigating Czar. . . .

The Senate probers singled out Defense Secretary McElroy as the real Missile Czar, but he, in turn, declared the real czar is President Eisenhower. This seems to put things back where they started.

Vice President Nixon declared he will not assume any czarist role. . . . (*Washington Post,* 2 Dec. 1957)

37. NSC-5724, "Deterrence and Survival in the Nuclear Age," Security Resources Panel of the Science Advisory Committee (the "Gaither Report"), 7 Nov. 1957, NASA HO. Quote from p. 1.

38. Allen Dulles to Andrew Goodpaster, 28 Oct. 1957, DDE Library, Office of the Special Assistant for Science and Technology, Box 1.

39. Samuel P. Huntington, *The Common Defense: Strategic Programs in National Politics* (New York, 1961), p. 113.

40. Eisenhower, *Waging Peace,* pp. 219–22.

41. Interview with Glen Wilson, NASA HO.

42. Gale, "Post-Sputnik Washington," p. 232.

43. The Preparedness Subcommittee included Johnson, defense heavyweights John Stennis (D., Miss.) and Stuart Symington (D., Mo.), Estes Kefauver (D., Tenn.), and three New England Republicans, Styles Bridges (N.H.), Leverett Saltonstall (Mass.), and Ralph E. Flanders (Vt.).

44. U.S. Senate, Committee on Armed Services, *Inquiry into Satellite and Missile Programs. Hearings Before the Preparedness Investigating Subcommittee,* 85th Cong., 1st and 2nd Sess. (Washington, D.C., 1958), vol. 1, pp. 1–2.

45. Ibid., pp. 21–23.

46. Ibid., p. 65.

47. Ibid., p. 122.

48. Ibid., pp. 208–09, 265, 273.

49. Ibid., p. 285.

50. Kurt Stehling, German engineer on the project, cited in Constance M. Green and Milton Lomask, *Vanguard. A History* (Washington, D.C., 1970), p. 209. Reporter Nate Haseltine wrote descriptively: "There was no explosion in the blast sense of the word, and no one was hurt—physically" ("Vanguard Fails . . ." *Washington Post,* 7 Dec. 1957).

51. U.S. Senate, *Inquiry into Satellite and Missile Programs,* vol. 1, pp. 464–78 (Brucker), 505–12 (Gavin), 539–73 (Medaris), 581–604 (von Braun).

52. Letter, LBJ to the President, 4 Dec. 1958; Letter, Eisenhower to LBJ, 21 Jan. 1958: both NASA HO.

53. U.S. Senate, *Inquiry into Satellite and Missile Programs,* vol. 1, p. 1004.

54. Ibid., vol. 2, 1678–79, 1710.

55. Ibid., vol. 3, 2428–30. The "organized brainpower" phrase was in an address to the Junior Chamber of Commerce, Wichita Falls, Texas, 29 Nov. 1957, LBJ Library, Senate Papers, Box 359.

56. Eric Hoffer, *Before the Sabbath* (New York, 1979), p. 55.

Chapter 7

1. Truman addressing the American Legion, Welch, W. Va., in *Washington Post,* 12 Nov. 1957.

2. Annual Message to the Congress on the State of the Union, 9 Jan. 1958, *PP of P, Dwight D. Eisenhower 1958,* pp. 2–15. Quotes from pp. 2–3. The notion of "total Cold War" was anticipated in the San Francisco speech of Vice President Nixon. On 6 Dec. 1957, Nixon iterated his assertion that the main Cold War threat was shifting to nonmilitary areas of competition and to the developing world. He conceded before the National Association of Manufacturers at the Waldorf-Astoria that "we are in the midst of a world conflict in which the sputniks are but a single episode. Call it a Cold War or contest for men's minds or a race for outer space. Call it whatever you will. It is as Mr. Khrushchev has bluntly told us, a war of many phases—military, political, economic, psychological. A total war" ("Texts of Nixon's N.A.M. Address and Satellite Statement," *New York Times,* 7 Dec. 1957).

3. Annual Message to the Congress on the State of the Union, 9 Jan. 1958, *PP of P, Dwight D. Eisenhower 1958,* pp. 3–7. Quotes from p. 7.

4. Ibid., p. 15.

5. Annual Budget Message to Congress—Fiscal Year 1959, 13 Jan. 1958, *PP of P, Dwight D. Eisenhower 1958,* pp. 17–74. Quotes from pp. 17, 27.

6. Executive Order #10521, "Administration of Scientific Research by Agencies of the Federal Government," 17 Mar. 1954.

7. All quotes from Memo, I. I. Rabi for Gordon Gray, 19 July 1957 (forwarded by Gray to the President, 5 Aug. 1957): DDE Library, Records, Central File/Official File, Box 674. Italics in original.

8. Barbara Barksdale Clowse, *Brainpower for the Cold War: The Sputnik Crisis and the National Defense Education Act of 1958* (Westport, Conn., 1981), pp. 28–39. (Killian quote, p. 30; Commission quote, p. 38.)

9. Mortimer B. Smith, *And Madly Teach: A Layman Looks at Public Education* (Chicago, 1949), p. 87, cited by Clowse, *Brainpower for the Cold War,* p. 31; Rickover cited by Clowse, pp. 34–35; von Braun in Senate hearings cited by Clowse, p. 84.

10. *Congressional Record,* 28 Mar. 1956, p. 5785; "Izvestia Gloats at U.S. Hysteria," *New York Times,* 21 Nov. 1957; House Hearings on Education Bill, cited by Clowse, *Brainpower for the Cold War,* pp. 26, 62–63, 77 (quote from lobbyist).

11. Pusey speech at Waldorf-Astoria, *New York Times,* 25 Oct. 1957; Conant in Dwight D. Eisenhower, *The White House Years,* 2 vols., vol. 2, *Waging Peace* (New York, 1965), pp. 241–42. See also Clowse, *Brainpower for the Cold War,* pp. 36–37.

12. Special Message to the Congress on Education, 27 Jan. 1958, *PP of P, Dwight D. Eisenhower 1958,* pp. 127–32. Quote from p. 132. A few educators rued the measure and were not soothed by the assurances about federal non-intervention. See, for instance, Russell Kirk, *Decadence and Renewal in the Higher Learning. An Episodic History of American University and College Since 1953* (South Bend, Ind., 1978), pp. 44–45.

13. This was done by "skillfully avoid[ing] the church-state issue and other issues that had earlier proved to be roadblocks to federal support of education" (James R. Killian, *Sputniks, Scientists, and Eisenhower. A Memoir of the First Special Assistant to the President on Science and Technology* [Cambridge, Mass., 1977], p. 196).

14. Within this hard-boiled group, "Rockefeller was the boyish one—highly theoretical in many of his approaches, not too familiar with the military set-up but wedded to certain concepts, more like a successful Ph.D. candidate than a mature man of great stature. But charming as an individual and companion." Oliver M. Gale, "Post-Sputnik Washington from an Inside Office," *Cincinnati Historical Society Bulletin* 31 (Winter 1973): 233.

15. Special Message to Congress on Reorganization of the Defense Establishment, 3 Apr. 1958, *PP of P, Dwight D. Eisenhower 1958,* pp. 274–90. On drafting the plan, see Douglas Kinnard, *President Eisenhower and Strategy Management: A Study in Defense Politics* (Lexington, Ky., 1977), pp. 89–93.

16. Eisenhower, *Waging Peace,* pp. 244–53. Quote from p. 251.

17. Kinnard, *Eisenhower and Strategy Management,* pp. 86–87.

18. Gale, "Post-Sputnik Washington," pp. 236–37.

19. Richard Hirsch and Joseph John Trento, *The National Aeronautics and Space Administration* (hereafter *NASA*) (New York, 1973), p. 7.

20. Alex Roland, *Model Research: A History of the National Advisory Committee for Aeronautics, 1915–1958*, NASA SP-4103 (Washington, D.C., 1985), chaps. 11–12; Hirsch and Trento, *NASA*, pp. 16–18; Arthur L. Levine, *The Future of the US Space Program* (New York, 1975), pp. 23–26.

21. Donald T. Rotunda, "The Legislative History of the National Aeronautics and Space Act of 1958," NASA HHN-125 (Sept. 1972), p. 25.

22. Army Ballistic Missile Agency, Development Operations Division, "Proposal—A National Integrated Missile and Space Vehicle Development Program," 10 Dec. 1957, DDE Library, White House Office of the Staff Secretary, DoD subseries, Box 8.

23. Memo for the Chief of Staff, USAF, "Vanguard—Jupiter C Firings," 22 Jan. 1958, NASA HO.

24. Enid Curtis Bok Schoettle, "The Establishment of NASA," in Sanford A. Lakoff, ed., *Knowledge and Power. Essays on Science and Government* (New York, 1966), pp. 162–270, quote from p. 187. Schoettle's essay, a NASA historical work, was originally entitled "Making American Space Policy: The Establishment of NASA," NASA 09-63 (Jan. 1963), NASA HO.

25. S.3126, a bill to create a Department of Science, was introduced by Senators Humphrey, Yarborough, and McClellan.

26. See Herblock cartoon, reproduced here, which first appeared in the *Washington Post*, 22 Nov. 1957.

27. John B. Medaris, *Countdown for Decision* (New York, 1960), pp. 222–24.

28. Eisenhower, *Waging Peace*, p. 256.

29. Letter, Anderson to C. H. Greenewalt, 5 Feb. 1958, Library of Congress (hereafter LC), Anderson Papers, Box 911.

30. On the committees, see Schoettle, "The Establishment of NASA," pp. 229–31. The Brooks quote is from Ken Hechler, *The Endless Space Frontier: A History of the House Committee on Science and Astronautics, 1959–1978*. AAS History Series, Vol. 4 (San Diego, 1982), an abridged version of Ken Hechler, *Toward the Endless Frontier: History of the Committee on Science and Technology, 1959–1979*, Committee Print, U.S. House of Representatives (Washington, D.C., 1980). Pagination will vary between the two.

31. "Statement by Senator Anderson to Special Subcommittee on Outer Space, Joint Committee on Atomic Energy," 20 Feb. 1958, LC, Anderson Papers, Box 910.

32. Memo, S. Paul Johnston for Dr. Killian, 21 Feb. 1958, forwarding "Preliminary Observations on the Organization for the Exploitation of Outer Space," NASA HO.

33. Preliminary Staff Draft, "Organization for Civil Space Programs," 22 Feb. 1958, NASA HO. Quotes from pp. 1, 5.

34. Ibid., pp. 8–10. For the presidential message incorporating these recommendations, 2 Apr. 1958, *PP of P, Dwight D. Eisenhower 1958*, pp. 269–73.

35. Comments on the draft National Aeronautics and Space Act: Memo, Quarles (DoD) to Stans (BoB), 1 Apr. 1958; Memo, Roy Johnson (ARPA) to Quarles, 16 Apr. 1958; Memo, Loftus Becker (counsel, State Department) to J. F. Dulles, 31 Mar. 1958; Memo, Doolittle (NACA) to Dryden, 28 Mar. 1958; all in NASA HO. "Motorcycle" quote from Schoettle, "The Establishment of NASA," p. 238.

36. PSAC, "Introduction to Outer Space" (Washington, D.C., 1958), is reprinted as appendix 4 of James R. Killian, *Sputniks, Scientists, and Eisenhower: A Memoir of the First Special Assistant to the President for Science and Technology* (Cambridge, Mass., 1976). It was drafted by Purcell, Land, and Francis Bello, a scientist-journalist then with *Fortune*.

37. PSAC, *Introduction to Outer Space* (Washington, D.C., March 26, 1958); quotes from p. 6.

38. The administration bill S.3609, "The National Aeronautics and Space Act," was introduced by Senators Lyndon Johnson and Styles Bridges. The House version was H.R. 11881, 14 April 1958.

39. Eilene Galloway, "The Problems of Congress in Formulating Outer Space Legislation," 7 Mar. 1958; Galloway, "Nature of the Task Confronting the Senate Special Committee on Astronautics and Outer Space"; Charles S. Sheldon II, "Congressional Intent on Outer Space Development," 25 Mar. 1958, all in NA, Papers of the Senate Committee on Aeronautical and Space Sciences (hereafter Papers of the Senate Space Committee), Box 9.

40. Committee staff, "Making the Record in Outer Space," 11 Apr. 1958, NA, Papers of the Senate Space Committee, Box 9.

41. Memo, Lt. Gen. Elmer J. Rogers, Jr., Acting Chief of Staff, to Assistant Secretary of the Air Force (R & D), 20 June 1958, NASA HO. On congressional concern over the role of the military, see Memo, Solis Horwitz, "National Aeronautics and Space Act of 1958—S. 3609," 3 May 1958, and Staff Memo, "The Conflicting Views of the House and Senate on Organization for the Conduct of Space Activities" and "Guidelines to the Hearings on Outer Space": both in NA, Papers of the Senate Space Committee, Box 9.

42. Killian, *Sputniks, Scientists, and Eisenhower,* pp. 135–36; Hirsch and Trento, *NASA,* pp. 22–27; Arthur L. Levine, *The Future of the US Space Program* (New York, 1975), pp. 21–27.

43. Quotes from "Reasons for Confusion over Outer Space Legislation and How To Dispel It," 11 May 1958, NA, Papers of the Senate Space Committee, Box 9. See also "Report of the Special Senate Committee on Space and Astronautics on S. 3609," LBJ Library, Senate Papers, Box 357.

44. Note, Victor Emanuel to Quarles, "Telephone Conversation with John McCormack," 1 June 1958 and Note, Donald L. Wilkins to Victor Emanuel, 3 June 1958: both NASA HO.

45. Khrushchev paraphrased by Donald W. Cox, *The Space Race. From Sputnik to Apollo . . . and Beyond* (Philadelphia, 1962), pp. 32–33.

46. Frank Gibney and George J. Feldman, *The Reluctant Space Farers: A Study in the Politics of Discovery* (New York, 1956), p. 68. Feldman was executive director of the House Space Committee, Gibney a consultant.

47. Killian, *Sputniks, Scientists, and Eisenhower,* p. 137. See also "Off the Record Meeting: The President, Senator Johnson," 7 July 1958, DDE Library, Ann Whitman File, Box 21.

48. On the compromise, see Schoettle, "The Establishment of NASA," pp. 259–60; Rotunda, "Legislative History of the NAS Act," pp. 34–35 (quote from p. 35); and Hechler, *The Endless Space Frontier,* p. 17.

Chapter 8

1. Joseph M. Goldsen and Leon Lipson, "Some Political Implications of the Space Age," RAND P-1435, 24 Feb. 1958. Italics in original.

2. Mansfield D. Sprague, "Proposal for a National Policy on Outer Space," International Security Affairs, Assistant Secretary of Defense, 25 Feb. 1958, NASA HO. Italics in original.

3. U.S. House of Representatives, Committee on Science and Astronautics, "U.S. Policy on the Control and Use of Outer Space," 86th Cong., 1st Sess. (Washington, D.C., 1959), p. 6.

4. These public exchanges may be found in many sources. See U.S. ACDA, *Documents on Disarmament 1945–1959,* vol. 2, pp. 938–39, 976–77. Eisenhower quote also in U.S. Senate, Committee on Aeronautical and Space Sciences, *Documents on International Aspects on the Exploration and Use of Outer Space, 1954–1962,* 88th Cong., 1st Sess. (Washington, D.C., 1963), pp. 52–53, Bulganin's reply (p. 54), Soviet proposal (p. 57), and Foster Dulles (p. 54), or in *Department of State Bulletin* 38 (3 Feb. 1958): 166–67.

5. "Soviet Proposal on the Question of Banning the Use of Cosmic Space for Military Purposes . . ." 15 Mar. 1958; Letter, Khrushchev to Eisenhower, 22 Apr. 1958; Letter, Eisenhower to Khrushchev, 28 Apr. 1958; U.S. Senate, *Documents on International Aspects . . . of Outer Space,* pp. 57–58, 62–64.

6. See Robert A. Divine, *Blowing on the Wind: The Nuclear Test Ban Debate 1954–1960* (New York, 1978), pp. 206–40; James R. Killian, *Sputniks, Scientists, and Eisenhower: A Memoir of the First Special Assistant to the President for Science and Technology* (Cambridge, Mass., 1976), pp. 150–68.

7. Discussion in the Cabinet, Minutes, 15 Aug. 1958, DDE Library, Ann Whitman File, Box 2.

8. NSC 5814/1, "Preliminary U.S. Policy on Outer Space," 18 Aug. 1958, p. 1: DDE Library, Office of the Special Assistant for National Security Affairs, Box 67.

9. Ibid., p. 5.

10. Ibid., pp. 8–9.

11. Ibid., p. 9.

12. Annex to NSC Action #1553, November 21, 1956, cited in ibid., pp. 11 ("effective inspection"), 12, and 13 ("imaginative position").

13. Ibid., pp. 13–14. Italics in original.

14. Ibid., p. 20.

15. Ibid., pp. 20–21.

16. Memo, Harr, ". . . relationship of the OCB to the question of U.S. policy toward Outer Space," prepared for briefing of NASC meeting of 24 Sept. 1958, NASA HO.

17. Memo of Meeting: OCB Working Group on Outer Space, 17 Sept. 1958, p. 2, NASA HO.

18. Memo of Meeting: OCB Working Group on Outer Space, 16 Oct. 1958, p. 2, NASA HO.

19. "Rocket to Moon Expected Aug. 17," *New York Times,* 27 July 1958.

20. "Minutes, OCB Working Group on Outer Space," 24 Oct., 31 Oct. 1958, NASA HO.

21. Memo, Robert O. Piland to Killian, 27 Oct. 1958, NASA HO.

22. Ibid. Also, Letter, Herter to Killian, 22 Aug. 1958, and Letter, Killian to Herter, 3 Sept. 1958: both in NASA 69 A 1729, Box 2.

23. Address by the Secretary to the General Assembly, *Department of State Bulletin* 38 (18 Sept. 1958): 528–29.

24. U.S. Senate, *Documents on International Aspects . . . of Outer Space,* pp. 84–86. Quote from p. 86.

25. UN General Assembly, 13th Sess., Agenda Item 60, A/4009, 28 Nov. 1958.

26. U.S. Senate, Special Committee on Space and Astronautics, *Space Law: A Symposium,* compiled by Eilene Galloway, 85th Cong., 2nd Sess. (Washington, D.C., 1958), p. vii.

27. John Cobb Cooper, "High Altitude Flight and National Sovereignty," *International Law Quarterly* (July 1951): 411–18, reprinted in U.S. Senate, *Space Law: A Symposium,* pp. 1–7. Quote from p. 7.

28. See the summary of proposals for defining "air space" in Myres S. McDougal, Harold D. Lasswell, and Ivan A. Vlasic, *Law and Public Order in Space* (New Haven, 1963), pp. 323–59.

29. Loftus Becker, "Major Aspects of the Problem of Outer Space," in U.S. Senate, *Space Law: A Symposium,* pp. 367–74. Quote from pp. 373–74.

30. George J. Feldman, "An American View of Jurisdiction in Outer Space," in U.S. Senate, *Space Law: A Symposium,* pp. 428–33. Quotes from pp. 431, 432.

31. Oscar Schachter, "Who Owns the Universe?" in U.S. Senate, *Space Law: A Symposium,* pp. 8–17. Quotes from pp. 10, 12.

32. Andrew G. Haley, *Space Law and Government* (New York, 1963), pp. 24–39.

33. Summary of McDougal's views in S. Bhatt, *Legal Controls of Outer Space: Law, Freedom, and Responsibility* (New Delhi, 1973), pp. 32–33. The literature on space law is extensive and highly repetitive. The most important early books on the subject are McDougal's, Haley's, Philip C. Jessup and Howard J. Taubenfeld, *Controls for Outer Space and the Antarctic Analogy* (New York, 1959), C. Wilfred Jenks, *Space Law* (New York, 1963), and John Cobb Cooper, *Explorations in Aerospace Law* (Montreal, 1968). See generally Irvin L. White, C. E. Wilson, and J. A. Vosburgh, *Law and Politics in Outer Space: A Bibliography* (Tucson, 1972), Kuo Lee Li, *World Wide Space Law Bibliography* (Toronto, 1978), the *Yearbooks of Air and Space Law* of the Institute of Air and Space Law, McGill University, and the *Journal of Space Law.*

34. Soviet jurisprudence for space activities is discussed at length in chapter 12.

35. Mary Shepard, Library of Congress, Legislative Reference Service, "An International Outer Space Agency for Peaceful Purposes," 20 Mar. 1958, NA, Papers of the Senate Space Committee.

36. *Congressional Record,* House, 2 June 1958, p. 9912.

37. *Congressional Record,* Senate, 23 July 1958, p. 14753.

38. Memo, Solis Horwitz to LBJ, "Briefing on the 'Discoverer' Satellite," 26 Nov. 1958, NA, Papers of the Senate Space Committee.

39. See, for instance, Goodpaster Notes, "Memo of a Conference with the President," 6 Feb. 1958, DDE Library, Ann Whitman File, Box 18. On ARPA, see Herbert York, *Race to Oblivion. A Participant's View of the Arms Race* (New York, 1970), pp. 117–24.

40. Letters, Secretary of Defense to the President, 30 Apr., 28 July, 29 Oct. 1959, DDE Library, Ann Whitman File, Box 28; Anthony Kenden, "US Reconnaissance Satellite Programs," *Spaceflight* 20 (July 1978): 243–62; ARPA, "Military Space Projects: Report of Progress," Quarterly, June 1959–May 1960, DDE Library, Office of Staff Secretary, DoD subseries, Box 9.

41. John Noble Wilford, *The Map Makers* (New York, 1981), describes the imperfect state of global surveys at the dawn of the Space Age. Wernher von Braun explained as early as 1951:

> The missile has a poor reputation due to the dispersion of the V-2s, but the missile art was still young. . . . One of the gravest handicaps in improving missile accuracy is the poor accuracy of the geodetic survey of a great portion of the globe. For example, the Eurasian landmass relative to the continent of the Americas is not known to more accuracy than 300 to 400 yards. But even within continents the national survey grids are often poorly linked together. The very good national surveys of Great Britain and France were so poorly linked that this alone accounted for error in the V-2s of approximately 150 yards. (Von Braun Speech, "Why Guided Missiles," LC, von Braun Papers, Box 46)

42. Memo, Stans for the President, 29 July 1958, NASA HO.

43. Memo of Meeting: OCB Working Group on Outer Space, 8 Jan. 1959, NASA HO, noting publicity restrictions on upcoming USAF Discoverer flights.

44. Itek Corporation, "Political Action and Satellite Reconnaissance," 24 Apr. 1959, DDE Library, Office of the Special Assistant for Science and Technology, Box 15.

45. Memo, Shapley to Stans, "Meetings of the NASC—UN Ad Hoc COPUOS," 24 Apr. 1959, NASA HO. See also "UN Ad Hoc COPUOS, Draft Position Paper on Topic 1(b)," 23 Apr. 1959, DDE Library, Office of the Special Assistant on National Security Affairs, Box 62.

46. UN General Assembly, 14th Sess., Agenda Item 25, A/4141, "Report of the Ad Hoc COPUOS," 14 July 1959.

47. "Report of the NSC Ad-Hoc Working Group on the Monitoring of Long-Range Rocket Test Agreement," 26 Mar. 1958, DDE Library, Office of the Special Assistant for Science and Technology, Box 7.

48. Letter, J. F. Dulles to Killian, 2 May 1958, DDE Library, Office of the Special Assistant for Science and Technology, Box 1.

49. Memo, Wiesner, "Urgency of a Complete Missile Ban," 20 Nov. 1959, DDE Library, Office of the Special Assistant for Science and Technology, Box 7.

50. Note, Rathjens, "Comments on Wiesner's Paper 'The Urgency of a Complete Missile Test Ban,'" 3 Dec. 1959, DDE Library, Office of the Special Assistant for Science and Technology, Box 7.

51. Letter, Killian to Kistiakowsky, 11 Dec. 1959, DDE Library, Office of the Special Assistant for Science and Technology, Box 7. In the same file see also the continued arguments against a missile freeze in Memo, Rathjens, 21 Dec. 1959, "Further Comments on Negotiations with the Soviets Relating to Testing and Control of Ballistic Missiles," and 18 Feb. 1960, Report entitled "Feasibility and National Security Implications of a Monitored Agreement to Stop or Limit Ballistic Missile Testing and/or Production," which states: "For a test ban to be effective in limiting missile development, it would be necessary that space research be abandoned or subject to rigid controls. . . . An absolute ban on production would be very dangerous to the U.S. if implemented as early as 1961" (p. 3).

52. On the transition from disarmament to arms "control" aimed at "stability" in the thinking of U.S. strategists, see Lawrence Freedman, *Evolution of Nuclear Strategy* (London, 1981), pp. 195–99.

Chapter 9

1. U.S. House of Representatives, Select Committee on Astronautics and Space Exploration, *Authorizing Construction for the NASA*, 85th Cong., 2nd Sess. (Washington, D.C., 1958), pp. 9, 12. On Dryden, see Alex Roland, *Model Research: A History of the National Advisory Committee for Aeronautics, 1915–1958*, NASA SP-4103 (Washington, D.C., 1985), chap. 11.

2. James R. Killian, *Sputniks, Scientists, and Eisenhower: A Memoir of the First Special Assistant to the President for Science and Technology* (Cambridge, Mass., 1976), pp. 139–41. Quotes from pp. 139, 140. Glennan admitted to being a Republican, but that was not the motive for his appointment, said Ike. Indeed, the President was chided by some Republicans for surrounding himself with so many scientific advisers of the other political persuasion. He mentioned this once to Herb York over breakfast, and York replied, "Well, Mr. President, don't you know that all scientists are Democrats?" Eisenhower laughed and said, "Well, I don't believe that, but it doesn't make a damned bit of difference to me" (James R. Killian, Jr., Oral History Interview, p. 12, 23 July 1974, NASA HO).

3. Glennan, National Air and Space Museum (hereafter NASM) Seminar, 28 Jan. 1982. The Glennan diary is now open for research at the DDE Library.

4. Herbert York, "Briefing on the Army Satellite Program," 19 Nov. 1957, DDE Library, Office of the Special Assistant for Science and Technology, Box 15.

5. On the institutional and psychological preference of the USAF for piloted manned flight over ballistic spaceflight, see the exquisite account of Tom Wolfe, *The Right Stuff* (New York, 1979).

6. Loyd S. Swenson, Jr., James M. Grimwood, and Charles C. Alexander, *This New Ocean: A History of Project Mercury,* NASA SP-4201 (Washington, D.C., 1966), p. 93.

7. Institute for Defense Analysis (ARPA contractor), "Strategic Space Force, Key to National Survival," 26 Aug. 1958, NASA HO.

8. Classified Address by Roy Johnson at Industrial College of the Armed Forces, 12 Dec. 1958; ARPA, "Long Range Plan for Advanced Research," 30 July 1959: both in NASA HO.

9. On military space programs generally, see the following popular works based on nonclassified sources: John F. Loosbrock, ed., *Space Weapons: A Handbook of Military Astronautics* (New York, 1959); Michael N. Golovine, *Conflict in Space: A Pattern of War in a New Dimension* (London, 1962); Eldon W. Downs, *The United States Air Force in Space* (New York, 1966); Robert Salkeld, *War and Space* (Englewood Cliffs, N.J., 1970); Bhupendra M. Jasani, *Space—Battlefield of the Future?* (Stockholm, 1978). The three most recent essays are G. Harry Stine, *Confrontation in Space* (Englewood Cliffs, N.J., 1981); David Ritchie, *Space War* (New York, 1982); and Thomas Karas, *The New High Ground. Strategies and Weapons of Space-Age War* (New York, 1983).

10. Glennan, NASM Seminar, 28 Jan. 1982.

11. Ibid. On the JPL transfer, see Clayton R. Koppes, *JPL and the American Space Program: A History of the Jet Propulsion Laboratory* (New Haven, 1982), pp. 96–99; on the ABMA transfer, see Robert L. Rosholt, *Administrative History of NASA, 1958–1963,* NASA SP-4101 (Washington, D.C., 1966), pp. 46–47; and especially U.S. Senate, Committee on Aeronautical and Space Sciences, NASA Authorization Subcommittee, *Transfer of Von Braun Team to NASA,* 86th Cong., 2nd Sess. (Washington, D.C., 1960).

12. Medaris had undergone an operation for cancer in 1956. The disease was caught in time by apparent accident, and Medaris believed he had been spared for a purpose higher even than his military calling. After retirement, he beat cancer on two more occasions and was ordained an Episcopal priest; see Gordon Harris, *A New Command: The Story of a General Who Became a Priest* (Plainfield, N.J., 1976).

13. U.S. Senate, *Transfer of Von Braun Team to NASA. Subcommittee Hearings on HJ Resolution 567,* 86th Cong., 2nd Sess. (Washington, D.C., 1960); John B. Medaris, *Countdown for Decision* (New York, 1960), pp. 257–69; Rosholt, *Administrative History of NASA,* pp. 109–15; Michael H. Armacost, *The Politics of Weapons Innovation. The Thor-Jupiter Controversy* (New York, 1969), pp. 238–44; Glennan, NASM Seminar.

14. Air Force Command History Office (AFCHO) K140, 11-6, "Documents on Air Force Space History"; AFCHO, K239, 04-61, W. A. Heflin, "NACA and NASA," pp. 1–18; Claude Witze, "How Our Space Policy Evolved," *Air Force/Space Digest* (Apr. 1962): 83–92; AFCHO, SHO-S-60/66, "ARDC History," 1-12/58, vol. 1, pp. 23–28; Bowen, "Air Force Space Activities 1945–1959," SHO-C-64/50, all in Air Force Systems Command History Office, Andrews Air Force Base.

15. Witze, "How Our Space Policy Evolved," p. 91.

16. Swenson, Grimwood, and Alexander, *This New Ocean,* pp. 101–06.

17. Ibid., p. 111.

18. "Dr. Killian commented that we may have a recurrence of the Sputnik hysteria if the

Soviets get a 'man in space' first" (Goodpaster Notes, "Memo of a Conference with the President," 17 Feb. 1959, DDE Library, Ann Whitman File, Box 24).

19. Glennan, NASM Seminar, 28 Jan. 1982: "Ike and I agreed that we were mature enough as a nation not to let some other country determine our behavior and policy. Hence we opposed a 'Space race,' and while we wanted to advance rapidly, not to do foolish things just because the Russians were doing them."

20. Ibid.

21. Goodpaster Notes, "Memo of a Conference with the President," 17 Feb. 1959, DDE Library, Ann Whitman File, Box 24.

22. Letter, Glennan to Killian, 27 May 1959, NASA HO.

23. Statement by Glennan before Senate Appropriations Committee, 13 July 1959, NASA HO.

24. Indeed, some of the most effective Soviet propaganda was inspired by the post-Sputnik breast beating of American writers. The vicious circle in which the West reacted with increasing alarm to Soviet playbacks of American self-criticism was observed by Arnold L. Horelick, "The Soviet Union and the Political Uses of Outer Space," RAND P-2480 (Nov. 1961).

25. Goodpaster Notes, "Memo of a Conference with the President," 23 Oct. 1959, DDE Library, Ann Whitman File, Box 29.

26. "Notes on Discussion of Question 'How Important in the Current Scheme of Things is the matter of competing in the Space Field aggressively and ultimately successfully with the Soviet Union?' " 23 Sept. 1959, NASA HO. See also George Kistiakowsky, *A Scientist at the White House: The Private Diary of President Eisenhower's Special Assistant for Science and Technology* (Cambridge, Mass., 1976), pp. 72–76, 104–05, 114.

27. Memo, Kistiakowsky for the Record, 2 Oct. 1959, NASA HO. See also *Wall Street Journal*, 5 Oct. 1959, citing General Medaris: "The first thing this country has to do is to make up its mind whether it is in a space race with Russia or not. It is my personal opinion that we are stupid if we don't make a race of it." See also "Space Race Called for by Medaris," *Washington Post*, 30 Oct. 1959.

28. Letter, Dryden to Greenewalt, 7 Oct. 1959, Milton S. Eisenhower Library, Johns Hopkins University, Dryden Papers; see also Letter, Glennan to Greenewalt, 7 Oct. 1959, NASA HO.

29. Rostow was then professor at MIT, Nitze was affiliated with Johns Hopkins and was president of the Foreign Service Educational Foundation. The others were Frank Stanton, president of CBS; James A. Perkins, vice-president of Carnegie; Mervin J. Kelly, a research management consultant; Professor Edward Purcell of Harvard; Raymond J. Saulnier of the Council of Economic Advisors; Lee A. DuBridge, president of Caltech; and Kistiakowsky.

30. Notes for Greenewalt Committee Meeting; Suggested Areas of Discussion; Background Paper and Questions (quote from this section, pp. 2–3), 10 Dec. 1959, NASA HO; also interview by the author with Paul Nitze, 16 Sept. 1980.

31. Interview, Nitze with the author, 16 Sept. 1980.

32. "Remarks by the VP," 10 Dec. 1959, in Notes on Greenewalt Committee, NASA HO.

33. *Newsweek*, 19 Oct. 1959; *New York Times*, 7 and 11 Oct. 1959; *Washington Star*, 7 Oct. 1959; *New York Herald Tribune*, 7 Oct. 1959.

34. NSC Memos, "Splits in U.S. Policy on Outer Space," 23 Dec. 1959; Meeting of NASC, "Report on Comparative Study of U.S. and USSR Capabilities in Space Science and Technology," 8 Jan. 1960: both in NASA HO.

35. NSC-5918, "U.S. Policy on Outer Space," 17 Dec. 1959 (quotes from pp. 1, 10, 11, 12), DDE Library, Office of the Special Assistant for National Security Affairs, Box 70.

36. Letter, Glennan to Eisenhower, 26 Apr. 1960, NASA HO.

37. Hailsham cited by Arnold W. Frutkin, *International Cooperation in Space* (Englewood Cliffs, N.J., 1965), p. 9.

38. Ibid., pp. 28–36.

39. The most indefatigable of American critics of policy for space cooperation was Leonard E. Schwartz. He wrote and spoke on every occasion, prompting one State Department official to remark to the author: "We just couldn't get him to shut up!" See Schwartz, "When Is International Space Cooperation International?" *Bulletin of the Atomic Scientists* 19 (June 1963): 12–18. See also Don Kash, *The Politics of Space Cooperation* (West Lafayette, Ind., 1967).

Chapter 10

1. On traditional civilian views of the American officer corps, see Samuel Huntington's classic, *The Soldier and the State: The Theory and Politics of Civil-Military Relations* (Cambridge, Mass., 1957).

2. See, for instance, Albert Wohlstetter, "Strategy and the Natural Scientists," and Bernard Brodie, "The Scientific Strategists," in Robert Gilpin, ed., *Scientists and National Policy Making* (New York, 1964), pp. 174–256; Philip Green, "Strategy, Politics, and Social Scientists" (pp. 39–68) and critical replies in Morton A. Kaplan, ed., *Strategic Thinking and Its Moral Implications* (Chicago, 1973). See also, Lawrence Freedman, *Evolution of Nuclear Strategy* (London, 1981), pp. 175–81.

3. Bernard Brodie, "Strategy as a Science," *World Politics* 1 (July 1949): 467–88.

4. Paul Peeters, *Massive Retaliation: The Policy and Its Critics* (Chicago, 1959).

5. Henry A. Kissinger, *Nuclear Weapons and Foreign Policy* (New York, 1957).

6. Bernard Brodie, *Strategy in the Missile Age* [1959] (Princeton, N.J., 1965), esp. pp. 348–57, 390–409. See also Thomas Schelling, *The Strategy of Conflict* (New York, 1960).

7. Hermann Kahn, *On Thermonuclear War* (Princeton, 1960), esp. pp. 556–76.

8. The challenge to military professionalism implicit in the industrialization of war (or militarization of industry) has not been adequately examined. Paul A. C. Koistinen, *The Military-Industrial Complex. A Historical Perspective* (New York, 1980), highlights the phenomenon (pp. 13–20), but traces its origins back to the nineteenth century and indeed to the nature of democratic capitalism. Michael Geyer, on the other hand, has traced the same loss of autonomy by the military under the impact of technological change and mass mobilization of society in Germany; see "Military Work, Civil Order, Militant Politics: The German Experience 1914–1945," Woodrow Wilson International Center for Scholars, International Security Studies Program, Working Paper (June 1982). Is it proper, therefore, to speak of this challenge as an American Cold War phenomenon? At least in terms of R & D, it is. Even Koistinen admits that "all the policies, techniques, and even violence of the Cold War abroad had come home to debauch and defile practically every institution in American life" (p. 18). His New Left perspective, however, prevents him from asking whether, given the Soviet challenge, another outcome was possible.

9. Maxwell D. Taylor, *The Uncertain Trumpet* (New York, 1959), esp. pp. 6–10, 19–22, 130–40, 165–80.

10. John B. Medaris, *Countdown for Decision* (New York, 1960), esp. pp. 270–95.

11. James M. Gavin, *War and Peace in the Space Age* (London, 1959), p. 212.

12. Ibid., pp. 232–44, 269–74. Quote from p. 269. See also Donald Cox and Michael Stoiko, *Spacepower* (New York, 1958).

13. Walt W. Rostow, *The Stages of Economic Growth: A Non-Communist Manifesto* (Cambridge, Eng., 1960).

14. Ibid., also M. F. Millikan and Walt W. Rostow, "Foreign Aid: The Next Phase," *Foreign Affairs* 36 (Apr. 1958): 418–36.

15. *Washington Post*, 11 Dec. 1957.

16. W. W. Rostow, "Notes on U.S. Space Policy" (Greenewalt Committee), 11 Jan. 1960, NASA HO.

17. Cited by Donald Cox, *The Space Race: From Sputnik to Apollo . . . and Beyond* (Philadelphia, 1962), p. 128.

18. In contrast to the Rockefeller (and Seaborg) recommendations for state-supported R & D, President Eisenhower's own commission recommended a "modest" increase in funding of *basic* science of some 70 percent—but "We should avoid like the plague the enticing danger of too much, and too concentrated planning of our national scientific development . . . the supposed benefits of centralized planning are an illusion"; see The President's Commission on National Goals, *Goals for Americans* (New York, 1960), p. 124.

19. PSAC, *Scientific Progress, the Universities, and the Federal Government* (Washington, D.C., 1960), pp. 1–4.

20. Ibid., pp. 10–11, 14, 28–31. Italics in original. The NSF had already argued the link between scientific R & D and economic growth; see Letter, Piore to Alan Waterman, 22 Apr. 1958, and Draft, Kreidler, "An Anti-Recession Program in Science and Technology,"

23 Apr. 1958; also Waterman, "The Challenge of Excellence," *Conference on the Mass Media and the Image of Science,* 6 Nov. 1959; all NASA HO.

21. Charles Van Doren scandalized the nation when he confessed to being "fed" the answers on a television quiz program, thereby blowing the lid off corruption in the whole genre.

22. Memo, Rostow to Senator Kennedy, "A Democratic Strategy for 1960," 2 Jan. 1960, pp. 1–4. NASA HO, document from John F. Kennedy Library, Waltham, Massachusetts (hereafter JFK Library).

23. Ibid., pp. 6–7. On space policy, see also Letter, Rostow to Archibald Cox, 11 Aug. 1960, NASA HO, document from JFK Library.

24. Cited by Oliver M. Gale, "Post-Sputnik Washington from an Inside Office," *Cincinnati Historical Society Bulletin* 31 (Winter 1973): 246.

25. See, for instance, the fictional account by William F. Buckley, Jr., *Marco Polo If You Can* (New York, 1981), premised on the notion that the CIA knew the U-2 would soon become vulnerable to Soviet air defense, and thus that the United States might "make use" of its last and predictably fatal flight in a novel way.

26. Gale, "Post-Sputnik Washington," p. 247.

27. Eisenhower's Secretary of Defense from 1957 to 1959 put on the record: "I don't like to be invidious to the Senators who were in this up to their necks—but I think relatively few of the Senators really thought there was much to [the missile gap], because we told them, of course, everything we knew" (Neil McElroy Oral History, p. 20, J. F. Dulles Papers, Firestone Library, Princeton). On the U-2 inquiry, see David Wise and Thomas B. Ross, *The U-2 Affair* (New York, 1962), pp. 174–75.

28. News Conference, 26 Jan. 1960, *PP of P, Dwight D. Eisenhower 1960,* p. 127.

29. Ibid., 3 Feb. 1960, pp. 144–47.

30. Ibid., 11 Feb. 1960, p. 172.

31. Kennedy speech at Portland, Oregon, 7 Sept. 1960. "Checkers" was Nixon's dog, made famous in the 1952 speech in which Nixon defended himself against charges of misuse of campaign funds.

32. Nixon speech at Philadelphia, 29 Oct. 1960.

33. *Missiles and Rockets,* 17 Sept. 1960.

34. "If the Soviets Control Space—They Can Control Earth—Kennedy," *Missiles and Rockets,* 10 Oct. 1960, and "Nixon: Military Has Mission to Defend Space," *Missiles and Rockets,* 31 Oct. 1960.

35. John Logsdon, *The Decision To Go to the Moon: Project Apollo and the National Interest* (Cambridge, Mass., 1970), pp. 55–57.

36. Robert L. Rosholt, *An Administrative History of NASA, 1958–1963.* NASA SP-4101 (Washington, D.C., 1966), pp. 130–31.

37. Roger E. Bilstein, *Stages to Saturn: A Technological History of the Apollo/Saturn Launch Vehicles,* NASA SP-4206 (Washington, D.C., 1980), pp. 45–50; John L. Sloop, *Liquid Hydrogen as a Propulsion Fuel 1945–1959,* NASA SP-4404 (Washington, D.C., 1978), pp. 230–43.

38. Philip J. Klass, *Secret Sentries in Space* (New York, 1971), pp. 79–99; Anthony Kenden, "U.S. Reconnaissance Satellite Programs," *Spaceflight* 20 (1978): 243–62; Herbert F. York and G. Allen Greb, "Strategic Reconnaissance," *Bulletin of the Atomic Scientists* 33 (Apr. 1977): 33–42. On CIA/USAF competition for control of early spy satellite programs, see Peter Pringle and William Arkin, *The SIOP. The Secret U.S. Plan for Nuclear War* (New York, 1983), pp. 92–98. Midas was not fully operational until 1972.

39. Wernher von Braun, "Speech to the Second Annual Meeting of the Association of the US Army," General Gavin presiding, 26 Oct. 1956, LC, Von Braun Papers, Box 46.

40. Jack Doherty, "Space in the 1960 Campaign," NASA HO.

41. See, for example Note, P. M. Chayes to JFK, "Space Flight—The Challenge of Our Times," 19 Aug. 1960; Memo, Chayes to Cox, "Appointment of Kennedy Scientific Advisory Group"; Memo, Gerald Siegel to Chayes, "Major Organizational and Policy Aspects of the National Space Program," 6 Sept. 1960: all in NASA HO (documents from JFK Library). Also, "Selected Statements by President Kennedy on Defense Topics, December 1957–August 1, 1962," NASA Historical Note.

42. Statement by Senator Lyndon B. Johnson, for release in AM's of Monday, October 31, 1960; see also the preparatory memo, Max Lehrer to LBJ, " 'White Paper' on Space,"

26 Oct. 1960; both in LBJ Library, Senate Papers, "Science, Space and Aeronautics," Box 1249.

43. PSAC, "Report of the Ad Hoc Panel on Man-in-Space," 14 Nov. 1960, NASA HO; see also Logsdon, *Decision To Go to the Moon*, pp. 34–36.

44. Goodpaster Notes, "Memo of a Conference with the President," 13 Oct. 1960, DDE Library, Ann Whitman File, Box 34.

45. Gerald M. Steinberg, "The Legitimization of Reconnaissance Satellites: An Example of Informal Arms Control Negotiation," Ph.D. diss., Cornell Univ., 1981, pp. 48–54. See also *Aviation Week and Space Technology*, 14 Nov. 1960, p. 26.

Part III Conclusion

1. *PP of P, Dwight D. Eisenhower, 1960–61*, pp. 689–90. See also Susan Miller et al., "Statements of Dwight D. Eisenhower on Space Exploration 1952–1964," NASA HHN-23 (July 1964), for a thorough compilation of Ike's *public* statements on the space program.

2. James R. Killian, *Sputniks, Scientists, and Eisenhower: A Memoir of the First Special Assistant to the President for Science and Technology* (Cambridge, Mass., 1976), p. 241.

3. Dwight D. Eisenhower, *The White House Years*, 2 vols., vol. 2, *Waging Peace, 1956–61* (New York, 1965), pp. 614–16; Killian, *Sputniks, Scientists, and Eisenhower*, pp. 237–39.

4. "Farewell Radio and Television Address to the American People," 17 Jan. 1961, *PP of P, Dwight D. Eisenhower 1960–61*, pp. 1035–40. Quote from 1038–39; "military-industrial complex" phrase on p. 1038.

5. News Conference, 18 Jan. 1961; ibid., pp. 1045–46.

6. Claude Witze, "How Our Space Policy Evolved," *Air Force/Space Digest* (Apr. 1962): 89.

7. Goodpaster Notes, "Memo of a Conference with the President," 18 June 1958, DDE Library, Ann Whitman File, Box 20.

Part IV

Headquotes: Korolev in Nicholas Daniloff, *The Kremlin and the Cosmos* (New York, 1972), p. 94; Alexandrov in *Izvestia*, 28 July 1962; "Harebrained scheming," an implicit condemnation of Khrushchev, in *Pravda*, 17 Oct. 1964; Khrushchev in *Khrushchev Remembers: The Last Testament* (Boston, 1974), p. 532; Voznesensky in James H. Billington, *The Icon and the Axe: An Interpretive History of Russian Culture* (New York, 1970), p. 571.

1. The literature on Khrushchev is large, but dates mostly from the 1960s and emphasizes by and large his background, rise to power, and circumstances of his fall. In addition to the works cited in other notes, see Edward Crankshaw, *Khrushchev. A Career* (New York, 1967); Mark Frankland, *Khrushchev* (New York, 1967); and Konrad Kellen, *Khrushchev, A Political Portrait* (New York, 1961).

2. *Pravda*, 28 Jan. ("many brilliant and new principles") and 4 Feb. ("complete abundance") 1959. On the Twenty-first Party Congress, see Carl A. Linden, *Khrushchev and the Soviet Leadership 1957–1964* (Baltimore, Md., 1966), pp. 72–89.

Chapter 11

1. But compare Leonid Vladimirov, *The Russian Space Bluff*, trans. David Floyd (London, 1971), pp. 69–71, who argues that the people were so accustomed to Stalinist propaganda about Russian and Soviet technology that Soviet victory in the satellite race seemed a "perfectly logical development," and not a surprise at all.

On Soviet space programs generally, see Martin Caidin, *Red Star in Space* (n.p.: Crowell-Collier, 1963); Nicholas Daniloff, *The Kremlin and the Cosmos* (New York, 1972); James E. Oberg, *Red Star in Orbit* (New York, 1981); G. I. Petrov, ed., *Conquest of Outer Space in the*

U.S.S.R.: Official Announcements by TASS, 1967–1970 (New Delhi, 1974); Evgeny Riabchikov, *Russians in Space* (Garden City, N.Y., 1971); Charles S. Sheldon II, *U.S. and Soviet Progress in Space: Summary Data Through 1979 and a Forward Look* (Washington, D.C., 1980); William Shelton, *Soviet Space Exploration. The First Decade* (New York, 1968); Peter L. Smolders, *Soviets in Space,* trans. Marian Powell (New York, 1973); Michael Stoiko, *Soviet Rocketry: Past, Present, and Future* (New York, 1970); Vladimirov, *The Russian Space Bluff.*

Soviet achievements in space science and technology are the subject of Valentin P. Glushko, *Development of Rocketry and Space Technology in the U.S.S.R.* (Moscow, 1973); Firmin J. Krieger, *Behind the Sputniks: A Survey of Soviet Space Science* (Washington, D.C., 1958); R. G. Perel'man, *Goals and Means in the Conquest of Space,* Israeli Scientific Translations (Jerusalem, 1970); I. A. Slukhai, *Russian Rocketry. A Historical Survey,* Israeli Scientific Translations (Jerusalem, 1968); V. N. Sokol'sky, *A Short Outline of the Development of Rocket Research in the U.S.S.R.,* Israeli Scientific Translations (Jerusalem, 1967).

2. Reston in *New York Times,* 9 Oct. 1957; Khrushchev on rockets in *Pravda,* 11 Oct. 1957. The following discussion is based heavily on the excellent examination of Soviet post-Sputnik propaganda by Arnold L. Horelick and Myron Rush, *Strategic Power and Soviet Foreign Policy* (Chicago, 1966). This work, based on the authors' RAND project, was the first to compile the Soviet pronouncements on missiles and space, many of which are cited in this chapter and the next.

3. *Pravda,* 19 Nov. 1957.

4. *Pravda,* 29 Nov. 1957.

5. "A Policy from Positions of Folly," *International Affairs* (Dec. 1957).

6. *Pravda,* 7 Dec. 1957 and 31 Mar. 1958.

7. L. Ilyichov, "The Sputniks and International Relations," *International Affairs* (Moscow) (Mar. 1958), pp. 8–9. Italics in original.

8. Ibid., pp. 9, 11.

9. Ibid., p. 11.

10. *Pravda,* 14 Nov. 1958. See also *Pravda,* 28 Jan. 1959.

11. *Pravda,* 23 May 1960.

12. *Pravda,* 24 Aug. 1960.

13. *Krasnaia zvezda* [Red Star], 4 and 22 Feb. 1959.

14. *Pravda,* 28 Jan. 1959.

15. *Pravda,* 9 May 1959.

16. On the strategic balance and the Berlin crisis, see Jack M. Schick, *The Berlin Crisis 1958–1962* (Philadelphia, 1972). The relaxation of Soviet propaganda prior to summit meetings has been demonstrated by Michael M. Milenkovitch, *The View from Red Square. A Critique of Cartoons From Pravda and Izvestia 1947–1964* (New York, 1966).

17. United States Information Agency (hereafter USIA), Research and Reference Service, "The Impact of Sputnik on the Standing of the US vs. the USSR," WE-52, Dec. 1957.

18. USIA, Office of Research and Intelligence, "Free World Views of the US-USSR Power Balance," R-54-60, Aug. 1960.

19. USIA, Research and Reference Service, "Western European Climate of Opinion on the Eve of the Paris Summit Conference," WE-63, Apr. 1960.

20. Vladimirov, *Russian Space Bluff,* pp. 73–75.

21. On the Nedelin catastrophe, see Oberg, *Red Star in Orbit,* pp. 39–49. According to Oberg, the first account to reach the West, in Oleg Penkovsky, *The Penkovsky Papers* (New York, 1965), is "highly distorted." See also Zhores A. Medvedev, *Soviet Science* (New York, 1976), pp. 97–99; N. S. Khrushchev, *Khrushchev Remembers: The Last Testament,* trans. Strobe Talbott (Boston, 1974), p. 51.

22. Vladimirov, *Russian Space Bluff,* pp. 86–97.

23. *Pravda,* 17 Apr. 1961.

24. *Pravda,* 12 May 1961.

25. Oberg, *Red Star in Orbit,* p. 53.

26. F. T. Orechov, "Forum: Space Exploration and International Relations," *International Affairs* (Moscow) (June 1961): 58.

27. See Oberg, *Red Star in Orbit,* a *tour de force* exposing official Soviet space deception and distilling the truth by inference from Soviet sources. The history of Soviet secrecy and propaganda in spaceflight proves *both* aphorisms: "Oh, what a tangled web we weave / When first we practice to deceive"; and "A lie goes halfway 'round the world before the truth can tie its shoes."

28. USIA, Research and Reference Service, "The Image of U.S. vs. Soviet Science in Western European Public Opinion," WE-3 (Oct. 1961).

29. USIA, Office of Research and Analysis, "Initial World Reaction to Soviet 'Man in Space,'" R-17-61, Apr. 1961.

30. The Soviets themselves did not expect decolonized states to adhere immediately to Communist models of development. After 1956, and especially after the Moscow Declaration of 1960, the Khrushchev regime followed the line of support for "national democracy" in the Third World, on the model of Nasser's Egypt, with the expectation of gradual shifts to the left by such societies as economic development proceeded. See Thomas Perry Thornton, *The Third World in Soviet Perspective: Studies by Soviet Writers on the Developing Areas* (Princeton, N.J., 1964), esp. pp. 1–29, 271–75, 301–04. Propaganda based on Soviet space triumphs was expected to dissuade new states from preserving links to the West and to encourage neutralism rather than to win immediate adherents to Marxism-Leninism.

The Berkeley Space Science Lab studied the impact of the space race on Third World policies. Beneath its sociological jargon, the study concluded that while the Soviets increased trade, aid, and treaty relations with developing countries at a greater rate after Sputnik than before, the Third World countries involved leaned toward Moscow for reasons other than perceived Soviet leadership in space. Third World opinion generally disparaged the "wasteful rivalry" in space regardless of which Superpower was ahead. Donald A. Strickland, "New States, Prestige, and the Space Age: Some Probable Connections," Internal Working Paper #2, Univ. of California, Berkeley, Space Sciences Laboratory, May 1964.

31. On the Khrushchev-Korolev relationship, see Vladimirov, *Russian Space Bluff,* pp. 107–09; Oberg, *Red Star in Orbit,* pp. 36–38; Daniloff, *Kremlin and the Cosmos,* pp. 108–09.

32. Khrushchev, *The Last Testament,* p. 53.

Chapter 12

1. See the discussions of H-bomb and ICBM development in chapter 2, and P. T. Astashenkov, *Akademik S. P. Korolev* (Moscow, 1969), p. 112.

2. Herbert York in *Scientific American* (Oct. 1975): 110–11, and Arnold Kramish, *Atomic Energy in the Soviet Union* (Stanford, Calif., 1959), pp. 125–26. Not until September 1955 (after the August test) did Khrushchev boast of a several megaton bomb based on "a relatively small quantity of fissionable material."

3. Ronald Amann, Julian Cooper, and R. W. Davies, *The Technological Level of Soviet Industry* (New Haven, 1977), p. 459; Thomas W. Wolfe, *Soviet Power and Europe 1945–70* (Baltimore, 1970), p. 182; Leonid Vladimirov, *The Russian Space Bluff,* trans. David Floyd (London, 1971), pp. 50–51; N. S. Khrushchev, *Khrushchev Remembers: The Last Testament,* trans. Strobe Talbott (Boston, 1974), pp. 47–48.

4. On Soviet deployment of ICBMs, see Amann et al., *Technological Level of Soviet Industry,* pp. 461–68; on Soviet space launches, see U.S. Senate, Committee on Aeronautical and Space Sciences, *Soviet Space Programs 1966–70,* ed. Charles S. Sheldon II (Washington, D.C., 1971), pp. 115–19.

5. See Amann, Cooper, and Davies, *Technological Level of Soviet Industry,* p. 465; William H. Schauer, *The Politics of Space. A Comparison of the Soviet and American Space Programs* (New York, 1976), p. 65. Missile deployments are also recorded in *Jane's Weapons Systems 1974–75* (London, 1974); International Institute for Strategic Studies, *The Military Balance 1975–76* (London, 1975); *SIPRI Yearbook 1974* (Stockholm, 1974).

6. The K-factor is equal to $\dfrac{y^{2/3}}{CEP^2}$ where y = yield and CEP = circular error probability. K must be increased, of course, as the enemy increases the hardness of his silos (the "psi factor," or pounds of pressure per square inch the silo can withstand).

7. Khrushchev, *The Last Testament,* p. 47.

8. Ibid., p. 48.

9. Alice Langley Hsieh, *Communist China's Strategy in the Nuclear Era* (Englewood Cliffs, N.J., 1962), chaps. 1–2.

10. *Ta Kung Pao,* 29 Aug. 1957; *Jen-min Jih-pao* [People's daily], 30 Aug. 1957; *Ta Kung*

Pao, 6 Oct. 1957; *Kuang-ming Jih-pao,* 7 Oct. 1957; "alarming" quotation from Chen Chihin, *Hsüeh-hsi* [Study], 3 Nov. 1957; on capitalist collapse in war, Chang-wen-tien in *People's Daily,* 2 Nov. 1957, and Mao Tse-tung in U.S. Consulate General, Hong Kong, *Current Background,* 13 Nov. 1957; all cited by Hsieh, *China's Strategy,* pp. 77–83.

11. Walt W. Rostow, *The Diffusion of Power: An Essay in Recent History* (New York, 1972), pp. 22–23.

12. Harold P. Ford, "The Eruption of Sino-Soviet Politico-Military Problems 1957–60," in Raymond L. Garthoff, ed., *Sino-Soviet Military Relations* (New York, 1966), pp. 102–3, cited by Rostow, *The Diffusion of Power,* p. 31.

13. Rostow, *The Diffusion of Power,* pp. 28–35; Hsieh, *China's Strategy,* pp. 94–109.

14. On the roots of the Sino-Soviet split, see Donald S. Zagoria, *The Sino-Soviet Conflict, 1956–1961* (Princeton, N.J., 1962); William E. Griffith, *The Sino-Soviet Rift* (Cambridge, Mass., 1964); and G. F. Hudson, Richard Lowenthal, and Roderick Mac Farquhar, *The Sino-Soviet Dispute* (New York, 1961).

15. Khrushchev, *The Last Testament,* p. 269.

16. Arnold L. Horelick and Myron Rush, *Strategic Power and Soviet Policy,* RAND P-434-PR (Santa Monica, Calif., 1965), pp. 105–09. This study was republished as *Strategic Power and Soviet Foreign Policy* (Chicago, 1966). The interpretation in this section is based largely upon it.

17. *Pravda,* 22 Nov. 1957.

18. *Pravda,* 16 Jan. 1960.

19. Khrushchev, *The Last Testament,* p. 54.

20. Ibid., p. 536.

21. *New York Times,* 26 Jan. 1958.

22. Nikita S. Khrushchev, *For Victory in Peaceful Competition with Capitalism* (New York, 1960), p. 133 (Russell letter, 5 Mar. 1958).

23. Ibid.

24. Ye. Korovin, "International Status of Cosmic Space," *International Affairs* (Jan. 1959): 53–59. See also Y. A. Pobedonostsev et al., "Conquest of Outer Space and Some Problems of International Relations," *International Affairs* (Nov. 1959): 88–96.

25. For instance, G. A. Osnitskaya, "International Law Problems of the Conquest of Space," from the Soviet Yearbook of International Law for the Year 1959, pp. 65–71; reprinted in U.S. Senate, Committee on Aeronautical and Space Sciences, *Legal Problems of Space Exploration. A Symposium,* 87th Cong., 1st Sess. (1961), 1088–94 ("[military] intentions are alien to the Soviet Union. We are not using the historic successes of Soviet science and technology to pursue a bellicose policy or to impose a diktat on other states . . . [but] in the fight for universal peace" (p. 1094).

26. Korovin, "International Status of Cosmic Space," pp. 58–59.

27. For instance, A. Kislov and S. Krylov, "State Sovereignty in Airspace," *International Affairs* (Moscow) (Mar. 1956): 35–44.

28. G. Zadorozhnyi, "The Artificial Satellite and International Law," *Sovetskaia Rossiia* (17 Oct. 1957), in U.S. Senate, *Legal Problems of Space Exploration,* pp. 1047–49.

29. A. Galina, "On the Question of Interplanetary Law," *Sovetskoe Gosudarstvo; Pravo* 7 (July 1958): 52–58, translated by RAND, reprinted in U.S. Senate, *Legal Problems of Space Exploration,* pp. 1050–58. Quotes from pp. 1054–55. See also E. Korovin, "International Status of Cosmic Space," and "On the Neutralization and Demilitarization of Outer Space," *International Affairs* (Moscow) (Dec. 1959): 82–83.

30. UN General Assembly, 14th Sess., Agenda Item 25, A/C.1/L.247, "General Debate and Consideration of the Draft Resolution on the Peaceful Uses of Outer Space," 11 Dec. 1959. On the origins of COPUOS and its procedures, see U.S. Senate, Committee on Aeronautical and Space Sciences, 89th Cong. 1st Sess., *International Cooperation and Organization for Outer Space* (Washington, D.C., 1965), pp. 163–216; Eilene Galloway, "The UN Ad Hoc COPUOS: Accomplishments and Implications for Legal Problems," 4 Sept. 1959 and "World Security and the Peaceful Uses of Outer Space," 16 Aug. 1960, NA Papers of the Senate Space Committee; Historical Office of the Department of State, "International Negotiations Regarding the Use of Outer Space, 1957–1961," Research Project no. 539, Feb. 1962.

31. *Pravda,* 8 May 1960.

32. Ibid., 8 July 1960.

33. Ibid., 22 June 1960.

34. Ibid., 10 May 1960.
35. Georgi P. Zhukov, "Space Espionage Plans and International Law," *International Affairs* (Oct. 1960).
36. On Soviet space law theory generally, see Robert D. Crane, "The Beginnings of a Marxist Space Jurisprudence," *American Journal of International Law* 57 (1963): 616–25, and the bibliography in U.S. Senate, *Legal Problems of Space Exploration* (1961), pp. 1022–37; the bibliographies and references listed in chapter 8, note 33; and the references in Schauer, *Politics of Space*, pp. 245–60.
37. On Resolution 1721 (XVI), see UN General Assembly, 16th Sess., 20 Dec. 1961. A/C.1/L.301/Rev. 1 and Corr. 1. On negotiations in COPUOS leading to Res. 1721, see U.S. Senate, Committee on Aeronautical and Space Sciences, *Soviet Space Programs: Organization, Plans, Goals, and International Implications* (Washington, D.C., 1962), pp. 163–73; also U.S. Senate, Committee on Aeronautical and Space Sciences, *International Cooperation and Organization for Outer Space* (Washington, D.C., 1965), pp. 183–205; and Report, Harlan Cleveland to Dean Rusk, "UN Outer Space Proposals for the Sixteenth General Assembly," 20 Nov. 1961, LBJ Library, Legislative Background to the Outer Space Treaty, Box 1.
38. UN General Assembly, 16th Sess., 7 Dec. 1961, Document A/C.1/SR. 1213, p. 263.
39. Committee on the Legal Problems of Outer Space of the Academy of Sciences of the USSR, *Kosmos i Mezhdunarodnoye Pravo* [Space and international law], Publishing Office of the Institute of International Relations (Moscow, 1962), p. 69; see also pp. 4–16, 31–34, 38, 53–55, 65–69.
40. Ibid., p. 53; see also pp. 171–82.
41. The "propagating war or hatred" clause—a reference to potential misuse of the coming communications satellites—and the insistence on prior agreement on all use of space were proposed by the USSR as additions to the principles in Resolution 1721 (XVI). The Soviets also attempted to append a ban on satellite intelligence-gathering. Thereafter they continually charged the United States with hypocrisy. See Morozov's speech in UN General Assembly, 17th Sess., Document A/C.1/PV.1289, 3 Dec. 1962, p. 57.
42. UN General Assembly, 17th Sess., Document A/RES/1802 (XVII), 19 Dec. 1962, article I.
43. UN General Assembly, 17th Sess., Document A/AC.105/PV.12, 12 Sept. 1962, p. 52.

Chapter 13

1. Roman Kolkowicz, *The Soviet Military and the Communist Party* (Princeton, N.J., 1967), pp. 31–34, 78–79, 117–19.
2. Ibid., pp. 71–79, and Raymond L. Garthoff, *Soviet Strategy in the Nuclear Age* (New York, 1958), pp. 62–63.
3. Garthoff, *Soviet Strategy,* p. 62.
4. On "new phenomena," *Voennaia mysl'* (Nov. 1953): 12. On "Military Science," *Voennaia mysl'* (Apr. 1955): 16–22, cited by Garthoff, *Soviet Strategy,* pp. 64, 67.
5. *Voennaia mysl'* (Mar. and May 1955), and "Voennye ideolog; kapitalisticheskikh stran . . ." in Garthoff, *Soviet Strategy,* pp. 72–73. The general staff insisted that "Wars are won only when the enemy's will to resist is broken and that can only be broken . . . when the armed forces of the enemy are destroyed. Therefore, *the objective of combat operations must be the destruction of the armed forces, and not strategic targets in the rear.*" Italics in original.
6. *Krasnaia zvezda [Red star],* 25 Sept. 1956; also Zhukov in *Krasnaia zvezda,* 23 Mar. 1957, and *Sovetskaia aviatsiia,* 17 Mar. 1957, in Garthoff, *Soviet Strategy,* pp. 76–77. One can speculate that this view set well with Soviet infantry generals whose only experience with massed bombing had been disastrous (1941) or merely tactical (1944–45), a far different experience from that of Americans in World War II. Italics in original.
7. *Pravda,* 9 Aug. 1953.
8. *Pravda,* 1 Jan. 1956.
9. Kolkowicz, *Soviet Military and Communist Party,* pp. 239–55. The Stalingrad Group was composed of officers who had shared in that dogged defense and counterattack. Khrushchev himself had been a Party leader on the Stalingrad front and shared the officers'

resentment when first Stalin, then Zhukov, claimed the role of mastermind in that holiest of battles.

10. *Pravda,* 15 Jan. 1960.

11. See Raymond L. Garthoff, *Soviet Military Policy: An Historical Analysis* (New York, 1966), pp. 110–11; Marshal Shulman, *Stalin's Foreign Policy Reappraised* (Cambridge, Mass., 1963), pp. 4–7; Carl A. Linden, *Khrushchev and the Soviet Leadership 1957–1964* (Baltimore, Md., 1966), pp. 90–93.

12. N. S. Khrushchev, *Khrushchev Remembers: The Last Testament,* trans. Strobe Talbott (Boston, 1974), pp. 48–52. Quotes from p. 52.

13. On economic factors in Soviet defense policy, see Lincoln P. Bloomfield, Walter C. Clemens, and Franklyn Griffiths, *Soviet Interests in Arms Control and Disarmament: The Decade under Khrushchev 1954–1964,* MIT Center for International Studies, C/65-1 (Cambridge, Mass., 1965), pp. 48–50.

14. Underestimation of this phenomenon makes dollar-ruble comparisons of relative defense spending problematical. If the cost of keeping one U.S. soldier under arms is estimated at, say, $40,000 per year and that figure is simply multiplied by the number of Soviets (or Chinese!) under arms, the Soviet defense estimate is inflated, and (in the Chinese case) ridiculous. The reverse process—costing Soviet hardware by American industrial prices—can result in Soviet defense estimates that are too low. See Franklyn Holzman, "Observations on the Military Spending Gap," Colloquium Paper, The Kennan Institute, 22 June 1982.

15. Bloomfield, Clemens, and Griffiths, *Soviet Interests in Arms Control,* pp. 46–48.

16. CIA Report, "Soviet Manpower 1960–1970," May 1960, pp. 1–3, 15.

17. One analysis of the Seven Year Plan reveals that Soviet planners expected military and R & D costs to more than double from 1959 to 1965, but that weapons and space spending were to rise only 65 to 85 percent. When missile and space R & D proved more costly and the army cuts were restored, the Khrushchev budget was busted. See Abraham Becker, *Soviet National Income and Product in 1965: The Goals of the Seven Year Plan,* RAND 3520-PR (Santa Monica, Calif., 1963), p. 19; also Fritz Ermath, "Economic Factors and Soviet Arms Control Policy: The Economic Burden of the Soviet Defense Policy," MIT, Center for International Studies (Cambridge, Mass., 1964), vol. 1, pp. 11–17.

18. This is Garthoff's conclusion, *Soviet Strategy,* p. 151.

19. Kolkowicz, *Soviet Military and the Communist Party,* pp. 265–78.

20. See Thomas W. Wolfe, *Soviet Strategy at the Crossroads,* RAND RM-4085-PR (Santa Monica, Calif., 1964), pp. 12–25. Note: in Harvard University Press edition (Cambridge, Mass., 1964), pp. 6–12.

21. V. D. Sokolovsky, ed., *Voennaia Strategiia,* 1st ed. (Moscow, 1962). All three editions have been translated and compiled in a single volume, with the additions and deletions of each edition, by the Stanford Research Institute, Harriet Fast Scott, ed., *Soviet Military Strategy* (New York, 1968). Following citations are from the Scott edition. Sections on the military uses of space include pp. 47–50, 84–89, 455–58.

22. Ibid., p. 456.

23. Ibid.

24. Ibid., pp. 457–58.

25. Ibid., p. 458.

26. See Herbert L. Sawyer, "The Soviet Space Controversy 1961 to 1963," Ph.D. diss. Fletcher School of Law and Diplomacy, 1969. The following citations (through note 34) on the military space lobby are drawn from Sawyer except where otherwise indicated.

27. Vladimir Orlov, "Kosmos-Kosmos," *Pravda,* 13 Aug. 1962.

28. Leonid Sedov, "Podvig Veka" [Exploit of the century], *Izvestia,* 13 Apr. 1963. See also A. A. Blaganravov, "Kaskad otkritii" [Cascade of discoveries], *Kosmomol'skaia Pravda,* 12 Apr. 1962.

29. V. Larionov, "Rakety i strategiia" and "Kosmos i strategiia," *Krasnaia zvezda,* 18 and 21 Mar. 1962.

30. See, for example, M. Kroshkin, "Na rubezhe kosmicheskogo vega" [On the frontier of the space age], *Kommunist vooruzhennykh sil* (July 1961): 16–21; N. Zaitsev, "Amerikanskie shpiony v kosmose" [American Spies in Space], *Aviatsiia i kosmonautika* (Jan. 1962): 92–94; M. Sakharov, "Kosmos v voennykh tseliakh zapada" [Outer space in the military aims of the west], *Mirovaia ekonomika i mezhdunarodnye otnosheniia* (Mar. 1962): 106–9; I. Anureev, "Imperialisticheskaia agressia v kosmose," [Imperialist aggression in space] *Kommunist*

vooruzkennykh sil (Aug. 1962): 17–23; M. Mil'shtein and A. Slobodenko, "O voennoi doktrine S ShA" [On the military doctrine of the USA], *Krasnaia zvezda* (6 Feb. 1963): 3.

31. See, for example, B. Alexandrov and V. Poliansky, "Pentagon i kosmos," *Krasnaia zvezda*, 5 July 1963, p. 4.

32. *Pravda*, 23 Feb. 1963; also Baryshev in *Krasnaia zvezda*, 2 Sept. 1962, cited by Wolfe, *Soviet Strategy at the Crossroads*, p. 262 (in Harvard edition, p. 206).

33. Colonel-General Shtememko in *Kommunist Vooruzhennykh sil* 3 (Feb. 1963), in Wolfe, *Soviet Strategy at the Crossroads*, p. 262 (in Harvard edition, p. 206).

34. G. Deborin, "Vo imia misa i bezopastnosti' naradov" [In the name of peace and the safety of peoples], *Krasnaia zvezda*, 21 Apr. 1961, p. 2; V. Liutyi, "Put' v kosmos" [The road to space], *Voenno-istoricheskii zhurnal* 8 (Aug. 1961): 25–36; G. Pokrovskii, "Chelovek v kosmos" [Man into space], *Kommunist* (Armenian SSSR), 7 Jan. 1962, p. 4; V. Larionov and V. Vaneev, "Strategiia i kibernetika," *Krasnaia zvezda*, 30 June (p. 6) and 3 July (p. 3) 1962; I. Baryshev, "Chto takoe protivokosmicheskaia oborona" [What is antispace defense], *Krasnaia zvezda*, 2 Sept. 1962, p. 3; P. Pliachenko, "Kosmos i iadernaia voina" [Outer space and nuclear war], *Aviatsiia i kosmonautika* 1 (Jan. 1963): 91–95; E. Malyshkin, "Voina orbit" [Orbital warfare], *Krasnaia zvezda*, 8 Aug. 1963, p. 3.

35. See Herbert S. Dinerstein, *The Making of a Missile Crisis: October 1962* (Baltimore, 1976).

36. Scott, ed., *Soviet Military Strategy*, pp. 454–55 (in the original, V. D. Sokolovsky, ed., *Voennaia Strategiia*, 2nd ed. (Moscow, 1963), p. 394.

37. TASS Release, 16 March 1962, on the Kosmos series, reprinted in U.S. Senate, Committee on Aeronautical and Space Sciences, *Soviet Space Programs, 1966–70*, 92d Cong., 1st Sess. (1971), p. 173.

38. U.S. Senate, *Soviet Space Programs 1966–70*, pp. 170–88; chart on p. 186. Also G. E. Perry, "The Cosmos Programme," *Flight International* (26 Dec. 1968): 1077–79.

39. Philip J. Klass, *Secret Sentries in Space* (New York, 1971), pp. 153–57.

40. U.S. Senate, *Soviet Space Programs 1966–70*, pp. 533–39.

41. UN General Assembly, 18th Sess., Document A/AC.105/PV.20, 10 Oct. 1963, p. 28.

42. On the reversal of Soviet policy on reconnaissance satellites at the UN, see Gerald M. Steinberg, "The Legitimization of Reconnaissance Satellites: An Example of Informal Arms Control Negotiation," Ph.D. diss. Cornell Univ., 1981, pp. 99–104; Klass, *Secret Sentries in Space*, pp. 123–29; William H. Schauer, *The Politics of Space. A Comparison of the Soviet and American Space Programs* (New York, 1976), pp. 252–55.

43. UN General Assembly, 18th Sess., Document A/RES/1962 (XVIII), 13 Dec. 1963.

Chapter 14

1. *Pravda*, 17 July 1955. The following discussion is based on Julian M. Cooper, "The Scientific and Technical Revolution in Soviet Theory," in Frederic J. Fleron, Jr., ed., *Technology and Communist Culture: The Socio-Cultural Impact of Technology under Socialism* (New York, 1977), pp. 146–79. See also Robert F. Miller, "The Scientific-Technical Revolution and the Soviet Administrative Debate," in Paul Cocks, Robert V. Daniels, and Nancy Whittier Herr, eds., *The Dynamics of Soviet Politics* (Cambridge, Mass., 1976), pp. 137–55.

2. *Programma Kommunisticheskoi partii sovetskogo soyuza* (Moscow, 1961), p. 27, cited by Cooper, "The Scientific and Technical Revolution," p. 154.

3. This definition became official in 1967, according to Cooper, "The Scientific and Technical Revolution," p. 147.

4. N. I. Bukharin, *Historical Materialism* [1925] (Ann Arbor, Mich., 1969), pp. 255–62, in ibid., p. 149.

5. A. Bogdanov, *A Short Course of Economic Science* (London, 1923), pp. 378–80 and L. Trotsky, *Marxism and Science* (Columbo, Sri Lanka, 1973), p. 29, both in ibid., pp. 151–53.

6. N. I. Bukharin, *Socialist Reconstruction and the Struggle for Technique* (Moscow, 1932), p. 10, and M. Rubinshtein, *O material'notekhnicheskoi baze perekhod ot sotsializma k kommunizmu* (Moscow, 1940), p. 51, in ibid., p. 150.

7. Ibid., p. 157.

8. *Sovremennaya nauchno-tekhnicheskaya revolyutsiya-istoricheskoe issledovanie,* 2nd ed. (Moscow, 1970), pp. 33–34, in ibid., p. 155.

9. Ibid., p. 158.

10. *Sovremennaya nauchno-tekhnicheskaya revolyutsiya-istoricheskoe issledovanie,* pp. 225–27, in ibid., p. 157.

11. See William Zimmerman, *Soviet Perspectives on International Relations 1956–1967* (Princeton, N.J., 1969), pp. 4–5.

12. Ibid., pp. 30–31.

13. Ibid., chap. 3.

14. "Gone with the wind" remark by V. Korionov, "The Crisis of the 'Position of Strength' Policy," *International Affairs* (Moscow) (Mar. 1958); on the three crises of capitalism, see Zimmerman, *Soviet Perspectives on International Relations,* pp. 131–35.

15. Zimmerman, *Soviet Perspectives on International Relations,* pp. 165–79.

16. M. Baturin, "Peace and the Status Quo," *International Affairs* (Moscow) (Jan. 1958). Italics in original.

17. On the perception of the new autonomy of the state in U.S. foreign policy, see Akademiia Nauk, SSSR, Institut Mirovoi Ekonomiki i Mezhdunarodnykh Otnoshenil (hereafter IMEMO), *Mezhdunarodnye otnosheniia posle vtoroi mirovoi voiny* [International relations since the second world war], 3 vols. (Moscow, 1962–65), vol. 1, introduction; P. Vladimirsky, "The Pentagon's Diplomacy," *International Affairs* (Moscow) (Nov. 1964); Boris Dmitriev [B. Piadyshev], *Pentagon i vneshniaia politika S ShA* [The Pentagon and U.S. foreign policy] (Moscow, 1961). On the development of "state monopoly capitalism," S. A. Dalin, *Voenno-gosudarstvennyi monopolisticheskii kapitalizm v S ShA* [Military-state monopoly capitalism in the USA] (Moscow, 1961); discussed by Zimmerman, *Soviet Perspectives on International Relations,* pp. 211–17.

18. Zimmerman, *Soviet Perspectives on International Relations,* pp. 227–29; see especially IMEMO, *Dvizhushchie sily vneshnei politiki S ShA* [The motive forces of U.S. foreign policy] (Moscow, 1965), pp. 20–26.

19. *Mezhdunarodnoe soveshchanie kommunistlicheskikh i rabochikh partii. Dokumenty i materialy* (Moscow, 1969), p. 303, cited by Cooper, "The Scientific and Technical Revolution," p. 146.

20. Dvorkin in *Mirovaya ekonomika i mezhdunarodnaya otnosheniya,* no. 2 (1967); cited by Cooper, "The Scientific and Technical Revolution," p. 173.

21. These figures are taken from the official Soviet statistical annuals. They exclude capital investment related to "science" as well as R & D buried in the military budget. Hence they are undoubtedly low. For discussions of Soviet R & D estimates, see Robert W. Campbell, *Reference Source on Soviet R & D Statistics 1950–1978,* NSF 78-SP-1023, and Alexander G. Korol, *Soviet R&D: Its Organization, Personnel, and Funds* (Cambridge, Mass., 1975), pp. 18–38.

22. Quote from N. S. Khrushchev, *Khrushchev Remembers: The Last Testament,* trans. Strobe Talbott (Boston, 1974), p. 61. On the Academy of Sciences, see Bruce Parrott, *Politics and Technology in the Soviet Union* (Cambridge, Mass., 1983), chap. 4, and Korol, *Soviet R&D,* pp. 18–38.

23. Parrott, *Politics and Technology,* pp. 139–42.

24. Zhores A. Medvedev, *Soviet Science* (New York, 1976), pp. 60–65.

25. Roy A. and Zhores A. Medvedev, *Khrushchev: The Years in Power* (New York, 1970), pp. 107–09; Arnold Kramish, *Atomic Power in the Soviet Union* (Stanford, 1959), pp. 187–96.

26. Roy and Zhores Medvedev, *Khrushchev: The Years in Power,* pp. 44–45. On Soviet computers, see Richard Judy, "The Case of Computer Technology," in S. Wasawski, ed., *East-West Trade and the Technology Gap* (New York, 1970), pp. 43–71; Martin Cave, "Computer Technology," in Ronald Amann, Julian Cooper, and R. W. Davies, *The Technological Level of Soviet Industry* (New Haven, 1977), pp. 377–406.

27. Cooper, "The Scientific and Technical Revolution," pp. 151–58.

28. Roy and Zhores Medvedev, *Khrushchev: The Years in Power,* pp. 143–58; Carl A. Linden, *Khrushchev and the Soviet Leadership 1957–1964* (Baltimore, Md., 1966), esp. chaps. 6–8; Michel Tatu, *Power in the Kremlin from Khrushchev to Kosygin* (New York, 1968), esp. pp. 364–98.

29. Marshal S. S. Biryuzov (formerly chief of rocket forces), "Voranyi inzhener—aktivnyi vospitatel" [The military engineer—an active educator] (Moscow, 1962), cited by Roman

Kolkowicz, *The Soviet Military and the Communist Party* (Princeton, N.J., 1967), p. 312.

30. Marshal M. V. Zakharov, *Krasnaia zvezda,* 12 Oct. 1962, cited in ibid.

31. *Krasnaia zvezda,* 30 Jan. and 10 Nov. 1963. Cited in ibid., p. 318.

32. A. O. Baranov, *Voennaia tekhnika i moral'no-boevye kachestva voina* (Moscow, 1961), pp. 5–7, cited in ibid., p. 315.

33. Ibid. See also Kolkowicz, "The Impact of Technology on the Soviet Military: A Challenge to Traditional Military Professionalism," RAND RM-4198-PR (August 1964).

34. Amann, Cooper, and Davies, *Technological Level of Soviet Industry,* pp. 63–66; Judy in Wasowski, ed., *East-West Trade and the Technology Gap,* pp. 63ff.; Philip Hanson, *Soviet Economics in a New Perspective* (Washington, D.C., 1976), pp. 786–811.

35. A. C. Sutton, *Western Technology and Soviet Economic Development,* vol. 3: *1945 to 1965* (Stanford, Calif., 1973), p. xxv, conclusion.

36. Zhores Medvedev, *Soviet Science,* pp. 62–77. For instance, a degree candidate was obliged to defend his thesis not only before his mentors but against two "official opponents." Korol, *Soviet R&D,* shows that for these and other reasons (including the delayed impact of the war on the generation that would have been reaching its thirties and forties) the average age of middle- and senior-level technicians rose steadily despite the big influx of young graduates.

37. See Andrei Sakharov, *Sakharov Speaks* (London, 1974), pp. 32–34; Khrushchev, *The Last Testament,* pp. 63–71; Zhores Medvedev, *Soviet Science,* pp. 89–90.

38. Zhores Medvedev, *Soviet Science,* pp. 88–101.

39. Roy and Zhores Medvedev, *Khrushchev: The Years in Power,* pp. 171–72.

40. This is the view of Tatu, *Power in the Kremlin,* pp. 391–98 ("the last straw" on p. 398).

41. CDSP XIV/5, p. 24, cited by Robert M. Slusser, *The Berlin Crisis of 1961: Soviet-American Relations and the Struggle for Power in the Kremlin* (Baltimore, 1973), p. 419.

42. *Izvestia,* 20 June 1961.

43. *Pravda,* 26 Aug. 1961.

44. Nicholas Daniloff, *The Kremlin and the Cosmos* (New York, 1972), pp. 108–09.

45. Among leading space officials were Anatoly A. Blagonravov, a lieutenant-general of artillery; Georgi I. Pokrovsky, a major-general of engineering; Yuri A. Pobedonostsev, a colonel/professor of artillery; Konstantin N. Runov, Mstislav Keldysh, Petr V. Dement'ev, and Vasily S. Emel'inaov all held high positions in military aviation and atomic energy; U.S. Senate, *Soviet Space Programs,* 1962, pp. 65–66.

46. On the organization of the Soviet program, see U.S. Senate, *Soviet Space Programs,* 1962, pp. 61–71; John D. Holmfled, "Organization of the Soviet Space Program," in U.S. Senate, Committee on Aeronautical and Space Sciences, *Soviet Space Programs, 1966–70,* 92d Cong., 1st Sess. (1971), pp. 69–105; William H. Schauer, *The Politics of Space. A Comparison of the Soviet and American Space Programs* (New York, 1976), pp. 25–30.

47. G. I. Pokrovsky, *Science and Technology in Contemporary War,* trans. Raymond Garthoff (New York, 1959), p. 164.

48. James E. Oberg, *Red Star in Orbit* (New York, 1981), p. 69.

49. Ibid., p. 70. On Tereshkova, see also Daniloff, *Kremlin and the Cosmos,* p. 109, and Leonid Vladimirov, *The Russian Space Bluff,* trans. David Floyd (London, 1971), pp. 113–15.

50. Vostok did have an exceptionally useful afterlife as an unmanned vehicle in the first recoverable spy satellite program.

51. Khrushchev, *The Last Testament,* p. 47.

52. Vladimirov, *Russian Space Bluff,* pp. 174–78.

53. The figure of 1.5 to 2.0 percent of GNP for Soviet space spending has been oft repeated and never challenged. See U.S. Senate, *Soviet Space Programs,* 1971, pp. 108–14. If the CIA has better (or just lower) estimates, it has not been talking.

54. The figures on Saturn development are from the *NASA Historical Data Book 1958–1968,* NASA SP-4012 (Washington, D.C., 1976), vol. 1: *NASA Resources,* p. 141. *Saturn 5* costs through 1968 were $5.36 billion out of total NASA R & D spending of $25.18 billion.

55. U.S. Senate, *Soviet Space Programs,* 1971, pp. 146–48; Oberg, *Red Star in Orbit,* pp. 121–24.

56. Khrushchev quoted in Daniloff, *Kremlin and the Cosmos,* p. 144. On Soviet remarks

about the moon race, see Schauer, *Politics of Space*, pp. 164–72; Oberg, *Red Star in Orbit*, pp. 111–24; Daniloff, *Kremlin and the Cosmos*, pp. 140–53.

57. Alexandr P. Romanov, *Designer of Cosmic Ships* (Moscow, 1969), pp. 109–11.

58. K. Sergeyev in *Pravda*, 1 Jan. 1964 and 1 Jan. 1965, cited by Daniloff, *Kremlin and the Cosmos*, pp. 116–17. Even poor Korolev's name was taboo. He published New Year's articles annually in *Pravda* under the pseudonym Professor K. Sergeyev. The same formula was used for his "deputy," Mikhail Kuzmich Yangel, who became in print "Vasily Mikhailovich."

59. For a summary of Soviet lunar flights, see U.S. Senate, *Soviet Space Programs, 1971*, pp. 205–09.

60. Vladimirov, *Russian Space Bluff*, pp. 123–27.

61. *Pravda* quote in Oberg, *Red Star in Orbit*, p. 77.

62. On *Voskhod I*, see Vladimirov, *Russian Space Bluff*, pp. 123–33; Oberg, *Red Star in Orbit*, pp. 74–78.

Part IV Conclusion

1. James H. Billington, *The Icon and the Axe: An Interpretive History of Russian Culture* (New York, 1970), p. 480.

2. Eugene Zamyatin, *We*, trans. G. Zilbourg (New York, 1959), cited in ibid., p. 509.

3. Roy and Zhores Medvedev, *Khrushchev: The Years in Power* (New York, 1970), chap. 16.

4. General M. V. Zakharov in *Krasnaia zvezda*, 4 Feb. 1965.

5. Brezhnev in *Moscow Domestic Service*, 22 Jan. 1969, and Sedov in *New Times*, 3 Mar. 1971, pp. 23–24, cited in U.S. Senate, Committee on Aeronautical and Space Sciences, *Soviet Space Programs 1966–70*, 92d Cong., 1st Sess. (Washington, D.C., 1971), pp. 36–37.

6. *Pravda*, 22 Oct. 1964.

7. Kendall E. Bailes, *Technology and Society in the Soviet Union Under Lenin and Stalin: Origins of the Soviet Technical Intelligentsia, 1917–1941* (Princeton, N.J., 1978), p. 414.

8. Aviation and Cosmonautics (Moscow) [1968]: 376, cited by Nicholas Daniloff, *The Kremlin and the Cosmos* (New York, 1972), p. 159.

9. Ibid.; also Leonid Vladimirov, *The Russian Space Bluff*, trans. David Floyd (London, 1971), pp. 139–40.

10. James E. Oberg, *Red Star in Orbit* (New York, 1981), pp. 87–89; Zhores A. Medvedev, *Soviet Science* (New York, 1976), pp. 87–88.

Part V

Headquotes: Pascal cited by Arthur O. Lovejoy, *Reflections on Human Nature* (Baltimore, Md., 1961), p. 240; Rickover to the National Press Club (1959) in Donald Cox, *The Space Race: From Sputnik to Apollo . . . and Beyond* (Philadelphia, 1962), p. 102; Kennedy in *PP of P, John F. Kennedy 1962*, p. 669; Stevenson, "Science and Technology in the Political Arena," *Science and Society: A Symposium* (Xerox Corp., 1965), p. 4.

1. Interviews of C. S. Draper by E. M. Emme, 13 Oct. 1971, and by J. M. Grimwood, 2 June 1974, NASA HO.

2. Hugh Sidey, *John F. Kennedy, President* (New York, 1964), p. 118.

3. The following quotes from Kennedy are all from "Special Message to the Congress on Urgent National Needs," 25 May 1961, *PP of P, John F. Kennedy 1961*, pp. 396–406.

4. Daniel J. Boorstin, *The Image; Or What Happened to the American Dream* (New York, 1962), p. 239.

5. On pseudo-events and the role of the media, see ibid., chaps. 1–2.

6. Telegram, Hearst, "On Trip to USSR," 5 Feb. 1955, DDE Library, Ann Whitman File, International Series, Box 45.

7. Boorstin, *The Image*, p. 243.

8. Ibid., pp. 246–47, 251–52. On the role of television in promoting the importance of "image," see Stephen J. Wayne, *The Road to the White House: The Politics of Presidential Elections* (New York, 1980), pp. 155–58.

Chapter 15

1. Ernest May, *"Lessons" of the Past: The Use and Misuse of History in American Foreign Policy* (New York, 1973) discusses the phenomenon of peak experiences in generations and their influence on future policy predispositions. See also Robert Jervis, *Perception and Misperception in International Politics* (Princeton, N.J., 1976).

2. David Halberstam, *The Best and the Brightest* (Greenwich, Conn., 1973), pp. 57, 153.

3. Inaugural Address, 20 Jan. 1961, *PP of P, John F. Kennedy 1961*, pp. 1–3.

4. Walt W. Rostow, *The Diffusion of Power: An Essay in Recent History* (New York, 1972), pp. 136–37. On Kennedy's economic policies, Keynesian proclivities, and desire for government-business cooperation, see Jim F. Heath, *John F. Kennedy and the Business Community* (Chicago, 1969); Hobart Rowan, *The Free Enterprisers: Kennedy, Johnson, and the Business Establishment* (New York, 1964); Seymour Harris, *Economics of the Kennedy Years* (New York, 1964).

5. "Summary of NASC Meetings 1958–1960" in NASA HO; also Letter, Glennan to Eisenhower, 16 Nov. 1959, NASA HO; Memo, "Proposed Amendments to the NAS Act of 1958," 8 Dec. 1959, DDE Library, White House Central File, Official File, NAS Council, 342A; Memo, BoB Military Division to Elmer Staats, "Status of the President's Plan To Amend the Space Act of 1958," 29 Sept. 1960, NASA HO.

6. John Logsdon, *The Decision To Go to the Moon: Project Apollo and the National Interest* (Cambridge, Mass., 1970), pp. 67–70. The discussion that follows on the evolution of the lunar landing decision relies heavily on Logsdon, pp. 64–130. Most of the documents cited from the NASA HO were first gathered, declassified, or consulted at what would become the JFK Library by Logsdon and the first NASA historian, Eugene M. Emme.

7. Hugo Young, Bryan Silcock, and Peter Dunn, "Why We Went to the Moon: From the Bay of Pigs to the Sea of Tranquility," *Washington Monthly* 2 (Apr. 1970): 25–58. Quote from p. 43. Kerr's rival for the chair of the Space Committee was Clinton Anderson, who bowed out after learning of the White House preference. Interview with Clinton P. Anderson, cited by Logsdon in *Decision To Go to the Moon*, p. 68. Anderson finally took the chair after Kerr's death on New Year's Day, 1963.

8. "Report to the President-Elect of the Ad-Hoc Committee on Space," 10 Jan. 1961 (quotes from pp. 16–17), NASA HO. Italics in original. The committee consisted of Wiesner as chairman, Kenneth BeLieu (staff director of the Senate Space Committee), Edwin Land, Maxwell Lehrer (BeLieu's assistant), Edward M. Purcell (chairman of the PSAC spaceflight panel and Harvard professor), Bruno B. Rossi (MIT physics professor), and Henry J. Walters (Land's assistant).

9. *New York Times,* 26 Jan. 1961.

10. "The President's News Conference of 8 Feb. 1961," *PP of P, John F. Kennedy 1961*, pp. 66–70.

11. *Congressional Record—Appendix*, 23 Mar. 1960, p. A2749.

12. James Webb Oral History, pp. 4–7, LBJ Library. On Webb's biography, see Jay Holmes, *America on the Moon: The Enterprise of the Sixties* (Philadelphia, 1962), pp. 190–92. Webb wrote a touching letter to his former boss, with whom he had always associated the phrase "Mr. President," describing his meeting with Kennedy on the NASA job: "I couldn't get off the hook." Letter, Webb to Harry S Truman, 8 Feb. 1961: HST Library, Post-Presidential Name File.

13. Zarem cited by Donald Cox, *The Space Race: From Sputnik to Apollo . . . and Beyond* (Philadelphia, 1962), pp. 72–73.

14. Webb, in foreword to Robert L. Rosholt, *An Administrative History of NASA, 1958–1963* (Washington, D.C., 1966), p. iv.

15. See, for instance, Letter, Glennan to Rep. James M. Quigley, 4 Apr. 1960, rebutting these views, and Memo, Hjornevik for the President, 2 Oct. 1959, reporting Glennan's and McElroy's concurrence on the need for a divided but coordinated program, NASA HO, as

well as the complaints of Medaris and Gavin cited in chapters 8 and 9 herein.

16. Memo, Thomas D. White (USAF Chief of Staff) to Generals Landon, Wilson, LeMay, Schriever, Dr. Perkins, and the Undersecretary of the USAF, 14 Apr. 1960, NASA HO:

I am convinced that one of the major long range elements of the Air Force future lies in space. It is also obvious that NASA will play a large part in the national effort in this direction and, moreover, inevitably will be closely associated, if not eventually combined with the military. It is perfectly clear to me that particularly in these formative years the Air Force must, for its own good as well as for the national interest, cooperate to the maximum extent with NASA.

17. Lt. Col. Kenneth F. Gantz, ed., *The U.S. Air Force Report on the Ballistic Missile: Its Technology, Logistics, and Strategy,* preface by General Thomas D. White, intro. by Major General Bernard Schriever (New York, 1958); Kenneth F. Gantz, *Man in Space: The United States Air Force Program for Developing the Spacecraft Crew* (New York, 1959); Editors of Air Force Magazine, *Space Weapons: A Handbook of Military Astronautics* (New York, 1959). *Man in Space* even argued for a moon program under military auspices. If the project proved to be impossible, "we shall know that no one else can succeed"; if it proved possible, "This outpost, under our control, would be the best possible guarantee that all of space will indeed be preserved for the peaceful purposes of man" (p. 253).

18. Letter, Schriever to Gardner, 11 Oct. 1960, NASA HO. Schriever compared the proposed study to the von Neumann committee of 1954. Gardner's associates included William O. Baker (vice president for research of Bell Labs), Harold Brown (of the University of California's Lawrence Radiation Lab), William C. Foster (vice president of Olin Mathieson Chemical Corp.), and eleven other luminaries from science, business, and the USAF. They were advised by a special study group of scientists and engineers convened at Los Alamos, chaired by Dr. Theodore B. Taylor (Senior Research Adviser of General Dynamics' Atomic Division).

19. Office of the Secretary of the USAF, Memo to commanders and contractors, "Air Force Competency in Space Operations," 1 Dec. 1960, in U.S. Congress, House Committee on Science and Astronautics, *Defense Space Interests,* 87th Cong. 1st Sess., 1961, pp. 93–96.

20. "Report of the Air Force Space Study Committee" ("Gardner Report"), HQ, ARDC, 20 Mar. 1961, NASA HO. Quotes from pp. 2, 3, 4, 54. Italics in original. The lunar base proposal drew on a previous USAF project calling for the integration of NASA and military requirements into one national lunar program, with establishment of a lunar base to be considered a military expedition (Directorate of Space Planning and Analysis, USAF, Ballistic Missile Division [BMD] "Military Lunar Base Program or S.R. 183 Lunar Observatory Study," Apr. 1960, NASA HO).

21. Memo, Overton Brooks, "Attitude of Committee on Science and Astronautics Relative to the National Space Program," 13 Feb. 1961, NASA HO.

22. Letter, Brooks to the President, 9 Mar. 1961, NASA HO.

23. U.S. Congress, House Committee on Science and Astronautics, *Defense Space Interests,* 87th Cong., 1st Sess. (March 1961), pp. 35 (Gilpatric quote), 92–93 (General White testimony). See also the committee's conclusions in U.S. Congress, Committee on Science and Astronautics, *Military Astronautics (Preliminary Report),* 87th Cong. 1st Sess. (May 1961), pp. 35–37.

24. Letter, Kennedy to Brooks, 23 Mar. 1961, NASA HO.

25. Schriever Speech before the Allegheny Conference on Community Development, 21 Nov. 1960, reprinted in *Congressional Record—Appendix,* 1961, pp. A93–94. On NASA-DoD relations generally, see also Richard H. Campbell, "The Impact of NASA on the Air Force Role in Space," Air Command and Staff College thesis, Maxwell AFB, Ala., 1965, and Stephen Ira Grossbard, "The Civilian Space Program: A Case Study in Civil-Military Relations," Ph.D. diss., University of Michigan, 1968.

26. Memo, Hyatt, "Some Notes on the Relationship of NASA-DoD Space Programs," 19 Apr. 1961, NASA HO.

27. NAS, Space Science Board, "Man's Role in the National Space Program," p. 1, undated (resulting from meetings of 10–11 Feb. 1961) and forwarded in Letter, Berkner to Waterman (NSF), 31 Mar. 1961, NASA HO. Italics in original. The document was re-released, with reference to the February meetings, on 7 Aug. 1961, and may be found in U.S. Senate, Committee on Aeronautical and Space Sciences, *National Goals for the Post-Apollo Period,* 89th Cong., 1st Sess. (Aug. 1965), pp. 242–43. On the Space Science Board's role in 1959–61, see also the brief accounts by Charles M. Atkins, "NASA and the Space

Science Board of the National Academy of Sciences" NASA, HHN-62, Office of Policy Analysis, Sept. 1966, and Logsdon, *Decision To Go to the Moon,* pp. 88–89.

28. NAS, Space Science Board, "Man's Role in the National Space Program," p. 3.

29. Homer E. Newell, Memo for Files, "Notes on the SSB Meeting," 10–11 Feb. 1961, NASA HO. Those who held manned spaceflight to be misdirected included Joshua Lederberg, Donald Hornig, and Willis Shapley (from BoB); Berkner on "negativism" (p. 3) and "pull together a statement" (p. 4). On Webb hearing of the SSB action a month early (in late February), see Logsdon, *Decision To Go to the Moon,* p. 89.

30. Memo, BoB, Military Division, to Bell (Director, BoB), "Meeting with Mr. James E. Webb," 15 Feb. 1961, NASA HO.

31. Jay Holmes, *America on the Moon: The Enterprise of the Sixties* (Philadelphia, 1962), p. 194; Logsdon, *Decision To Go to the Moon,* pp. 89–90.

32. Webb Speech to the American Astronautical Society, 17 Mar. 1961, cited by Logsdon, *Decision To Go to the Moon,* p. 90.

33. Memo, Bell for the President, "National Aeronautics and Space Administration budget problem," undated (describes 22 Mar. 1961 meeting), NASA HO.

34. Quoted both by Holmes, *America on the Moon,* p. 194, and Logsdon, *Decision To Go to the Moon,* p. 91.

35. Robert Jastrow (NASA-Goddard Space Flight Center), "Evaluation of USSR vs. US Output in Space Physics," 27 Feb. 1961, NASA HO.

36. "Administrator's Presentation to the President," 21 Mar. 1961, NASA HO.

37. Interviews with David Bell and Edward Welch by John Logsdon, *Decision To Go to the Moon,* p. 100.

38. *Washington Post,* 13 Apr. 1961; *New York Times,* 16–17 Apr. 1961.

39. U.S. Congress, House Committee on Science and Astronautics, *Hearing on H.R. 6169—A Bill to Amend the National Aeronautics and Space Act of 1958,* 87th Cong., 1st Sess. (April 1961), Fulton quote from p. 5.

40. U.S. Congress, House Committee on Science and Astronautics, *Discussion of Soviet Man-in-Space Shot,* 87th Cong., 1st Sess. (April 1961), Fulton quote from p. 7, Anfuso quote from p. 13. On "tired of coming in second-best," see Ken Hechler, *The Endless Space Frontier. A History of the House Committee on Science and Astronautics, 1959–78* (Washington, D.C., 1980), p. 82.

41. Hechler, *The Endless Space Frontier,* p. 83.

42. The President's News Conference, 12 Apr. 1961, *PP of P, John F. Kennedy 1961,* pp. 262–63.

43. NASA Press Conference on Russian Space Shot, 12 Apr. 1961, cited by Logsdon, *Decision to Go to the Moon,* p. 105.

44. Hugh Sidey, *John F. Kennedy, President* (New York, 1964), pp. 121–23.

45. Interview with James Webb by Logsdon, *Decision To Go to the Moon,* p. 109. The Bay of Pigs itself was apparently not a factor in the lunar deliberations, at least in the documentary record. Ed Welsh insists that it was never brought up by anyone in the Kennedy counsels (Edward C. Welsh Oral History, JFK Library, p. 20).

46. Letter, Webb to Glennan, 12 Apr. 1961, NASA HO.

47. Memo, Kennedy for the Vice President, 20 Apr. 1961, NASA HO.

48. The President's News Conference, 21 Apr. 1961, *PP of P, John F. Kennedy 1961,* pp. 309–10.

49. Memo, NASA for the Vice President, 22 Apr. 1961, NASA HO.

50. Logsdon, *Decision To Go to the Moon,* pp. 115–16. Cook was Executive Vice President of American Electric Power Corporation.

51. Memo, Johnson for the President, "Evaluation of the Space Program," 28 Apr. 1961, NASA HO.

52. Richard Hirsch and Joseph John Trento, *The National Aeronautics and Space Administration* (New York, 1973), p. 108.

53. Letter, Webb to LBJ, 26 Aug. 1966, LBJ Library, White House Central Files, Outer Space; "close to demanding," from Logsdon, *Decision To Go to the Moon,* p. 120.

54. Letter, Webb to LBJ, 4 May 1961, NASA HO.

55. Memo, Colonel Burris to the Vice President, "Views of Secretary McNamara on Acceleration of Booster Capabilities and Manned Exploration in Space," 23 Mar. 1961, NASA HO.

56. Bright, *The Jet Makers,* pp. 169–73; William L. Baldwin, *The Structure of the Defense*

Market 1955–1964 (Durham, N.C., 1967); Tae Saeng Shin, "A Financial Analysis of the Airframe-Turned-Aerospace Industry," Ph.D. diss., Univ. of Illinois, 1969 (on business and financial risk, p. 146).

57. Charles D. Bright, *The Jet Makers* (Lawrence, Kans., 1978), pp. 70–73. The breakdown of aerospace profits and losses from 1958 to 1960 was found in *Aviation Week and Space Technology*, 12 June 1961, p. 115. Only three major manufacturing sectors in the economy had lower composite profit margins than aerospace.

58. Major-General Ostrander (USAF Director, Launch Vehicle Programs) had already gotten word of a planned moon program and recommended that it be given to NASA on the condition that responsibility for all other launcher programs, meteorological and communication satellites, and orbital operations be given to the USAF. Memo, Ostrander for Robert Seamans, "Reflections on the Present American Posture in Space," 21 Apr. 1961, HST Library, Webb Papers, Box 103.

59. Willis Shapley, National Air and Space Museum Seminar, 11 Feb. 1982. Shapley reports that no one was more stunned than young Berkeley physicist Harold Brown, summoned to discuss the moon his first day on the job at the Pentagon! Uncomfortable in his California sportcoat amid all the gray suits, he said nothing. Brown became Secretary of Defense under Jimmy Carter. Overton Brooks was also invited to the meeting but did not attend. He expressed his views in a long memo, "Recommendations re the National Space Program," 4 May 1961, attached to Letter, Brooks to LBJ, 5 May 1961, NASA HO.

60. Logsdon, *Decision To Go to the Moon*, p. 124.

61. Willis Shapley, NASM Seminar, 11 Feb. 1982.

62. "Recommendations for Our National Space Program: Changes, Policies, Goals" (the Webb-McNamara Report), May 1961, JFK Library, Presidential Office Files. Quotes from pp. 10, 11, 18, 25.

63. Memo, Shapley to Bell, "Staff Review of Proposed Increase in Space Programs," 18 May 1961, NASA HO.

64. McGeorge Bundy Interview by Logsdon, *Decision To Go to the Moon*, p. 126; Jerome Wiesner Interview by E. M. Emme and Alex Roland, NASA HO.

65. Letter, Webb to LBJ, 23 May 1961, NASA HO.

66. See, for instance, Howard Margolis, "Technological and Political Models of the Decision To Go to the Moon," Institute for Defense Analyses—International and Social Studies Division (Sept. 1972), who argued (once the jargon is penetrated) that *both* of his models of decision making account for Apollo. "The key question was how a decision of this magnitude could have been made without a serious technical review" (p. 174).

Chapter 16

1. On the Kennedy-McNamara relationship, see William W. Kaufmann, *The McNamara Strategy* (New York, 1964), pp. 3, 300; on the whiz kids and civil management of defense, pp. 31–33; on the "unfinished business" inherited from Eisenhower, see Alain Enthoven and K. Wayne Smith, *How Much Is Enough? Shaping the Defense Program, 1961–1969* (New York, 1971), pp. 1–30.

2. See Richard Fryklund, *100 Million Lives: Maximum Survival in a Nuclear War* (New York, 1962), pp. 1–16; Peter Pringle and William Arkin, *The SIOP. The Secret U.S. Plan for Nuclear War* (New York, 1983), pp. 101–8 (Eisenhower reference, pp. 107–8); David Alan Rosenberg, "The Origins of Overkill: Nuclear Weapons and American Strategy, 1945–1960," *International Security* 7 (Spring 1983).

3. Desmond Ball, *Politics and Force Levels: The Strategic Missile Program of the Kennedy Administration* (Berkeley, 1980), pp. 34–38. This is the latest and best researched history of American strategy and missile deployment in the 1960s. The following discussion depends on it heavily. But see also Elie Abel, *The Missile Crisis* (New York, 1966); Richard J. Barnet, *The Economy of Death* (New York, 1969), a leftist attack on the military-industrial complex; C. W. Borklund, *The Department of Defense* (New York, 1968); Edgar M. Bottome, *The Missile Gap: A Study of the Formulation of Military and Political Policy* (Cranbury, N.J., 1971); Enthoven and Smith, *How Much Is Enough?* by the chiefs of McNamara's Systems Analysis Office; Roger Hilsman, *To Move a Nation* (New York, 1964), and *The Politics of Policy-Making in Defense and Foreign Affairs* (New York, 1971); Samuel P. Huntington, *The Common*

Defense (New York, 1966); Edmund S. Ions, ed., *The Politics of John F. Kennedy* (London, 1964); Kaufmann, *The McNamara Strategy*, an apologia by one of the whiz kids; George E. Lowe, *The Age of Deterrence* (Boston, 1964); Robert S. McNamara, *The Essence of Security: Reflections in Office* (London, 1968), not a memoir (he has never produced one), but a self-justifying essay on defense policy; Seymour Melman, *Pentagon Capitalism: The Political Economy of War* (New York, 1970), another New Left attack on the military-industrial complex and American militarism; Robert L. Perry, *The Ballistic Missile Decisions*, RAND P-3686 (Oct. 1967), a highly professional and documented narrative by an air force historian; George H. Quester, *Nuclear Diplomacy: The First Twenty-Five Years* (Boston, 1971); James M. Roherty, *Decisions of Robert S. McNamara* (Coral Gables, Fla., 1970); Walt W. Rostow, *The Diffusion of Power: An Essay in Recent History* (New York, 1972); Theodore C. Sorensen, *Decision-Making in the White House* (New York, 1963); Henry L. Trewhitt, *McNamara: His Ordeal in the Pentagon* (New York, 1971); Herbert York, *Race to Oblivion. A Participant's View of the Arms Race* (New York, 1970).

4. Perry, *The Ballistic Missile Decisions*, p. 14; Ball, *Politics and Force Levels*, pp. 43–46.

5. "Special Message to the Congress on the Defense Budget," 28 Mar. 1961, *PP of P, John F. Kennedy 1961*, pp. 229–40.

6. Ball, *Politics and Force Levels*, p. 119.

7. McNamara testimony in U.S. Senate, Committee on Government Operations, Subcommittee on National Policy Machinery, *Organization for National Security*, 87th Cong., 1st Sess., 1961, vol. 1, p. 1197.

8. Maxwell D. Taylor, *The Uncertain Trumpet* (New York, 1959), p. 158.

9. Ball, *Politics and Force Levels*, pp. 59–87; "not get murdered" comment made to Wiesner, 9 Dec. 1961 (p. 87).

10. Ibid., pp. 132–35.

11. *Washington Post*, 7 Feb. 1961.

12. "The President's News Conference of 8 Feb. 1961," *PP of P, John F. Kennedy 1961*, pp. 67–68.

13. Ball, *Politics and Force Levels*, pp. 94–95.

14. *New York Times*, 27 Nov. 1961.

15. McNamara Speech, "U.S. Nuclear Strategy," in San Francisco, 18 Sept. 1967, *Department of State Bulletin* 57 (9 Oct. 1967): 443–51.

16. U.S. Senate, Appropriations Committee, *Department of Defense Appropriations for 1964*, 88th Cong., 1st Sess., 1963, p. 40.

17. U.S. Senate, Armed Services Committee, *Military Procurement FY 1963*, 88th Cong., 1st Sess., 1963, p. 50. McNamara also told Representative Gerald R. Ford (R., Mich.) that downgrading of the Soviet ICBM threat by as much as 25 percent would have some, but not much, impact on U.S. deployment programs; see U.S. Congress, Appropriations Committee, *Department of Defense Appropriations for 1962*, 87th Cong., 1st Sess., 1961, Pt. 3, pp. 111–12, cited by Ball, *Politics and Force Levels*, pp. 167–68.

18. Kaufmann, *The McNamara Strategy*, p. 53.

19. Ball, *Politics and Force Levels*, pp. 182–83.

20. Ibid., pp. 190–91.

21. Kaufmann, *The McNamara Strategy*, pp. 114–20. Quote from p. 116.

22. Trewhitt, *McNamara*, pp. 115–16.

23. Ball, *Politics and Force Levels*, pp. 198–205.

24. McNamara, *The Essence of Security*, esp. chap. 4.

25. "Shifting sands" remark by James R. Schlesinger, "Uses and Abuses of Analysis," in U.S. Senate, Committee on Government Operations, *Planning, Programming, Budgeting*, 91st Cong., 2nd Sess., 1970, cited by Ball, *Politics and Force Levels*, pp. 265–66.

26. See Ball, *Politics and Force Levels*, pp. 212–31 (on the B-70 and Skybolt), pp. 240–52 (on military pressures), and pp. 252–63 (on industrial pressures and the recession).

27. McNamara, *The Essence of Security*, pp. 94–95. Quote from p. 95. On the PPBS, see Enthoven and Smith, *How Much Is Enough?* pp. 32–53. See also Charles J. Hitch, "Plans, Programs, and Budgets in the Department of Defense," *Operations Research* 11 (Jan.–Feb. 1963): 1–17; and Samuel A. Tucker, ed., *A Modern Design for Defense Decision: A McNamara-Hitch-Enthoven Anthology* (Washington, D.C., 1966).

28. McNamara, *The Essence of Security*, pp. 96–100.

29. Kaufmann, *The McNamara Strategy*, pp. 292–93.

30. Ibid., p. 240.

31. Alain C. Enthoven, Address before the Naval War College, Newport, R.I., 6 June 1963. See also Enthoven and Smith, *How Much Is Enough?* chap. 3.

32. Ibid.

33. Kaufmann, *The McNamara Strategy,* p. 181.

34. On systems analysis, its delights and its ultimate futility, see C. West Churchman, *Challenge to Reason* (New York, 1968), and *The Systems Approach* (New York, 1968). Churchman exposes the artificialities of systems analysis in chap. 2 and the solipsistic and "closed loop" nature of all systems in chap. 8, of *Challenge to Reason.* Systems analysis is discussed at more length in chapter 21 herein.

35. Top Secret Presentation by General Thomas S. Power, CINCSAC, on SAC's Strategic Role, the Pentagon, 28 Apr. 1964. A classic case study of conflict between civilian analysts and the brass in these years is Robert J. Art, *The TFX Decision: McNamara and the Military* (Boston, 1968).

36. This formula appears to have originated in the NASC report of Vice President Johnson in 1961 and is to this day the U.S. foreign policy line on the use of outer space.

37. In his March 1961 defense message to Congress, *PP of P, John F. Kennedy 1961,* pp. 230–31.

38. York testimony in U.S. Congress, Committee on Science and Astronautics, *Science, Astronautics, and Defense,* 87th Cong., 1st Sess., 1961, p. 8. The USAF share rose to 91 percent when its provision of launch services for army and navy spacecraft was taken into account. The other services made last-ditch appeals for a single, unified DoD space agency to forestall the USAF (General Trudeau in U.S. Congress, House Committee on Science and Astronautics, *Research and Development for Defense,* 87th Cong., 1st Sess., Feb. 1961, pp. 51–64, and Navy Memo, "The Organization of the US National Space Effort," in House Committee on Science and Astronautics, *Equatorial Launch Sites—Mobile Sea Launch Capability,* May 1961, pp. 37–49), but to no avail. McNamara made USAF military space primacy official in an executive order of 6 Mar. 1961.

Conflict among the services for larger pieces of the cosmic pie inspired this forensic gem from Congressman James G. Fulton (R., Pa.), interrogating Elvis J. Stahr, Secretary of the Army:

> Another point I would like to make: the Army is trained for certain purposes. The Moon is in space, obviously. You might have the Navy take you out to a launch platform at sea, you might have the Air Force get you up so high, or launch you. But when you get to the Moon, the people who might walk around on the Moon and who might be in tanks on the Moon when some adverse enemy power might have gotten there first, are going to be who?
>
> Are they going to be Air Force people who then don the Army uniform and become tank experts and artillery experts? Is it going to be a Marine landing, under Navy auspices?
>
> Is Rickover going to circle the Moon in a submarine? You see, it isn't quite as logically simple as saying that the so-called Air Force will then be in a place where there is no air. (U.S. Congress, House Committee on Science and Astronautics, *Defense Space Interests,* 87th Cong., 1st Sess. March 1981, p. 135).

39. See Michael N. Golovine, *Conflict in Space: A Pattern of War in a New Dimension* (London, 1962), p. 302.

40. U.S. Congress, House Committee on Science and Astronautics, *Research and Development for Defense,* Exec. Sess., 87th Cong., 1st Sess., 1961, pp. 34, 178.

41. Lt. General James Ferguson (Deputy Chief of Staff, USAF Research and Technology): "The Air Force must exploit space to retain U.S. military superiority in order to insure the peaceful uses of space" (House Armed Services Committee, 19 Feb. 1962); General Curtis E. LeMay (Chief of Staff, USAF): "War in the future may be waged and decided without a weapon being applied against an earth target. A nation that has maneuverable space vehicles and revolutionary armaments can indeed control the world, for peace or for aggression" (Speech, 26 Oct. 1961); General Thomas S. Power (CINCSAC): "Absolute superiority in space is essential to the future welfare and security of the Free World. In order to reserve space for peaceful purposes we must achieve the capability to defend ourselves against the wartime use of space" (Speech, 10 Apr. 1962); Power again:

> The space experiments so far are but a first step toward the goal of conducting manned space missions on a routine operational basis in support of national security needs. The U.S. space military requirements call for (1) rendezvous capability; (2) improved

satellite acquisition, tracking, and identification systems; (3) nuclear propulsion developments; (4) large manned and unmanned space stations in orbit (29 Mar. 1962 in House Committee on Science and Astronautics).

These and the quotations in the following note can be found in Research and Analysis Division, SAFAA, *The Military Mission in Space: A Summary of Published Views by Prominent Americans,* 24 Aug. 1962, pp. 3, 9, 13, NASA HO.

42. Harold Brown, 28 March 1962, in the House Committee on Science and Astronautics; McNamara, 7 Mar. 1962, in a speech to the Advertising Council; ibid., pp. 1, 12.

43. U.S. Congress, Committee on Appropriations, *DoD Appropriations for 1959,* 85th Cong., 2nd Sess., 1958, p. 391. See also Robert H. Puckett, "The Military Role in Space— A Summary of Official, Public Justifications," RAND P-2681 (Aug. 1962).

44. Brockway McMillan (Assistant Secretary of the USAF for R & D), 18 July 1962, in a speech to the American Rocket Society; ibid., p. 11. See also Letter, Webb to Gilpatric, 8 June 1962, NASA HO, making the same point: "The President's policy is to conduct the space effort with a civil, peaceful, international orientation as long as is possible and to the fullest extent possible, but always to develop the technology and preserve the ability to move rapidly to a military emphasis should this be required."

45. Philip J. Klass, *Secret Sentries in Space* (New York, 1971), pp. 130–39; Anthony Kenden, "U.S. Reconnaissance Satellite Programs," *Spaceflight* 20 (1978): 247–50.

46. Klass, *Secret Sentries in Space,* pp. 142–46. The Big Bird launch, on the mighty Titan 3D, was delayed until 15 June 1971. The Midas early-warning system ran into trouble in the early 1960s. Its sensors did not perform well, and doubts remained about the optimal orbital strategy: many satellites circling randomly, or a few sophisticated ones perched in geosynchronous orbit? Barren lines of development forced Harold Brown to admit in June 1963 that half of the $423 million spent on Midas had been wasted. The program was ultimately downgraded in favor of a new-generation system. In 1968 the first geosynchronous early warning satellite was launched and stationed permanently south of Western Russia (Kenden, "U.S. Reconnaissance Satellite Programs," p. 254).

47. Dr. L. L. Kavanau (Special Assistant, Space, to the DDR&E), "Department of Defense Views on the Military Mission in Space," 6 Nov. 1961, NASA HO.

48. General Ferguson statement before the House Committee on Armed Services, 19 Feb. 1962, NASA HO. See also Schriever's testimony, 20 July 1961, to the Preparedness Subcommittee of the Senate Armed Services Committee, cited in Letter, Schriever to LeMay, NASA HO; and Ferguson, "The Air Force Space Plan" and "USAF Studies US, Soviet Space Potential," *Aviation Week and Space Technology,* 26 Feb. and 5 Mar. 1962.

49. Memo, Hyatt to Webb, Dryden, and Seamans, 3 Nov. 1961, NASA HO.

50. "The Gleeful Conspiracy," *Aviation Week and Space Technology,* 23 Apr. 1962.

51. See the nostalgic account of Orion by Freeman Dyson, *Disturbing the Universe* (New York, 1979), pp. 109–15, 127–29. Interest in nuclear rocket propulsion dated from 1955, when it was shown that neutrons emitted by nuclear fission could heat and expel hydrogen at exhaust velocities greater than those achieved by chemical combustion rockets. Project Rover began at that time under AEC-USAF sponsorship to exploit this technique for military missiles. In 1958 it became a NASA-AEC research program. First tests of the Kiwi reactor in 1959 spawned a budget increase and the Rift (Reactor in Flight) Program. Congress added more money in 1962 to prime the pump of Project Nerva, a full-fledged nuclear rocket engine. Rift was cancelled in 1963, as was Dyson's Orion. Nerva was downgraded in 1968 and cancelled in 1972. At its peak, the nuclear rocket program consumed $100 million per year.

52. Air Research and Development Command, "Abbreviated Systems Development Plan. System 464L: Hypersonic Strategic Weapons System," C7-115361, 10 Oct. 1957, Andrews AFB, Air Force Systems Command (AFSC) HO; Memo of Understanding, "Principles for Participation of NACA in Development and Testing of the Air Force System 464L Hypersonic Boost Glide Vehicle (DynaSoar 1)," 20 May 1958, NASA HO.

53. The Dyna-Soar's growing weakness also stemmed from a loss of congressional support. In 1961 Senator Kerr had already expressed severe doubts about "duplication"— NASA, after all, was in his committee's bailiwick, not the USAF. His successor Clinton Anderson expressed the same doubts, eliciting from Deputy Secretary of Defense Gilpatric the admission that "In their present form, neither Gemini nor Dyna-Soar can perform a genuine military mission. Neither provides much room for military equipment" (Letter,

Gilpatric to Anderson, "Some Thoughts on Relationships of NASA and DoD," 27 Sept. 1963, LC, Anderson Papers, Box 908).

54. AFSC Historical Publications Series 63-50-I, *History of the X-20A Dyna-Soar,* by Clarence J. Geiger, Oct. 1963, Andrews Air Force Base, AFSC HO.

55. Letter, Webb to McNamara, 16 Jan. 1963, NASA HO. Reputedly Webb and McNamara also fell into personal acrimony by late 1962 that left them "not speaking to each other" (Shapley, National Air and Space Museum Seminar, 11 Feb. 1982). Given the size of their talents and egos, their rival claims to be administrative prima donna, and the inescapable conflicts between their agencies, it is rather likely that Washington proved not big enough for both of them.

56. By joint letter of 21 Jan. 1963, the NASA and USAF established a Gemini Program Planning Board to plan experiments to meet the objectives of both agencies. See Colonel Daniel D. McKee, "The Gemini Program," in Eldon W. Downs, ed., *The. U.S. Air Force in Space* (New York, 1966), pp. 6–15.

57. Letter, Anderson to Gilpatric, 30 Sept. 1963, LC, Anderson Papers, Box 908.

58. Clarence J. Geiger, *Termination of the X-20A Dyna-Soar,* AFSC Historical Publications Series 64-51-III (1964). McNamara's news briefing of 10 Dec. 1963 referred to the limited objectives and great expense of Dyna-Soar. Its main purpose, in his view, was to "explore means of controlling re-entry more precisely." That objective would now be pursued through the unmanned Asset Program, in which a small "lifting body," a craft with aerodynamic qualities, would be guided back to earth from very high altitudes. *Asset 1* was launched in September 1963 and *Asset 2* followed in March 1964. Both were lost in reentry.

59. William J. Coughlin, editor of *Missiles and Rockets,* "The Air Force and Gemini," NASA HO (clipping file).

60. Even at this point, the administration worried about the economic costs of any large cutback in aerospace R & D. Thousands of workers lost their jobs (see Arms Control and Disarmament Agency 29, *A Case Study of the Effects of Dyna-Soar Contract Cancellation upon Employees of the Boeing Company in Seattle,* July 1965); some lost their dreams as well. Their passion and humor found expression in Andy Oberta's send-up of Poe's "The Raven" in *Missiles and Rockets,* 27 Jan. 1964. A sample:

Ah, distinctly I remember it was early last December.
It was felt that very shortly we would be employed no more.
Every day we feared the morrow; vainly we had sought to borrow
Funds to budget us tomorrow for our work on Dyna-Soar—
On the sleek and wingèd spacecraft that they called the Dyna-Soar—
Cancelled now forever more.

. . . .

From off the duct I pulled the shutter, when, with many a flirt and flutter
Out there flew a stately raven of the saintly days of yore;
Not the least obeisance made he; not a minute stopped or stayed he;
But with mien of lord or lady, perched beside my office door—
Upon a bust of Eugen Sänger on the bookcase by the door—
Perched, and sat, and nothing more.

. . . .

"Prophet!" said I, "Thing of evil! Tell me, agent of the devil,
Whether McNamara axed the program or just cut it back some more?
Will he make a presentation to Congress for appropriation?
Does he plan continuation after Fiscal '64?
Is there funding in the budget? Tell me, tell me, I implore."
Quoth the raven, "Nevermore."

61. The Johnston Island A-SAT launches are listed in the chronology of U.S. space missions in U.S. Senate, Committee on Aeronautical and Space Sciences, *U.S. Civil Space Programs, 1958–1978* (Washington, D.C., 1980), pp. 127ff. See also Robert Salkeld, *War and Space* (Englewood Cliffs, N.J., 1970), pp. 61, 147.

62. Duncan L. Clarke, *Politics of Arms Control: The Role and Effectiveness of the U.S. Arms Control and Disarmament Agency* (New York, 1979). See esp. chap. 2 on the passage of the Arms Control and Disarmament Act of 1961, and chap. 3 on the agency's image.

63. Ibid., p. 32.

64. Jet Propulsion Laboratory, Arms Control Study Group Memo 40-1, "A Basis for a Separate Treaty for Banning Weapons in Space," by Richard P. Schuster, Jr. (20 Nov.

1962), p. 3. See also Alton Frye, "Space Arms Control: Trends, Concepts, and Prospects," RAND P-2873 (Feb. 1964). When NASA headquarters got word that such policy studies were being conducted at JPL, it scolded Director William Pickering and recalled all copies of the report (Memo, F. Dixon to Seamans, 28 Dec. 1962).

65. G. E. Hlavka and G. R. Pitman, Jr., "Clandestine Production of Missile Components," ACDA Study ST-13, 10 Apr. 1963, p. 2. ACDA Library, Rosslyn, Virginia.

66. Memo, Walter D. Sohier (NASA General Counsel) for Webb, "Disarmament," 27 Feb. 1964, NASA Archives, 71A4083.

67. Letter, Sohier to Herbert Scoville, "Impact on NASA Programs of Banning Certain Strategic Delivery Vehicles Developments," 18 Mar. 1964, NASA Archives, 71A4083.

68. Ibid.

69. A. R. Hibbs Draft Report, "Effects of Arms Control and Disarmament on Future NASA Programs," JPL, 24 Apr. 1964, p. 3, NASA Archives, 71A4083.

Chapter 17

1. David Halberstam, *The Best and the Brightest* (New York, 1969), pp. 153–54.

2. See, for instance, the speculations (mostly optimistic) by social scientists and government officials in the host of compilations published in the early Space Age: American Assembly, *Outer Space: Prospects for Man and Society*, ed. Lincoln Bloomfield (Englewood Cliffs, N.J., 1962); Brookings Institution, *Proposed Studies on the Implications of Peaceful Space Activities for Human Affairs* (Washington, D.C., 1961); W. E. Frye, ed., *Impact of Space Exploration on Society* (Washington, D.C., 1966); Joseph M. Goldsen, *International Political Implications of Activities in Outer Space* (New York, 1959), and *Outer Space in World Politics* (New York, 1963); Arnold L. Horelick, "Outer Space and Earthbound Politics," *World Politics* 13 (1961): 323–29; Klaus Knorr, "On the International Implications of Outer Space," *World Politics* 12 (1960): 564–84; Lillian Levy, ed., *Space: Its Impact on Man's Society* (New York, 1965); Frederick J. Ossenbeck and Patricia C. Kroeck, eds., *Open Space and Peace: A Symposium on Effects on Observation* (Stanford, Calif., 1964); RAND Corporation, *International Political Implications of Activities in Outer Space: Report of a Conference, Oct. 22–23, 1959*, RAND R-362-RC, 1960; Oscar H. Rechtshaffen, ed., *Reflections on Space: Its Implications for Domestic and International Affairs* (Colorado Springs, 1964); Howard J. Taubenfeld, ed., *Space and Society: Studies for the Seminar on Problems of Outer Space Sponsored by the Carnegie Endowment for International Peace* (Dobbs Ferry, N.Y., 1964). Even the Democratic National Committee speculated on such matters: "[Space] will have a profound effect in making people more 'one world' conscious. In short, that cooperation on space exploration and technology may be an especially fruitful way to weaken the barriers of nationalism" ("Comments on International Agreements in Space," JFK Library, Democratic National Committee, Box 208).

3. State Department, Bureau of Research and Intelligence, "Main Trends in Soviet Policy, 1960–1965," 31 Oct. 1960, NA, State Research & Analysis Reports, No. 8362.1. On the Kennedy transition, see Philip J. Klass, *Secret Sentries in Space* (New York, 1971), pp. 104–07.

4. See, for example, *Washington Post*, 13 Oct. 1960, and *Christian Science Monitor*, 14 Sept. 1960.

5. Memo, Arthur Sylvester (Assistant Secretary of Defense) for the President, 26 Jan. 1961, JFK Library, National Security Files—Departments and Agencies, cited by Gerald M. Steinberg, "The Legitimization of Reconnaissance Satellites: An Example of Informal Arms Control Negotiation," Ph.D. diss., Cornell Univ., 1981, pp. 60–62.

6. State Department Draft, "Foreign Policy Aspects of Outer Space," 3 Apr. 1961 (quotes from pp. 1, 4, 12), and its implementation in Memo, Harland Cleveland to the Secretary of State, "U.S. Outer Space Proposals for the Sixteenth General Assembly," 20 Nov. 1961, LBJ Library, Legislative Background—Outer Space Treaty History, Box 1.

7. Abraham Chayes Oral History, JFK Library, pp. 215–16.

8. Memo, Rusk for the President, 2 Feb. 1961, JFK Library, Primary Office Files, Departments and Agencies—State 2/61; Memo, John McCone, 14 Mar. 1961, JFK Library, White House Central Files, Box 653; cited by Steinberg, "The Legitimization of Reconnaissance Satellites," pp. 69–71.

9. "Summary of Foreign Policy Aspects of the U.S. Outer Space Program," 5 June 1962, Annex C: "Development of Procedures for U.S. Reports for the UN Registration of Space Launches," pp. 1–2, LBJ Library, Legislative Background—Outer Space Treaty History, Box 1.

10. Chayes Oral History, JFK Library. See also Steinberg, "The Legitimization of Reconnaissance Satellites," p. 88. The other members of the Task Force were Joseph Charyk, Gilpatric, Bundy, Wiesner, and Herbert Scoville.

11. Memo for the Record, "Space Policy Statement," E. C. Welsh, 14 May 1962 (source of quote) and Memo, Welsh, "Space Policy," 28 May 1962; also Draft NASC, "United States Policy on Outer Space," 28 May 1962; all LBJ Library, Legislative Background—Outer Space Treaty History, Box 1. This Space Council document was planned as an unclassified policy statement, hence a brief, generalized reaffirmation of American commitment to leadership in all aspects of space technology, cooperation as well as competition, and "the maximum degree of openness compatible with national security considerations." It showed less candor than Eisenhower's public declarations, and its drafting simultaneous to secret policy papers demonstrated that substantive policy issues, despite official openness, were scarcely even to be acknowledged before the American public.

12. NSAM-156 as cited by Steinberg, "The Legitimization of Reconnaissance Satellites," pp. 89–90. See also State Department, "U.S. Policy on Outer Space," 25 Oct. 1962, LBJ Library, Legislative Background—Outer Space Treaty History, Box 1. Directives on the handling of military space activities included National Security Action 2454 and NSAM 183. For legal and academic opinions jibing with U.S. policy at this time, see John A. Johnson (NASA General Counsel), "The Future of Manned Space Flight and the 'Freedom of Outer Space,'" Speech to the American Bar Association, 4 Aug. 1962, and Robert D. Crane, "Law and Strategy in Space," *ORBIS* 6 (Summer 1962): 281–300. Ossenbeck and Kroeck, eds., *Open Space and Peace*, a Hoover Institution colloquium, was also supportive of administration policy. See especially Carl H. Amme, Jr., "The Implications of Satellite Observation for U.S. Policy," *Open Space and Peace*, pp. 105–25. The book's motto: "*Si vis pacem, para scientiam.*"

13. Albert Gore, 3 Dec. 1962, in U.S., ACDA, *Documents on Disarmament, 1962* (Washington, D.C., 1963), vol. 2, p. 1121.

14. On orbital bombardment see Donald G. Brennan, "Arms and Arms Control in Outer Space," in American Association, *Outer Space: Prospects for Man and Society*, pp. 129–32.

15. J. P. Lorenz, Draft Historical Project, "OUTER SPACE. Negotiation of Outer Space Treaty. Background," Oct. 1968, p. 4, citing Policy Paper #4052, Rollefson to McGee, 6 Nov. 1962, LBJ Library, Legislative Background—Outer Space Treaty History, Box 1.

16. Gilpatric Speech to a Meeting of Midwestern Industry and University Representatives, 5 Sept. 1962; U.S. Senate, Committee on Aeronautical and Space Sciences, *Documents on International Aspects of the Exploration and Use of Outer Space, 1954–1962*, 88th Cong., 1st Sess., May 1963, pp. 317–18.

17. Sedov quoted by Dryden after a personal conversation at a scientific conference, Letter, Dryden to Glennan, 7 Sept. 1960, Milton Eisenhower Library, Johns Hopkins Univ., Dryden Papers.

18. Memo, Frutkin for Glennan, 11 Sept. 1960, NASA HO; Arnold W. Frutkin, *International Cooperation in Space* (Englewood Cliffs, N.J., 1965), pp. 34–35.

19. See, for example, Don Kash, *The Politics of Space Cooperation* (West Lafayette, Ind., 1967), pp. 10–22, 44–48; Leonard E. Schwartz, "When Is International Space Cooperation International?" *Bulletin of the Atomic Scientists* (June 1963): 12–18, and "Control of Outer Space," *Current History* (July 1964): 39–46.

20. "An Interview with John F. Kennedy," *Bulletin of the Atomic Scientists* (Nov. 1960): 347.

21. Inaugural Address, 20 Jan. 1961, *PP of P, John F. Kennedy 1961*, p. 2.

22. State of the Union Address, 30 Jan. 1961, *PP of P, John F. Kennedy 1961*, pp. 26–27.

23. Webb Testimony, U.S. Congress, Committee on Appropriations, *Independent Offices Appropriations for 1963*, 87th Cong., 3rd Sess., 1962, vol. 3, pp. 418–419.

24. Leon Lipson, "Current Problems of Space Control and Cooperation: A Summary," RAND RM-2805-NASA, July 1961, NASA HO. See also Robert H. Pickett, "International Control and Limitation of Space Activities: The American National Interest," Ph.D. diss., University of Chicago, 1962.

25. "Presidential Meeting with Khrushchev in Vienna, June 3–4 [1960], SCIENTIFIC COOPERATION," NASA HO.

26. See Frutkin, *International Cooperation in Space*, pp. 92–93; Dodd L. Harvey and Linda C. Ciccoritti, *U.S.-Soviet Cooperation in Space* (Washington, D.C., 1974), pp. 86–87; and Letter, Kennedy to Khrushchev, 7 March 1962, in U.S. Senate, Committee on Aeronautical and Space Sciences, *International Cooperation and Organization for Outer Space*, 89th Cong., 1st Sess. (Washington, D.C., 1965), pp. 137–40.

27. Letter, Khrushchev to Kennedy, 20 Mar. 1962, in U.S. Senate, Committee on Aeronautical and Space Sciences, *Documents on International Aspects of the Exploration and Use of Outer Space*, 88th Cong., 1st Sess. (Washington, D.C., 1963), pp. 248–51. Quote from p. 251.

28. "The President's News Conference, 17 July 1963," *PP of P, John F. Kennedy 1963*, p. 568. The Dryden-Blagonravov negotiations have been described more often than their results warrant. See especially the accounts in U.S. Senate, *International Cooperation and Organization for Outer Space*; Harvey and Ciccoritti, *U.S.-Soviet Cooperation in Space*; and James A. Malloy, Jr., "U.S.-U.S.S.R. Space Negotiations and Cooperation, 1958–1965," NASA HHN-113 (Sept. 1971).

29. Frutkin, *International Cooperation in Space*, pp. 100–01.

30. Ibid., pp. 133–37. European space programs and problems of cooperation with the United States are discussed in chapter 20 herein. On international cooperation in this period generally, see Lincoln P. Bloomfield, "Outer Space and International Cooperation," *International Organization* 19 (1965): 603–21; Eugene M. Emme, Jr., *Statements by the Presidents of the United States on International Cooperation in Space* (Washington, D.C., 1971); Richard Gardner, "Cooperation in Outer Space," *Foreign Affairs* 41 (1963): 344–59.

31. Arthur C. Clarke, "A Short Pre-History of Comsats, Or: How I Lost a Billion Dollars in My Spare Time," in Clarke, *Voices from the Sky: Previews of the Coming Space Age* (New York, 1965), pp. 119–28.

32. Pierce first suggested a comsat system in 1952, but, fearing ridicule, wrote under a pseudonym in *Amazing Science Fiction.* He came out of the closet on comsats in 1955 and urged his fellow scientists to promote rocketry and communications technology so as to realize "orbital radio relays" in the near future. See Delbert D. Smith, *Communications via Satellite: A Vision in Retrospect* (L Boston, 1976), pp. 18–19.

33. Smith, *Communications via Satellite*, pp. 58–59. The Hughes engineers were Harold Rosen, Donald Williams, and Thomas Hudspeth.

34. In this the United States was unique. In most other countries governments owned a monopoly of communications. But while the U.S. government had aided the spread of telegraph and telephone, radio and undersea cables (even racing the British for an "all-blue" trans-Pacific cable in the ninteenth century), it had not performed R & D for civilian innovations in these fields.

35. Smith, *Communications via Satellite*, pp. 60–77. On projected communications volume, see also Booz-Allen and Hamilton, "Business Planning Study for a Commercial Telecommunications Satellite," prepared for Lockheed Aircraft Corporation, 31 Oct. 1960.

36. Glennan Speech, "Space Exploration and Exploitation," Portland, Ore., 12 Oct. 1960, NASA HO.

37. "Statement by President Eisenhower on Communications Satellites," 31 Dec. 1960, *Department of State Bulletin*, 16 Jan. 1961, p. 77.

38. Memos, Nunn to Webb, 1 and 11 May 1961, NASA Nunn Files, cited by Smith, *Communications via Satellite*, p. 78; Special Message to the Congress on Urgent National Needs, 25 May 1961: *PP of P, John F. Kennedy 1961*, p. 404. Letter, Kennedy to Vice President Johnson, 15 June 1961, *PP of P, John F. Kennedy 1961*, p. 472.

39. Lloyd D. Musolf, ed., *Communications Satellites in Political Orbit* (San Francisco, 1968), pp. 40–43.

40. "Statement by the President on Communications Satellite Policy," 24 July 1961, *PP of P, John F. Kennedy 1961*, pp. 529–31. Quote from p. 530. See also U.S. Congress, House Committee on Interstate and Foreign Commerce, *Communications Satellites*, 87th Cong., 2nd Sess., Mar. 1962, Pt. 2, pp. 616–17. On the military interest, see Memo, Webb for Dryden, 12 June 1961, HST Library, Webb Papers, Box 30: "I informed them [communications executives] very directly that the President had urged the bringing into being of a capability at the earliest possible time, that the Defense Department had a very real interest in this although it was important not to discuss this. . . ."

41. Smith, *Communications via Satellite*, pp. 94–100.

42. See Jack Posner, "The Implementation of the Communications Satellite Act of 1962," Ph.D. diss., American Univ., 1967, pp. 107–08; and Smith, *Communications via Satellite*, p. 104. Michael E. Kinsley, *Outer Space and Inner Sanctums: Government, Business, and Satellite Communication* (New York, 1976), speculates that Kerr's bill was a ploy that, combined with his influence over the fate of other presidential programs, obliged the White House to compromise with him (p. 5).

43. Testimony of James E. Dingman (Executive vice president of American Telephone and Telegraph), U.S. Senate, Committee on Aeronautical and Space Sciences, *Communications Satellite Legislation*, 87th Cong., 2nd Sess., Feb.–Mar. 1962, pp. 308–16.

44. Testimony of Estes Kefauver, U.S. Senate, Committee on Aeronautical and Space Sciences, *Communications Satellite Legislation*, 87th Cong., 2nd Sess., Feb.–Mar. 1962, pp. 332–78. Quotes from pp. 345, 351.

45. U.S. Senate, Committee on Commerce, *Communications Satellite Act of 1962*, 87th Cong., 2nd Sess., 11 June 1962, pp. 49–55.

46. The cloture vote was a moment of high drama in the Senate, for the upper chamber had tolerated filibusters for decades and had recently seen them used to block civil rights legislation. When a number of senators resigned themselves to voting cloture against liberals on behalf of the President's space policy, they could no longer in good conscience uphold filibusters against integration: how could one vote cloture for AT&T, but not for the NAACP? as one observer put it. Some conservatives voted against cloture regardless of their views on comsats, however, in order to uphold the Senate's "unlimited debate" tradition. They included John Sparkman (D., Miss.), John Stennis (D., Miss.), Strom Thurmond (D., S.C.), John Tower (D., Tex.), and Barry Goldwater (R., Ariz.), thus making the nays list a particolored spackling of left-wing and right-wing names.

47. On the benefits accruing to AT&T as a result of the Comsat Act, see Kinsley, *Outer Space and Inner Sanctums*. Data cited here are from pp. 17–25.

48. Richard N. Gardner (International Organizations Division, State Department), "Space Meteorology and Communications: A Challenge to Science and Diplomacy," *Department of State Bulletin*, 13 May 1964, pp. 743–45. For deeper foreign policy considerations, see RAND's Murray L. Schwartz and Joseph M. Goldsen, *Foreign Participation in Communications Satellite Systems: Implications of the Communications Satellite Act of 1962*, RM 3484-RC (Feb. 1963).

49. Smith, *Communications via Satellite*, pp. 132–35.

50. U.S. Congress, Committee on Government Operations, *Satellite Communications*, 88th Cong., 2nd Sess., 1964, Pt. 2, pp. 661–65. Kinsley says that "The State Department made it impossible for the Soviet Union to join the new international organization" (*Outer Space and Inner Sanctums*, p. 112), but the Soviets' own political hostility to such a system and their technical inferiority were just as important in rendering a "truly global system" impossible.

51. Jonathon F. Galloway, *The Politics and Technology of Satellite Communications* (Lexington, Mass., 1972), p. 87.

52. British Postmaster-General, cited in U.S. Congress, Committee on Government Operations, *Hearings on Satellite Communications*, 88th Cong., 2nd Sess., 1964, Pt. 1, p. 28. On European organization, see Smith, *Communications via Satellite*, pp. 135–37.

53. On the negotiations leading to INTELSAT, see U.S. Senate, *International Cooperation and Organization for Outer Space*, pp. 117–20; Smith, *Communications via Satellite*, pp. 135–41; Joseph N. Pelton, *Global Communications Satellite Policy: INTELSAT, Politics and Functionalism* (Mt. Airy, Md., 1974), pp. 54–60. On the subsequent history of INTELSAT, see esp. Galloway, *The Politics and Technology of Satellite Communications*; also Judith Tegger Kildow, *INTELSAT: Policy Maker's Dilemma* (Lexington, Mass., 1973), and Joseph N. Pelton and Marcellus S. Snow, eds., *Economic and Political Problems in Satellite Communications* (New York, 1977).

54. Gardner, "Space Meteorology and Communications," pp. 743–44.

55. *New York Times*, 22 Apr. 1965, cited by Kinsley, *Outer Space and Inner Sanctums*, p. 114.

Chapter 18

1. See, for example, Bruce Mazlish, ed., *The Railroad and the Space Program; An Exploration in Historical Analogy* (Cambridge, Mass., 1965); John Logsdon, *The Decision To Go to the Moon: Project Apollo and the National Interest* (Cambridge, Mass., 1970); Vernon Van Dyke, *Pride and Power: The Rationale of the Space Program* (Urbana, Ill., 1964); Amitai Etzioni, *The Moon-Doggle, Domestic and International Implications of the Space Race* (Garden City, N.Y., 1964); and Norman Mailer, *Of a Fire on the Moon* (New York, 1969) for varying interpretations of why Apollo was somehow very "American." It is interesting that, through the years, almost no commentators, whether they were for or against it, thought Apollo to be a rather "un-American" novelty in our history.

2. Jane Van Nimmen and Leonard C. Bruno with Robert L. Rosholt, *NASA Historical Data Book, 1958–1968*, 2 vols., vol. 1: *NASA Resources* (Washington, D.C., 1976).

3. "Reappraising NASA's Organizational Structure to Achieve the Objectives of an Accelerated Program," 12 June 1961, NASA HO. See also Robert L. Rosholt, *An Administrative History of NASA, 1958–1963*, NASA SP-4101 (Washington, D.C., 1966), pp. 209–11, and Arnold S. Levine, *Managing NASA in the Apollo Era*, NASA SP-4102 (Washington, D.C., 1982), pp. 1–4, 34–43.

4. Memo, Webb for the Vice President, 23 May 1961, pp. 1–2, NASA HO.

5. Ibid., p. 2.

6. Ken Hechler, *The Endless Space Frontier. A History of the Committee on Science and Astronautics, 1959–1978* (Washington, D.C., 1980), p. 178. On the location of NASA facilities, see Thomas P. Murphy, *Science, Geopolitics, and Federal Spending* (Lexington, Mass., 1971), pp. 197–223. The biggest congressional tiff was touched off by the abrupt NASA decision to locate a new Electronics Research Center on Route 128 outside of Boston, the preferred choice of a number of prominent Democrats (Murphy, chap. 8).

7. Hugo Young, Bryan Silcock, and Peter Dunn, "Why We Went to the Moon," *Washington Monthly* 2 (Apr. 1970): 48. Their allegations are corroborated in the memoirs of Bobby Baker himself (with Larry L. King), *Wheeling and Dealing* (New York, 1978), pp. 103–04, 169–70. On the selection of Houston for the Manned Spaceflight Center, see the comparison of eighteen cities in Mary Holman, *The Political Economy of the Space Program* (Palo Alto, Calif., 1974), p. 50, and Webb's Memo to Dryden and Seamans, 31 Jan. 1963, citing the reasons for their judgment in favor of North American, HST Library, Webb Papers, Box 33. Houston tied for first on objective criteria with Los Angeles, San Diego, San Francisco, and New Orleans. On politics in the selection of NASA sites and award of contracts, see also Young, Silcock, and Dunn, "Why We Went to the Moon," and Hechler, *The Endless Space Frontier*, p. 185. The North American Aviation contract became a favorite of scandalmongers in 1967 when astronauts Grissom, White, and Chafee were incinerated in the test Apollo command module. See U.S. Congress, Committee on Science and Astronautics, Subcommittee on NASA Oversight, *Investigations into the Apollo 204 Accident*, 90th Cong., 1st Sess., 1967.

8. The public image of NASA and the astronauts is the commonest theme of popular writing about the space program. In most cases the authors aggressively "humanize" the astronauts and engineers (not a difficult task, their being human), only to reveal qualities more endearing than those concocted by the public relations men themselves. See especially Tom Wolfe, *The Right Stuff* (New York, 1979), and James Michener, *Space* (New York, 1982).

9. On the 1961 managerial reforms, see Rosholt, *An Administrative History of NASA*, pp. 209–39; Levine, *Managing NASA*, pp. 34–46. These 1961 reforms were not wholly successful. The functional lines of authority meant that one center might be involved in projects divided among five different program offices. The burden fell on Seamans, who could not untangle all the strings by himself. In 1963 Webb decentralized management, promoting project managers to autonomy as associate administrators for their own offices.

10. Levine, *Managing NASA*, pp. 68–93; quotation from p. 53. On PERT, see pp. 156–57. But Harvey M. Sapolsky, *The Polaris System Development. Bureaucratic and Programmatic Success in Government* (Cambridge, Mass., 1972), considers PERT overrated even in its debut with the navy (pp. 94–130).

11. On the heated eighteen-month debate on the lunar mode, see John Logsdon, "Selecting the Way to the Moon: The Choice of the Lunar Orbital Rendezvous Mode," *Aerospace Historian* 18 (June 1971): 66–70; Holman, *Political Economy of the Space Program,* pp. 98–102; Roger E. Bilstein, *Stages to Saturn: A Technological History of the Apollo/Saturn Launch Vehicles,* NASA SP-4206 (Washington, D.C., 1980), pp. 60–68; Courtney Brooks, James Grimwood, and Loyd S. Swenson, Jr., *Chariots for Apollo: A History of Manned Lunar Spacecraft,* NASA SP-4205 (Washington, D.C., 1979), pp. 61–86. The motives of George Low and Wernher von Braun in finally opting for LOR are still obscure. Each had held tenaciously to the other modes. Von Braun himself thought it important to develop a capacity for orbital rendezvous, important as it was to the future of space development. On the other hand, von Braun *wanted* to continue building ever larger rockets. If he had known that choosing LOR meant the end not only of Nova but even of the *Saturn 5* after 1972, perhaps he would have continued to fight. There is speculation that a deal might have been struck, for von Braun had remained wedded to EOR until the climactic Huntsville meeting itself (7 June 1962), when he suddenly rose to endorse LOR, thereby undercutting his own people: "You could hear a pin drop." It is also possible, however, that von Braun simply wearied of the fight: "We are already losing time in our overall program," he said, "as a result of lacking a mode decision."
12. The heated public exchange in September 1962 has been oft described. Von Braun was hosting the presidential party around Marshall Spaceflight Center and explaining the LOR sequence when Kennedy interrupted, "I understand Dr. Wiesner doesn't agree with this. Where is Jerry?" Wiesner approached and was obliged to register his objection. But the debate occurred out of earshot of reporters, and Kennedy soon called it to a halt.
13. Webb acknowledged that ancillary benefits of potential military value to be derived from EOR versus LOR were an issue of great importance that required detailed discussion with the DoD (Memo, Webb to Brainerd Holmes, 8 June 1962, NASA HO).
14. Holman, *Political Economy of the Space Program,* p. 99; Harold L. Nieburg, *In the Name of Science* (Chicago, 1966), p. 29; U.S. House, Committee on Science and Astronautics, *NASA Lunar Orbit Rendezvous Decision,* 87th Cong., 2nd Sess., July 1962.
15. Loyd S. Swenson, Jr., James M. Grimwood, and Charles C. Alexander, *This New Ocean. A History of Project Mercury.* NASA SP-4201 (Washington, D.C., 1966), p. 477.
16. Ibid., pp. 492 ("trust in God"), 420 ("flying camera"), 503 ("dust on highway").
17. Van Nimmen, Bruno, with Rosholt, *NASA Historical Data Book,* vol. 1, p. 134. Total space spending is recounted annually in the rear of the *Aeronautics and Space Report of the President,* published annually by the NASA HO.
18. Letter, Webb to the President, 29 Oct. 1962, NASA HO.
19. Letter, Webb to the President, 30 Nov. 1962, NASA HO.
20. Memo, Kennedy for the Vice President, 9 Apr. 1963, NASA HO.
21. Letter, Webb to the Vice President, 3 May 1963, HST Library, Webb Papers, Box 34.
22. Address by James Webb, Institute of Foreign Affairs, 24 Jan. 1963, p. 2, NASA HO.
23. James Webb, *Space Age Management: The Large Scale Approach* (New York, 1969), pp. 15, 17, 22, 29, 33–37.
24. Address by Webb, Milwaukee Press Club Gridiron Dinner, 17 Apr. 1963, in NASA, *The American Space Program. Its Meaning and Purpose,* HHN-47 (Dec. 1964), p. 25.
25. Webb, *Space Age Management,* p. 113.
26. Ibid.; see also Interview with Webb by Dr. Harvey et al., 16 Mar. 1968, pp. 16–17, NASA HO.
27. Interview with Webb by Dr. Harvey et al., 16 Mar. 1968, p. 16, NASA HO.
28. Webb, *Space Age Management,* p. 117.
29. Ibid.
30. Raymond Bauer, *Second Order Consequences: A Methodological Essay on the Impact of Technology* (Cambridge, Mass., 1969), pp. 3–4. Webb cites Bauer in *Space Age Management,* p. 121.
31. Webb, *Space Age Management,* pp. 123–24.
32. Ibid., p. 151.
33. Ibid., pp. 16–17.
34. Webb in "Conference on Space, Science, and Urban Life," Dunsmuir House, Oakland, Calif., 30 Mar. 1963, in NASA, *The American Space Program. Its Meaning and Purpose,* p. 33.

35. U.S. Senate, Committee on Appropriations, *Independent Offices Appropriations, 1962,* 87th Cong., 1st Sess., 1961, p. 649.

36. *Congressional Record,* 17 July 1963, p. A4508.

37. Welsh Speech, "Benefitting from Aerospace Revolution," Institute of Electrical and Electronics Engineers, Schenectady, N.Y., 17 May 1969, pp. 1–2. See also Welsh Speech, "Space and the National Economy," Conference on Space Science and Space Law, Norman, Okla., 20 June 1963.

38. National Science Foundation, *Federal Funds for R & D,* NSF 79-310, vol. 27, appendices.

39. Holman, *Political Economy of the Space Program,* esp. chaps. 5 and 9.

40. Letter, Webb to Vannevar Bush, 15 May 1961, HST Library, Webb Papers, Box 67.

41. Van Dyke, *Pride and Power. The Rationale of the Space Program,* pp. 224–30; Homer E. Newell, *Beyond the Atmosphere: Early Years of Space Science,* NASA SP-4211 (Washington, D.C., 1980), pp. 225–28.

42. Letter, Webb to DuBridge, 29 Aug. 1961, NASA HO.

43. Letter, Webb to Killian, 10 Jan. 1962, Box 31; Memo, Webb for Dryden, Seamans, and Simpson, 20 May 1964, Box 164: both HST Library, Webb Papers. Note: Most of Webb's outgoing correspondence from his tenure as NASA administrator can also be found in the Chron File at the NASA HO.

44. Memo, Webb for Dryden, 19 June 1964, HST Library, Webb Papers, Box 165.

45. Letter, Webb to Mose L. Harvey (Director, Center for Advanced International Studies, Univ. of Miami, Fla.), 29 June 1964, HST Library, Webb Papers, Box 36.

46. Newell, *Beyond the Atmosphere,* pp. 232–34; quotation from p. 233.

47. Ibid.

48. Ibid.; see also Webb, "Education in the Space Age," *Educational Record* (Winter 1964): 33–40.

49. Kerr Speech, "The Challenge of Outer Space to the Legal Profession," Norman, Okla., 27 Apr. 1961, NA, Papers of the Senate Space Committee, Box 102.

50. James Reston, "Senator Johnson's Move," *New York Times,* 8 Jan. 1958.

51. Von Braun quoted by Enid Curtis Bok Schoettle, "The Establishment of NASA," in Sanford A. Lakoff, *Knowledge and Power. Essays on Science and Government* (New York, 1966), p. 185.

52. Memo, Webb for Horace Busby (Special Assistant to the President), 23 Mar. 1965, p. 3, LBJ Library.

53. Ibid. Italics in original.

54. Ibid., p. 2.

55. Walter Prescott Webb quoted by James Webb in draft of a speech to be delivered at Cornell University, 3 Mar. 1965, manuscript in LBJ Library. See also James E. Webb, "Education in the Space Age," *Educational Review* (Winter 1964): 33–40 (on pp. 38–39 Webb cites Frederick Jackson Turner, *The Frontier in American History* [New York, 1921], p. 283, on the link between the frontier, universities, and science). On the other frontier theorist, "the sage of West Texas," see Necah Stewart Furman, *Walter Prescott Webb. His Life and Impact* (Albuquerque, 1976).

Chapter 19

1. Editorial from *Territorial Enterprise and Virginia City News,* Nevada, Jan. 1960, copy in NASA HO.

2. Testimony of Dr. Philip Abelson, in U.S. Senate, Committee on Aeronautical and Space Sciences, *Scientists' Testimony on Space Goals,* 88th Cong., 1st Sess., 1963, pp. 3–24. Quote from p. 16.

3. Ibid., pp. 62–73, 244–46.

4. Donald A. Strickland, "Physicists' Views of Space Politics," NASA General Grant NSG-243-62 (1962), NASA HO.

5. Testimony of Dr. Lloyd Berkner and Dr. Frederick Seitz, in U.S. Senate, *Scientists' Testimony on Space Goals,* pp. 88–137. Quotes from p. 89.

6. Walter Lippmann, "Money and the Moon," *Washington Post,* 2 Apr. 1962.

7. Interview with Eisenhower, *Saturday Evening Post,* 6 Aug. 1962, p. 24.

8. "GOP Calls for Military Direction of Space Program," including Representative Miller's reaction, *Space Daily,* 7 Jan. 1963, p. 2.

9. Goldwater Speech at Maxwell AFB, Alabama, 14 Nov. 1960, in "Statements of Barry M. Goldwater on Space Exploration," NASA HHN-46, 27 July 1964.

10. Ibid.

11. Goldwater Speech at National Rocket Club, Washington, D.C., 17 July 1962, in *Congressional Record, Senate,* 1962, pp. A5481–83.

12. Goldwater Speech at Maxwell AFB, Alabama, 19 July 1963, in *Congressional Record, Senate,* 1963, pp. 12382–85.

13. Staff Report of Senate Republican Policy Committee, "A Matter of Priority," 17 May 1963, and Republican Task Force on Space and Aeronautics, Subcommittee on the Military Role in Space, "The Military Role in Space," 27 Sept. 1963, both in LC, Anderson Papers, Boxes 908 and 914.

14. *Reader's Digest* (Aug. 1963): "While we race to the moon, Soviets stride toward military conquest of near-space."

15. William F. Buckley, Jr., "The Moon *and* Bust?" 1 June 1963, reprinted in Buckley, *The Jeweler's Eye* (New York, 1968), pp. 75–78.

16. For USAF opinions, see, for example, William Leavitt, "U.S. Power in the Space Cold War," *Air Force/Space Digest* (Apr. 1962): 39–46; "The Gilpatric Catalyst," *Space Daily,* 26 June 1962, pp. 751–52; "Secretary of the Air Force on Space Tasks," *Air Force Information Policy Letter for Commanders* 16 (1 June 1962): 1; Sen. Howard W. Cannon (D., Nev.), "Needed: A Space Platform for US Strength," *Space Digest* 5 (Oct. 1962): 51–53; "Air Force Renews Its Plea for Better Space Role," *Space Daily,* 4 Mar. 1963; "Air Force Space Program Gains," *Missiles and Rockets,* 19 Nov. 1962, p. 18; "Thinking Matures on Military's Space Role," *Aviation Week and Space Technology,* 22 July 1963: 209. For views of civilian strategists, see: Alton Frye, "The Military Danger: Our Gamble in Space," *Atlantic* (Aug. 1963): 46–50, and Frye, "Space Denial—An Amplification," *Space Digest* 5 (Oct. 1962): 54–56; Robison's views expressed later in "Self-Restrictions in the American Military Use of Space," *ORBIS* (Spring 1965): 116–39. The Goldwater quotation is from *Congressional Record, Senate,* 17 Oct. 1963, p. 18772. On Kennedy and the military relevance of Apollo, see Walt W. Rostow, *The Diffusion of Power: An Essay in Recent History* (New York, 1972), p. 184. The studies continued at the JCS level (Letter, W. F. Boone [NASA Deputy Associate Administrator for Defense Affairs] to LeMay, 16 Jan. 1964, NASA HO) even beyond the 1964 election, since Webb came to rely on the purported military value of Apollo more and more in NASA budget battles.

17. Edwin Diamond, *The Rise and Fall of the Space Age* (Garden City, N.Y., 1964), p. 1ff.

18. Amitai Etzioni, *The Moon-Doggle, Domestic and International Implications of the Space Race* (Garden City, N.Y., 1964), pp. xv, 99–112.

19. Ibid., pp. 107, 185, 191, *xv,* and 152. For another critique of Apollo, see William L. Crum, *Lunar Lunacy and Other Commentaries* (Philadelphia, 1965).

20. Fulbright in *Congressional Record,* 17 Oct. 1963, pp. 18277, 18777–78; Clark in *Congressional Record,* 20 Nov. 1963, p. 21344; Proxmire in *Congressional Record,* 4 Nov. 1963, pp. 19990–91. These and other citations are collected in Harold W. Babbit, "Priorities, Frugality, and the Space Race: A Preliminary Analysis of Congressional Criticism of Project Apollo," NASA Historical Note #41, Sept. 1964.

21. The President's News Conference of 24 Apr. 1963, *PP of P, John F. Kennedy 1963,* p. 350.

22. Remark made at a breakfast meeting of Republican congressmen, Washington, D.C., 12 June 1963.

23. Letter, Lovell to Dryden, 23 July 1963, NASA HO. Partial text of the letter together with replies from Webb and Dryden are published in Dodd L. Harvey and Linda C. Ciccoritti, *U.S.-Soviet Cooperation in Space* (Washington, D.C., 1974), pp. 114–19.

24. The President's News Conference of 17 July 1963, *PP of P, John F. Kennedy 1963,* pp. 567–68.

25. Dissent registered by Representative Thomas M. Pelly (R., Wash.).

26. The President's Address to the UN General Assembly, 20 Sept. 1963, *PP of P, John F. Kennedy 1963,* p. 695.

27. Dryden Memo for the Record, "Luncheon with Academician Blagonravov in New York, 11 Sept. 1963," 17 Sept. 1963, and McGeorge Bundy, Memo for the President, "Your 11 A.M. Appointment with James Webb," 18 Sept. 1963; both in NASA HO. On the origins of the Kennedy joint moon mission proposal, see Arthur M. Schlesinger, Jr., *A Thousand Days: John F. Kennedy in the White House* (Boston, Mass., 1965), pp. 919–21; Edward Welsh and Hugh Dryden Oral Histories, JFK Library.

28. James Webb Oral History, LBJ Library.

29. Kennedy's UN speech appeared in *Department of State Bulletin* (7 Oct. 1963), pp. 532–33. See also *PP of P, John F. Kennedy 1963,* pp. 693–98 (Moon, p. 695). The UN speech episode is ably recounted in Harvey and Ciccoritti, *U.S.-Soviet Cooperation in Space,* pp. 120–32; but see also James A. Malloy, Jr., "U.S.-USSR Space Negotiations and Cooperation, 1958–1965," NASA HHN-113, Sept. 1971, pp. 79–96; and Alton Frye, "The Proposal for a Joint Lunar Expedition: Background and Prospects," RAND P-2808 (Jan. 1964). In November, Kennedy issued NSAM #271, which instructed Webb to take personal initiative and responsibility for substantive cooperation in space with the USSR. NASA made studies, but it took two to tango.

30. Malloy, "U.S.—U.S.S.R. Space Negotiations and Cooperation, 1958–1965," p. 93.

31. Khrushchev's remarks in *Izvestia,* 26 Oct. 1963, but see also *New York Times* front page story, 27 Oct. 1963. On American reaction, see Joseph G. Whelan, "The Press and Khrushchev's 'Withdrawal' from the Moon Race," *Public Opinion Quarterly* 32 (Summer 1968): 233–50.

32. Congressional Record, 10 Oct. 1963, p. 18298, and U.S. Senate, Report 641, 88th Cong., 1st Sess., 13 Nov. 1963, p. 21. The campaign against a joint lunar mission was led by Congressman Thomas M. Pelly (R., Wash.). See Whelan, "The Press and Khrushchev's 'Withdrawal,' " pp. 247–49, and Babbit, "Priorities, Frugalities, and the Space Race," pp. 13–17.

33. Letter, Bush to Glennan, 18 Apr. 1960, NASA HO.

34. Letter, Bush to Webb, 11 Apr. 1963, HST Library, Webb Papers, Box 67.

35. Eisenhower, "Why I Am a Republican," *Saturday Evening Post,* 11 Apr. 1964, pp. 17ff. Quote from p. 18.

36. The latter suspicion was that of the political pathologist of the United States, Theodore H. White, in *The Making of the President, 1964* (New York, 1965), esp. pp. 359–60, 388–89.

37. Harold Brown Oral History; Clark Clifford Oral History; Eugene Rostow Oral History (Rusk cited by Rostow); all in LBJ Library.

38. Annual Message to the Congress on the State of the Union, 8 Jan. 1964, *PP of P, Lyndon B. Johnson 1964,* Doc. 91.

39. New York World's Fair 1964–65 Corporation, *New York World's Fair 1964–1965: Progress Reports* (New York, 1962–64), frontispiece to first pamphlet.

40. Ibid. Various pamphlets.

41. Daniel J. Boorstin, *The Image; Or What Happened to the American Dream* (New York, 1962), p. 294, referring to William J. Lederer and Eugene Burdick, *The Ugly American* (New York, 1958); C. S. Lewis, *Reflections on the Psalms* (New York, 1958), pp. 69–70.

42. Barry M. Goldwater, "A Realistic Space Program for America," *Science and Mechanics* (June 1964): 44–47, 100 ("too ludicrous" on p. 44; "lasers" on p. 100). See also Goldwater in Congressional Record—Senate, July 22, 1963, p. 12383: "We are moonstruck. And, to be sure, the Moon is most romantic. It's had sex appeal for centuries. But while our eyes are fixed upon it, we could lose the earth or be buried in it."

43. Memo, Horace Busby for Ed Welsh, 13 May 1964, LBJ Library, Aides Files—Horace Busby, Box 1304.

44. Memo, Welsh for the President, "Public Opinion Poll re Space," 24 Oct. 1964, LBJ Library, White House Central Files—Outer Space.

45. Remarks in St. Louis, Missouri, 21 Oct. 1964, *PP of P, Lyndon B. Johnson 1963–1964,* pp. 1396–1401. Quotes from pp. 1397–99.

46. Remarks at City Hall, Los Angeles, 28 Oct. 1964, *PP of P, Lyndon B. Johnson 1963–1964,* pp. 1492–99. Quotes from pp. 1493–94, 1495–96.

47. Ibid.

Part V Conclusion

1. Address at Rice University in Houston on the Nation's Space Effort, 12 Sept. 1962, *PP of P, John F. Kennedy 1962*, pp. 668–71. Quote from p. 669.
2. James Webb Oral History, LBJ Library.
3. Letter, Webb to the President, 30 Nov. 1964, HST Library, Webb Papers, Box 36.
4. Ibid.
5. Adlai E. Stevenson, "Science and Technology in the Political Arena," *Science and Society: A Symposium* (Xerox Corp., 1965), p. 4.
6. Lyndon B. Johnson, Interview with Walter Cronkite, "Man on the Moon: The Epic Journey of Apollo 11," CBS News, 21 July 1969.

Part VI

Headquotes: LBJ in *PP of P, Lyndon B. Johnson 1965*, 4 Dec. 1965, vol. 2, p. 1135; Kosygin in *Izbrannye rechi* (Moscow, 1969), p. 438; Price in *The Scientific Estate* (Cambridge, Mass., 1965), p. 278; Rubashov in Arthur Koestler, *Darkness at Noon* [1941] (Bantam Books ed. 1981), p. 153; Neil Armstrong's first words upon landing on the moon, 20 July 1969 (NASA quickly edited his sentence to read "small step for [a] man").

1. For Webb's knowledge of LBJ's plans to step down, see James Webb Oral History, p. 48, LBJ Library. For the vivid depiction of *Apollo 8*, see Richard S. Lewis, *From Vinland to Mars: A Thousand Years of Exploration* (New York, 1976), pp. 201–06.
2. *New York Times*, 8 Dec. 1965, interview with James Reston, cited by Nicholas Daniloff, *The Kremlin and the Cosmos* (New York, 1972), p. 151.
3. Norman Mailer, *Of a Fire on the Moon* (New York, 1969), esp. pp. 436–41. Quote from p. 441.
4. Text of Anderson remarks in the Senate, 10 Mar. 1970, LC, Anderson Papers, Box 921.

Chapter 20

1. Memo, Charles E. Johnson for Hayes Redmon (White House), 5 May 1966, LBJ Library, Legislative Background, Outer Space Treaty History.
2. Memo, Moyers to the President, 6 May 1966, including "Draft Statement for the President," and Memo, Rusk to the President, "An Approach to the Soviet Union on a Celestial Bodies Treaty": both in LBJ Library, White House Central File, Confidential File, Box 59.
3. Statement of Herbert Reis, Legal Adviser U.S. Mission to UN, in U.S. House of Representatives, Subcommittee on Space Science and Applications, *International Space Law*, 94th Cong., 1st Sess., July 1976, pp. 17–20. See also Telegram, Goldberg to Rusk, 22 Sept. 1965, LBJ Library, Legislative Background, Outer Space Treaty History; John Michael Kemp, "Evolution Toward a Space Treaty: An Historical Analysis," NASA HHN-64, Sept. 1966, pp. 98–105.
4. Memo, OSD, attached to Memo, State Department, "Principles for Inclusion in a Treaty on the Exploration of Celestial Bodies," 10 Nov. 1965, LBJ Library, Office Files of Joseph Califano, Celestial Bodies Treaty Negotiating History (hereafter Califano Treaty History).
5. Memo, Frutkin (NASA) to Meeker (State), 16 Nov. 1965; Memo, Wheeler (JCS) to McNamara, 24 Nov. 1965; and Letter, McNamara to Rusk, 11 Jan. 1966: all in LBJ Library, Califano Treaty History.
6. Letter, Leonard Meeker (State) to Frutkin (NASA), 27 Jan. 1966; Letter, Arthur W. Barber (Dep. Asst. Sec. of Def.) to Meeker, 14 Feb. 1966; Letter, Meeker to Frutkin, 16 Feb. 1966; Letter, Reis (US Mission to UN) to Meeker, 17 Feb. 1966; Memo, Vladimir Toumanoff

(State Dept. Soviet Desk) to Reis, 11 Mar. 1966; Telephone Message, Duane Anderson (DoD) to Reis, 11 Mar. 1966; Memo, Meeker to Rusk, "A United States Initiative for a Treaty Governing Activities on Celestial Bodies—Action Memo," 1 Apr. 1966; Memo, Walt Rostow for LBJ, 5 Apr. 1966, all in LBJ Library, Califano Treaty History.

7. For example, *New York Times,* 9 and 10 May 1966; *Washington Evening Star,* 10 May 1966.

8. Hines in *Washington Evening Star,* 12 May 1966.

9. Memo, Galloway, "Proposed Treaty on the Peaceful Exploration of the Moon and Other Celestial Bodies," 12 May 1966, NA, Senate Committee on Aeronautical and Space Sciences.

10. Gromyko's announcement in UN Document A/6341, 31 May 1966; Soviet draft treaty in UN Document A/6352, 16 June 1966; U.S. speculations in Letter, Goldberg to David H. Popper (State Dept.), 19 May 1966, LBJ Library, Legislative Background, Outer Space Treaty History.

11. Memo, State Dept., "Double Scenario for Negotiating the Celestial Bodies Treaty," 16 May 1966; Memo, Frutkin to Meeker, "Draft Treaty Governing the Exploration of the Moon and Other Celestial Bodies," 17 May 1966; Letter, Meeker to Frutkin, 18 May 1966; Memo, Thomas L. Hughes (State Dept., Dir. of Intell.) to Rusk, "Soviets Table Space Exploration Treaty Designed To Be Acceptable to US," 17 June 1966: all in LBJ Library, Legislative Background, Outer Space Treaty History. Instructions to Goldberg (fourteen in all) described in Draft Telegram, Meeker and Reis to Goldberg, circulated for executive branch clearance 22 June 1966, transmitted 6 July 1966; see also "omnibus text" for use of U.S. delegation, 3 July 1966, LBJ Library, Califano Treaty History.

12. Prior to my declassification of the original sources cited in subsequent notes, available references on the negotiation of the space treaty included Goldberg's Press Release #491, 16 Sept. 1966; also, Kemp, "Evolution Toward a Space Treaty," pp. 122–26; U.S. Senate, Committee on Aeronautical and Space Sciences, "Treaty on Principles Concerning the Activities of States in the Exploration and Use of Outer Space, Including the Moon and Other Celestial Bodies. Analysis and Background Data," 90th Cong., 1st Sess., Mar. 1967; Paul G. Dembling and Daniel M. Arons, "The Evolution of the Outer Space Treaty," *Journal of Air Law and Commerce* 33 (Jan. 1967): 419–56.

13. Telegram, U.S. Mission, Geneva to Rusk, #325, 19 July 1966; Telegram, Rusk to Geneva, #672, 20 July 1966; Telegrams, U.S. Mission, Geneva, to Rusk, #341 and #355, 20–21 July 1966. Agreements and remaining "splits" as of the morning of 21 July are listed, article by article, in Telegram, U.S. Mission, Geneva (Meeker) to Rusk, #354, 21 July 1966: all in LBJ Library, Califano Treaty History.

14. Telegram, Rusk to Goldberg, #836, 23 July 1966, LBJ Library, Califano Treaty History.

15. Telegrams, U.S. Mission, Geneva to Rusk, #571, #603, #624, 1–2 Aug. 1966, LBJ Library, Legislative Background, Outer Space Treaty History.

16. Letter, Douglas MacArthur II (Asst. Sec. of State for Congress. Relations) to Sen. Anderson, 17 Aug. 1966, LC, Anderson Papers, Box 915; State Dept. Memo, "Background Papers for Forthcoming Outer Space Talks in New York," 29 Aug. 1966; also Memo, Reis to Meeker, 8 Sept. 1966, suggesting "three 'final' proposals which we might make to the Soviets on access, reporting, and 'equal conditions'": all in LBJ Library, Califano Treaty History.

17. Telegrams, Goldberg to Rusk, #714, #725, #736, #737, #767, 12–14 Sept. 1966; also State Dept. Memo, "Soviet Proposal for a Tracking Facilities Provision in the Space Treaty and U.S. Counter-proposals," 16 June–7 Oct. 1966: all in LBJ Library, Califano Treaty History. See also Reis testimony in U.S. Senate, Committee on Aeronautical and Space Sciences, *International Cooperation in Outer Space: A Symposium,* 92nd Cong., 1st Sess., Dec. 1971, pp. 255–58.

18. Telegrams, Goldberg to Rusk, #1251 and #1256, 4 Oct. 1966, LBJ Library, Califano Treaty History. Reference to "subjects lying outside space field" from #1256, p. 4.

19. Telegrams, Goldberg to Rusk, #2185 and #2489, 9 and 19 Nov. 1966, LBJ Library, Califano Treaty History; Telegram, Moyers to the President, 8 Dec. 1966; Letter, Goldberg to Johnson, 20 Dec. 1966, and LBJ's minutes, all in LBJ Library, White House Central Files, U.S. Mission to UN, Box 59.

20. U.S. Senate, Committee on Foreign Relations, *Treaty on Outer Space. Hearings,* 90th Cong., 1st Sess., April 1967, pp. 108, 112.

21. Memo, Oscar E. Anderson, Jr. (NASA) to Frutkin, "Plans for Presentation of UN

Outer Space Treaty to the Senate Foreign Relations Committee," 5 Jan. 1967, reporting a briefing by Asst. Sec. State Joseph Sisco; also Memo, Frutkin to Meeker, "Comments on Substantive Articles, UN Outer Space Treaty," 12 Jan. 1967: both in NASA 71A 4083, Box 5; briefing booklet attached to "Statement for presentation by James E. Webb," Jan. 1967, NASA 71A 4083, Box 2.

22. U.S. Senate, *Treaty on Outer Space. Hearings,* pp. 1–14. Quotes from pp. 4, 12–13.

23. Ibid., pp. 16–17, 23–26.

24. Ibid., pp. 29–36. Quote from p. 30. When Fulbright tried to interpret Goldberg's words on this issue, Senator Gore laughingly interjected, "I must say, Mr. Chairman, I am a little more cautious now. I remember I accepted the Tonkin Bay [sic] resolution upon your explanation." The Tonkin Gulf resolution was used by the administration to justify sending ground forces to Vietnam.

25. Ibid., pp. 69–70.

26. Ibid., pp. 82–98. Quotes from pp. 82, 96.

27. Letters, James J. Gehrig to Senator Howard W. Cannon (D., Nev.) and Senator Anderson, 31 Mar. 1967, reporting the Executive Session of the Foreign Relations Committee of 6 Mar. 1967, LC, Anderson Papers, Box 919; Memo, John T. McNaughton (Asst. Sec. Def.) to Dep. Sec. Def., "The Space Treaty and Observation Satellites," 11 Mar. 1967, reporting that prohibition of spy satellites was never part of the treaty negotiations; Memo, Meeker to McNaughton, 23 Mar. 1967: both in LBJ Library, Legislative Background, Outer Space Treaty History; Letter, Gehrig to Anderson, 13 Mar. 1967, detailing U.S. positions on the meaning of "benefits and interests of all countries," the lack of a definition of outer space, the continued freedom for military R & D and the freedom of each signatory to define "threats to its security" as it wished. "The hearing," concluded Gehrig, "failed to bring out what I believe to be the most important aspect of the Treaty—that most of the policies established are U.S. policies. . . . In not too many years (remembering the Space Age is not yet 10 years old), many nations will be in space. It is to the advantage of the US to establish space law now." LC, Anderson Papers, Box 733.

28. "Statement on Treaty on Outer Space 1967—Elizabeth A. Kendall" in U.S. Senate, *Treaty on Outer Space. Hearings,* p. 159.

29. This view was eloquently expressed by Herbert Reis, Chief Counsel, U.S. Delegation to the UN (interview with author, 4 Apr. 1979), and by Michel Bourely, legal counsel to European Space Agency (interview with author, 12 Apr. 1979).

30. Arnold S. Levine, *Managing NASA in the Apollo Era,* NASA SP-4102 (Washington, D.C., 1982), p. 242.

31. For example, Letter, Webb to LBJ, 2 Oct. 1968, warning of imminent Soviet leadership in space if NASA budgets continued to slide (HST Library, Webb Papers, Box 91). On applications of NASA technology in other fields, see Letter, Webb to Maxwell Taylor, 16 Jan. 1964; Letter, Earl D. Hilburn (Dep. Assoc. Admin., NASA) to the NASA Centers, 8 Feb. 1966, with "Problems for Presentation to Symposium To Consider Technical Solutions for Vietnam War Problems"; Memo, NASA Office of Defense Affairs, "NASA Technology Which May Support DoD Limited Warfare Problems," 21 July 1966; Memo, R. D. Ginter (Defense Projects Support Office) to Colonel David V. Armstrong (Army Materiel Command), 25 Sept. 1968; Memo, James M. Beggs to H. Julian Allen (Director, Ames Laboratory), "Application of Sensor Technology," 9 Oct. 1968; Draft Report, C. P. Cabell, "NASA Benefits to National Security," 2 May 1967: all in NASA HO. On NASA's role in solving social and economic problems, see, for example, Memo, Webb for Seamans, 7 Dec. 1967, HST Library, Webb Papers, Box 42; and Address, Edward C. Welsh (Exec. Sec. NASC), "The Aerospace Revolution," 10 Apr. 1967, NASA HO.

32. President's Science Advisory Committee, "The Space Program in the Post Apollo Period," Feb. 1967; Space Science and Technology Panel, "U.S. Strategy for Space Research and Exploration," 19 Dec. 1967, cited by Levine, *Managing NASA,* p. 253.

33. President's Space Task Group, "The Post-Apollo Space Program: Directions for the Future," Sept. 1969, NASA HO. For a narrative of the decline in NASA funding, see Thomas P. Murphy, *Science, Geopolitics, and Federal Spending* (Lexington, Mass., 1971), pp. 363–89.

34. Richard S. Lewis, "The Kennedy Effect," *Bulletin of the Atomic Scientists* (Mar. 1968): 2–5.

35. Memo for Senator Anderson, 1 Nov. 1971, LC, Anderson Papers, Box 921.

36. John Logsdon, "The Policy Process and Large-Scale Space Efforts," *Space Humanization*

Series 1 (1979): 65–79, and "The Space Shuttle Decision: Technology and Political Choice" (consulted in ms.), Aug. 1978; David Baker, "Evolution of the Space Shuttle," *Spaceflight* (June–Sept. 1963); Claude E. Barfield articles in *National Journal*, 31 Mar., 12 and 19 Aug. 1972; U.S. House of Representatives, Committee on Science and Technology, Subcommittee on Space Science and Applications, *U.S. Civilian Space Programs 1958–1978*, 97th Cong., 1st Sess., 1981, pp. 445–56.

37. For de Gaulle's ideas on politics and technology, see especially his *Memoirs of Hope: Renewal and Endeavor* (New York, 1976), the collections of speeches in Ambassade de France, *Major Addresses, Statements, and Press Conferences of General Charles de Gaulle* (New York, 1964), and André Passeron, ed., *De Gaulle parle*, 2 vols. (Paris, 1962–66); also the following works on his foreign policy: Paul-Marie de la Gorce, *De Gaulle entre deux mondes* (Paris, 1964), and *La France contre les empires* (Paris, 1969); W. W. Kulski, *De Gaulle and the World* (Syracuse, 1966); John Newhouse, *De Gaulle and the Anglo-Saxons* (New York, 1970); and Paul Reynaud, *The Foreign Policy of Charles de Gaulle*, trans. Mervyn Savill (New York, 1964).

38. For example, de Gaulle declared early in 1958 that "I would quit NATO if I were running France.... NATO is no longer an alliance, it is a subordination," in C. L. Sulzberger, *The Last of the Giants* (New York, 1970), pp. 61–62. Although the official justification of the *force de frappe* was to provide France with a modern deterrent under its own control, Gaullist ministers invariably spoke of it as the only way for France to rejoin the ranks of the Great Powers, make itself heard in world councils, receive equal treatment in the Western alliance, and qualify for American nuclear aid. Even Raymond Aron, the influential political analyst and friend of NATO, saw the *force de frappe* as a "political trump" in dealings with the United States.

39. On French rocket development, see U.S. Congress, Committee on Science and Technology, Subcommittee on Space Science and Applications, *World Wide Space Programs*, 95th Cong., 1st Sess., 1977, pp. 142–57. On the origins and development of the Force de Frappe, see Wilfrid L. Kohl, *French Nuclear Diplomacy* (Princeton, N.J., 1971); Wolf Mendl, *Deterrence and Persuasion: French Nuclear Armament in the Context of National Policy, 1945–1969* (London, 1970); Bertrand Goldschmidt, *L'aventure atomique* (Paris, 1962); Lawrence Scheinmann, *Atomic Energy Policy in France under the Fourth Republic* (Princeton, N.J., 1965); Charles Ailleret, *L'aventure atomique française* (Paris, 1968). French nuclear research was well advanced in 1940 when the German conquest halted the work of the Curies. The Fourth Republic established the French atomic energy commission, which worked steadily toward the fabrication of weapons-grade plutonium, often without official blessing. In 1954 the cabinet of Pierre Mendès-France approved continuation of work leading to a bomb test. By that time French military thinkers already justified going nuclear as an economy move, "more force for the franc," in imitation of Eisenhower's massive retaliation; see, for example, Charles Ailleret, "L'arme atomique, arme à bon marché," *Revue de défense national*, 10 (1954): 315–25. Hence, when de Gaulle assumed power in 1958, he had only to go public with France's intention to build its own nuclear force—and vastly increase the funding.

On the social effects of Gaullist technocracy, see above all Robert Gilpin, *France in the Age of the Scientific State* (Princeton, N.J., 1968), and Michael L. Martin, *Warriors to Managers. The French Military Establishment Since 1945* (Chapel Hill, N.C., 1981).

40. De Gaulle, *Memoirs of Hope*, p. 135.

41. On the strategy and execution of French space programs in the 1960s, see Walter A. McDougall, "Space Age Europe: Gaullism, Euro-Gaullism, and the American Dilemma," *Technology and Culture* 26 (April 1985); U.S. Congress, *World Wide Space Programs*, pp. 139–70; Georges L. Thomson, *La Politique Spatiale de l'Europe*, 2 vols. (Dijon, 1976), vol. 1: *Les Actions Nationales*, chap. 1. The importance of military applications of space ancillary to a nuclear missile force was clearly expressed by Colonel Petkovsek, "L'utilisation militaire des engins spatiaux," *Revue militaire générale* (July 1961). French competitive strategy, especially within Europe, was admitted by Robert Aubinière, "Réalisations et projêts de la recherche spatiale française," *Revue de défense nationale* (Nov. 1967): 1736–49. See also "Programme spatial français jusqu'en 1965," *Le Figaro*, 28 July 1961; "Le débat sur le Centre d'Études spatiales va s'engager à l'Assemblé Nationale," *Le Monde*, 19 Oct. 1961; "More French Satellites after 'Diamond,'" *Daily Telegraph*, 1 June 1962; Kenneth Owen, "France's Space Programme: The Reasons Why," *Flight International*, 12 July 1962; Le Commandant L. Germain, "Le recherche spatiale en France," *Revue militaire d'information*

(Jan. 1963); Charles Cristofini (President of SEREB), "Planned Cooperation Is France's Aim," *Financial Times*, 10 June 1963.

42. De Gaulle, *Memoirs of Hope*, p. 136.

43. John L. Hess, "The Last Countdown," *New York Times*, 11 Feb. 1967.

44. *Le Monde* headline, 28 Nov. 1965.

45. C. Freeman and A. Young, *The Research and Development Effort in Western Europe, North America, and the Soviet Union* (Paris, 1965), p. 71; "Le programme pour la recherche scientifique," *Le Figaro*, 4 May 1961.

46. The idea of using Blue Streak as the first stage of a European space launcher seems to have originated with De Haviland engineer G. K. C. Pardoe: "European Cooperation in Space," *Spaceflight* (Jan. 1961): 5–8. David Price, "Political and Economic Factors Relating to European Space Cooperation," *Spaceflight* (Jan. 1962): 6–15 (Quote from p. 10); Kenneth Owen, "Europe's Future in Space," *Flight*, 6 July 1961, pp. 5–9.

47. SEREB and Hawker-Siddeley Aviation, *L'Industrie et l'Espace* (Industry and Space) (Paris, 1961), NASA 64A664, Box 9; Jean Delorme in EUROSPACE, *Proposals for a European Space Program* (Fontenay-s-bois, 1963), pp. 11–13, 96–97. Quote from p. 11.

48. Yves Demerliac, "European Industrial Views on NASA's Plans for the '70s," AAS Goddard Memorial Symposium, Washington, D.C., Mar. 1971 (Quotes from pp. 3, 4); see also the testimony of Eilene Galloway after interviews with European space officials in April 1967, LC, Anderson Papers, Box 919. On proposed European space research, see EUROSPACE, *Proposals for a European Space Program*, pp. 17–67; and U.S. Senate, Committee on Aeronautical and Space Sciences, *International Cooperation and Organization for Outer Space*, 89th Cong., 1st Sess., 1965, pp. 123–28. A fascinating proposal for a reusable "space transporter" or "shuttle" was spelled out in detail in EUROSPACE, *Aerospace Transporter* (Fontenay-s-bois, 1964). The preface, written by the aging Eugen Sänger, imagined the uses of such a vehicle as passenger transport, satellite launching and recovery, construction and supply of large, manned space stations and way stations between the earth and the moon, and the policing of orbital space. Thus Sänger anticipated the building of the Space Shuttle by a decade.

49. A. V. Cleaver, "European Space Activities Since the War: A Personal View," British Interplanetary Society Paper, 8 Mar. 1974. The United States welcomed ESRO but tried to persuade the Europeans to abandon ELDO as a duplication of American efforts. See also Lord Caldecote, "Problems of American European Space Cooperation," *Space Digest* 8 (1965): 66–69, and Jean Delorme, "European Space Policy," in the same issue, pp. 59–61.

50. See ELDO, *1960–1965: First Annual Report* (Brussels, 1965); ESRO, *First General Report, 1964–1965* (Paris, 1966); and *Annual Reports* of ESRO and ELDO (1966–), European Space Agency Library, Paris. European works on the frustrations of the 1960s include Jacques Tassin, *Vers l'Europe spatiale* (Paris, 1970); Thomson, *La politique spatiale de l'Europe*, vol. 2; Orio Giarini, *L'Europe et l'Espace* (Lausanne, 1968); Alain Dupas, *La lutte pour l'espace* (Paris, 1977), chap. 10.

51. EUROSPACE, *Towards a European Space Program* (Fontenay-s-bois, 1966), esp. pp. 32, 114–15; Michel Drancourt, *Les Clés de pouvoir* (Paris, 1964); Jean-Jacques Servan-Schreiber, *Le Défi américain* (Paris, 1967); Atlantic Institute, *The Technology Gap: United States and Europe* (London, 1970); articles by Hubert Humphrey, Richard Morse, and Harry G. Johnson, all entitled "The Technology Gap and the Brain Drain," in Edwin Mansfield, ed., *Defense, Science, and Public Policy* (New York, 1968), pp. 175–91; Richard R. Nelson, *The Technology Gap: Analysis and Appraisal*, RAND P-3694-1 (Santa Monica, Calif., 1967); Klaus-Heinrich Standke, *Europäische Forschungspolitik im Wettbewerb* (Baden-Baden, 1970); Pierre Vellas, *L'Europe face à la Révolution technologique américaine* (Paris, 1969); Norman J. Vig, *Science and Technology in British Politics* (Oxford, 1968); Roger Williams, *European Technology: The Politics of Collaboration* (London, 1973). McNamara reference in Williams, *European Technology*, p. 25; Brzezinski quotation in Memo, Webb to Frutkin, 22 June 1967, NASA 71A 4083, Box 5.

52. Events in Britain determined the ultimate fate of ELDO. Despite the role of Price, Massie, Thorneycroft, A. V. Cleaver, and other Britons in the founding of the organization, British governments remained cool toward space research, a symptom of Britain's general confusion about science and technology. Historically, the United Kingdom had the world's third highest R & D budget, while its industrial decline periodically raised alarms about the need for modern technology. Nevertheless, budgetary pressures and bungling seemed always to prevent a coherent R & D effort on the French model. Officials responsible for

space programs grew dizzy from a bureaucratic Irish jig that shifted them among nine different ministries at one time or another. Having cancelled its own missile programs, the British government revived scientific rocket research in 1964 and succeeded in launching a single homemade satellite on the Black Knight in 1971. Britain depended on the United States for strategic missiles and was increasingly willing to rely on NASA for access to space, earning the British representatives to European space conclaves the epithet "the delegates from the United States."

On British confusion over R & D policy after Sputnik, see Norman J. Vig, *Science and Technology in British Politics* (Oxford, 1968).

53. On the transition to ESA in European space politics, see the interim reports and discussions: Michel Bourely, *La Conférence spatiale européenne* (Paris, 1970); J. Henrici, *An Overall Coherent and Long Term European Space Program* (Munich, 1969); Théo Lefevre, *Europe and Space* (Brussels, 1972); Laurence Reed, *Ocean Space—Europe's New Frontier* (London, 1969); C. R. Turner, "A Review of the Third EUROSPACE U.S.-European Conference," *Spaceflight* (Jan. 1968): 7–9; T. H. E. Nesbitt, "Future US-European Cooperation in Space: Possibilities and Problems," *Spaceflight* (May 1969): 150–52; A. V. Cleaver, "The European Space Program, DISCORDE," *Aeronautics and Astronautics* (Oct. 1968): 80–85; Burl Valentine, "Obstacles to Space Cooperation: Europe and the Post-Apollo Experience," 21 May 1971, NASA HO; Kenneth Gatland, "A European Space Agency," *Spaceflight* (Oct. 1973): 92–93; ESRO, *Europe in Space* (Paris, 1974); ESA, *Space—Part of Europe's Environment* (Paris, 1979), and *Europe's Place in Space* (Paris, 1981), both in ESA Library; also the summary in U.S. Congress, *World Wide Space Programs*, pp. 285–314.

54. The growing competition in space markets attracted the attention of Congress in the early 1980s. See Office of Technology Assessment, *Civilian Space Policy and Applications* (Washington, D.C., 1982), and *Competition and Cooperation in Outer Space* (Washington, D.C., 1984). On competition between Ariane and the Shuttle, see Alain Dupas, *Ariane et la Navette Spatiale* (Paris, 1981).

55. The origins of Chinese missile research are shrouded, but Tsien Hsue-shen, the brilliant CalTech and JPL scientist, is assumed to have played a leading role after his departure for China in 1955. He was labeled a security risk after the Communist victory in China and denied an exit visa until his information on American rocketry was no longer current. After falling out with Khrushchev, the Chinese pushed their own research, testing ICBMs by the mid-1960s and launching their first satellite in 1970. Since that time Chinese missile and space programs appear to be devoted primarily to national defense.

Japan initially confided rocketry and space science to an academic group at the University of Tokyo, which also succeeded in orbiting a satellite, on a pencil-shaped Vanguard-type rocket, in 1970. But Japanese industry prevailed on the government to inaugurate a commercially motivated space program. Thanks to the purchase of American Delta technology and the fastest-growing space budget in the world in the 1970s, Japan has emerged as a growing competitor in space applications.

India is the third Asian country to opt for "vertical expansion," cooperating extensively with the USSR and the United States (on a prototype system for educational television by satellite). But India also insisted on developing its own rockets and launched its first satellite in 1980.

56. On Soviet space programs in the 1970s, see U.S. Senate, Commerce Committee, Subcommittee on Space Science and Applications, *Soviet Space Programs 1971–1975* and *1976–1980*, ed. Charles S. Sheldon II (Washington, D.C., 1976, 1981), and James E. Oberg, *Red Star in Orbit* (New York, 1981).

57. On the campaign against ABM beginning in 1964, see Freedman, *Evolution of Nuclear Strategy*, pp. 252–53. A succinct summary of the anti-ABM position was published by Herbert York and Jerome Wiesner, "National Security and the Nuclear Test Ban," *Scientific American* (Oct. 1964). The best treatment of SALT-I is John Newhouse, *Cold Dawn: The Story of SALT* (New York, 1973). Soviet publications on issues in space law dating from this period include A. S. Piradov, *International Space Law*, Science Translation Service (Santa Barbara, Calif., 1974), and E. G. Vasilevskaya, *Legal Problems of the Conquest of the Moon and Planets*, Science Translation Service (Santa Barbara, Calif., 1974).

58. All major agreements in space law are reproduced in U.S. Senate, Committee on Commerce, Science, and Transportation, *Space Law: Selected Basic Documents*, 95th Cong., 2nd Sess., 1978. On the ASTP, see Edward Clinton Ezell and Linda Newman Ezell, *The Partnership: A History of the Apollo-Soyuz Test Project*, NASA SP-4209 (Washington, D.C.,

1978). On détente in space generally, see J. C. D. Blaine, *The End of an Era in Space Exploration: From Competition to Cooperation* (San Diego, Calif., 1976).

59. Data on strategic systems from David Holloway, *The Soviet Union and the Arms Race* (New Haven, 1983), pp. 58–60; see also Stockholm International Peace Research Institute (SIPRI), *World Armaments and Disarmament: SIPRI Yearbook 1980* (London, 1980), pp. xlii–xlvi.

60. Bruce Parrott, *Politics and Technology in the Soviet Union* (Cambridge, Mass., 1983), pp. 187, 203. For a recent, insightful discussion of Soviet attitudes toward nuclear war, see Freeman Dyson, *Weapons and Hope* (New York, 1984), chaps. 15 and 20.

61. A. N. Kosygin, *Izbrannye rechi i stat'i* (Moscow, 1974), p. 438, and *Izvestia Akademii nauk, Seriia ekonomicheskaia* 4 (1970): 7, cited by Parrott, *Politics and Technology*, pp. 232, 247.

62. NSF, *Soviet R & D Statistics 1977–1980*, #80-SP-0727 (Washington, D.C., 1980), pp. 5, 14, 31–34; Holloway, *Soviet Union and the Arms Race*, p. 134. Various estimates of the military share of total Soviet R & D range from 40 to 80 percent. Holloway considers the 50 percent figure judicious.

63. *Materialy XXIV s''yezda KPSS* (Moscow, 1971), p. 46, cited by Holloway, *Soviet Union and the Arms Race*, p. 171.

64. On the importance of space systems in C^3I, see Paul Bracken, *The Command and Control of Nuclear Forces* (New Haven, 1983), esp. chap. 2. Discussions of the prospects for A-SAT weapons, space-based ABM, and earth-based beam weapons, their strategic and legal implications, include David A. Andelman, "Space Wars," *Foreign Policy* (Fall 1981): 94–106; Thomas Blau and Dan Gouré, "A Preface to Space Defense," *Comparative Strategy* 3 (1981): 135–49; B. Anthony Fessler, *Directed Energy Weapons: A Juridical Analysis* (New York, 1979); Lt. General Daniel O. Graham (Ret.), *High Frontier: A New National Strategy* (Washington, D.C., 1982), a Heritage Foundation proposal; Jerry Grey, *Beach-heads in Space. A Blueprint for the Future* (New York, 1983); Thomas Karas, *The New High Ground. Strategies and Weapons of Space-Age War* (New York, 1983); Geoffrey J. H. Kemp in Institute for Contemporary Studies, ed. W. Scott Thompson, *National Security in the 1980s: From Weakness to Strength* (San Francisco, 1980), pp. 69–88; Walter A. McDougall, "How Not To Think about Space Lasers," *National Review*, 13 May 1983; 550–56, 580–81; David Ritchie, *Space War* (New York, 1982); G. Harry Stine, *Confrontation in Space* (Englewood Cliffs, N.J., 1981); Barry L. Thompson, "'Directed Energy' Weapons and the Strategic Balance," *ORBIS* (Fall 1979): 697–709; Kosta Tsipis, "Laser Weapons," *Scientific American* (Dec. 1981): 51–57; United States Air Force Academy, *The Great Frontier: Military Space Doctrine*, 4 vols. (Colorado Springs, 1981), esp. articles by Barry J. Smernoff, Barry Watts, and Lance Lord in vol. 4. In *Weapons and Hope*, chaps. 6–7, Freeman Dyson argues the potential usefulness of ABM, space-based or otherwise, in an overall campaign to reduce dependence on nuclear deterrence. After all, disarmament and ABM ultimately are directed toward the same goal.

65. Current issues in space law may be followed in the *Journal of Space Law* published by the Institute of Air and Space Law, McGill Univ., Montreal. Summaries include Seyom Brown, et al., *Regimes for the Ocean, Outer Space, and Weather* (Washington, D.C., 1977), a Brookings Institution study; Center for Research of Air and Space Law, *Space Activities and Implications: Where From and Where To at the Threshold of the '80s* (Montreal, 1981); Nicholas M. Matte, *Aerospace Law* (Montreal, 1977); Gijsbertha C. M. Reijnen, *Legal Aspects of Outer Space* (Utrecht, 1977); and Irvin L. White, *Decision-Making for Space: Law and Politics in Air, Sea, and Outer Space* (West Lafayette, Ind., 1970). See also U.S. Congress, Office of Technology Assessment, *UNISPACE '82: A Context for International Cooperation and Competition,* OTA-TM-ISC-26 (Washington, D.C., 1983); *Report of the Second United Nations Conference on the Exploration and Peaceful Uses of Outer Space,* A/CONF.101/10 (1983); U.S. Congress, Committee on Science and Technology, *United States Civilian Space Programs,* 97th Cong., 1st Sess. (January 1981), pp. 5–28.

On the moon treaty, initialed under Jimmy Carter but unratified, see U.S. Senate, Committee on Commerce, Science, and Transportation, *Agreement Governing the Activities of States on the Moon and Other Celestial Bodies,* 96th Cong., 2nd Sess., 1980, Pt. 4.

Chapter 21

1. Daniel Bell, ed., and the Commission on the Year 2000, *Toward the Year 2000* (Boston, 1968); Hermann Kahn and Anthony J. Wiener, *The Year 2000: A Framework for Speculation on the Next Thirty-Three Years* (New York, 1967). On the origins, evolution, and pretensions of futurology, see Edward Cornish, *The Study of the Future: An Introduction to the Art and Science of Understanding and Shaping Tomorrow's World* (Washington, D.C., 1977), a publication of the World Future Society. The roots of futurology (as opposed to the occasional concoction of utopias) seem to rest in the 1920s, when S. Colum Gilfallin, Bertrand Russell, and others contemplated the prospects of taming technology after the excesses of World War I destroyed the prior expectation of natural human progress. Gilfallin coined the term "mellontology" for the study of the future; H. G. Wells dreamed of a science of the future and inspired the British *Tomorrow: The Magazine of the Future.* After World War II French existentialists, especially Jean-Paul Sartre, preached perfect human freedom to choose and shape one's own future without regard for any system of ethics. The past was dead; history had failed. Physicist Ennis Gabor later capsulized this attitude in the phrase "inventing the future," which futurists then adopted as their own. Clearly the RAND Corporation and Hudson Institute, among others, represented an official departure into futurology, but in the technocratic 1960s self-styled futurist writers and organizations proliferated. Chief among them was the Institute for the Future in Menlo Park, California, but traditional think tanks joined the movement and helped to give it a modicum of respectability, as well as a more or less accepted methodology. Major works dating from this short period give an indication of the popularity of the genre: Robert U. Ayres, *Technological Forecasting and Long-Range Planning* (New York, 1969); Kurt Baier and Nicholas Rescher, eds., *Values and the Future* (New York, 1969); Burnham P. Beckwith, *The Next 500 Years: Scientific Predictions of Major Social Trends* (New York, 1967); Adrian Berry, *The Next Ten Thousand Years: A Vision of Man's Future in the Universe* (New York, 1974); James R. Bright, ed., *Technology: Forecasting for Industry and Government* (Englewood Cliffs, N.J., 1968); Arthur B. Bronwell, ed., *Science and Technology in the World of the Future* (New York, 1970); Marvin J. Cetron, *Technological Forecasting: A Practical Approach* (New York, 1969); Stuart Chase, *The Most Probable World* (New York, 1968); Gerald Feinberg, *The Prometheus Project: Mankind's Search for Long-Range Goals* (New York, 1969); Victor Ferkiss, *The Future of Technological Civilization* (New York, 1971); Jay W. Forrester, *World Dynamics* (Cambridge, Mass., 1971); Dennis Gabor, *Inventing the Future* (New York, 1964); Olaf Helmer, *Social Technology* (New York, 1966); Robert Jungk and Johan Galtung, eds., *Mankind 2000* (London, 1969); Hermann Kahn and B. Bruce-Briggs, *Things To Come: Thinking about the Seventies and Eighties* (New York, 1972); John McHale, *The Future of the Future* (New York, 1969); Donald N. Michael, *The Unprepared Society: Planning for a Precarious Future* (New York, 1968); Harvey Perloff, ed., *The Future of the U.S. Government: Toward the Year 2000* (New York, 1971); Fred L. Polak, *Prognostics: A Science in the Making Surveys and Creates the Future* (Amsterdam, 1971); Robert W. Prehoda, *Designing the Future: The Role of Technological Forecasting* (Philadelphia, 1967); Alvin Toffler, ed., *The Futurists* (New York, 1972).

2. On the "Big L-systems" and continental systems integration, see Peter Pringle and William Arkin, *SIOP, The Secret U.S. Plan for Nuclear War* (New York, 1983), pp. 209–19. Analysis and critique of these systems is the theme of Paul Bracken, *The Command and Control of Nuclear Forces* (New Haven, 1983). On Big L-systems, see pp. 15–17, 186–88.

3. Memo, Seamans, "AFSC Briefing (by General Schriever)," 13 Mar. 1964, NASA HO; "Schriever Urges Bold Approach to Future (Project Forecast)," *Armed Forces Management* (May 1965) (source of quote).

4. U.S. Congress, Committee on Aeronautical and Space Sciences, *The National Space Program—Its Values and Benefits*, 90th Cong., 1st Sess., 1967, p. 2.

5. Richard J. Barber, *The Politics of Research* (Washington, D.C., 1966), pp. 1–24; H. L. Nieburg, *In the Name of Science* (Chicago, 1966), pp. 114–16; and Don K. Price, "The Scientific Establishment," in Robert Gilpin and Christopher Wright, eds., *Scientists and National Policy Making* (New York, 1964), pp. 19–40, all mourn that money for science was appropriated almost on faith. The "R & D Cult" is Nieburg's phrase.

6. Barber, *Politics of Research,* pp. 24–31. Econometricians were still unhappy with efforts to construct models to measure the impact of R & D. See Z. Griliches, "Issues in Assessing the Contributions of R & D to Productivity Growth," *Bell Journal of Economics* 10 (Spring 1974): 92–116.

7. Nieburg, *In the Name of Science,* pp. 71–74.

8. On computers, see James E. Tomayko, "NASA's Impact on Computer Development," Wichita State Univ. (1980), consulted in manuscript, NASM; on the studies, see Mary A. Holman and Theodore Suranyi-Unger, Jr., "The Political Economy of American Astronautics," in Frederick C. Durant III, ed., *Between Sputnik and the Shuttle: New Perspectives on American Astronautics* (San Diego, 1981), pp. 176–78, and Mary A. Holman, *The Political Economy of the Space Program* (Palo Alto, Calif., 1974), p. 107; on the difficulties of implementing vast reforms (for instance, agricultural or ecological) based on satellite data, see Sir Bernard Lovell, *The Origins and International Economics of Space Exploration* (Edinburgh, 1973), pp. 83–84. The 1976 study was done by Chase Econometrics Associates and the 1967 space station study by the Planning Research Corporation. Some idea of the vagaries of measuring the spin-off phenomenon is conveyed by the earnest efforts of the Denver Research Institute, *Mission-Oriented R&D and the Advancement of Technology: Impact of NASA Contributions, Final Report* (Denver, 1974). Its authors foreswore the usual practice of starting with NASA-derived innovations, then tracing their commercial applications. Instead they worked backward from a sample of major technical developments in the general economy to their origins (if any) in the space program. This method, no worse than any other, still required subjective judgments in the selection of fields of technology to study, identification of "major" developments, identification and characterization of NASA contributions, and, finally, assessment of the importance of those contributions and of the technology itself for the economy.

The authors stated up front that no satisfactory solution had ever been devised to the problem of measuring innovation and transfer. They drew on cost-benefit analysis but ran into "fundamental conceptual problems" in estimating how much of a given "benefit" could be credited to NASA's account. Chief among these was that no person, group, or organization worked in a vacuum. Rather, several groups were likely to be at work on aspects of a research problem in an interlocking, mutual relationship. This led the authors into network theory and marginal analysis. But results were so imprecise that the team could in good conscience speak only of "high, moderate, and low levels of influence." These terms were themselves subjective, of course, and based on "weighted average for actual and potential impact," derived in part from "least squares regression analysis."

Not surprisingly, the authors reached timid conclusions. NASA contributions seemed broader, more complex, and more indirect than previously thought. They were concentrated in military and aerospace applications—only one-quarter found commercial application. The economic impact of NASA contributions was judged to be "moderate," the scientific impact "moderate-to-low," and the direct social impact "low."

9. Holman, *Political Economy of the Space Program,* chaps. 5 and 9; Nieburg, *In the Name of Science,* pp. 77–84. Panofsky quote from p. 78.

10. U.S. Senate, Select Small Business Subcommittee, *Hearings on Economic Aspects of Government Patent Policies,* 88th Cong., 1st Sess., 1963, p. 23. Economist Walter Adams termed this "neomercantilism" and "industrial feudalism" in "The Military-Industrial Complex and the New Industrial State," *American Economic Review* 58 (May 1968): 652–65. But was it new? The "corruption" of the free market by large military expenditures is a phenomenon dating back to the Revolutionary War itself. In the 1930s the so-called Nye Commission (Senator Gerald P. Nye) exposed war profiteering in World War I but feared that a legislative cure would, in case of another war mobilization, prove worse than the disease by supplanting free enterprise altogether. See Paul Koistinen, *The Military-Industrial Complex. A Historical Perspective* (New York, 1980), pp. 54–61. What then is novel about the Space Age? It is the "bastardization" of the economy in *peacetime* and the active role of government in "fathering the bastard" through contract R & D on an unprecedented scale.

11. See Robert E. Bickner, *The Changing Relationship Between the Air Force and the Aerospace Industry,* RAND RM-4101-PR (July 1964); M. J. Peck and F. M. Scherer, *The Weapons Acquisition Process: An Economic Evaluation* (Boston, 1962); Herman O. Stekler, *The Structure and Performance of the Aerospace Industry* (Berkeley, Calif., 1965); Stanford Research Institute, *The Industry-Government Aerospace Relationship* (Menlo Park, Calif.,

1963); Editors of *Fortune, The Space Industry: America's Newest Giant* (Englewood Cliffs, N.J., 1962).

12. J. Fred Weston, ed., *Defense-Space Market Research* (Cambridge, Mass., 1964), chaps. 1 and 4; Murray Weidenbaum in Edwin Mansfield, ed., *Defense, Science, and Public Policy* (New York, 1968), p. 32; Charles D. Bright, *The Jet Makers* (Lawrence, Kans., 1978), pp. 59–60, 70–75; J. Stefan Dupre and Sanford A. Lakoff, *Science and the Nation: Policy and Politics* (Englewood Cliffs, N.J., 1962), p. 174. Nieburg provides a lengthy "catalog of atrocities," including examples of "duplication, confusion, and do-nothing exercises, overstated costs, unconscionable hidden profits, instances of deliberate fraud, collusive price-fixing, overruns, over-complication of products, egregious technical errors, and company—as well as government—mismanagement"; see Nieburg, *In the Name of Science*, pp. 268–87.

13. Bright, *The Jet Makers*, pp. 133–47; Denver Research Institute, *Defense Industry Diversification: An Analysis*, by John S. Gilmore (Denver, 1964). Two documents from Webb's NASA papers illustrate the federal agency's role. In Letter, Webb to Attorney General Ramsay Clark, 25 Apr. 1967, he discusses the condition of the aerospace industry, its need for work, and the necessity for mergers such as the current McDonnell-Douglas one (HST Library, Webb Papers, Box 41). In Letter, Webb to J. S. McDonnell, 21 Aug. 1967, Webb suggests to the firm's chairman the capabilities NASA would like to see McDonnell-Douglas develop (HST Library, Webb Papers, Box 50).

14. H. L. Nieburg, "R & D and the Contract State: Throwing Away the Yardstick," *Bulletin of Atomic Scientists* (Mar. 1966): 20–24; Nieburg, *In the Name of Science*, chaps. 10–12. See also the relatively uncritical view of the government-industry relationship in Thomas P. Murphy, *Science, Geopolitics, and Federal Spending* (Lexington, Mass., 1971), pp. 51–64.

15. U.S. Congress, Committee on Science and Astronautics, *Hearings, Space Posture*, 88th Cong., 1st Sess., 1963, p. 7.

16. Lloyd V. Berkner, *The Scientific Age: The Impact of Science on Society* (New Haven, 1964), pp. 29–30.

17. Michael Reagan, *Science and the Federal Patron* (New York, 1969), p. 322, table 4; Dupre and Lakoff, *Science and the Nation*, chap. 3; Murphy, *Science, Geopolitics, and Federal Spending*, pp. 111–16.

18. Amitai Etzioni, *The Moon-Doggle. Domestic and International Implications of the Space Race* (Garden City, N.Y., 1964), p. 68; Reagan, *Science and the Federal Patron*, pp. 64–70. Denver Research Institute, *Effects of a National Space Program on Universities* (Denver, 1968), a NASA-sponsored study, praised federal largesse but found it more valuable to the universities than to government! See also Thomas W. Adams and Thomas P. Murphy, "NASA's University Research Programs: Dilemma and Problems at the Government-University Interface," *Public Administration Review* 27 (Mar. 1967).

19. U.S. Congress, Committee on Government Operations, Research and Technology Programs Subcommittee, *The Use of Social Research in Federal Domestic Programs*, 90th Cong., 1st Sess., 1967, Pt. IV, pp. 34–35, 60, cited by Reagan, *Science and the Federal Patron*, p. 168.

20. Cited by Reagan, *Science and the Federal Patron*, p. 188. Of course, the Soviets had already preached the same message. The party program of the Twenty-Second Party Congress noted: "There must be intensive development of research work in the *social sciences*, which constitute the scientific basis for the guidance of the development of society." See N. S. Khrushchev, *Program of the Communist Party of the Soviet Union* (New York, 1963), p. 129.

21. Bertrand de Jouvenel, "The Political Consequences of the Rise of Science," *Bulletin of the Atomic Scientists* (Dec. 1963): 2–8. Quote from p. 8.

22. For instance, Ralph E. Lapp, *The New Priesthood* (New York, 1965); Don K. Price, *The Scientific Estate* (Cambridge, Mass., 1965); Sanford Lakoff, *Knowledge and Power: Essays on Science and Government* (New York, 1971); Harvey Brooks, *The Government of Science* (Cambridge, Mass., 1968); and Nieburg, *In the Name of Science*, chaps. 7 and 13.

23. Nieburg, *In the Name of Science*, p. 118.

24. Robert C. Wood, "Scientists and Politics: The Rise of an Apolitical Elite," and Werner R. Schilling, "Scientists, Foreign Policy, and Politics," in Gilpin and Wright, *Scientists and National Policy-Making*, pp. 41–72, 144–73.

25. Hanson W. Baldwin, "Slow Down in the Pentagon," in Mansfield, *Defense, Science, and Public Policy*, p. 62.

26. Nieburg, *In the Name of Science*, pp. 340–44.

27. Ibid., pp. 101–02; Reagan, *Science and the Federal Patron*, p. 279; National Academy of Sciences, "Technology Assessment," in John G. Burke, ed., *The New Technology and Human Values*, 2nd ed. (Belmont, Calif., 1972), pp. 78–90.

28. The Princeton study, headed by Klaus Knoor and Oskar Morgenstern, cited by Charles Hitch, "The Case for Cost-Effectiveness Analysis," in Mansfield, *Defense, Science, and Public Policy*, p. 94; RAND analysis cited in Bickner, *Changing Relationship between the Air Force and Aerospace*, chap. 3.

29. Interview with James Webb by John Logsdon, 1 Sept. 1970, NASA HO.

30. Carl Kaysen, "Improving the Efficiency of Military R&D," in Mansfield, *Defense, Science, and Public Policy*, pp. 114–22; Nieburg, *In the Name of Science*, pp. 78, 226; Bernard Brodie, "The Scientific Strategists," in Gilpin and Wright, *Scientists and National Policy-Making*, pp. 240–56; John S. Foster, Jr., "Basic Research and the Military," in Burke, *New Technology and Human Values*, pp. 153–58.

31. A. Hunter Dupree, "A New Rationale for Science," *Saturday Review*, 7 Feb. 1970; Holman and Suranyi-Unger, "Political Economy of American Astronautics," pp. 168–69.

32. National Science Foundation, *Federal Funds for R&D*, NSF 79-310, vol. 32 (Washington, D.C., 1979); National Science Board, *Scientific Indicators 1978* (Washington, D.C., 1979); National Science Foundation, *Soviet R&D Statistics 1977–1980* (Washington, D.C., 1980); Merrill Lynch, Pierce, Fenner, & Smith, "Aerospace/Defense 1982: A Merrill Lynch Industry Review" (New York, 1982).

33. The perception of a New Class consisting of bureaucrats, consultants, teachers, professors, journalists, social workers, federal, state and municipal office workers, professional advocates for "victimized" minorities, many doctors and lawyers, and all others engaged in "knowledge industries" that rely on government expenditure or regulation rather than private risk-taking is most often associated with the name of Daniel Bell, *The Coming of Post-Industrial Society* (New York, 1973), and the *Cultural Contradictions of Capitalism* (New York, 1976). The problem is capsulized in his essay "The New Class: A Muddled Concept," in *The Winding Passage: Essays and Sociological Journeys, 1960–1980* (New York, 1980), pp. 144–64.

34. For a lament and critique of the backlash, see Samuel Florman's eloquent briefs on behalf of the engineering profession, *The Existential Pleasures of Engineering* (New York, 1976) and *Blaming Technology: The Irrational Search for Scapegoats* (New York, 1981).

35. Nieburg, *In the Name of Science*, p. 62.

36. David Ehrenfeld, *The Arrogance of Humanism* (New York, 1978), pp. 44–54, 104–25, describes the myths and realities of environmental engineering and includes examples of counterproductive applications of technology. The conflict of values between economic growth and environmental health is described in Burke, *New Technology and Human Values*, pp. 47–90.

37. Alvin Toffler, *Future Shock* (New York, 1970); Donella Meadows, et al., *The Limits to Growth: A Report for the Club of Rome's Project on the Predicament of Humanity* (New York, 1972); Jeremy Rifkin, *Entropy: A New World View* (New York, 1980).

38. Ehrenfeld, *Arrogance of Humanism*, pp. 116–18.

39. C. West Churchman, *The Systems Approach* (New York, 1968), p. 11.

40. Ehrenfeld, *Arrogance of Humanism*, p. 122.

41. Churchman, *The Systems Approach*, pp. 11–15.

42. Ibid., p. 31. How can a "value-free" systems approach exist? It cannot, but Churchman also recognizes "this century's strenuous effort to keep the deeper problems of values out of social system design"; see C. West Churchman, *The Systems Approach and Its Enemies* (New York, 1979), p. 47. The appeal of analytical models to the technocratic statesmen is precisely its gift of an apparently scientific rationale for action.

43. Franz Kafka, *The Castle* [1926] (New York, 1930); Mary Wollstonecraft Shelley, *Frankenstein, or the Modern Prometheus* [1818] (London, 1831); Karel Čapek, *R.U.R.* (Rossum's Universal Robots) [1920] (New York, 1930); George Orwell, *1984* (London, 1948); Aldous Huxley, *Brave New World* [1932] (New York, 1946); C. S. Lewis, *That Hideous Strength* [vol. 3 of a trilogy including *Out of the Silent Planet* and *Perelandra*] (London, 1946); J. R. R. Tolkien, *The Lord of the Rings*, 3 vols. (London, 1954–55).

44. Waddington in Philip M. Boffey, "American Science Policy: OECD Publishes a Massive Critique," *Science*, 12 Jan. 1968, pp. 176–78, cited by Reagan, *Science and the Federal Patron*, p. 273.

45. Churchman, *Systems Approach and Its Enemies*, p. 113, illustrates this with Kant's Fourth Antimony: "Thesis: there exist absolutely necessary conditions, some of which can be known, for there to be a betterment of the human condition. Antithesis: The necessary conditions for the betterment of the human condition either cannot be known or do not exist."

46. Ibid., pp. 32–47, 97–99.

47. Bukharin wrote in 1931:

> The crisis of bourgeois consciousness goes deep, and traces out marked furrows; on the whole front of science and philosophy we have gigantic dislocations which have been excellently formulated (from the standpoint of their basic orientation) by O. Spann: the main thing is a war of destruction against materialism. This is the great task of culture, in the opinion of the warlike professor, who protests against knowledge without God and knowledge without virtue.... It is not surprising, therefore, that from any scientific hypothesis quasi-philosophic (essentially religious) conclusions are being drawn, and on the extreme and most consistent wing there is openly being advanced the watchword of *a new medievalism.* (Essays by N. I. Bukharin, et al., *Science at the Crossroads. Papers Presented to the Second International Congress of the History of Science and Technology* [1931] (London, 1971), pp. 32–33. Italics in original.)

48. Ehrenfeld, *Arrogance of Humanism*, pp. 16–17.

49. This line is from B. F. Skinner's *Walden Two*, cited by Ehrenfeld, *Arrogance of Humanism*, p. 30.

50. Mailer, *Of a Fire on the Moon*, p. 441.

51. Churchman, *Systems Approach and Its Enemies*, pp. 21–22, 111–12.

52. Lewis Mumford, *The Myth of the Machine*, vol. 2: *The Pentagon of Power* (New York, 1970), pp. 303–11.

Chapter 22

Headquotes: C. S. Lewis, *The Chronicles of Narnia*, vol. 3, *The Voyage of the "Dawn Treader"* (New York, 1952), p. 152; Psalm 9:20 from the Authorized (King James) Version.

1. Alexander Gerschenkron, *Economic Backwardness in Historical Perspective. A Book of Essays* (Cambridge, Mass., 1962), pp. 16–21.

2. "If the arrangement of society is bad (as ours is), and a small number of people have power over the majority and oppress it, every victory over nature will inevitably serve only to increase that power and that oppression." So wrote Leo Tolstoy in "The Superstitions of Science," *The Arena*, 20 (1898): 52–60 (Quote from p. 59). Wrote Bertrand Russell:

> Brief and powerless is Man's life; on him and all his race the slow, sure doom falls pitiless and dark. Blind to good and evil, reckless of destruction, omnipotent matter rolls on its relentless way; for Man, condemned to-day to lose his dearest, to-morrow himself to pass through the gate of darkness, it remains only to cherish, ere yet the blow falls, the lofty thoughts that ennoble his little day; disdaining the coward terrors of the slave of Fate, to worship at the shrine that his own hands have built; undismayed by the empire of chance, to preserve a mind free from the wanton tyranny that rules his outward life ... to sustain alone, a weary but unyielding Atlas, the world that his own ideals have fashioned despite the trampling march of unconscious power (Bertrand Russell, *Mysticism and Logic* [London, 1903], cited by Louise B. Young, ed., *Exploring the Universe* [New York, 1963], pp. 344–45).

Aldous Huxley perceived that the "belief in all-round progress" was "based upon the wishful dream that one can get something for nothing." Its underlying assumption is that gains in one field do not have to be paid for by losses in other fields. For the ancient Greeks, hubris, or overweening insolence, whether directed against the gods, one's fellow men, or nature, was sure to be followed, sooner or later, in one way or another, by avenging Nemesis. Unlike the Greeks, we of the twentieth century believe that we can be insolent with impunity; see Aldous Huxley, *Science, Liberty, and Peace* (New York, 1946), pp. 419–20.

Similar quotations could be gleaned from Hilaire Belloc, G. K. Chesterton, C. S. Lewis, and others.

3. An interesting panel discussion on this topic is *Why Man Explores,* NASA EP-125, a symposium held at the Beckman auditorium, Caltech, 2 July 1976, Donald P. Hearth, moderator, featuring James Michener, Norman Cousins, Philip Morrison, Jacques Cousteau, and Ray Bradbury.

4. Daniel J. Boorstin, *The Republic of Technology: Reflections on Our Future Community* (New York, 1978); John William Ward, "The Meaning of Lindbergh's Flight," *American Quarterly* 10 (1958): 3–16.

5. See Herbert Butterfield, *The Origins of Modern Science* (New York, 1957), pp. 197–202. For an introduction to this much-debated subject, see Lynn White, Jr., *Machina ex Deo: Essays in the Dynamism of Western Culture* (Cambridge, Mass., 1968), and Lewis Mumford, *The Myth of the Machine,* vol. 1: *Technics and Human Development* (New York, 1967).

6. Dean E. Wooldridge, *Mechanical Man: The Physical Basis of Intelligent Life* (New York, 1968), p. 167.

7. Ibid., pp. 183 (free will), 190 (personal God), 192–93 (government), 204 (final quote).

8. Lewis Mumford, *The Myth of the Machine,* vol. 2: *The Pentagon of Power* (New York, 1964), p. 56, citing Fuller.

9. Bruce Mazlish, "The Fourth Discontinuity," *Technology and Culture* 8 (Jan. 1967): 1–15.

10. C. West Churchman, *Systems Approach and Its Enemies* (New York, 1979), p. 214.

11. Edouard Bornecque-Winandye, *Droit de l'Impérialisme Spatial* (Paris, 1962), and Alex Roudène, *Contre la Conquête de l'Espace* (Nancy, 1967).

12. Huxley, *Science, Liberty, and Peace,* p. 421. Emphasis added.

13. Leo J. Haigerty, ed., *Pius XII and Technology* (Milwaukee, 1962).

14. Pope John XXIII, "Pacem in Terris," reprinted in *Congressional Record—Senate,* 11 Apr. 1963, pp. 6116–25.

15. Wernher von Braun, text of Commencement Address, St. Louis Univ., 3 June 1958, pp. 9–10, LC, Von Braun Papers, Box 46.

16. Editors of *Fortune, The Space Industry,* pp. x–xi.

17. James A. Pike, "Religious Responsibility in the Space Age," in Lillian Levy, ed., *Space: Its Impact on Mankind* (New York, 1965), pp. 185–87.

18. NASA Historical Note #25, "The Impact of Space on Religion," comp. Martha Wheeler George, Aug. 1963, NASA HO.

19. Anis Mansour, editor of *Al Akhbar* (Cairo), cited in the *Washington Post,* 15 Oct. 1962.

20. Editors of *Fortune, The Space Industry,* p. x.

21. Carl Sagan, *The Cosmic Connection. An Extraterrestrial Perspective* (New York, 1973), p. viii. The cosmic perspective that conjured for many the image of earth as of infinitesimal significance was popularized as much as anyone by French astronomer and popular science writer Camille Flammarion. See, for instance, his *La fin du monde,* translated as *Omega: The Last Days of the World* (New York, 1894).

Kant surmised in 1755 that the fainter points of light in the night sky, infinitely remote, must be elliptical galaxies on the order of our own Milky Way. In the 1920s Edwin Hubble and Harlow Shapley confirmed this with the new 100-inch telescope at Mt. Wilson, California, and popularized the notion that the earth was "nothing but" a grain of sand circling a humdrum star out in the suburbs of the galaxy, and thus of minute importance in the material universe.

22. Sagan, *The Cosmic Connection,* pp. 3–8ff.; Arthur C. Clarke, *Report on Planet Three and Other Speculations* (New York, 1972), p. 40.

23. W. Norris Clarke, S.J., "Technology and Man: A Christian Vision," *Technology and Culture* 3 (Fall 1962): 422–42. Clarke was Professor of Philosophy at Fordham University.

24. Martin J. Heinecken, *God in the Space Age* (Philadelphia, 1959), pp. 188–89.

25. "Canterbury Thinks God May Doom Man," *New York Times,* 14 July 1958. An American cleric, Monsignor Ronald Knox, writing in 1946 from an orthodox position, also saw the dangers of great and concentrated power in the hands of a fallen race. The promise of technology (he was thinking primarily of atomic power) was in fact its greatest danger in a world that "has very largely gone barbarian"; see Ronald Knox, *God and the Atom* (New York, 1945), p. 163.

26. Eugene S. Ferguson, "Toward a Discipline of the History of Technology," *Technology and Culture* 15 (Jan. 1974): 13–30. Quotes from pp. 20, 23.

27. Emmanuel G. Mesthene, "Some General Implications of the Research of the Harvard

University Program on Technology and Society," *Technology and Culture* 10 (Oct. 1969): 489–513. Quote from p. 498.

28. James R. Killian, *Sputniks, Scientists, and Eisenhower: A Memoir of the First Special Assistant to the President for Science and Technology* (Cambridge, Mass., 1976), pp. 261–62.

29. William Grosvenor Pollard, *Chance and Providence: God's Action in a World Governed by Scientific Law* (New York, 1958).

30. Mumford, *The Pentagon of Power*, p. 57.

31. William James, *The Varieties of Religious Experience* (New York, 1902), p. 519.

32. Bruce Mazlish, *The Railroad and the Space Program: An Exploration in Historical Analogy* (Cambridge, Mass.), introduction, esp. pp. 11–14.

33. Cox cited by Mesthene, "Some General Implications," p. 503.

34. Homer E. Newell, *Beyond the Atmosphere. Early Years of Space Science*, NASA SP-4211 (Washington, D.C., 1980), pp. 362–63.

35. William Grosvenor Pollard, *Physicist and Christian* (New York, 1961), pp. 104–06.

36. Ibid., p. 110. But University of Chicago philosopher Mortimer J. Adler argues in *How To Think about God. A Guide for the 20th Century Pagan* (New York, 1980) that the "radical contingency" of the universe—its demonstrable need for an outside force sustaining its continued existence at every moment—proves the existence of a God even if one sets aside the problem of Creation.

37. Dante, *The Divine Comedy*, vol. 2: *Purgatory*, canto xxxiii, ll. 142–45, Dorothy Sayers translation [1955] (New York, 1979), p. 335.

38. "Phaedo," *Great Dialogues of Plato*, trans. W. H. D. Rouse (New York, 1956), p. 514.

Index